Einstein Studies

Editors: Don Howard John Stachel

Published under the sponsorship
of the Center for Einstein Studies,
Boston University

Volume 1: Einstein and the History of General Relativity
Don Howard and John Statchel, editors

Volume 2: Conceptual Problems of Quantum Gravity
Abhay Ashtekar and John Stachel, editors

Volume 3: Studies in the History of General Relativity
Jean Eisenstaedt and A.J. Kox, editors

Volume 4: Recent Advances in General Relativity
Allen I. Janis and John R. Porter, editors

Volume 5: The Attraction of Gravitation: New Studies in the History of General Relativity
John Earman, Michel Janssen, and John D. Norton, editors

Volume 6: Mach's Principle: From Newton's Bucket to Quantum Gravity
Julian B. Barbour and Herbert Pfister, editors

Volume 7: The Expanding Worlds of General Relativity
Hubert Goenner, Jürgen Renn, Jim Ritter, and Tilman Sauer, editors

Hubert Goenner
Jürgen Renn
Jim Ritter
Tilman Sauer
Editors

The Expanding Worlds of General Relativity

Birkhäuser
Boston • Basel • Berlin

Hubert Goenner
Institut für Theoretische Physik
Universität Göttingen
D-3400 Göttingen
Germany

Jim Ritter
Département de Mathématiques,
Histoire et Philosophie des Sciences
Université de Paris VIII
F-93526 Saint-Denis
France

Jürgen Renn
Max-Planck-Institut für
Wissenschaftsgeschichte
D-10117 Berlin
Germany

Tilman Sauer
Institut für Wissenschaftsgeschichte
Universität Göttingen
D-37073 Göttingen
Germany

Library of Congress Cataloging-in-Publication Data
The expanding worlds of general relativity / Hubert Goenner, editor
... [et al.].
p. cm. — (Einstein studies ; v. 7)
Includes bibliographical references and index.
ISBN 0-8176-4060-6 (hardcover : alk. paper)
1. General relativity (Physics) I. Goenner, Hubert. II. Series.
QC173.6.E97 1999
530.11—dc21 98-29451
CIP

AMS Subject Classifications: 83C05, 83C02, 8301, 8306

Printed on acid-free paper.

The Einstein studies series is published under the sponsorship of the Center for Einstein Studies, Boston University.

Birkhäuser

ISBN 0-8176-4060-6
ISBN 3-7643-4060-6

Typeset by the editors.
Printed and bound by Braun-Brumfield, Inc., Ann Arbor, MI.
Printed in the United States of America.

9 8 7 6 5 4 3 2 1

Contents

Part III *Relativity at Large*

Part IV *Relativity in Debate*

Preface

The past decade has seen a considerable surge of interest in historical and philosophical studies of gravitation and relativity, due not only to the tremendous amount of world-wide research in general relativity and its theoretical and observational consequences, but also to an increasing awareness that a collaboration between working scientists, historians and philosophers of science is, in this field, particularly promising for all participants. The expanding activity in this field is well documented by recent volumes in this Einstein Studies series on the History of General Relativity as well as by a series of international conferences on this topic at Osgood Hill (1986), Luminy (1988), and Pittsburgh (1991). The fourth of these conferences, hosted by the Max Planck Institute for the History of Science, was held in Berlin from 31 July to 3 August 1995, with a record attendance of some 80 historians and philosophers of science, physicists, mathematicians, and astronomers. Based on presentations at the Berlin conference, this volume provides an overview of the present state of research in this field, documenting not only the increasing scope of recent investigations in the history of relativity and gravitation but also the emergence of several key issues that will probably remain at the focus of debate in the near future.

Relativity in the Making

The papers of this section deal with the origins and genesis of relativity theory. Chronologically, they cover the comparatively short period from the end of the last century to about 1915, but thematically they range from experimental work on gravitational absorption, via Minkowski's role in developing special relativity, to the formulation of the field equations of general relativity by Einstein. Precisely because of this wide thematic range, these essays illustrate a feature of the creation of general relativity that tends to be neglected in the usual historical accounts: the scope of knowledge that was required as the basis for what appears to be Einstein's lonely contribution and the disciplinary space that had to be conquered for this knowledge.

The paper by Roberto DE ANDRADE MARTINS, first of all, reminds the reader that the task of formulating a field theory of gravitation was on the agenda of physics long before the advent of special relativity in 1905. The success of Maxwellian electrodynamics suggested not only the possibility of a field theory of gravitation constructed within the same framework, but also basic qualitative models for conceiving of a physical field and its effects: induction, wave propagation, and absorption. Martins traces the attempts at pursuing experimental and observational efforts to measure gravitational absorption and sets them into the context of the evolving theory of general relativity and also into that of the growing technical potential for realizing these attempts. Although these attempts never succeeded in establishing gravitational absorption as a stable effect and although gravitational absorption remained alien to general relativity, the suggestiveness of the qualitative model alone was sufficient to inspire ever new attempts to measure it.

Special relativity had emerged from an integration of the knowledge of mechanics and electrodynamics; the electrodynamics of moving bodies played the role of a borderline problem which brought these two domains of knowledge in contact with each other. The newly established kinematics of relativity then represented a borderline problem of a new kind, between the physical knowledge supporting it and the mathematical knowledge available for its elaboration. Obviously, borderline problems tend to represent challenges to the disciplinary organization of science and provide an occasion for a contest between disciplines, comprising imperialistic claims and rebuttals. Scott WALTER portrays the development of the four-dimensional representation of the kinematics of special relativity as such a contest, in this case between physicists and mathematicians. He reconstructs in detail how Minkowski, a mathematician, came as an outsider to the newly created theory of relativity of 1905, developed the mathematical representation named after him, and attempted to present it as a decisive conceptual innovation due to mathematicians.

Einstein's long and winding path to the creation of the general theory of relativity from the formulation of the equivalence principle in 1907 to the gravitational field equations of 1915 has traditionally appeared as a comedy of errors. This path was indeed marked by brilliant insights, such as that into the role of the metric tensor for representing the gravitational potential, as well as by major blunders, such as Einstein's misconception of the Newtonian limit at the beginning of his search. On the basis of a research project on the emergence of general relativity, pursued jointly with Michel Janssen, John Norton, and John Stachel, Jürgen RENN and Tilman SAUER present an account of Einstein's heuristics that throws new light on this bewildering trajectory. Their analysis is based on a detailed reconstruction of Einstein's calculations and scattered remarks in a research notebook dating from the period between 1912 and 1913, the "Zurich notebook." In this period Einstein collaborated with his mathematician friend Grossmann in constructing candidates for the left-hand side of the gravitational field equations which Einstein then examined according to a number of physical criteria. Einstein attempted to embed insights from classical and special relativistic physics into a generalized mathematical framework, with the aim of fulfilling as far as possible the requirements of an extended principle of relativity. Renn and Sauer argue that Einstein's search for the gravitational field equations can be reconstructed as the consequence of systematically pursuing a "double strategy" by which he related his exploration of a mathematical formalism to the available physical knowledge of gravitation.

An important heuristic role in the formulation of general relativity was played by the idea of the "relativity of rotation." According to Einstein's view of the Machian critique of Newtonian mechanics, it was legitimate to consider a rotating frame to be at rest and to describe the forces occurring in it as being due to a gravitational interaction. This consideration became part of Einstein's "double strategy" as one of the physical principles guiding his search for the gravitational field equations. How decisive the heuristic role of this idea actually was is revealed by the study of Michel Janssen. The study is based on several hitherto neglected historical documents and, in particular, on a careful reconstruction of some calculations in another research manuscript, containing entries by Einstein as well as by his friend Besso dating to 1913. In this manuscript they had jointly attempted to determine the perihelion shift on the basis of Einstein's non-generally-covariant *Entwurf* theory and failed to obtain the correct value. These efforts led them, however, to develop techniques allowing them to also check whether the *Entwurf* theory implies that rotation may be considered as a state of rest. But due to a calculational error they missed the opportunity to find out that it actually does not satisfy this requirement; Einstein's belief in the *Entwurf* theory was consequently strengthened. Only two years later, in the fall of 1915, did Einstein come back to this problem and discover his error.

Relativity at Work

Even after the theory of relativity was in place, some of the tensions and conflicts that had marked its birth continued to characterize also its further development. As the essays collected in this section illustrate, these tensions can take a number of forms and operate on a number of levels: the personal level, within the community of physicists; the conceptual level, between the theory's theoretical innovations and the heuristic role which traditional concepts from classical physics continue to play in its further elaboration; the disciplinary level, between the mathematical elaboration of general relativity and its physical interpretations.

Peter Havas's contribution on the Austrian relativity community shows that the early growth of general relativity was shaped in an important way by a scientific community which had its center in Vienna and was there fostered by a number of specific local, cultural as well as personal, conditions. Among these conditions are the links between Austrian relativists and the Social Democratic movement and, more generally, the cultural context of "Red Vienna" with its university, the Social Democratic Party and its workers' educational institutions, but also with the Vienna Circle philosophical movement. This context was, for instance, closely related to an early Austrian interest in the popularization of general relativity. Such insights are not just the result of considering an isolated example but of Havas' comprehensive longitudinal study which has made it possible to delineate essential features of an important scientific community in the early history of general relativity. Even within the numerically small and apparently homogeneous scientific community the spectrum of debate, conflict and opposition is significant. Of course, ultimately, a conflict infinitely more tragic, that associated with the rise of fascism, put an end to all debate. It is not the least merit of this essay that its poignant evocation of this destruction is that of an eyewitness.

The issue of gravitational radiation has been a ticklish one from the start, not the least because it was coupled both with expectations from classical physics and the often unexpected and sometimes even ambivalent results of its exploration in the context of general relativity. Classical physics suggested two basic approaches, both relying on analogies with Maxwell's equations: either one succeeded in obtaining an exact or approximate solution for the Einstein field equations which could plausibly be interpreted as free radiation; or one tried to determine the decrease in the energy of a material system, which one could then attribute to the effect of the transport out of the system by gravitational radiation. Daniel KENNEFICK takes up the story of the second of these two approaches. After first situating the rather confused situation in the pre-World War II period—initial certainty about the existence of graviational radiation had turned to scepticism following the publication of Einstein and Rosen's 1937 paper on cylindrical waves in general relativity—he then concentrates on the debate between proponents and sceptics at the Bern and Chapel Hill Conferences of the mid-1950s and thereafter. The growing deception with both slow and fast-motion approximation techniques in the 1960s, coupled with the impetus afforded by Weber's apparent detection of gravitational waves, led to attempts to decide the issue by other means. The controversy over the correct quadrupole formula finally ended when it became possible to decide this theoretical ambivalence by taking into account new data from binary pulsar observations.

John EARMAN has chosen a case where another apparently marginal question, that of singularities, eventually led to a remodeling of the very boundaries of the theory, as a result of the evolving tension between the mathematical elaboration of the general theory of relativity and its physical interpretation. He delineates several clearly marked periods: in the 1930s there existed agreement within the relativity community that singularities exist but their definition was still subject to controversy. While the Schwarzschild solution provided a terrain of conflict over the nature of the possible singularities at $r = 2M$ and at $r = 0$, the earliest work on gravitational collapse failed to elicit either recognition of its importance or concern for the possible impact of such results on the theory itself. In the next two decades the question of singularities receded from view. Only in the mid-1950s, the work of Raychaudhuri and Komar marked a significant change in the status of the question. The polemical nature which the debate about singularities acquired in the sequel of their contribution—due in part to the initial negative reaction from the Soviet school—was instrumental in focusing attention on this issue. In this way, a dynamics was initiated that was to culminate in the singularity theorems of Penrose, Hawking and Geroch. As so often when scientific development involves conceptual changes, the question—What is a singularity?—was eventually not so much resolved as displaced, giving rise to whole families of questions: What is the significance of singularities? What are the questions they allow us to pose about space-time? To what other kind of global questions can the new methods, introduced to resolve the singularity debate, be applied?

Relativity at Large

The enduring fascination with cosmology, attested in earlier volumes of the *Einstein Studies* series as well as in this one, is perhaps a reflection of the peculiar nature of the subject. The difficulty of obtaining solid empirical information, coupled with the conceptual problems of a science which treats a unique object, the 'universe as a whole,' have formed a field in which the initial metaphors and analogies which guide the construction of a theory remain more powerful and suggestive than is usual elsewhere. Thus philosophical issues, such as the completeness and coherency of a physical theory, the primacy of matter or space, causality, etc., have often played a direct role in the creation of models, as the following essays show.

The universe as a whole was an issue even in Newtonian physics and gave rise there to considerations which may still be relevant for understanding general relativistic cosmology. This is made strikingly evident by John Norton's essay which traces the attempts to escape from the paradox between infinite total mass and vanishing average matter density in Newtonian cosmology and to cope with the problems of the instability or indeterminacy of finite and infinite matter distributions. He analyzes the various escape-routes from these problems, such as modifications of Newton's law, the introduction of negative masses, or the admission of non-Euclidean geometry, as they were debated by Seeliger, Neumann, Kelvin, Arrhenius, Föppl, Charlier—and Einstein. Norton pays special attention to the ingenious, though little-studied, hierarchic model of Charlier and Selety which removed many of the difficulties within Newtonian cosmological models.

The application of general relativity to cosmological issues required, from the beginning, not only the identifying of appropriate solutions but also providing these solutions with an adequate physical interpretation. The paper by Silvio Bergia and Lucia Mazzoni focuses on the intervention of Hermann Weyl in a debate of the nineteen twenties which split the pioneer relativistic cosmological community. They were puzzled by the lack of gravitating matter and thus of a preferred rest system in the De Sitter solution. In order to transform this solution into a cosmological model some additional principle had to be introduced. The authors show Weyl's prolonged effort to formulate and bring into play a concept of global causal interrelation of matter as just such a means of grounding the De Sitter model. But far from being generally accepted by the cosmological community, Weyl's principle became the basis for an extension of the scope of debate with the participation of Milne, Robertson, Silberstein and Tolman.

By the end of the nineteen thirties the debate seemed to be over; a standard cosmological model held sway, aside from some sniping from Milne's kinematic relativity. But the conflict about the proper model was to resume in the postwar period with the challenge of the steady-state cosmological theory. In the first of two papers on the subject, George GALE and John URANI relate Bondi's steady-state cosmology precisely to the philosophical and methodological debates of the preceding decade. Bondi's work, they argue, was directly inspired by Milne's axiomatic approach, in direct opposition to the empirico-inductivist thinking of mainstream cosmology. The important debate in the 1937 special issue of *Nature* is treated in detail and its relation to the later ideas of the founders of the steady-state theory underscored.

A complementary view on these issues is provided by the final contribution in this section. In his paper Helge KRAGH concentrates on the technical and scientific aspects of the steady-state versus relativistic cosmological model dispute. In particular, the contributions of McCrea, McVittie and Pirani to the steady-state model are discussed as well as the attempts by some physicists to find a common meeting ground between the rival theories. A central role is shown to be played by attempts of fitting matter creation into such a theory, a concept unacceptable to most physicists. This occasionally acrimonious debate ended with the defeat of the steady-state theory—due essentially to the arrival of new empirical evidence in the nineteen sixties.

RELATIVITY IN DEBATE

In the last section questions around the meaning and nature of space-time theories and of general relativity in particular, are the subjects of the controversies under examination. The revolutionary conceptual implications of general relativity, in fact, made it hard for some physicists to fully accept this theory and, even for those who did, to agree on its nature and epistemological consequences.

An example of the long and sometimes painful process of adopting to Einstein's new theory is discussed by José SÁNCHEZ-RON in the context of Larmor, Lodge, and a generation of British physicists firmly rooted in the conceptual traditions of nineteenth-century physics. As he shows, it was not necessarily any difficulty in grasping the new mathematics of tensor calculus; even well-trained and capable mathematical physicists like Larmor had difficulties in accepting the new theory. Sánchez-Ron illustrates this by recounting how attempts were made to explain the anomalous advance of the perihelion of Mercury (Lodge) and the gravitational bending of light rays (Larmor) from within the conceptual framework of classical electrodynamics and the "electromagnetic worldview." Drawing on the correspondence between Larmor and Lodge, Sánchez-Ron analyses how these two physicists discussed their problems of understanding Einstein's theory, with Lodge never quite accepting it at all, while Larmor after many years began to realize that his own attempts had failed and accepted Einstein's theory.

The debate over the conceptual problems raised by the general theory of relativity even among its defenders is the subject of the final two contributions. Robert RYNASIEWICZ's paper deals with Erich Kretschmann's famous argument, expounded in 1917, that the mathematical representation of a theory by means of generally covariant equations does not necessarily entail that the theory also satisfies a generalized principle of relativity, and the controversy with Einstein that followed. The debate raised important questions regarding the distinction between the physical content of a theory and its mathematical representation—in what sense does a theory satisfy a principle of relativity if it is not by being expressed by means of equations of a certain covariance group, and in what sense can the general theory of relativity be called truly generally relativistic? To explore this point Kretschmann advanced a version of what is now referred to as the point-coincidence argument and argued that all physical theories may be expressed by means of generally covariant equations. Based on a new English translation of Kretschmann's difficult paper, Rynasiewicz gives a detailed interpretation of Kretschmann's analysis of the issues in question, one informed by a reconstruction of the original argument in terms of modern intrinsic geometry. The paper concludes with an intriguing confrontation of Kretschmann's understanding of the point-coincidence argument with that of Einstein.

The philosophical dimension of the conceptual implications of the general theory of relativity is also the subject of Don HOWARD's contribution. He takes issue with the thesis that the point-coincidence argument is at the historical origin of the empiricist and verificationist conceptions of science characteristic of later positivism. Howard takes a close look at the actual historical role the argument played in the discussions, notably by Einstein, Schlick, and Reichenbach. A key distinction observed by Howard is whether the point coincidences are taken to be real, by means of their invariance, or observable. In order to disentangle the respective implications, Howard carefully distinguishes between what he calls infinitesimal "point coincidences" between world lines and extended "pointer coincidences" between pointers of measuring instruments such as, typically, coincidences between the hands of a clock and points on its dial. While the latter are observable by definition, the former need not be. Discussing Einstein's philosophical reading of Mach, Duhem and Petzold, Howard argues that Einstein did not simply take point coincidences to be observable, and that the later verificationist identification of the real with the observable is based on a serious misreading of the debate.

The wide variety of subjects and of methodological approaches to which this fourth collection of essays bears witness is a most encouraging sign of the spirit of cooperation and understanding in this interdisciplinary group of working scientists, historians and philosophers of relativity. As this volume goes to press, the next international conference is in preparation and promises that this spirit will continue and grow in the expanding field of the history and philosophy of general relativity.

Hubert Goenner — Jürgen Renn
Jim Ritter — Tilman Sauer

Acknowledgements

It is a pleasure for the editors to acknowledge the support, assistance, and encouragement of the following people and organizations:

For the financial assistance that made possible the Fourth International Conference on the History of General Relativity we thank the Deutsche Forschungsgemeinschaft, the Max-Planck-Gesellschaft, and the Max-Planck-Institut fur Wissenschaftsgeschichte in Berlin. We are particularly grateful to the staff of the Max-Planck-Institut for their enthusiastic participation in the organization and smooth running of the conference. We also wish to thank the Einstein Forum of Potsdam for making possible an excursion to Einstein's summer house in Caputh and the round table discussion that enlivened the conference.

The preparation of this book, as is the case for all others in the Einstein Studies Series, owes much to the help of the Series Editors Don Howard and John Stachel and to the patience of the staff of Birkhäuser Boston.

A Note on Sources

In view of the frequent citations of unpublished correspondence or other items in the Einstein Archive, we have adopted a standard format for such citations. For example, the designation "EA 26-107" refers to item number 26-107 in the Control Index to the Einstein Archive. Copies of the Control Index can be consulted at the Jewish National and University Library (The Hebrew University), Jerusalem, where the Archive is housed; and at Mudd Manuscript Library, Princeton University, and Mugar Memorial Library, Boston University, where copies of the Archive are available for consultation by scholars.

Part I

Relativity in the Making

The Search for Gravitational Absorption in the Early Twentieth Century

Roberto de Andrade Martins

> *Unlike any other known physical influence, it [gravitation] is independent of medium, it knows no refraction, it cannot cast a shadow. It is a mysterious power, which no man can explain; of its propagation through space, all men are ignorant.*
>
> Charles Boys (1894)

THE RECEIVED HISTORICAL VIEW OF GENERAL RELATIVITY tells us that, at the end of the nineteenth century, physicists accepted Newton's law of gravitation: there is an attractive force between any pair of bodies in the world, and this force depends only on the masses of those bodies and on their mutual distance. Newton's gravitational law had been able to explain very well the motion of the planets. It had also been confirmed in the laboratory. Something, however, could not be explained: there was an anomaly in the motion of Mercury that seemed incompatible with this theory.

Soon after the development of special relativity (in 1905), Albert Einstein began to study gravitation. In 1916 he was able to formulate the so-called "general theory of relativity." This theory explained the anomalous precession of Mercury's perihelion. Besides that, it predicted two new effects: a wavelength increase of the spectral lines of the Sun due to its gravitational field; and the deflection of star light when it passes near the limb of the Sun, during eclipses. Those effects were confirmed, and hence in the early 1920s general relativity replaced Newtonian gravitation theory.

That, of course, is just a textbook version of what happened. Historians of physics know that things did not really happen that way. General relativity was indeed the successful gravitational theory but many other theories were proposed in the late nineteenth and early twentieth centuries. The three 'classical tests' of general relativity are very significant, but there were several other anomalous (non-Newtonian) astronomical and terrestrial gravitational effects that deserved attention in the early twentieth century. Indeed, if one consults scientific journals of the late nineteenth and early twentieth century, one finds a large number of revolutionary studies on gravitation. Many of them proposed alternative gravitational

H. Goenner, J. Renn, J. Ritter, T. Sauer (eds.)
The Expanding Worlds of General Relativity
(Einstein Studies, volume 7) pp. 3–44

theories. Besides theoretical papers, one finds a lot of experimental work guided by those alternative theories. Those works can be collectively called 'the search for non-Newtonian gravitational effects.'

Since the publication of Whittaker's book (1951–53), historians have been well aware of alternative gravitational theories in the early twentieth century. Two more recent book-length studies covering this period have been published: Woodward (1972) studied the history of gravitational models from Newton's time to the 1920s; Roseveare (1982) discussed gravitational theories from the late nineteenth century to general relativity. Gillies (1987) published a fairly complete bibliography of experimental studies of gravitational force up to the time of his compilation, including most relevant empirical work of the turn of the century.[1] Of course, it is impossible to describe all gravitational research of that period in a single paper. Among all those works searching for non-Newtonian effects, this paper will discuss only *gravitational absorption*.

1. The meaning of gravitational absorption

During the eighteenth century, after the general acceptance and diffusion of Newtonian theory, gravitation was generally regarded as an inexplicable phenomenon. It was accepted as an immediate *action at a distance*.[2] It did not depend on anything that might exist between the attracting bodies. Newton's law of gravitation was very simple, and there was only a single kind of relevant gravitational experiment: to test whether that law was valid or not in the laboratory, measuring the force between two bodies. Of course, this state of affairs did not stimulate any other gravitational experiment.

However, it is also possible to conjecture that gravitation is an interaction requiring some mediate cause in the intervening space. It is well known that Newton himself speculated about the mechanism of gravitation.[3] In his Trinity notebook, after sketching the ether-stream model of gravitation, Newton suggested several relevant experiments that should be tried to test this model:

> Try whether the weight of a body may be altered by heat or cold, dilatation or condensation, beating, powdering, transferring to several places or several heights, or placing a hot or heavy body over it or under it, or by magnetism. Whether lead or its dust spread abroad is heaviest. Whether a plate flat ways or edge ways is heaviest. Whether the rays of gravity may be stopped by reflecting or refracting them.[4] (Newton, cited McGuire & Tamny 1983: 431)

The precise way of understanding and testing the existence of gravitational absorption depends on the chosen gravitational model. According to Newton's

---

[1] The same author has also compiled a more popular bibliography (Gillies 1990)

[2] See Fink 1982. There were exceptions, of course, see Woodward 1972: 58–89.

[3] Aiton 1969; Hawes 1968; Rosenfeld 1969.

[4] It is not known whether Newton checked all those consequences of his early ether model or just thought about them. In his mature gravitational work, nothing of this sort can be found.

early (pre-gravitational) ether stream model, gravity would be produced by a steady flow of ether towards the centre of the Earth. In that case, the weight of a body under a thick roof would be expected to be smaller than outside the cover (Figure 1).[5] On the other side, a thick slab of matter below a body would not change its weight. The total weight of two bodies placed side by side would be larger than that of the same two bodies placed one over the other, because in the second case the upper body would reduce the weight of the lower body.

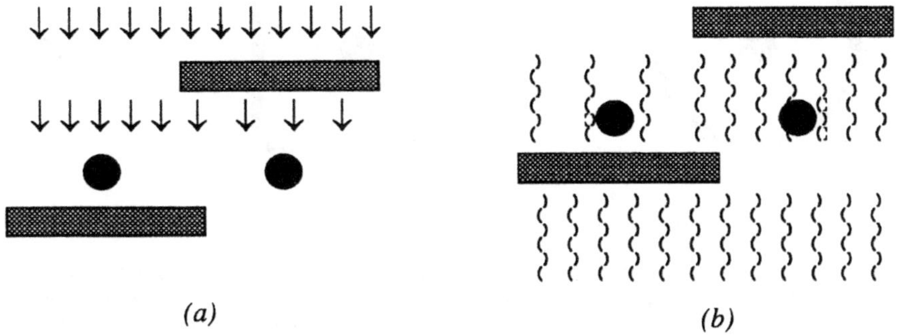

Figure 1. Two models of gravitational interaction.

The way of testing the existence of gravitational absorption might depend on the model of gravitational interaction that is assumed. *(a)*. If gravity was produced by something coming from space towards the Earth, and if this something could be absorbed by matter, the weight of a body would be reduced by a shield placed above the test body. *(b)*. If, on the other side, gravity was produced by "something" flowing from the Earth towards space (for instance, some kind of gravitational wave), the weight could be reduced by a shield placed below the test body.

Descartes' vortex model, on the other side, explained gravity as being a hydrostatical thrust due to the motion of a subtle matter circulating around the Earth.[6] In this case, a thick roof would produce no weight change, but thick walls *around* a test body would reduce its weight.

Any experimental test of gravitational absorption should take into account the gravitational attraction of the slab of matter that will be checked for its absorption effect. According to the standard (post-*Principia*) Newtonian theory of gravitation, the weight of a body under a thick roof would be reduced, of course—due to the gravitational attraction produced by the roof, and not because of any absorption effect. Therefore, in order to detect the absorption effect it is necessary either to compute (or measure) and take into account such attraction effects, or to eliminate them by a suitable symmetrical disposition of matter around the test

5 Those predictions, of course, do not take into account the gravitational attraction produced by the matter slabs. In his early speculations on gravity, Newton did not suppose that there was such an attraction.

6 The models proposed by Huygens and Leibniz, after publication of the *Principia*, also invoked the motion of matter around the Earth.

body (Figure 2). For instance: a hollow box (of spherical, cylindrical or cubic shape) will produce a null gravitational attraction upon a test body placed at its centre. However, if gravity were due to the flow of 'something' between the Earth and the test body, or to something that flows from the space towards the Earth, the walls of the box could reduce that flux and decrease the weight of the test body.

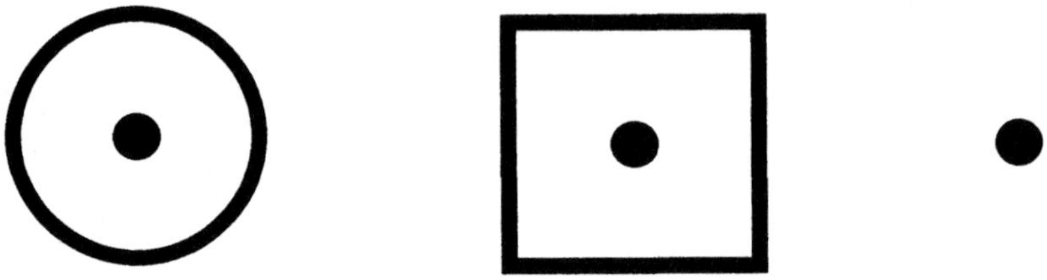

Figure 2. Gravitational shielding.

Independently of the specific model of gravitation that is assumed, if gravitational absorption by matter exists, the gravitational attraction acting upon a body should decrease when it is surrounded by a symmetrical shield. The gravitational attraction of the shield itself upon the test body will be null, in this case.

In the late nineteenth century, there appeared many hypotheses about the cause of gravitation. It was suggested that gravitation could be due to very fast particles traveling throughout all space, or to penetrating ether waves, or to a secondary effect of electromagnetic forces, and so on. In this context, there appeared many suggestions about possible new gravitational phenomena—and speculation led to a rich *experimental* investigation of all sorts of suggested effects.[7] There was no predominant alternative to the action-at-a-distance theory of gravitation. In some cases it was impossible to identify the specific model or hypothesis that suggested the experimental work.

This paper will shortly discuss the main experimental and observational work of this period on gravitational absorption and related effects, with stronger emphasis on the decade following 1910.[8] Laboratory experiments will be described first (Sections 2 to 5), followed by studies of gravitational absorption related to the study of the anomalous motion of the Moon (Sections 6 to 11).

[7] In 1881, Preston urged the search for new gravitational effects that were suggested by the kinetic theory of gravitation; he wondered "if certain small specific variations may not have escaped notice, owing to their *not having been searched for*, on account of the bias of preconceived ideas" (Preston 1881: 393). For a general view of the astronomical and experimental situation in the early twentieth century, one may consult Poynting 1900, Zenneck 1901 and Oppenheim 1920.

[8] For a general review of theories of gravitation at the end of the nineteenth century, see Drude 1897 and Taylor 1877.

2. Early experiments (to 1910)

Louis Winslow Austin and Charles Burton Thwing (University of Wisconsin) made the first known experimental attempt to test the existence of a change of gravitational force due to interposed matter (Austin & Thwing 1897). They did not present any specific theory of gravitation as a motivation of their search for a new effect. It seems that they were guided by a bare analogy to electromagnetism: electric and magnetic attractions between two bodies are affected by an intervening medium. Could there exist a similar effect for gravitation?

Austin and Thwing studied the effect of interposing screens of different materials between the attracting bodies in a torsion balance (Figure 3). They tested lead, zinc, mercury (because of their high densities), water, alcohol and glycerin (for their high dielectric constants) and iron (because of its high magnetic permeability). No significative change was observed. The experimental error was about 2×10^{-3}. Those experiments were not very sensitive, but they were significant because the authors were looking for something similar to the influence of the medium on electromagnetic forces. By analogy, one could expect that any effects of this kind would be much larger than 1/500.

At the beginning of the twentieth century, some researchers investigated a hypothetical relation between gravitation and radioactivity. The discovery of radioactivity led to the suspicion that several basic physical laws should be changed (Lodge 1912). It was very hard to explain the continuous emission of energy by radioactive substances. One of the several suggested explanations was that radioactive bodies obtain their energy from the gravitational field; therefore their weight should exhibit some kind of anomaly.[9]

The Curies and several other researchers of the time believed that the energy emitted by radioactive bodies came from outside. In their report to the International Congress of Physics of 1900, Pierre and Marie Curie suggested the following explanation:

> According to what has just been said, one could consider the Becquerel rays as a secondary emission due to rays analogous to X rays, that travel through all space and through all bodies.[10] (Curie & Curie 1900: 114)

A similar view was entertained by Lord Kelvin:

> It seems to me, therefore, absolutely certain, that if emission of heat [by radioactive bodies] ... can go on for month after month, energy must somehow be supplied from without to give the energy of the heat getting into the material of the calorimetric apparatus. I venture to suggest that somehow ethereal waves may supply energy to the radium while it is giving out heat to the ponderable matter around it. (Thomson 1903: 537)

---

[9] Other explanations included violation of the conservation of energy, or absorption of thermal energy of the environment and emission of that energy under a new form, in violation to the second law of thermodynamics.

[10] "Conformément à ce qui vient d'être dit, on pourrait considérer les rayons de Becquerel comme une émission secondaire due à des rayons analogues aux rayons X traversant tout l'espace et tous les corps."

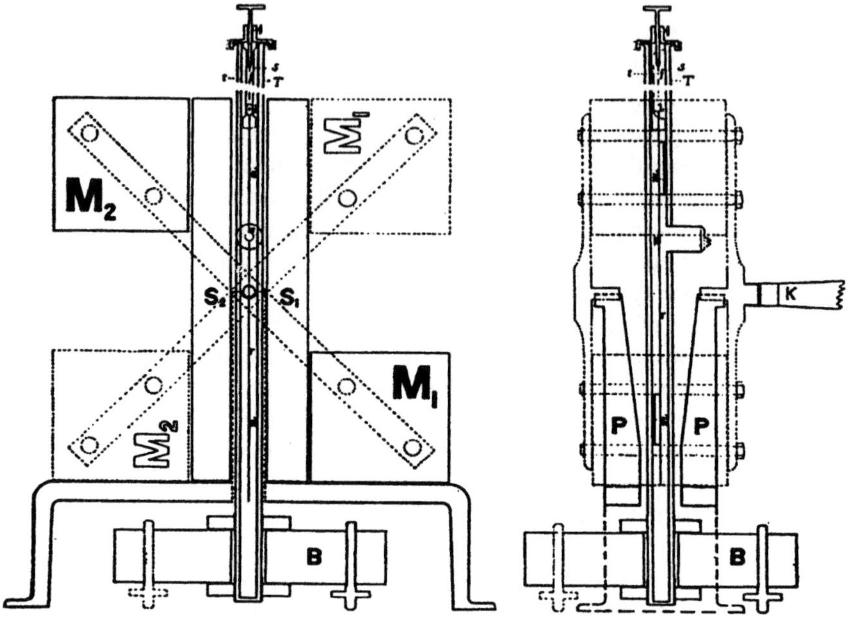

Figure 3. Two vertical sections of Austin and Thwing's apparatus (1897).

Two gold attracted masses (m_1 and m_2) were attached to a glass tube (r). This system was enclosed in a double metallic tube, with paraffin between its walls, to prevent air currents. The attracting masses were lead cubes (**M_1** and **M_1**) that could be moved so as to invert the direction of attraction. The interposed screens (**S_1** and **S_1**) were 3 cm thick, 10 cm wide and 29 cm high.

A few authors connected radioactivity to gravitation. Arthur Schuster (1903) recalled Le Sage's theory of gravitation. In this theory, space is filled with very fast corpuscles that push bodies towards one another. This hypothesis had been revived by Lord Kelvin, but had been objected to by Maxwell, who pointed out that the collisions of corpuscles with matter should produce heat. Schuster remarked that in radioactive bodies we do observe a rise of temperature that could not be explained by known causes. Perhaps it was the effect of Le Sage's corpuscles.

In this context, it was natural that a study was made of the gravitational properties of radioactive bodies.

Adolf Heydweiller (1856–1926) was the first to test the constancy of weight of radioactive substances (Heydweiller 1902). He described a significative weight change of radioactive substances in a few weeks (0.5 mg weight reduction of a 5 g sample). Heydweiller explained the effect as due to the transformation of gravitational energy into radiation. The effect was not confirmed by Dorn (1903).

Robert Geigel conjectured that radioactive bodies might exhibit a strong absorption of gravitational energy, transforming it into radiation. He tested this hypothesis and detected a significative reduction of the weight of a test body (40 μg in 6.5 g) when he put a radioactive material below it (Geigel 1903). The experiment was repeated (Forch 1903a, 1903b and Kaufmann 1903) and was not confirmed.

A few years later, a series of investigations on gravitational absorption was developed in Zurich, under the guidance of Alfred Kleiner. Two of his students (Fritz Laager and Theodor Erismann) and Kleiner himself tried to detect changes of gravitational attraction due to a shield.

Their work was motivated by the previous investigation of Austin and Thwing. Kleiner noticed that the geometry used by Austin and Thwing was not very convenient because the screens could produce forces upon the test bodies when they were not exactly in the middle position. In order to avoid this problem, Laager (1904) used screens in the form of cylindrical shells (Figure 4).

Laager initially observed noticeable effects, but soon detected experimental problems that made his experiment inconclusive. The experiment was repeated by Kleiner (1905) and by Erismann (1908, 1911) using spherical shields (Figure 5). No effect greater than experimental errors (of about 10^{-3}) was observed.

In 1905, Victor Crémieu reported the first positive result (Crémieu 1905a, 1905b, 1905c). He compared the gravitational attractions between bodies in air and in water (Figure 6). There seemed to be an increase of the gravitational attraction in water, of about 7%. Further repetitions of the experiment in improved conditions led to smaller effects of about 2–5% (Crémieu 1906, 1907). After several years of delicate researches, however, Crémieu himself detected a major source of error in his experiments (Crémieu 1910, 1913, 1917a, 1917b). He finally concluded that there was no measurable effect.

All those early attempts to detect a change in gravitational attraction tried to observe what could be described as large, easily detectable changes (10^{-2} to 10^{-3}, except for the radioactivity experiments). They could not detect smaller effects.

Around 1909, Roland von Eötvös also investigated the absorption of gravitation by matter.[11] He used a very sensitive method using a torsion balance and a device called a "gravitational compensator" (Figure 7).[12] He observed that a thickness of 10 cm of lead produced a relative gravitational absorption smaller than 2×10^{-11} (1/50 000 000 000). Eötvös also tried to detect any gravitational anomaly associated with radioactivity. He searched for absorption of gravitation and violation of the principle of equivalence, by radioactive substances. The outcome was negative.[13]

---

[11] Those investigations, made with the collaboration of Eötvös's students Jenö [Eugen] Fekete and Dezsö [Desiderius] Géza Sándor Pekár, were published posthumously (Eötvös, Pekár & Fekete 1922). The introduction to the article states that its contents had been written in 1909, and that it had obtained a prize from Göttingen University.

[12] This instrument had already been described by Eötvös in a previous paper (Eötvös 1896). Its original aim was to increase the sensitivity of the balance.

[13] Zeeman (1918) also used a torsion balance to test the same effect, with a similar result. He observed no violation of the principle of equivalence between inertial and gravitational mass for uranila nitrate, with a sensibility of 5×10^{-6}.

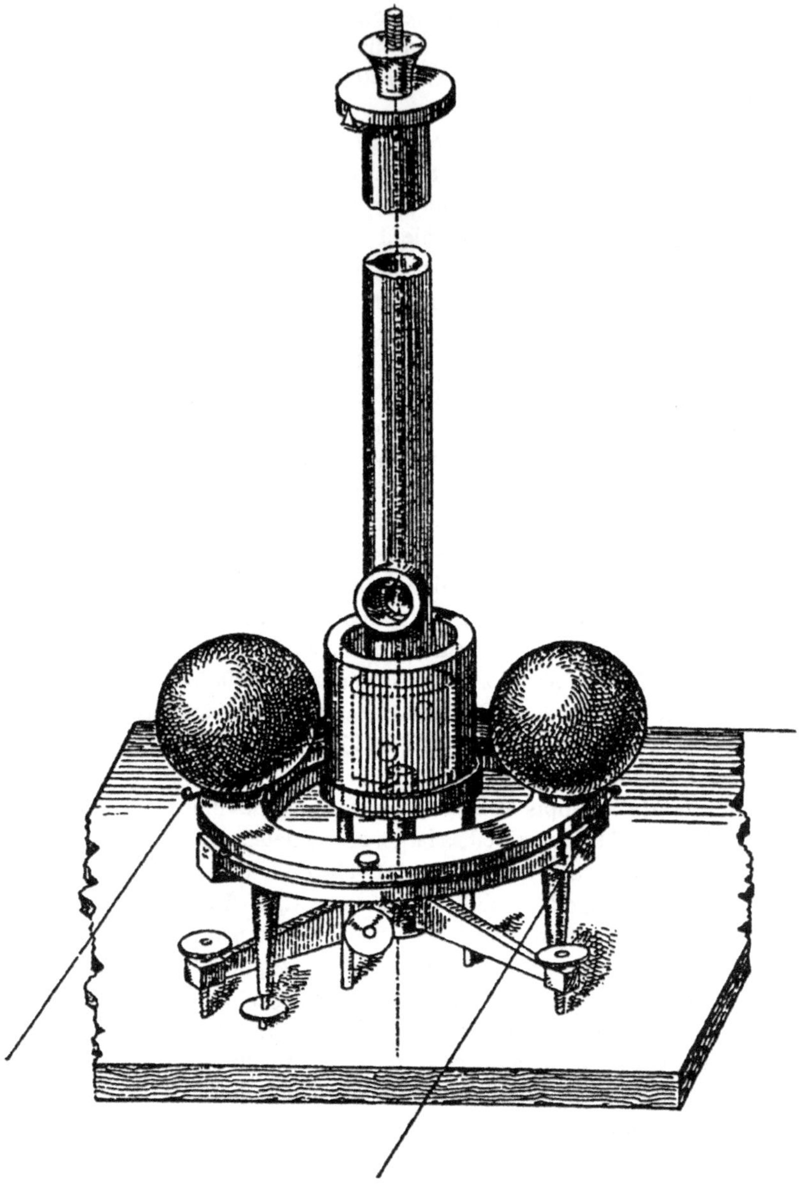

Figure 4. Laager's apparatus (1904).

The torsion balance was enclosed inside a vertical cylinder. The experiment tested the screening influence of this cylinder on the attraction between external bodies and test bodies. The external attracting masses could be moved to two different positions in order to produce a variation of the direction of the gravitational force. Attraction was measured with and without the screen.

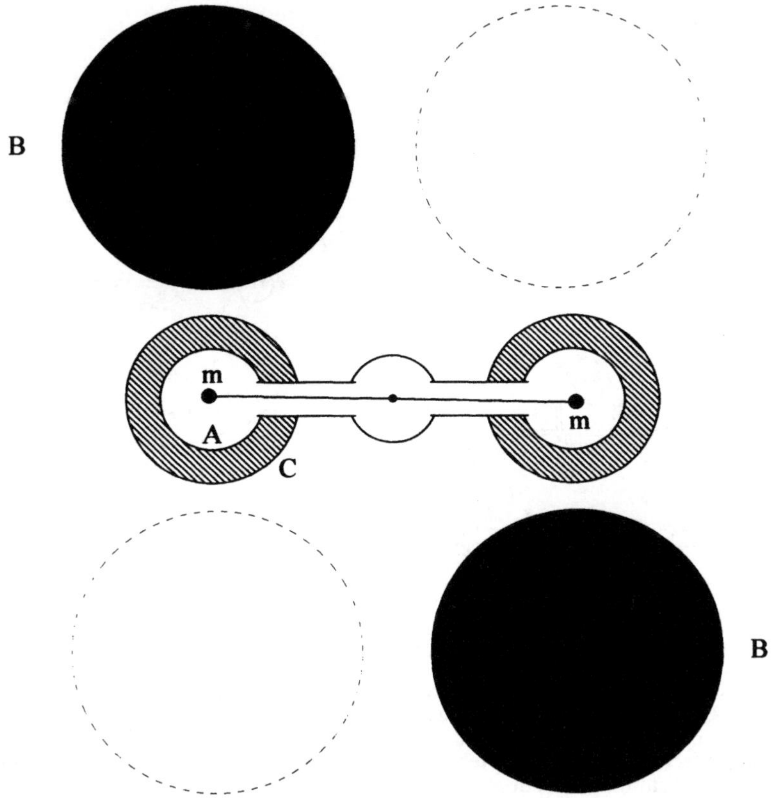

Figure 5. Erismann's apparatus (1908–1911).

This was a modified version of Laagers balance. Each test body (m) was surrounded by a double spherical shell. The inner spherical shell (A) was made of aluminum, the outer one (C) was made of copper. The space between the two metal sheets could be filled with different liquids. Inside the inner shell, a vacuum was made. The attracting spheres B, B could be moved to a different position.

3. Majorana's experiments: positive results

For about ten years, there were no further experiments on the subject. At the end of the 1910s, the Italian physicist Quirino Majorana (1871–1957) brought the search for gravitational absorption back to the laboratory. His experiments on gravitational absorption seem the best ever made. Majorana published the details of his work in several publications in Italian scientific journals (Majorana 1918/9, 1919/20a, 1919–20b, 1921–22). He also published shorter accounts of his researches in French (Majorana 1919a, 1919b, 1921, 1930) and in English (Majorana 1920).[14]

---

[14] For a modern description of Majoranas work, see Dragoni & Maltese 1994.

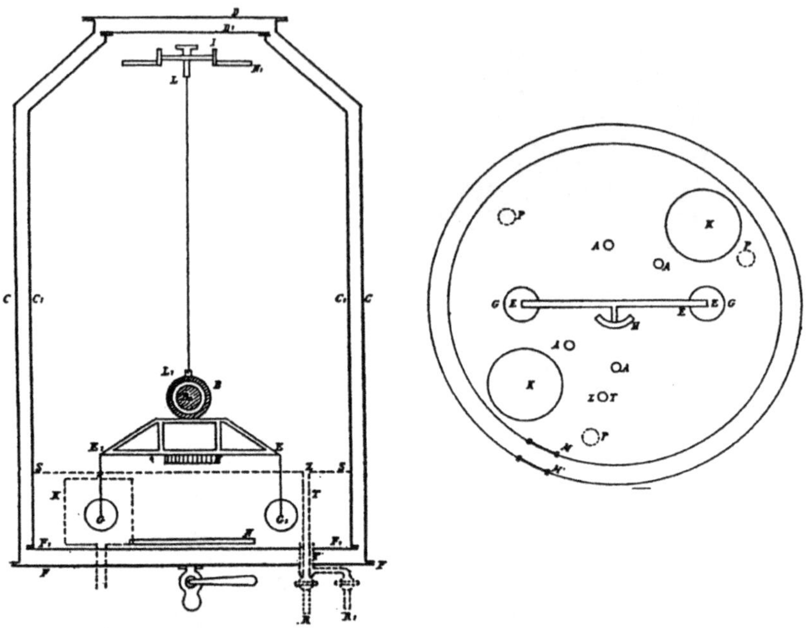

Figure 6. Crémieu's balance (1905–1917).

Test bodies G, G_1 attached to the balance beam EE_1 moved inside water that attained the level SS. Attraction produced by cylinders K, K was measured. Those cylinders were hollow and could be filled with mercury. A double metallic screen CC_1 filled with water helped to protect the balance from temperature changes. The experiments were made in a cave, far from disturbing vibrations.

Majorana conjectured that gravitation was due to the flow of gravitational energy from all bodies to the surrounding space. He also supposed that matter is not transparent to gravitational flux. According to him, gravitational energy can be absorbed by matter and transformed into heat. Therefore, each body would undergo a "spontaneous" heating. This effect would be noticeable only for large bodies, because the generation of heat would be proportional to the volume, and emission of heat proportional to the surface of the body. This process would account for stellar energy.[15]

Majorana knew that previous experimenters had tried to detect large variations of gravitation due to interposed matter and had found nothing.[16] For that reason he believed that only the search for a very weak gravitational absorption could lead to positive results.

[15] This idea was not developed in Majorana's early works. It was discussed, however, many years later (Majorana 1954).

[16] Majorana was aware of the experiments of Austin and Thwing, Kleiner, Laager, Crémieu and Erismann.

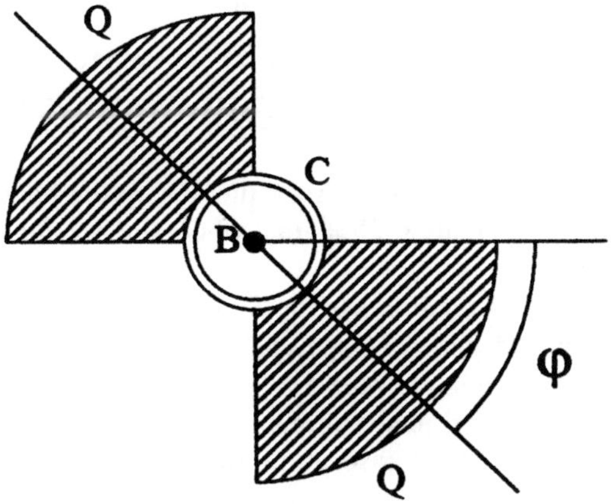

Figure 7. Eötvös' gravitational compensator (1909).

The apparatus was built of two lead masses of equal form, size and weight, symmetrically placed around each of the test bodies of a torsion balance. The test body (B) oscillated inside a hollow cylinder (C) with a diameter of 5 cm. Outside this cylinder, there were two opposite cylindrical quadrants ($Q\,Q$) of cast lead. The angle between the horizontal direction and the line that bisects the quadrants (φ) could be changed. By turning the gravitational compensator around a horizontal axis, the lead bodies could either be interposed between the test body and the Earth, or leave a free way between the test body and the ground. The experiment tried to detect any change of the weight of the test body when the gravitational compensator was turned.

He developed a theoretical analysis to evaluate the upper order of magnitude of the effect that was to be searched for (Majorana 1919/20a, 1919–20b). Let us suppose a homogeneous material medium. According to the simplest absorption hypothesis, a body of mass M placed in this medium would produce at the distance r a gravitational field g equal to:

$$g = \left(\frac{GM}{r^2}\right) e^{-Hr},$$

where H is the characteristic gravitational absorption constant of the medium. Majorana assumed that H does not depend on the chemical composition of the medium. It would be proportional to its density.

Suppose now a homogeneous sphere of radius R. Due to self-absorption of gravitation, the external field of this sphere would correspond to an apparent active gravitational mass M_a different from the sum of the gravitational masses of its parts. If ρ_v is the 'real' density of the sphere, its 'real' mass M_v is simply:

$$M_v = \frac{4}{3}\pi \rho_v R^3,$$

but its 'apparent' or 'effective' active gravitational mass M_a will be equal to ψM_v, where ψ is a factor that takes into account the self-absorption of gravitation.

Majorana computed this factor for a homogeneous sphere and found:[17]

$$\psi = \frac{3}{4}\left\{\frac{1}{RH} - \frac{1}{2}\frac{1}{(RH)^3} + \left[\frac{1}{(RH)^2} + \frac{1}{2}\frac{1}{(RH)^3}\right]e^{-2RH}\right\}.$$

As described above, Majorana assumed the absorption constant H to be proportional to the density of matter, $H = h\rho_v$, where the parameter h was supposed to be a universal constant.

Applying the above computations to the Sun, Majorana was able to evaluate an upper limit to h. The effective or apparent active gravitational mass of the Sun is known from its effect upon the planets. From this effective gravitational mass, it is easy to compute that the medium effective density of the Sun is about 1.41 $\mathrm{g\,cm^{-3}}$. If there is gravitational absorption, the Sun's real mean density must be greater than the above value.

Although the Sun is not homogeneous, Majorana applied the model of a homogeneous sphere to this case. Using values of true density larger than 1.41, he computed by successive approximations the corresponding values of h:

ρ_v ($\mathrm{g\,cm^{-3}}$)	ρ_a/ρ_v	h ($\mathrm{cm^2\,g^{-1}}$)
1.41	1.000	0
2.0	0.705	3.81×10^{-12}
5.0	0.281	7.08×10^{-12}
10	0.141	7.49×10^{-12}
15	0.094	7.63×10^{-12}
20	0.070	7.64×10^{-12}

This computation led to an unexpected result: if the true density of the Sun is supposed to increase to infinity, the absorption constant h approaches a finite value: $7.65 \times 10^{-12}\ \mathrm{cm^2\,g^{-1}}$. That is, if a simple model is applied to the Sun, its known apparent active gravitational mass imposes an upper limit to the value of gravitational absorption. Of course, the Sun is not a homogeneous sphere. However, even with this simple model, it is remarkable that Majorana could reach an upper limit for the constant of gravitational absorption.

Could such a small effect be detected? A simple computation will show that in laboratory conditions the effect would be very small indeed. As a first approximation, the weight of a body inside a spherical shell would suffer a relative reduction of about $hr\rho$, where r is the thickness of the shell. As an instance, take $\rho = 13.6\ \mathrm{g\,cm^{-3}}$ (mercury), $r = 10$ cm and $h = 10^{-11}\ \mathrm{cm^2\,g^{-1}}$. The relative weight reduction would amount to 1.36×10^{-9}, that is, a reduction of about 1 μg

[17] In one of his papers, Majorana presented a different result (Majorana 1919–20b: 314). The equation presented here was that published in his other articles (Majorana 1919/20a: 75, 1919–20b: 420, 1919a: 648, 1920: 494). Poincaré had already studied this theoretical problem and reached an equivalent equation (Poincaré 1906/7: 188).

for a 1 kg body. In order to measure such an effect, it would be necessary to attain a sensitivity 10 times better, that is, to detect changes of 0.1 μg in 1 kg (10^{-10}).

Therefore, if the effect existed, it should be undetectable for laboratory-size bodies using the previously attempted techniques. Instead of using a Cavendish torsion balance as former researchers had done, Majorana decided to use the common analytical balance and to look for weight variations when a test body was surrounded by a thick shield of dense matter, a method that had already been tried by Kleiner, but with much lower sensitivity.

No balance of that time could measure such a small change of weight. Majorana adapted the best analytical balance he could find, taking special care to avoid effects due to temperature change, air currents, etc. The whole experiment was controlled from a distance of 12 m from the apparatus to avoid any influence of the observer upon the balance (heat, air currents, and gravitational attraction). The gravitational shield was placed around the test body in such a way that the shield's resultant Newtonian attraction would be null (Figure 8).

It was necessary to control the position of the test body and of the shield around it, because any change of position between them would produce an effect larger than the hypothetical gravitational absorption. The effects of the shield upon the body attached to the other arm of the balance, and upon the balance itself, could not be neglected. All this was taken into account in Majorana's work.

In his first series of experiments, Majorana used a test body surrounded by a cylinder with about one hundred kilograms of liquid mercury. His balance was so sensitive that measurements could only be made in the first hours after midnight, to avoid vibrations due to street traffic. The best measurement conditions were obtained during a general strike.

It was not necessary to touch the test bodies during the experiment. The two masses attached to the two arms of the balance were kept in their place and the balance was equilibrated. The balance did not maintain a constant equilibrium position, however; its zero point exhibited a slow but detectable drift. The drift was regular, and therefore Majorana supposed that it would be possible to detect weight changes even though the drift could not be eliminated. The observations tried to detect minute changes of the beam position, when the shield was put around the test body or withdrawn. It was necessary to make several measurements in sequence, in order to take into account the equilibrium drift.

Majorana observed a weight reduction of about one part in one thousand million when the test body was surrounded by mercury (Majorana 1919/20a: 83; 1919–20b: 93). After taking into account several systematic errors, he obtained a value for the constant of gravitational absorption h compatible with his theoretical analysis:

$$h = (6.7 \pm 1.1) \times 10^{-12} \mathrm{cm}^2\,\mathrm{g}^{-1}.$$

Two years after the first series of measurements, Majorana repeated the experiment surrounding the test body with nine thousand kilograms of lead. For practical reasons, the lead shield had a cubic form, instead of the cylindrical form used in the case of mercury (Figure 9). He anticipated that the effect now would

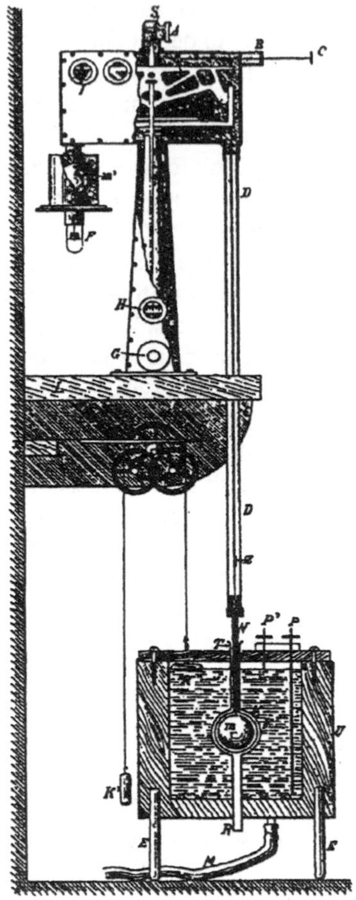

Figure 8. Majorana's first experimental arrangement for the measurement of gravitational absorption (1919–1920).

The balance and test bodies were enclosed in metallic vessels where a vacuum was produced. It was possible to manipulate the balance and the rider (of 10 mg) from outside (*C*). The oscillations of the balance were measured by a beam of light reflected from a mirror (*S*) at the top of the balance, through a strong glass wall (*A*). A deflection of 170 mm of the light beam corresponded to one mg, and it was possible to measure 0.1 mm, corresponding to 1/1 700 mg. Attached to the left side of the balance there was a 1 274 g sphere of lead (m'). Attached to the right side by a long brass thread (about 80 cm) there was a second lead ball (m) of equal mass, enclosed in a brass hollow sphere (V'), and this enclosed inside another brass hollow sphere (V). The second sphere could be surrounded by liquid mercury, contained in a strong wood cylindrical vessel (U). During the measurements, mercury was first introduced in the wooden vessel and then taken out, and any change of equilibrium of the balance was observed. Measurement and control were made at a distance of 12 m from the balance. The balance and vessel were covered by a threefold thick cover made of camel hair, to avoid changes of temperature.

be about 5 times larger than in the earlier experiment. There were, however, new significant experimental problems. The motion of the large mass of lead produced deformations of the whole building where the experiment was made. The deformation produced an angular tilting of about 10″ of the balance. It was necessary to measure and to try to compensate or evaluate all such changes.

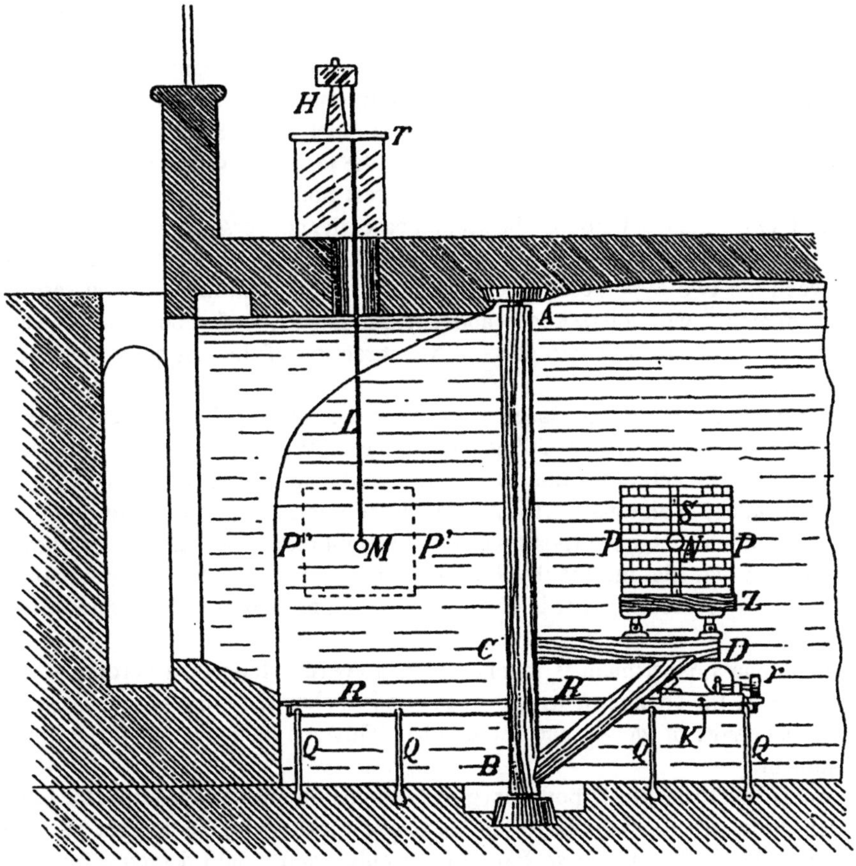

Figure 9. Majorana's second experiment (1921–1922).

The absorption of gravity was produced by a lead cube with dimensions of 95 cm and total weight of 9 603 kg. The cube was formed by two separate half-cubes. They could be moved 3 m away from the test body (M), by rotating them around the axis (AB) of their supports. The lead cube was mounted in the basement of the building. The balance (H) was on the ground floor.

After several corrections, the measured decrease of weight, ascribed to the absorption of gravity, was about half the expected value (Majorana 1921–22: 144). Therefore, in the case of lead, Majorana's measurements led to a different value for the constant h:

$$h = (2.8 \pm 0.1) \times 10^{-12} \mathrm{cm}^2\, \mathrm{g}^{-1}.$$

It would be possible to ascribe this difference either to experimental error, or to a dependence of gravitational absorption on the chemical composition of the absorbing body. Majorana did not however choose any of those alternatives. He assumed that his experiments would be reproduced by other scientists, and that those new experiments would elucidate his results.

In this second series of experiments, Majorana tried to decide whether gravity was due to something emitted from the Earth (such as Seeliger's "gravitational rays"), or coming to the Earth through space (such as Le Sage's "ultra mundane corpuscles"). In the first case, the weight of the test body would be decreased by a screen placed *between* the Earth and the test body, but not if the screen were placed *above* the test body. In the second case, the converse would be true.

The experiment with the test body inside the lead cube could not distinguish between the two hypotheses. For this reason, Majorana made a new series of measurements, with the test body above and below the lead cube. Now, however, it was necessary to take into account the attraction of the test body by the lead cube.

When the test body was placed 5 cm above the upper surface of the cube (position 1), a weight *increase* of 200 g was observed. When the test body was placed 5 cm *below* the cube (position 3), he measured a weight *decrease* of 204 g (see detailed data in Majorana 1921–22: 224). For comparison, when the test body was in the middle of the cube (position 2), the weight reduction was 2 g. Majorana's conclusion was that the first hypothesis is the correct one, that is, gravitation is produced by something that is emitted by attracting bodies (Majorana 1921–22: 79)[18]

Majorana's experiment is inconclusive, however. Indeed, according to both hypotheses, the change of weight of the body below the cube should be greater than its change of weight above the cube. This can be shown by the following argument:

According to the first hypothesis (gravitational influence emitted from the Earth), when the test body was above the lead cube (position 1), its weight W would increase by F (the attractive force of the cube) and would decrease by f (the absorption by the cube of the gravitational attraction of the Earth). When the test body was below the lead cube (position 3), its weight W would decrease by F (the attraction of the cube).

According to the second hypothesis (gravitational influence coming from space), when the test body was above the lead cube, its weight W would increase by F (the attraction of the cube). When the test body was below the lead cube, its weight W would decrease by F (the attraction of the cube) and would decrease by f (the absorption of the gravitational attraction of the Earth).

---

[18] The hypothetical emission of gravitational particles by matter was later further developed by him (see Majorana 1955).

	Test body above the cube	*Test body below the cube*
First hypothesis	$W + F - f$	$W - F$
Second hypothesis	$W + F$	$W - F - f$

Suppose that $F = 200$ g and $f = 4$ g. In this case, the changes of weight would be:

	Test body above the cube	*Test body below the cube*
First hypothesis	196	−200
Second hypothesis	200	−204

In both cases, therefore, the change of weight with the test body below the cube should be greater than with the test body above the cube. Majorana's test could not distinguish between the two hypotheses. This aspect of Majorana's work was not criticized by contemporary scientists, however.

4. **Reactions to Majorana's work**

After the publication of Majorana's first papers, Albert Michelson wrote to him and stated his intention to repeat his experiments in the Mount Wilson Observatory (Majorana 1921–22: 77). Majorana agreed enthusiastically, but Michelson never reproduced the experiment.

Majorana's first work attracted the attention of astronomers. Henry Norris Russell discussed some consequences of gravitational absorption (Russell 1921). As Eddington had formerly remarked, absorption of gravitation would lead to violation of the principle of equivalence between inertial and gravitational masses. This would produce a violation of Kepler's third law. Russell computed that In the case of Jupiter, the effect would correspond to an increase of 1% in the greater semi-axis. Astronomical measurements, however, showed that the largest possible deviation would be five hundred times smaller than that. Russell concluded that Majorana's measurements of gravitational absorption were incompatible with the motion of Jupiter. The only way out of those problems would be to assume that both the inertial and the gravitational masses were influenced by gravitational absorption.

Russell's final conclusion was that either Majorana had committed systematic errors, or that he had measured another phenomenon: a real mass variation due to the surrounding matter.[19] Russell even tried to link this suggestion to the general theory of relativity, where gravitational effects are not additive.

[19] Several decades after Majorana's work, an attempt was made by H. Grayson and Collin Williams to detect a change of mass induced by nearby matter (Grayson 1978). Their experiment was inconclusive, but their suggested improvements were remarkably similar to Majorana's experimental setup.

Russell's article was discussed by Arthur Eddington. Unexpectedly, Eddington defended Majorana (Eddington 1922), using a semi-relativistic argument. According to the strong principle of equivalence, a freely falling body could not absorb gravitational interaction, since, relative to this body, the gravitational field disappears. For that reason, the effect observed by Majorana in the laboratory would not have the consequences predicted by Russell in the case of the planets.

According to Eddington, the only odd consequence of Majorana's effect would be the possibility of building a gravitational perpetual motion machine, because the gravitational field would no longer be conservative.[20]

Majorana's *experimental method* was never criticized. Indeed, when one reads the detailed account of his measurements, it is very difficult to suggest any source of error that was not taken into account by Majorana himself. Discussion of Majorana's work focused on its consequences and compatibility with other accepted results. Majorana himself always stressed the importance of reproducing his experiments in order to check his results, but nobody ever did.

Majorana's experiments had been performed in the Physics Laboratory of the Turin Polytechnic. At the end of 1921, however, Majorana assumed the chair of Physics at the University of Bologna, as a successor to Augusto Righi. It seems that the new laboratory was better equipped than the former (see Perucca 1954: 359). There Majorana began a new series of experiments on absorption of gravity.

The main difficulty Majorana had found in his experiments was the deformation of the building resulting from displacement of about 10 tons of lead. In order to avoid this problem, Majorana reduced, in Bologna, the weight of lead to only 380 kg. The arrangement of the balance was also different: a cylindrical lead shield was successively placed around each of the two test bodies attached to the balance, in order to double the effect. Majorana stated that there were new sources of error and that it was impossible to derive any reliable value for the coefficient of absorption of gravitation from those measurements (Majorana 1930: 321).

At Bologna, Majorana also tried to improve his mercury experiments. In this case, a new arrangement of the mercury vessels was chosen, so that its whole weight was always applied to the same point of the floor. In 1930, Majorana was still improving the suspension of his balance and could present no quantitative results from this new arrangement:

> The few measurements that have already been carried out seem to give results that confirm the sense of the previously established effect, that is, an absorption of gravitational force. Although I cannot provide today quantitative results on the sought-for effect, I am confident that, with the new apparatus now under test, I will be able, after some time, to say my definitive word *on the subject.*[21] (Majorana 1930: 321)

[20] It is interesting to remark that Newton himself had sketched two different forms of a perpetual motion machine that could be built if gravitation were absorbed or refracted (McGuire & Tamny 1983: 431).

[21] "Les quelques mesures dejà faites semblent donner des résultats qui confirment le sens de l'effet autrefois établi, c'est-à-dire l'absorption de la force de gravitation. Bien que je ne puisse donner aujourd'hui des résultats quantitatifs sur l'effet recherché, j'ai confiance que avec le dernier appareil en cours d'expérimentation je serai à même, dans quelque temps, de dire, pour mon compte, un mot

Majorana's new measurements were never published. What happened? It seems that the results were not completely satisfactory and coherent, and other interests had captured his attention. Around 1930, Majorana was deeply involved in the development of communication by ultraviolet and infrared radiation, for military purposes (see Majorana 1941: 81–82). It seems that his gravitational experiments were successively postponed and never finished. Indeed, in 1941 Majorana was still referring to his Bologna attempts, remarking:

> The effect is of the same order of magnitude as that already observed in Turin. However it was impossible for me to establish its precise value in a definitive way. There are many perturbative causes that act in an erratic way when the experiment is varied. Notwithstanding this, hitherto the existence of the effect has always been confirmed. These are highly delicate researches that require months and years of accurate work for their preparation. If they are developed, they may in the future provide the last word on this interesting subject.[22] (Majorana 1941: 80)

That future time never arrived. Before his death in 1957, Majorana had published several works that refer to gravitational absorption, but he was unable to repeat his experiments (Majorana 1957a, 1957b).

Even with current techniques it would be difficult to attain the necessary sensitivity to repeat Majorana's experiments. It is remarkable, however, that no one has even *attempted* to repeat them.

5. Further laboratory experiments: Brush, Schlomka

Close to the time when Majorana reported his experiments, Charles Francis Brush (1849–1929) also published the description of anomalies ascribed to gravitational absorption. He claimed the detection of large violations of the proportionality between gravitational and inertial mass.

Brush made a series of measurements with a Cavendish balance and found that equal weights of different metals produced different attractions. Aluminum showed the greatest attraction, and bismuth the smallest. Compared to zinc, the attraction produced by aluminum was 30% higher, and that of bismuth was 28% lower (Brush 1921: 50). Brush also made pendulum experiments and measured a difference of about 1/35 000 between the ratio of inertial and gravitational mass for zinc and bismuth (Brush 1921: 56). Measurement in an inertial balance showed to him that the inertial mass of equal weights of those substances differed by one part in 1 300 (Brush 1921: 61).

There was a negligible chance that effects such as those claimed by Brush could have escaped former researchers. His results using the torsion balance

définitif *sur le sujet*."

[22] "L'entità dell'effetto è dello stesso ordine di grandezza di quello già osservato a Torino. Ma non mi è stato possibile fissarne in modo definitivo, il preciso valore. Molte sono le cause pertubatrici che agiscono in modo incostante al variare delle modalità di esperimento. Ma comunque, mi è apparsa finora confermata sempre la esistenza dell'effetto stesso. D'altra parte si tratta di delicatissime ricerche per la cui preparazione occorrono mesi ed anni di accurato lavoro. Esse, se sviluppate potranno in avvenire dire l'ultima parola sull'interessante argomento."

conflicted with previous results, such as those obtained in measurements of the gravitational constant by Baily, who had tested several different substances and had found no significative difference between them (Baily 1843). Brush's pendulum experiments conflicted both with Bessel's well-known pendulum results (Bessel 1833) and with Eötvös's torsion balance measurements. Although Brush's work was not regarded as a challenge to Newtonian physics, it was easy to check some of his results. His pendulum experiments were repeated and were soon disconfirmed by other authors (Potter 1922 and Wilson 1922). Most of his experiments were never repeated, however—perhaps because they were not as easy to reproduce as the pendulum experiments, and because the disconfirmation of a single set of measurements was enough to show that Brush's work did not deserve further investigation.

Brush's work has been largely ignored by the scientific community. He was regarded as an amateur and his experimental work was not of the same level as Majorana's, for instance.

The latest experimental researches on gravitational absorption that were developed in the period that concerns us here were done by Teodor Johannes Hermann Schlomka. His work was highly original and delicate. Schlomka was aware of earlier works on gravitational absorption. He used a Eötvös torsion balance to search for this effect, but in a new way (Schlomka 1927–30). He measured the gravitational effect of a large cubic mass of iron (1 200 kg) on the torsion balance and tested whether it depended on interposed matter (water). While previous researchers had taken the utmost care to avoid the gravitational attraction of the shield, Schlomka simply measured the effect of the iron cube both with and without intervening water, and the effect of water without the iron cube (Figure 10). He then checked whether the gravitational effects of iron and water were additive or not.

Schlomka reported that the gravitational effect of the iron mass was indeed influenced by its passage through the water prism (Schlomka 1927: 399). He stated that the results were reproducible and were not influenced by temperature. Schlomka claimed that the magnetic properties of iron could not have influenced the results. The anomalous effect was about ten times larger than the experimental errors, in the single measurement reported by him.

One might question whether the use of the Eötvös balance was acceptable in the experimental situation chosen by Schlomka. The theory of the torsion balance requires that, within the dimensions of the instrument, the gravitational field and its first derivatives should be uniform. This condition was not satisfied in Schlomka's experiment and it is difficult to analyse the consequences of this violation of Eötvös's requirements.

At the end of his article, Schlomka promised to reproduce and improve those measurements. It seems, however, that the promise was not fulfilled. It also seems that no other researcher reproduced his experiments.[23]

---

[23] Schlomka's experiments are certainly easier to reproduce than Majorana's, and it would be interesting to check them.

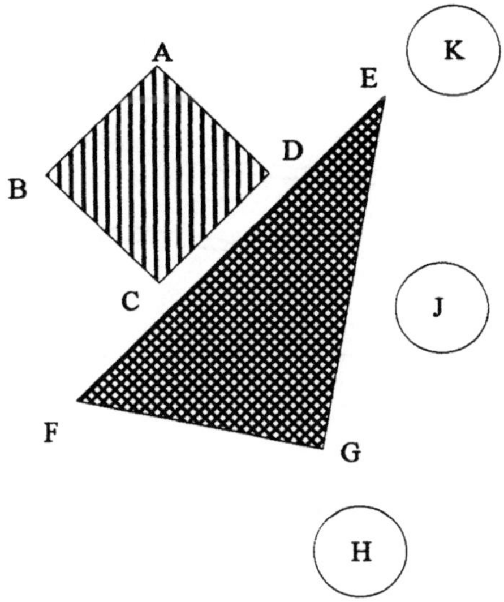

Figure 10. Schlomka's investigation of the effect of intervening matter on gravitation (1927–1930).

ABCD is a large iron mass whose effect was measured by a torsion balance. *EFG* is a triangular iron water tank with vertical walls. When full, it contained 825 liters of water. Measurements we made with the torsion balance at three different places (*H*, *J*, *K*). Keeping the torsion balance at the same place, the gravitational field was measured with the four different combinations of presence/absence of the iron mass and of water.

6. **Fluctuations of the motion of the moon**

In the 1910s, independently of the laboratory search for the absorption of gravitation, Kurt Bottlinger and other authors investigated the relation between the motion of the Moon and gravitational absorption.

The reasons that led Bottlinger to study gravitational absorption were some unexplained fluctuations in the motion of the Moon and a theoretical conjecture due to Seeliger.

Although the general behaviour and most of the details of the motion of the Moon were explained by Newtonian gravitational theory, there were a few problems that were detected towards the end of the nineteenth century. Simon Newcomb devoted the last decade of his life to the study of these anomalies.

According to Newcomb, the existing gravitational theory was unable to explain all features of the Moon's motion. After taking into account all conceivable gravitational influences, there was a difference between observed positions and theoretical predictions (Figure 11). There was a great fluctuation in the longitude of the Moon, with a period between 250 and 300 years, and other minor

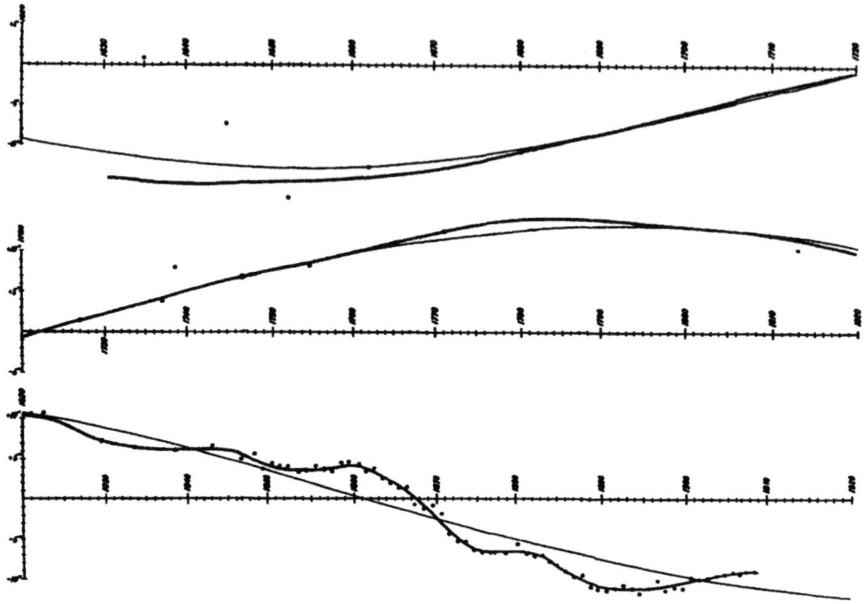

Figure 11. Newcomb's curves for the difference between observed and theoretical longitudes of the Moon from 1610 to 1908.

The whole time interval exhibited a large period fluctuation. From about 1820 onwards, when the number of observations is greater, it was possible to discern smaller period fluctuations.

fluctuations, with periods of about 60 and 20 years.[24]

Notice that the anomalous motion of the Moon is seldom cited in the traditional accounts of the rise of general relativity. As a matter of fact, the anomalous precession of Mercury's perihelion was just *one* of several unexplained astronomical phenomena at the beginning of the twentieth century.

Newcomb summarized the possible explanations of the fluctuations of the motion of the Moon. The first explanation that could be suggested was that the inequalities were only apparent, being due to fluctuations in the speed of rotation of the Earth. Indeed, astronomical observations measured the position of the Moon as a function of time; but astronomical time measurements used the rotation of the Earth as a clock.

[24] The "great fluctuation" was described by Newcomb as corresponding to a term:

$$12''.95 \sin[100°.6 + 131°.00(T - 1800.0)].$$

After Newcomb's death, the short- and medium-period oscillations were described by his co-worker Franck Elmore Ross by two terms with periods of 57 and 23 years (Ross 1911):

$$2''.9 \sin[350°.6 + 6°.316(T - 1900.0)] \quad \text{and} \quad 0''.8 \sin[313° + 15°.65(T - 1900.0)].$$

Pendulum clocks were used to measure short time periods, but those mechanical clocks were periodically checked and corrected by astronomical observations of meridian transit of standard stars. For long time periods, time was measured assuming that the angular speed of the Earth (that is, the apparent angular speed of the stars revolving around the Earth) was constant. Therefore, any anomaly of the angular speed of the Earth would falsify time computations and would result in observational anomalies in the speed of celestial bodies. The effect would be especially observable in the case of the Moon, because its motion can be studied much more acurately than those of other bodies of the solar system.

Could the observed fluctuations of the motion of the Moon be explained as mere fluctuations of astronomical time? Newcomb discussed and rejected that explanation.[25] If there were significant fluctuations in the speed of rotation of the Earth, there would be observable consequences on the motion of the planets, and observations of Mercury seemed to preclude the existence of these changes in the speed of the Earth. Besides, it would be very difficult to explain oscillations in the speed of the Earth of the required magnitude and period. The rotation of the Earth is influenced by tidal friction, but this effect cannot produce *fluctuations* of its angular velocity. If the Earth could undergo significant periodic changes in its moment of inertia, its angular velocity would also change and it would be possible to explain the apparent fluctuations of the motion of the Moon. However, no cause was known that could produce changes of the required magnitude.

This is the reason why, after describing these residual fluctuations of the motion of the Moon, Newcomb remarked:

> I regard these fluctuations as the most enigmatical phenomenon presented by the celestial motions, being so difficult to account for by the action of any known causes, that we cannot but suspect them to arise from some action in nature hitherto unknown. (Newcomb 1909: 168)

7. Bottlinger's theory of the Moon

In 1909, Hugo von Seeliger suggested that the attraction between the Moon and the Sun could decrease during lunar eclipses, due to absorption of gravity by the Earth (Seeliger 1909: 12).[26] One of his students, Kurt Felix Ernst Bottlinger (1888–1934),[27] used this hypothesis to explain the anomalies of the motion of the Moon (Bottlinger 1912a, 1912b). Bottlinger assumed that gravitation was produced by

---

25 Newcomb had presented the comparative data on the Moon and Mercury in a previous paper, where he concluded: "The evidence seems almost conclusive that the very improbable deviations in the Earth's rotation inferred from the observation of the Moon are unreal, and that the motion of our satellite is really affected by causes which have, up to the present time, eluded investigation" (Newcomb 1903: 318).

26 In former papers, Seeliger had already studied astronomical consequences—especially perihelion precession—due to hypothetical modifications of Newton's law of gravitation. He had particularly studied a modification with an exponential term, representing gravitational absorption, and its suitability for cosmological theories (Seeliger 1895, 1896).

27 For an account of Bottlinger's life and work, see Schneller 1934.

"gravitational rays" emited by all bodies. Some of the gravitational rays from the Sun would be absorbed by the Earth during lunar eclipses. That would affect the motion of the Moon (Figure 12).

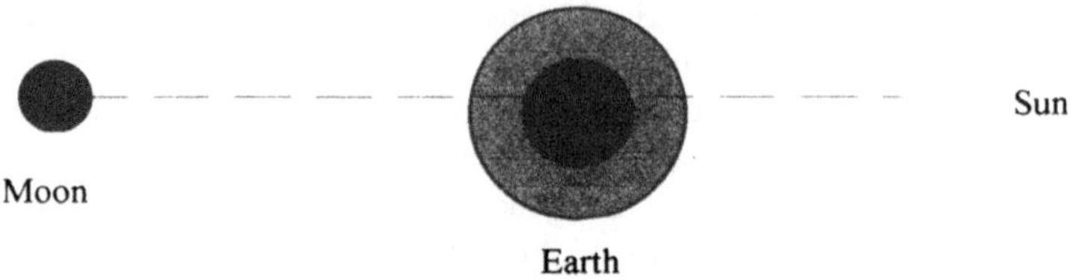

Figure 12. Seeliger's hypothesis (1909).

When the Earth is between the Moon and the Sun, during lunar eclipses, the gravitational attraction between Sun and Moon could suffer a reduction due to gravitational absorption by the Earth. Bottlinger studied the influence of eclipse details upon the supposed effect. It was necessary to take into account that the Earth is not homogeneous and that at each eclipse the Moon traverses a different path.

Bottlinger assumed the absorption of gravitational force to be proportional to the amount of matter between attracting bodies, according to an exponential law similar to that of light or X-rays.

$$F = F_0 e^{-\lambda d},$$

where F_0 is the value of the force computed according to Newtonian theory, λ is a coefficient of absorption of gravitation (proportional to the density of matter) and d is the distance traversed by gravitation in the absorbing medium.

Using data about duration and position of lunar eclipses during one century, and assuming a simple inner model of the Earth, Bottlinger computed the mass interposed between Sun and Moon in the case of each eclipse. He was able to develop a quantitative prediction of perturbations that would be produced by gravitational absorption, with a single adjustable parameter—the coefficient of absorption of gravitation by matter. In this way, Bottlinger computed the effect of all lunar eclipses, from 1830 to 1910.

The main effect, according to Bottlinger's theory, would be a fluctuation in lunar longitude, instead of an accumulative secular effect. Bottlinger compared his results to Newcomb's residues (Figure 13).

There was a nice qualitative agreement. The maxima and minima occurred at about the same years. Quantitative comparison allowed Bottlinger to compute that the maximum relative decrease of gravitational attraction between the Moon and the Sun was about 1/60 000. This ocurred when the gravitational rays travelled through the centre of the Earth. He also computed the corresponding value of the absorption constant for a substance with unit density (1 $g\,cm^{-3}$): $\lambda = 3 \times 10^{-15}$ cm^{-1}.[28]

[28] Bottlinger was probably unaware of Eötvös's unpublished results. It is relevant to remark, however, that the gravitational absorption computed by Bottlinger is much smaller than the sensitivity of Eötvös's experiment; therefore, it is compatible with Eötvös's results.

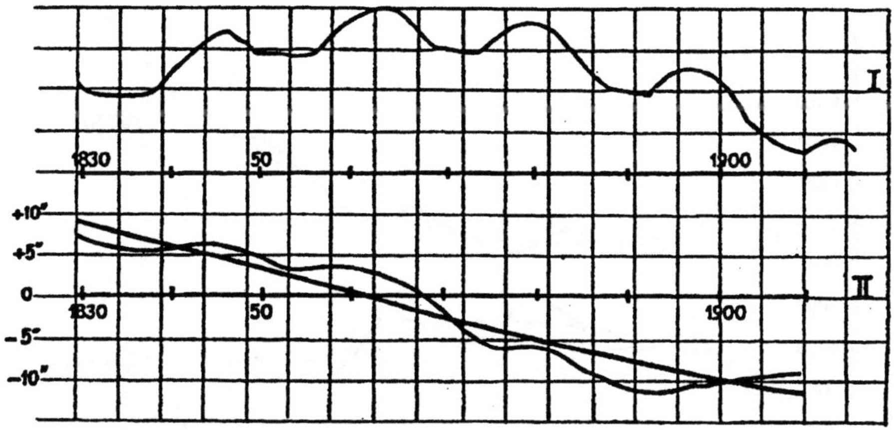

Figure 13. Bottlinger's comparison of theoretical and observed fluctuations in lunar latitude (1912).

Bottlinger compared the theoretical fluctuations of the longitude of the Moon due to gravitational absorption (I) to Newcomb's fluctuations (II). There was a general agreement between the positions of the maxima and minima.

Bottlinger's results were soon discussed by Willem De Sitter (De Sitter 1912). De Sitter had already studied gravitational absorption, but had not published his results. De Sitter did not criticize the basic hypothesis used by Bottlinger, but rather details of the theory. He stressed a few delicate points of Bottlinger's computation. The effects of successive lunar eclipses tend to cancel each other and therefore the effect computed over a large period of time is the sum of a series of positive and negative terms that do not differ much from one another. For that reason, it is necessary to compute highly accurate values for the absorption at each eclipse. De Sitter detected delicate aspects of some approximations in Bottlinger's theory that could lead to significant cumulative errors.

De Sitter compared his own previous unpublished studies to Bottlinger's. He noticed that their hypotheses and approximations were slightly different. There was a general agreement between the fluctuations found by Bottlinger and those computed by De Sitter. In both cases, the maxima and minima showed an agreement with those of the residues of Newcomb's analysis, for the nineteenth century.

There was, however, an important difference: from 1870 onwards, the perturbation computed by Bottlinger produced increasingly *negative* results (Figure 14). On the contrary, the effect computed by De Sitter led to increasingly *positive* values. This showed that slight changes in the assumptions produced relevant changes in the results.

It seems that in 1912 De Sitter believed that absorption of gravity could turn out to be the solution of the problem of lunar motion and that it could explain both small-period fluctuations and Newcomb's long-period term. Indeed, at one point of his paper, De Sitter remarked:

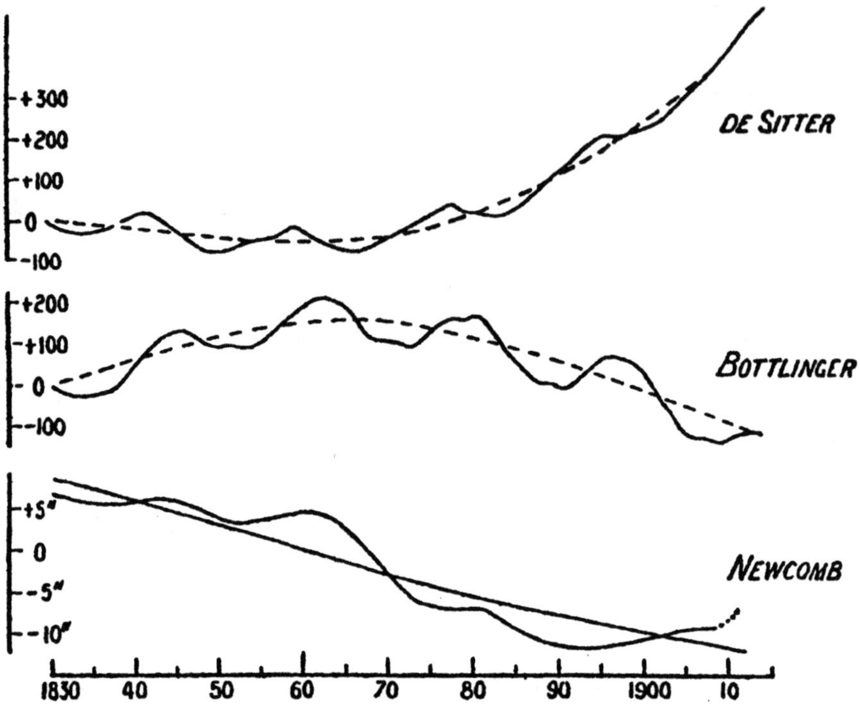

Figure 14. De Sitter's theory vs. Bottlinger's (1912).

De Sitter computed the effect of the supposed gravitational absorption using a different method. He also obtained fluctuations that roughly corresponded to those obtained by Bottlinger. However, there was a different secular (or long period) effect that was completely different in De Sitter's and Bottlinger's works.

> And perhaps we may entertain a slight hope that the slow 'undercurrent' will not only prove to be not incompatible with the observations, but may even be the explanation of the great fluctuation of 273 years' period. (De Sitter 1912: 393)

The final paragraph of De Sitter's paper is:

> However this [may] be, whether Dr. Bottlinger's results are confirmed or not, he must be congratulated on having completed a fine piece of work, which may ultimately prove to be of great importance for our intelligent understanding of natural phenomena. (De Sitter 1912: 393)

In a second paper presented in November of the same year, De Sitter presented a new detailed study of the effect of gravitational absorption on the motion of the Moon (De Sitter 1913). In his new work, De Sitter developed a more elaborate theory of lunar motion and used two different hypotheses about the internal structure of the Earth.

The results confirmed that any non-periodic effect was strongly dependent on the detailed assumptions about the structure of the Earth. Besides, it showed that the periodic fluctuations predicted by the theory were essentially the same, notwithstanding the use of different models of the Earth.

De Sitter compared the new theoretical predictions to observations and found significant differences between theory and the observed fluctuations when the whole period from 1700 to 1910 was used. Even over a shorter period, the concordance between predicted and observed fluctuations seemed to him poor (Figure 15). He concluded there was no reason to accept the existence of gravitational absorption by the Earth.

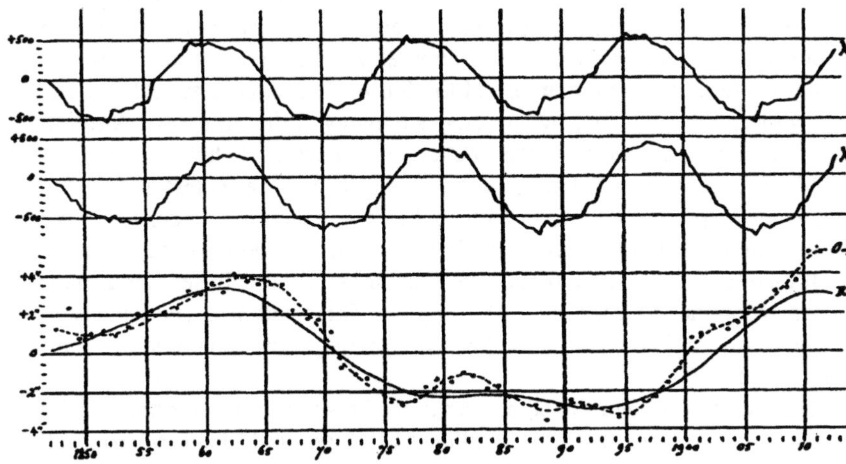

Figure 15. De Sitter's two hypotheses (1913).

In his second work, De Sitter improved the method of computing the effect of gravitational absorption on the Moon's motion and used two different hypotheses about the distribution of matter inside the Earth. In both cases the short period fluctuations (λs and λ′s) were very similar. De Sitter concluded that those theoretical fluctuations did not agree with observed fluctuations ($0\text{-}N_2$).

De Sitter was the only astronomer who took the trouble to discuss in detail and re-analyse Bottlinger's work. Other astronomers presented only short comments on the issue.

William Campbell regarded Bottlinger's work as a competitive solution of the problem of lunar theory.

> Bottlinger and De Sitter have recently investigated the hypothesis that the mutual gravitational attractions of two bodies may be influenced by the passing of a third body between the first two. ... There is some evidence that this hypothesis is an approximation to a fact of nature. (Campbell 1913a: 47)

The general opinion, however, was that the hypothesis would not provide an answer to the difficulties. Arthur Eddington was sceptic regarding the possibility of gravitational absorption (Eddington 1915). He presented another difficulty: if that effect existed, there would be no proportionality between inertial and gravitational mass and Kepler's laws would be violated. However, such a violation did not seem to exist.

In a later work (Bottlinger 1914), Bottlinger studied the consequences of two new aspects of the hypothetical absorption of gravitation:

(a) influence of solar eclipses upon the rotation of the Earth;
(b) influence of the absorption of gravitation on the motion of Mars's satellites.

In solar eclipses, the shadow of the Moon (and the corresponding assumed absoption of gravitation) traverses a small path on the surface of the Earth. When the eclipse is not a central one, the assymmetry of the solar eclipse will result in a change of the angular momentum of the Earth. That change would be so small, that it could not be detected by any clocks existing at that time. As astronomical time was reckoned according to the rotation of the Earth, the change of this rotation would be interpreted as a change in the speed of the other celestial bodies. The effect would be particularly observable in the case of the Moon, because its motion can be studied much more accurately than that of other bodies of the solar system.

Bottlinger computed the effects of both lunar and solar eclipses. He added their effects and compared his theory to observation. There was now a better agreement, and a new and smaller coefficient of absorption of gravitation could be computed ($\lambda = 1.3 \times 10^{-15}$ cm^{-1}).

Bottlinger also applied the same theory to the study of the motion of Phobos. The small distance between this satellite and Mars would lead to strong effects due to gravitational absorption. Anomalies in the motion of Phobos could amount to 36′ in longitude, with a period of 11 years. Available astronomical data did not allow the checking of this consequence of the theory. Bottlinger suggested the necessity of new observations, but it seems that the comparison was never made.

Majorana never referred to Bottlinger's studies. It is relevant to remark, however, that Bottlinger had obtained a coefficient of gravitational absorption about 500 times smaller than Majorana's and hence compatible with the limit computed by Russell from the behaviour of the planets.[29]

8. See's explanation of long-period fluctuations

Bottlinger's work could only account for short-period fluctuations. A few years later, the American astronomer Thomas Jefferson Jackson See tried to improve Bottlinger's work and to explain all fluctuations.

See is known for his 'capture theory' of the formation of the solar system, published in 1910. He might be characterized as a crank astronomer, as his work

[29] An anonymous reviewer of *The Observatory* (1920) remarked that Majorana's value for the absorption of gravitation was much higher than Bottlinger's and so there would be a difficulty in reconciling the former value with the observed motion of the Moon.

contains unorthodox views on every aspect of astronomy (Ashbrook 1962). Shortly after publication of Bottlinger's researches, he began to work on new ideas about electromagnetism and gravitation, including the hypothesis of gravitational absorption. His first results on this subject were published in 1917 as a series of 6 independently printed bulletins, afterwards collected as a book named *Electrodynamic Wave-Theory of Physical Forces* (See 1917).

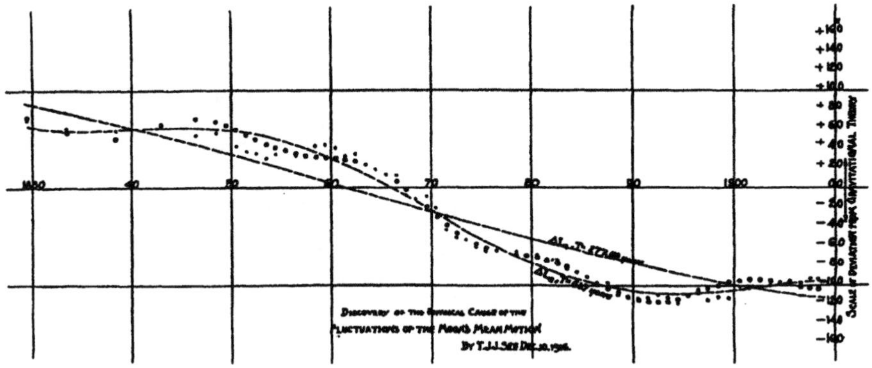

Figure 16. Fluctuations in the Moon's mean motion according to See (1917).

See's comparison between theoretical and observed fluctuations of the motion of the Moon. The small circles represent the difference between observed longitudes and Newtonian predictions (Newcomb's residues). The larger, open circles represent See's theoretical fluctuations. The curves corresponding to the longer period (277.59 years) and medium period (61.7 years) fluctuations are also shown.

Without elucidating the details of his work, See presented an impressive graphical comparison between his theory and observed fluctuations of the motion of the Moon (Figure 16). The agreement is striking. Besides, See presents exact values for the periods of the fluctuations: 18.0293 years, 61.7006 years and 277.590 years (See 1917: 4). The amplitudes and phases of the fluctuations were also exactly presented in his formulae:

$$\Delta L_1 = 1''.0 \sin[19^\circ.9675(t - 1800.0) + 239^\circ.42]$$

$$\Delta L_2 = 3''.0 \sin[5^\circ.83597(t - 1800.0) + 126^\circ.35]$$

$$\Delta L_3 = 13''.0 \sin[1^\circ.29691(t - 1800.0) + 100^\circ.6]$$

To arrive at these terms, See used vague analogies between light and gravitation. He assumed that the hypothetical gravitational waves would suffer refraction—"and perhaps absorption"—when they passed through the Earth (Figure 17). This would produce a weaker gravitational attraction between the Sun and the Moon during eclipses (See 1917: 87–91). Qualitatively, the hypothetical effect is similar to the one envisaged by Bottlinger, and See reproduces many of his equations.

However, it would be impossible to work out See's hypothesis in a quantitative way, because the effect would depend on several unknown details: refraction would depend on the precise constitution of the Earth, on the spectral composition of gravitational waves from the Sun, on the index of refraction and dispersion of each kind of gravitational wave through matter, on the absorption of gravitational waves, etc. Besides, the effect would not be confined to the Earth's shadow, but would also affect nearby regions.

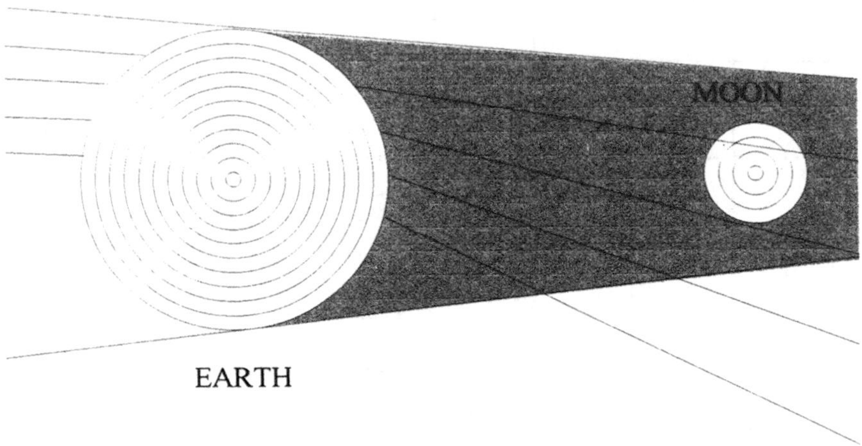

Figure 17. See's gravitational waves.

According to See, the gravitational waves emitted by the Sun would suffer refraction and absorption inside the Earth. This would affect the force between Sun and Moon even when the Moon is not in the shadow of the Earth. The effect could not be computed without detailed knowledge of the spectral composition of gravitational waves, of the inner constitution of the Earth, the index of refraction (and dispersion) of gravitational waves, and coefficient of absorption of gravitation by matter.

How, then, did See arrive at his exact theoretical evaluation of the Moon's fluctuations? That was very simple. He revised all known cycles associated with the motion of the Moon (See 1917: 101–107) and chose the periods that best suited the observed fluctuations. The amplitude and phase were taken from Newcomb's empirical representation of the fluctuations.

On close analysis, See's work was rejected as groundless. Its apparent success was just due to the use of Newcomb's empirical data (Jeffreys 1918). Instead of improving Bottlinger's work, it probably helped to decrease the interest of astronomers on this subject.

9. Einstein's explanation of the Moon's fluctuations

Shortly after the time when Bottlinger developed his theory, Albert Einstein formulated the general theory of relativity. It is well known that one of the successes of Einstein's theory was to explain the anomalous precession of Mercury's perihelion. However, there were other important anomalies, such as the fluctuations of the Moon. The theory of relativity was incompatible with gravitational absorption; although it was a non-linear theory, nothing similar to absorption appears in the field equations. Therefore, Einstein could not accept Bottlinger's theory and he was naturally led to look for another explanation of the fluctuations in the Moon's longitude.

Einstein's attempt to explain this phenomenon was a classical one (Einstein 1919a). He tried to ascribe it to fluctuations of the rotation of the Earth. The changes of rotation of the Earth were due, in Einstein's opinion, to tidal effects.

Tides produced by the Moon and the Sun produce a change of the moment of inertia of the Earth. That, in its turn, must produce a change in the speed of rotation of the Earth. If the rotation of the Earth is not uniform, astronomical observations of the Moon will exhibit anomalies, because astronomers measure time assuming that the rotation of the Earth is uniform. Periodic changes of the moment of inertia of the Earth will lead to periodic apparent fluctuations of the motion of the Moon.

Einstein's argument presupposes that there is no irregularity in the motion of the Moon, that is, it assumes that the Moon follows Newton's gravitational theory. At any given time t, the Moon would be in its 'correct' place. Suppose, now, that the rotation of the Earth fluctuates around its median angular motion. The angular position of a reference point on the surface of the Earth at time t is $\varphi = \omega t + \delta\varphi$. At this time t, if the Earth rotated with an uniform angular speed, the longitude of the Moon would be φ. However, due to the irregularity of the motion of the Earth, the longitude of the Moon measured relative to the Earth at time t will be $\varphi' = \varphi - \Delta\varphi$, where $-\Delta\varphi$ is the difference between the mean motion and the actual motion of the Earth at time t. Therefore, relative to the Earth, the motion of the Moon will fluctuate around its theoretical (Newtonian) value with an amplitude equal to the amplitude of the fluctuation of the motion of the Earth.

Einstein computed the effect of the tides. In his theory, there was only one adjustable parameter: the mean amplitude of tides. He assumed a value of 1.5 m and obtained fluctuations of the Moon's motion with amplitude of $1''$. He also computed the times of maxima and minima of those fluctuation, and they agreed with observed maxima and minima. The theoretical values of the fluctuations were smaller than observed ones, but Einstein ascribed the difference to the value he used for the moment of inertia of the Earth. Therefore, it seemed to Einstein probable that this was the correct explanation of the fluctuations of the Moon.

Einstein was not the first to try to explain the fluctuations of the Moon in this way. However, all previous attempts had obtained too small effects. Indeed, Einstein was wrong.

Immediately after the publication of Einstein's work, an astronomer named Albert von Brunn detected his mistake (Brunn 1919). Einstein had not taken into account the methods really used by astronomers in their measurements: "The explanation seems to be founded upon a mistake concerning the method of time determination in astronomy."[30] If things worked as Einstein supposed, then *all* celestial bodies (including the stars, Sun and planets) would exhibit the same fluctuations, with the same period and amplitude.[31] These fluctuations do not exist. By correcting Einstein's assumptions, Brunn showed that the periodical fluctuations of the Moon would have an amplitude 27 times smaller than those computed by Einstein. Therefore, Einstein's proposal could not explain the observed fluctuations.

In a note to Brunn's article, Einstein acknowledged his mistake:

> Herr von Brunn's criticism is completely well founded. Since my mistake is not devoid of a certain objective interest, I want to characterize it shortly too. My reflection would be correct, if the astronomers used the Earth itself as a spatial reference body, in connection with a particular clock for time measurement. In fact, the astronomers use the stellar heaven as a coordinate system for spatial measurements, and the rotation of the earth relative to the stars as a clock. Therefore, an irregularity of the rotation of the Earth relative to the time measurement can only be displayed in the way shown by Herr Brunn.[32] (Einstein 1919b: 711)

10. Einstein's mistake

What was wrong in Einstein's argument? The whole problem was the use of *absolute time t* in the argument. Einstein implicitly assumed that there was a way of measuring time *independently of the rotation of the Earth*. Of course, nowadays we can use atomic clocks, but in 1919 astronomers measured time using the rotation of the Earth relative to the stars as their clock. Even at that time, it was possible, of course, to *think* about some absolute (Newtonian) time and deduce consequences, but it would be necessary to analyse what *observable* effects could be measured by astronomers.

Suppose there exist a sidereal astronomer and a terrestrial astronomer. Suppose further that both use the same theory for computing the motion of the Moon, that is, they use the same formula $\phi = \phi(t)$ to compute its longitude. Both measure angles relative to the same set of 'fixed' stars. The sidereal astronomer uses an

---

[30] "Die Erklärung scheint auf einem Irrtum über die Methode der Zeitbestimmung in der Astronomie zu beruhen" (Brunn 1919: 710).

[31] "If this understanding was correct, the apparent right ascension of the stars, as also the longitudes of the Sun and the planets, would all exhibit the same essential periodicity as the Moon's longitude" ["Wäre diese Auffassung richtig, so würden so würden offenbar die Rektaszensionen aller Gestirne und damit auch die Längen der Sonne und der Planeten alle im wesentlichen die gleiche Periodizität zeigen wie die Mondlänge"] (Brunn 1919: 710).

[32] "Herrn von Brunns Kritik ist durchaus begründet. Da mein Irrtum nicht ohne ein gewisses objektives Interesse ist, will auch ich ihn noch einmal kurz charakterisieren. Meine Betrachtung wäre richtig, wenn sich die Astronomen der Erde als räumlichen Bezugskörpers in Verbindung mit einer besonderen Uhr als Zeitmaß bedienten. In Wahrheit dient den Astronomen der Fixsternhimmel als Koordinatensystem für die räumlichen Messungen, die Drehung der Erde relativ zu den Fixsternen als Uhr. Deshalb kann eine Ungleichmäßigkeit der Erddrehung nur Fehler bezüglich der Zeitmessung herbeiführen, wie Herr Brunn zutreffend ausgeführt hat."

absolute clock and the terrestrial astronomer uses the rotation of the Earth as a clock. That is, he assumes that the rotation of the Earth obeys a simple law:

$$\varphi = \omega t',$$

where the angular speed ω of the Earth is assumed to be constant and t' is the time measured by the terrestrial astronomer.

The sidereal astronomer observes that the rotation of the Earth is not uniform: its rotation is described by this astronomer as:

$$\varphi = \omega t + \Delta\varphi(t),$$

where ω is the *mean* angular velocity of the Earth. There will be a difference between the times assigned by the sidereal and terrestrial astronomers to any event, since $\omega t' = \omega t + \Delta\varphi(t)$; therefore $t' = t + \Delta\varphi(t)/\omega$.

Suppose that the sidereal astronomer computes the position of the Moon at a time $t = T$ and obtains the longitude ϕ that agrees with observation. What will the terrestrial astronomer find at this same time?

When the terrestrial astronomer computes the position of the Moon at time T, he obtains the same value $\phi(T)$ as the sidereal astronomer, because both use the same formula. However, when the Moon passes by this position $\phi(T)$, the terrestrial clock will not measure time $t' = T$, but a time $t' = T + \Delta\varphi(T)/\omega$, because of the irregularity of the rotation of the Earth. The terrestrial astronomer will conclude that the Moon is behind (or ahead) of its theoretical position, because at time $t' = T + \Delta\varphi(T)/\omega$ the Moon should be in the position

$$\phi\left[T + \frac{\Delta\varphi(T)}{\omega}\right] \approx \varphi(T) + \frac{\Delta\varphi(T)}{\omega}\frac{d\phi}{dt}.$$

If the motion of the Earth fluctuates with an amplitude a and the angular velocity of the Moon is $d\phi/dt = \omega'$, then the observed longitude of the Moon will fluctuate around its theoretical position with an amplitude $a' = a\omega'/\omega$, when observed by the terrestrial astronomer.

Notice that the fluctuation of the position of the Moon, $\Delta\phi$, will be *different* from the fluctuation of the Earth, $\Delta\varphi$, because the angular speed of the Moon, ω' is different from the angular speed of the Earth, ω. As ω' is about 27 times smaller than ω, the amplitude of the observable fluctuation of the Moon would be about 27 times smaller than the fluctuation in the Earth's rotation. If the amplitude of the oscillations of the Earth's motion is $2''$, the corresponding fluctuations of the Moon would amount to only $2''/27$.

If we apply the same argument to other celestial bodies (the Sun and planets), it will be easily perceived that their observable fluctations due to the fluctuation of the rotation of the Earth will be much smaller that that of the Moon, because, relative to the Earth, their angular velocities are always much smaller than ω'.

The argument presented here is a didactic reconstruction of Brunn's very short correction of Einstein's mistake. It seems that Brunn did not care to discuss the argument in detail because astronomers were well aware of all those distinctions. Einstein's mistake was due to his lack of acquaintance with astronomical methods of measurement.

11. Final explanation of lunar fluctuations

In the long run, lunar fluctuations were explained away. A general consensus was reached around 1940: the motion of the Earth is irregular—but the changes of the rotation of the Earth were not ascribed to the tides (Spencer Jones 1939). As a result of this interpretation, the rotation of the Earth could not be retained as the standard of time determination. A new measurement of time was introduced in astronomy: so-called *ephemeris time*, defined as the time parameter that complies with gravitational theory. It was adopted by the International Astronomical Union in 1955 (Spencer Jones 1955). In principle, it was associated to the motion of the Earth around the Sun (or the apparent motion of the Sun). In practice, however, ephemeris time was determined from observations of the Moon. The adopted definition used a correction equation that included a parameter B which was the empirical fluctuation in the Moon's longitude that is, the difference between its observed position and the motion predicted by gravitational theory (Spencer Jones 1956: 22).

Therefore, the fluctuations of the motion of the Moon disappeared *by definition*: the adoption of the new definition of time used the motion of the Moon itself as a clock. According to that clock, of course, the motion of the Moon is completely regular.

De Sitter himself greatly contributed to the establishment of this standard interpretation. He studied the motions of the Sun, Mercury and Venus and showed that all of them exhibited longitude fluctuations similar to those of the Moon, and proportional to their mean motions (De Sitter 1927, 1928). The agreement was especially good for the long- and medium-period terms and was better in the cases of Mercury and Venus than in the case of the Sun (Figure 18).[33] Harold Spencer Jones always supported this explanation of the fluctuations (Spencer Jones 1926, 1939). However, up to 1932, this explanation was not free from problems (Fotheringham 1927, 1932).

Around 1920 there was no quantitative explanation for the Moon's short-period fluctuations better than Bottlinger's. In that year, Ernest Brown, one of the leading authorities in lunar theory, reviewed the question (Brown 1920). Up to that time, a correlation had been found between the irregularities of the lunar motion and those of Venus and Mercury, but these seemed smaller than expected. Besides that, no cause was known that could produce fluctuations in the rotation of the Earth of the required magnitude: "There is, I think, a growing conviction that the Earth's average rate of rotation has not sensibly changed within historic times"

[33] See also Dyson & Cullen 1929. The correspondence between the fluctuations of the Sun and the Moon was at most as good as the correspondence between Bottlinger's theory and the observed lunar fluctuations.

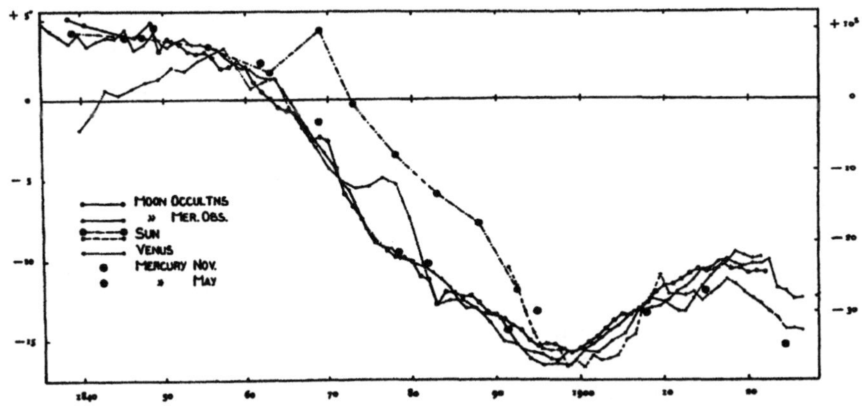

Figure 18. De Sitter on fluctuations of the Moon, Sun, Venus and Mercury (1927).

De Sitter's comparison between fluctuations of the motions of the Moon (measured by occultations and meridian observations), Sun, Venus, and Mercury, from 1840 to 1920. In the case of the Sun, agreement is poor. There was, however, a general agreement between the fluctuations of Venus, Mercury and Moon specially for the long period fluctuation.

(Brown 1920: 100–101). Brown's conclusion was that some unknown cause was producing real changes in the motions of the Moon and the planets.

> The irregular character and comparatively great magnitude of these fluctuations suggests that there is some set of forces acting on the bodies of the solar system which are related to the known irregular changes in the condition of the Sun. (Brown 1920: 101)

In the decade 1910–1920, Bottlinger's theory was an interesting, quantitative attempt to explain short period fluctuations of the Moon. It did not account for all known facts, but no other explanation did, either. However, most astronomers dismissed it without a detailed analysis and kept waiting for a classical explanation of the phenomenon.

12. **Conclusion**

Nowadays, we believe that gravitational absorption does not exist. From the concordance between gravitational theory and motions observed within the solar system,[34] as also from geophysical measurements,[35] an upper limit was reached for the constant of absorption of gravitation: it must be smaller than Bottlinger's value: $\lambda < 10^{-15}\ g^{-1}cm^2$.

According to present scientific knowledge, the similarity between Bottlinger's theoretical curve and the observed fluctuations of the Moon was due to chance.

[34] See Steenbeck & Treder 1984, especially pp. 16–25).

[35] See Bocchio 1971 and Groten 1972. See also Cook 1988: 719.

Also, according to current knowledge, Majorana measured nothing but experimental errors. Indeed, both in the old gravitational experiments and in recent ones, it is usual to find unexplained systematic effects (Cook 1987, 1988). As Cook has put it, "it is difficult to attain an adequate understanding of experiments at the limit of available techniques" (Cook 1987: 76). Majorana was certainly pushing the sensibility of weight measurements to its limit. Although he was a careful experimenter, some systematic error might be responsible for his results.

From the historical point of view, it is relevant to understand why those investigations of gravitational absorption did not receive much attention at the time they were published. Some of the attempts to detect anomalous effects produced null results and were not very exciting—they were forgotten. Some of the research that produced anomalous results was soon reproduced and errors were detected. This was the case with Brush's enormous violations of the principle of equivalence and Heydweiller's transformation of gravitation into radioactive energy. Some of the authors of these results could be classified as cranks and their works could be dismissed without detailed analysis. This was the case of See's wave theory of gravitation.

It is not so easy to understand why Bottlinger's and Majorana's explorations were also dismissed or did not receive much attention. They were high-quality work, but were not linked to the mainstream of gravitational research of the time. Their motivation was an old-fashioned corpuscular (or wave) theory of gravitation. That style of theory had been very popular among outstanding scientists at the end of the nineteenth century, but now had been replaced by another kind of theory. Since the beginning of the twentieth century, Poincaré, Lorentz, Abraham, Einstein, Nordström, Mie and several other physicists were striving to develop a relativistic theory of gravitation.[36] These new theories of gravitation did not appeal to mechanical models or analogies: they used sophisticated mathematics and their primary aim was to provide a unified relativistic description of gravitation and electromagnetism. From the point of view of this line of research, gravitational experiments or explanations grounded upon old models and analogies were mere child's play.

Majorana was unfortunate enough to publish his results at the time when all the world was celebrating Einstein's successful prediction of the bending of light rays near the Sun.[37] Besides the three classical tests of general relativity, there seemed to be no new phenomenon that could be observed. As an effect of widespread acceptance of general relativity in the 1920s, for some decades gravitational research was transformed into a mathematical subject and experimental gravitation came close to extinction. This might explain both the lack of reproduction of anomalous results, and the general oblivion of these interesting investigations.

---

[36] For a contemporary statement of theoretical gravitational research before the full development of general relativity, consult Abraham 1914.

[37] Of course, the relation between eclipse observations and theory was not as simple as usually assumed (see, for instance, Moyer 1978, Earman & Glymour 1980), but for the general scientific and non-scientific public the 1919 eclipse observations seemed a crucial test of general relativity.

ACKNOWLEDGEMENTS

This work was produced while the author was a visiting scholar of the Department of History and Philosophy of Science, University of Cambridge, and visiting fellow of Wolfson College. The author is also grateful to the State of São Paulo Science Foundation (FAPESP) for support of this research.

REFERENCES

ABRAHAM, Max. (1914). "Neuere Gravitationstheorien." *Jahrbuch der Radioaktivität und Elektronik* **11**: 470–520.

AITON, E..J. (1969). "Newton's Aether-Stream Hypothesis and the Inverse Square Law of Gravitation." *Annals of Science* **25**: 255–260.

[ANONYMOUS]. (1920). "Experimental Researches on Gravitation." *The Observatory* **43**: 286–287.

ASHBROOK, Joseph. (1962). "The Sage of Mare Island." *Sky and Telescope* **24**: 193, 202.

AUSTIN, Louis Winslow & THWING, Charles Burton. (1897). "An Experimental Research on Gravitational Permeability." *Physical Review* **5**: 294–300.

BAILY, Francis. (1843). "An Account of Some Experiments with the Torsion Rod, for Determining the Mean Density of the Earth." *Memoirs of the Royal Astronomical Society* **14**: 1–120, i–cclviii.

BESSEL, Friedrich Wilhelm. (1833). "Versuche über die Kraft, mit welcher die Erde Körper von verschiedener Beschaffeneheit anzieht." *Astronomische Nachrichten* **10** (223): col. 97–108.

BOCCHIO, Franco. (1971). "Old and New Approaches in the Study of Gravitational Absorption." *Geophysical Journal of the Royal Astronomical Society* **24**: 101–102.

BOTTLINGER, Kurt Felix Ernst. (1912a). "Die Erklärung der empirischen Glieder der Mondbewegung durch die Annahme einer Extinktion der Gravitation im Erdinnern." *Astronomische Nachrichten* **191**: 147–150.

——— (1912b). *Die Gravitationstheorie und die Bewegung des Mondes*. Freiburg: C. Trömer.

——— (1914). "Zur Frage der Absorption der Gravitation." *Sitzungsberichte der Königlichen Bayerischen Akademie der Wissenschaften zu München. Mathematische-physikalische Klasse* **44**: 223–239.

BOYS, Charles Vernon. (1894). "On the Newtonian Constant of Gravitation." *Nature* **50**: 330–334, 366–368, 417–419, 571.

BROWN, Ernest William. (1920). "The Problem of the Moon's Motion." *Publications of the Astronomical Society of the Pacific* **32**: 93–104.

VON BRUNN, Albert. (1919). "Zu Hrn. Einsteins Bemerkung über die unregelmässigen Schwankungen der Mondlänge von der genäherten Periode des Umlaufs der Mondknoten." *Sitzungsberichte der Preussischen Akademie der Wissenschaften* (Berlin): 710–711.

BRUSH, Charles F. (1921). "Discussion of a Kinetic Theory of Gravitation II, and Some New Experiments in Gravitation." *Proceedings of the American Philosophical Society* **60**: 43–61. [Continued in **61**: 167–183; **62**: 75–89; **63**: 57–61; **64**: 36–50].

CAMPBELL, William Wallace (1913). "A Possible Absorption of Gravitational Energy." *Publications of the Astronomical Society of the Pacific* **25**: 46–47.

COOK, Alan (1987). "Experiments on Gravitation." In *Three Hundred Years of Gravitation*. Stephen Hawking and Werner Israel, eds. 51–79. Cambridge: Cambridge University Press.

——— (1988). "Experiments on Gravitation." *Reports of Progress in Physics* **51**: 707–757.

CRÉMIEU, Victor. (1905a). "Attraction observée entre gouttes liquides suspendues dans un liquide de même densité." *Comptes Rendus des Séances de l'Académie des Sciences de Paris* **140**: 80–83.

——— (1905b). "Recherches sur la gravitation." *Comptes Rendus des Séances de l'Académie des Sciences de Paris* **141**: 653–656, 713–715.

——— (1905c). "Recherches expérimentales sur la gravitation." *Bulletin de la Société Française de Physique* **3**: 485–99, 103. [Reprinted in *Journal de Physique Théorique et Appliquée* (4) **5** (1906): 25–39].

——— (1906). "Recherches sur la gravitation." *Comptes Rendus des Séances de l'Académie des Sciences de Paris* **143**: 887–889.

——— (1907). "Recherches comparées sur les forces de gravitation dans les gaz et les liquides." *Bulletin de la Societé Française de Physique* **1**: 33–47, 14–15. [Reprinted in *Journal de Physique Théorique et Appliquée* (4) **6** (1907): 284–298].

——— (1910). "Sur une erreur systématique qui limite la précision de l'expérience de Cavendish. Méthode nouvelle pour l'étude de la gravitation." *Comptes Rendus des Séances de l'Académie des Sciences de Paris* **150**: 863–866.

——— (1913). "Effects de la flexion aux points d'attache du fil d'une balance de torsion." *Comptes Rendus des Séances de l'Académie des Sciences de Paris* **156**: 617–620.

——— (1917a). "Recherches expérimentales sur la gravitation." *Comptes Rendus des Séances de l'Académie des Sciences de Paris* **165**: 586–589, 688.

——— (1917b). "Nouvelles recherches expérimentales sur la gravitation." *Comptes Rendus des Séances de l'Académie des Sciences de Paris* **165**: 670–672.

CURIE, Pierre & CURIE, Marie (1900). "Les Nouvelles Substances radioactives et les rayons qu'elles émettent." In *Rapports présentés au Congrès International de Physique réuni à Paris en 1900 sous les auspices de la Société Française de Physique*. Vol. 3: 79–114. Charles-Édouard Guillaume and Lucien Poincaré, eds. Paris: Gauthier-Villars.

DE SITTER, Willem. (1912). "Absorption of Gravitation." *The Observatory* **35**: 387–393.

——— (1913). "On Absorption of Gravitation and the Moon's Longitude." *Proceedings of the Royal Academy of Amsterdam* **15**: 808–824, 824–830.

——— (1927). "On the Secular Accelerations and the Fluctuations of the Longitudes of the Moon, the Sun, Mercury and Venus." *Bulletin of the Astronomical Institutes of the Netherlands* **4**: 21–38.

——— (1928). "On the Rotation of the Earth and Astronomical Time." *Nature* **121** (supplement): 99–106.

DORN, Ernst. (1903). "Versuch über die zeitliche Gewichtsänderung von Radium." *Physikalische Zeitschriften* **4**: 530–531.

DRAGONI, Giorgio & MALTESE, Giulio. (1994). "Quirino Majorana e l'assorbimento gravitazionale." *Giornale di Fisica* **35**: 245–292.

DRUDE, Paul. (1897). "Über Fernewirkungen." *Annalen der Physik* **62**: i–xlix, 693.

DYSON, Frank W. & CULLEN, R. T. (1929). "Variability of the Earth's Rotation." *Monthly Notices of the Royal Astronomical Society* **89**: 549–551.

EARMAN, John & GLYMOUR, Clark. (1980). "Relativity and Eclipses: the British Eclipse Expeditions of 1919 and Their Predecessors." *Historical Studies in the Physical Sciences* **11**: 49–85.

EDDINGTON, Arthur Stanley. (1915). "Some Problems of Astronomy. XIX: Gravitation." *The Observatory* **38**: 93–98.

——— (1922). "Majorana's Theory of Gravitation." *Astrophysical Journal* **56**: 71–72.

EINSTEIN, Albert. (1919a). "Bemerkung über periodische Schwankungen der Mondlänge, welche bischer nach der Newtonschen Mechanik nicht erklärbar schienen." *Sitzungsberichte der Preussischen Akademie der Wissenschaften* (Berlin): 433–436.

——— (1919b). "Bemerkung zur vorstehenden Notiz." *Sitzungsberichte der Preussischen Akademie der Wissenschaften* (Berlin): 711.

von Eötvös, Roland. (1896). "Untersuchungen über Gravitation und Erdmagnetismus." *Annalen der Physik und Chemie* (3) **59**: 354–400.

von Eötvös, Roland, Pekár, Desiderius & Fekete, Eugen. (1922). "Beiträge zum Gesetze der Proportionalität von Trägheit und Gravität." *Annalen der Physik* (4) **68**: 11–66.

Erismann, Theodor. (1908). "Zur Frage nach der Abhängigkeit der Gravitationskraft von Zwischenmedium." *Vierteljahresschrift der Naturforschenden Gesellschaft zu Zürich* **53**: 157–185.

——— (1911). "Sur la dépendance de la force de gravitation du milieu intermédiaire à travers lequel elle s'exerce." *Archives des Sciences Physiques et Naturelles* **31**: 36–45.

Fink, Karl J. (1982). "*Actio in distans*, Repulsion, Attraction: the Origin of a 18th Century Fiction." *Archiv für Begriffsgeschichte* **25**: 69–87.

Forch, Carl. (1903a). "Bewirken radioaktive Substanzen eine Absorption von Gravitationsenergie?" *Physikalische Zeitschrift* **4**: 318–319.

——— (1903b). "Weitere Versuche zur Frage: Bewirken radioaktive Substanzen eine Absorption von Gravitationsenergie?" *Physikalische Zeitschrift* **4**: 443–445.

Fotheringham, J. K. (1927). "Changes in the Length of the Day." *Nature* **119**: 318–319.

——— (1932). "The Determination of the Accelerations and Fluctuations in the Motions of the Sun and Moon." *The Observatory* **55**: 305–316.

Geigel, Robert. (1903). "Über Absorption von Gravitationsenergie durch radioaktive Substanz." *Annalen der Physik* (4) **10**: 429–435.

Gillies, George T. (1987). "The Newtonian Gravitational Constant." *Metrologia—International Journal of Scientific Metrology* **24** (supplement): 1–56.

——— (1990). "Resource Letter MNG-1: Measurements of Newtonian Gravitation." *American Journal of Physics* **58**: 525–534.

Grayson, H. W. (1978). "An Experiment to Test the Gravitational Field of a Mass at its Center." In *The Theory of Relativity Revisited.* H. W. Grayson, ed. 253–261. Philadelphia: Dorrance.

Groten, Erwin. (1972). "The Problem of Gravitational Absorption." *Geophysical Journal of the Royal Astronomical Society* **27**: 447–448.

Hawes, Joan L. (1968). "Newton's Revival of the Aether Hypothesis and the Explanation of Gravitational Attraction." *Notes and Records of the Royal Society of London* **23**: 200–212.

Heydweiller, Adolf. (1902). "Zeitliche Gewichtsänderungen radioaktiver Substanz." *Physikalische Zeitschrift* **4**: 81–82.

Jeffreys, Harold. (1918). "A Theory of Physical Forces." *The Observatory* **41**: 217–220.

Kaufmann, Walter. (1903). "Bemerkungen zu der Arbeit des Hrn. R. Geigel: Über die Absorption von Gravitationsenergie durch radioaktive Substanz." *Annalen der Physik* (4) **10**: 894–896.

Kleiner, Alfred. (1905). [Influence des milieux sur la gravitation]. *Archive des Sciences Physiques et Naturelles* (4) **20**: 420–423.

Laager, Fritz. (1904). *Versuch mit der Drehwage die Abhängigkeit der Gravitation von Zwischermedium nachzuweisen.* Bern: Haller.

Lodge, Oliver Joseph. (1912). "The Discovery of Radioactivity and Its Influence on the Course of Physical Science." *Journal of the Chemical Society* **101**: 2005–2031.

McGuire, J. E. & Tamny, Martin. (1983). *Certain Philosophical Questions: Newton's Trinity Notebook.* Cambridge: Cambridge University Press.

MAJORANA, Quirino. (1918/9). "Nuove ipotesi cosmogoniche e nuovo fenomeno gravitazionale." *Atti della Reale Accademia delle Scienze di Torino* **54**: 667–669.

——— (1919/20a). "Sulla gravitazione." *Atti della Reale Accademia delle Scienze di Torino* **55**: 69–88.

——— (1919–20b). "Sulla gravitazione." *Rendiconti della Reale Accademia dei Lincei. Classe di Scienze Fisiche, Matematiche e Naturali* (5) **28** (2° Semestre): 165–174, 221–223, 313–317, 416–421, 480–489; **29** (1° Semestre): 23–32, 90–99, 163–169, 235–240.

——— (1919a). "Sur la gravitation." *Comptes Rendus des Séances de l'Académie des Sciences de Paris* **169**: 646–649.

——— (1919b). "Expériences sur la gravitation." *Comptes Rendus des Séances de l'Académie des Sciences de Paris* **169**: 719–721.

——— (1920). "On Gravitation. Theoretical and Experimental Researches." *London, Edinburgh and Dublin Philosophical Magazine* (6) **39**: 488–504.

——— (1921). "Sur l'absorption de la gravitation." *Comptes Rendus des Séances de l'Académie des Sciences de Paris* **173**: 478–479.

——— (1921–22). "Sull'assorbimento della gravitazione." *Rendiconti della Reale Accademia dei Lincei. Classe di Scienze Fisiche, Matematiche e Naturali* (5) **30** (2° Semestre): 75–79, 289–294, 350–354, 442–446; **31** (1° Semestre): 41–45, 81–86, 141–146, 221–226, 343–346.

——— (1930). "Quelques recherches sur l'absorption de la gravitation par la matière." *Journal de Physique et le Radium* (7) **1**: 314–323.

——— (1941). "Le mie ricerche scientifiche." *Nuovo Cimento* **18**: 71–86.

——— (1954). "Su di un'ipotesi cosmogonica." *Rendiconti della Accademia Nazionale dei Lincei. Classe de scienze fisiche, matematiche e naturali* (8) **17**: 150–157.

——— (1957a). "Sull'ipotesi dell'assorbimento gravitazionale." *Rendiconti della Academia Nazionale dei Lincei. Scienze fisiche, matematiche e naturali* (8) **22**: 392–397.

——— (1957b). "Ipotetiche conseguenze dell'assorbimento gravitazionale." *Rendiconti della Academia Nazionale dei Lincei. Scienze fisiche, matematiche e naturali* (8) **22**: 397–402.

MOYER, Donald Franklin. (1978). "Revolution in Science: the 1919 Eclipse Test of General Relativity." In *On the Path of Albert Einstein*. Arnold Perlmutter and Linda F. Scott, eds. 55–102. New York: Plenum.

NEWCOMB, Simon. (1903). "On the Desirableness of a Re-investigation of the Problems Growing Out of the Mean Motion of the Moon." *Monthly Notices of the Royal Astronomical Society* **63**: 316–324.

——— (1909). "Fluctuations in the Moon's Mean Motion." *Monthly Notices of the Royal Astronomical Society* **69**: 164–169.

OPPENHEIM, S. (1920). "Kritik des Newtonschen Gravitationsgesetzes." In *Encyklopädie der mathematischen Wissenschaften*. Band VI.2.B *Astronomie*. K. Schwarzschild, S. Oppenheim and W. Dyck, eds. 80–158. Leipzig: B. G. Teubner.

PERUCCA, Eligio. (1958). "Commemorazione del socio Quirino Majorana." *Rendiconti delle Sedute della Accademia Nazionale dei Lincei* **25**: 354–362.

POINCARÉ, Jules Henri. (1906/7). "Les Limites de la Loi de Newton." *Bulletin Astronomique* **17**: 121–269 (1953).

POTTER, Harold Herbert. (1922). "Note on the Gravitational Acceleration of Bismuth." *Physical Review* **19**: 187–188.

POYNTING, John Henry. (1900). "Recent Studies in Gravitation." *Nature* **62**: 403–408.

PRESTON, Samuel Tolver. (1881). "On the Importance of Experiments in Relation to the Mechanical Theory of Gravitation." *London, Edinburgh and Dublin Philosophical Magazine* (5) **11**: 391–393.

Rosenfeld, Léon. (1969). "Newton's View on Aether and Gravitation." *Archive for History of Exact Sciences* **6**: 29–37.

Roseveare, N. T. (1982). *Mercury's Perihelion from Le Verrier to Einstein.* Oxford: Clarendon Press.

Ross, Franck Elmore. (1911). "Empirical Short-Period Terms in the Moon's Mean Longitude." *Monthly Notices of the Royal Astronomical Society* **72**: 27–28.

Russell, Henry Norris. (1921). "On Majorana's Theory of Gravitation." *Astrophysical Journal* **54**: 334–346. [Reprinted in *Contributions of the Mount Wilson Observatory* No. 216: 1–13].

Schlomka, Teodor. (1927–30). "Über die Abhängigkeit der Schwerkraft von Zwischenmedium." *Zeitschrift für Geophysik* **3**: 397–400; **6**: 392–396.

Schneller, H. (1934). "Nekrolog. Kurt Felix Bottlinger." *Vierteljahresschrift der astronomischen Gesellschaft zu Leipzig* **69**: 350–353.

Schuster, Arthur. (1903). "Cosmical Radio-activity." *Chemical News* **88**: 166–167.

See, Thomas Jefferson Jackson. (1917). *Electrodynamic Wave-Theory of Physical Forces.* Vol. I. Bulletins 1 to 6 inclusive. ["Announcing the discovery of the physical cause of magnetism, of electrodynamic action, and of universal gravitation. With proof that these fundamental forces of nature are due to the interpenetration of waves propagated with the velocity of light through the free aether, but more slowly through the solid masses, whence arises also the refraction, dispersion and perhaps absorption of part of the wave energy, and thus the hitherto unexplained fluctuations of the Moon's mean motion established by Newcomb in 1909, and justly pronounced the most enigmatical phenomenon presented by the celestial motions."] Lynn (MA): Thos. P. Nichols & Sons.

von Seeliger, Hugo. (1895). "Über das Newton'sche Gravitationsgesetz." *Astronomische Nachrichten* **137**: 129–136.

——— (1896). "Über das Newton'sche Gravitationsgesetz." *Sitzungsberichte der Bayerischen Akademie der Wissenschaften zu München* **26**: 373–400.

——— (1909). "Über die Anwendung der Naturgesetze auf das Universum." *Sitzungsberichte der Königlichen Bayerischen Akademie der Wissenschaften zu München. Mathematisch-physikalische Klasse* **39** (4. Abhandlung): 1–25.

Spencer Jones, Harold. (1926). "The Rotation of the Earth." *Monthly Notices of the Royal Astronomical Society* **87**: 4–31.

——— (1939). "The Rotation of the Earth, and the Secular Acceleration of the Sun, Moon and Planets." *Monthly Notices of the Royal Astronomical Society* **99**: 541–558.

——— (1955). "Definition of the Second of Time." *Nature* **176**: 669–670.

——— (1956). "The Rotation of the Earth." In *Encyclopedia of Physics [Handbuch der Physik].* Vol. 47: 1–23. S. Flügge, ed. Berlin: Springer.

Steenbeck, Max & Treder, Hans-Jürgen. (1984). "Möglichkeiten der experimentellen Schwerkraftforschung." *Veröffentlichungen des Forschungbereichs Kosmische Physik* (Heft 11): 1–46.

Taylor, W. B. (1876). "Kinetic Theories of Gravitation." *Annual Report of the Board of Regents of the Smithsonian Institution*: 205–282.

Thomson, William (Lord Kelvin). (1903). [Discussion on the nature of the emanations from radium]. *Report of the Meeting of the British Association for the Advancement of Science* **73**: 535–537.

Whittaker, Edmund. (1951–53). *A History of the Theories of Aether and Electricity.* 2 vols. London: T. Nelson and Son.

Wilson, H. A. (1922). "Note on the Ratio of Mass to Weight for Bismuth and Aluminium." *Physical Review* **20**: 75–77.

WOODWARD, James F. (1972). "The Search for a Mechanism: Action-at-a-Distance in Gravitational Theory." Ph.D. dissertation, University of Denver, Denver (CO) [UMI 72-33,077].

ZEEMAN, Pieter. (1918). "Some Experiments on Gravitation. The Ratio of Mass to Weight for Crystals and Radioactive Substances." *Proceedings of the Royal Academy of Amsterdam* **20**: 542–553.

ZENNECK, J. (1901). "Gravitation." In *Encyklopädie der mathematischen Wissenschaften, mit Einschluß an ihrer Anwendung*. Band V.1. *Physik*. Arnold Sommerfeld, ed. 25–67. Leipzig: Teubner.

Minkowski, Mathematicians, and the Mathematical Theory of Relativity

Scott Walter

THE IMPORTANCE OF THE THEORY OF RELATIVITY for twentieth-century physics, and the appearance of the Göttingen mathematician Hermann Minkowski at a turning point in its history have both attracted significant historical attention. The rapid growth in scientific and philosophical interest in the principle of relativity has been linked to the intervention of Minkowski by Tetu Hirosige, who identified Minkowski's publications as the turning point for the theory of relativity, and gave him credit for having clarified its fundamental importance for all of physics (Hirosige 1968: 46; 1976: 78). Lewis Pyenson has placed Minkowski's work in the context of a mathematical approach to physics popular in Göttingen, and attributed its success to the prevalence of belief in a neo-Leibnizian notion of pre-established harmony between pure mathematics and physics (Pyenson 1985, 1987: 95). The novelty to physics of the aesthetic canon embodied in Minkowski's theory was emphasized by Peter Galison (1979), and several scholars have clarified technical and epistemological aspects of Minkowski's theory.[1] In particular, the introduction of sophisticated mathematical techniques to theoretical physics by Minkowski and others is a theme illustrated by Christa Jungnickel and Russell McCormmach.[2]

In what follows, we address another aspect of Minkowski's role in the history of the theory of relativity: his disciplinary advocacy. Minkowski's 1908 Cologne lecture "Raum und Zeit" (Minkowski 1909) may be understood as an effort to ex-

[1] On Minkowski's role in the history of relativity see also Illy 1981 and Pyenson 1987. Many references to the primary and secondary literature on the theory of relativity may be found in Miller 1981 and Paty 1993. Pauli 1958 remains an excellent guide to the primary literature.

[2] McCormmach 1976; Jungnickel & McCormmach 1986: II, 334–347.

H. Goenner, J. Renn, J. Ritter, T. Sauer (eds.)
The Expanding Worlds of General Relativity
(Einstein Studies, volume 7) pp. 45–86

tend the disciplinary frontier of mathematics to include the principle of relativity. We discuss the tension created by a mathematician's intrusion into the specialized realm of theoretical physics, and Minkowski's strategy to overcome disciplinary obstacles to the acceptance of his work. The effectiveness of his approach is evaluated with respect to a selection of responses, and related to trends in bibliometric data on disciplinary contributions to non-gravitational theories of relativity through 1915.

1. Minkowski's authority in mathematics and physics

At the time of the meeting of the German Association in late September 1908, Minkowski was recognized as an authority on the theory of relativity nowhere outside of the university town of Göttingen. The structure and content of Minkowski's lecture, we will see later, was in many ways a function of a perceived deficit of credibility. In order to understand this aspect of Minkowski's lecture, we first examine how Minkowski became acquainted with the electrodynamics of moving bodies.

Around 1907, Minkowski's scientific reputation rested largely upon his contribution to number theory.[3] Yet Minkowski was also the author of an article on capillarity (1906) in the authoritative *Encyklopädie der mathematischen Wissenschaften*, granting him a credential in the domain of mechanics and mathematical physics. In addition, Minkowski had lectured on capillarity, potential theory, and analytical mechanics, along with mathematical subjects such as Analysis Situs and number theory at Zurich Polytechnic, where Einstein, Marcel Grossmann and Walter Ritz counted among his students; he also lectured on mechanics and electrodynamics (among other subjects) in Göttingen, where he held the third chair in mathematics, created for him at David Hilbert's request in 1902.[4]

In Göttingen, Minkowski took an interest in a subject strongly associated with the work of many of his new colleagues: electron theory. An early manifestation of this interest was Minkowski's co-direction of a seminar on the subject with his friend Hilbert, plus Gustav Herglotz and Emil Wiechert, which met during the summer semester of 1905.[5] While Lorentz's 1904 paper (with a form of the transformations now bearing his name) was not on the syllabus, and Einstein's 1905 paper had not yet appeared, one of the students later recalled that Minkowski had hinted that he was engaged with the Lorentz transformations.[6]

Minkowski was also busy with his article on capillarity, however, and for the next two years there is no trace of his engagement with the theory of relativity. In October 1907, Minkowski wrote to Einstein to request an offprint of his *Annalen*

---

[3] Minkowski published his lectures on Diophantine analysis in Minkowski 1907a.

[4] Copies of Minkowski's manuscript notes of these lectures are in the Niels Bohr Library, Minkowski Papers, Boxes 7, 8 and 9.

[5] On the Göttingen electron theory seminar, see Pyenson 1985: 102.

[6] Undated manuscript, Niedersächsische Staats- und Universitätsbibliothek, Hilbert *Nachlaß* 570/9; Born 1959: 682.

article on the electrodynamics of moving bodies, for use in his seminar on the partial differential equations of physics, jointly conducted by Hilbert.[7] During the following Easter vacation, he gave a short series of lectures on "New Ideas on the Basic Laws of Mechanics" for the benefit of science teachers.[8]

In what seem to be notes to these holiday lectures, Einstein's knowledge of mathematics was subject to criticism. Minkowski reminded his audience that he was qualified to make this evaluation, since Einstein had him to thank for his education in mathematics. From Zurich Polytechnic, Minkowski added, a complete knowledge of mathematics could not be obtained.[9]

This frank assessment of Einstein's skills in mathematics, Minkowski explained, was meant to establish his right to evaluate Einstein's work, since he did not know how much his authority carried with respect to "the validity of judgments in physical things," which he wanted "now to submit." A pattern was established here, in which Minkowski would first suggest that Einstein's work was mathematically incomplete, and then call upon his authority in mathematics in order to validate his judgments in theoretical physics. While Minkowski implicitly recognized Einstein's competence in questions of physics, he did not yet appreciate how much Europe's leading physicists admired the work of his former student.[10] Even in his fief of Göttingen, Minkowski knew he could not expect any authority to be accorded to him in theoretical physics, yet this awareness of his own lack of credentials in physics did not prevent him from lecturing on the principle of relativity.

While the scientific world had no real means of judging Minkowski's competence in theoretical physics due to the paucity of relevant publications, Minkowski himself did not consider his knowledge in physics to be extensive. It is for this reason that he sought an assistant capable of advising him on physical matters, and when Max Born—a former student from the electron theory seminar—wrote him from Breslau (now Wrocław, Poland) for help with a technical problem, he found

---

[7] Minkowski to Einstein, 9 October 1907 (Einstein *CP5*: doc. 62); course listing in *Physikalische Zeitschrift* **8** (1907): 712. Fragmentary notes by Hermann Mierendorff from this seminar show a discussion of Lorentz's electrodynamics of moving media, see Niedersächsische Staats- und Universitätsbibliothek, Hilbert *Nachlaß* 570/5; Pyenson 1985: 83. During the same semester, Minkowski introduced the principle of relativity into his lectures on the theory of functions ("Funktionentheorie." Minkowski Papers: Box 9, Niels Bohr Library).

[8] "Neuere Ideen über die Grundgesetze der Mechanik," held in Göttingen from 21 April to 2 May, see *L'Enseignement Mathématique* **10** (1908): 179.

[9] Undated manuscript, Niedersächsische Staats- und Universitätsbibliothek, Math. Archiv 60: 4, 52. Minkowski's uncharitable assessment of mathematics at Zurich Polytechnic belied the presence on the faculty of his friend Adolf Hurwitz, a member of the mathematical elite, and a lecturer of great repute. Graduates included Marcel Grossmann, L.-Gustave du Pasquier and Minkowski's doctoral student Louis Kollros, all of whom were called upon to teach university mathematics upon completion of their studies. In recollections of his years as Einstein's classmate, Kollros wrote that there was "almost too much mathematics" at Zurich Polytechnic (Kollros 1956: 273). Minkowski's remark that Einstein's mathematical knowledge was incomplete may have been based on the fact that, unlike his classmates, Einstein did not elect to pursue graduate studies in mathematics, after obtaining the diploma from Polytechnic.

[10] In a letter of 18 October 1908, Minkowski wrote to Robert Gnehm of his satisfaction in learning—during the Cologne meeting of scientists and physicians—how much Einstein's work was admired by the likes of Walther Nernst, Max Planck and H. A. Lorentz (Seelig 1956: 131–132).

a suitable candidate.

Initially attracted to mathematics, Born heard lectures by Leo Königsberger in Heidelberg, and Adolf Hurwitz in Zurich, and later considered Hurwitz's private lectures as the high point of his student career. In Göttingen, Born obtained a coveted position as Hilbert's private assistant, and began a doctoral dissertation on Bessel functions under Hilbert's direction. When he abandoned the topic, as Born recalled in old age, Hilbert laughed and consoled him, saying he was much better in physics.[11] In the same year, Born attended Hilbert and Minkowski's electron theory seminar, along with Max Laue and Jakob Laub, among others (Born 1959: 682; Pyenson 1985: 102). Profoundly influenced by what he learned in this seminar, and deeply devoted to both Hilbert and Minkowski, Born was not permitted to write a dissertation on electron theory, although the idea appealed to him (Born 1959: 684). Felix Klein obliged him to write a dissertation on elasticity theory, but in order to avoid having "the great Felix" as an examiner, Born took up Karl Schwarzschild's suggestion to prepare for the oral examination in astronomy (Born 1906, 1968: 20–21). After defending his doctoral dissertation on 14 January 1907, Born spent six months in Cambridge with Joseph Larmor and J. J. Thomson before returning to Breslau, where the young theoretical physicists Stanislaus Loria and Fritz Reiche brought Einstein's 1905 *Annalen* paper on relativity to his attention (Born 1959: 684).

In studying relativity with Reiche, as Born recounted later, he encountered some difficulties. He formulated these in a letter to Minkowski, seeking his former teacher's advice. Minkowski's response to Born's letter was a great surprise, for instead of the requested technical assistance, Minkowski offered him the possibility of an academic career. Minkowski wrote that he had been working on the same problem as Born, and that he "would like to have a young collaborator who knew something of physics, and of optics in particular" (Born 1978: 130).[12] Besides mathematics, Born had studied physics in Göttingen, attending Woldemar Voigt's "stimulating" lectures on optics and an advanced course on optical experimentation (Born 1968: 21). It was just this background in optics that Minkowski lacked, and he looked to Born to guide him through unknown territory. In return, Minkowski promised Born he would open the doors to an academic career. The details were to be worked out when they met at the meeting of the German Association of Scientists and Physicians, later that year in Cologne (Born 1978: 130).[13]

In April 1908, Minkowski published a technically accomplished paper on the electromagnetic processes in moving bodies ("Die Grundgleichungen für die electromagnetischen Vorgänge in bewegten Körpern," hereafter *Grundgleichungen*).

---

[11] Transcript of an oral interview with Thomas S. Kuhn, 18 October 1962, Archives for History of Quantum Physics, p. 5.

[12] According to another version, the manuscript sent to Minkowski showed a new way of calculating the electromagnetic mass of the electron, described by Born as a combination of "Einstein's ideas with Minkowski's mathematical methods" (Born 1968: 25).

[13] Minkowski's premature death prevented him from personally fulfilling his obligation to Born, but his Göttingen colleagues accorded Born the venia legendi in theoretical physics, on Voigt's recommendation (Born 1978: 136).

In this essay, Minkowski wrote the empty-space field equations of relativistic electrodynamics in four-dimensional form, using Arthur Cayley's matrix calculus. He also derived the equations of electrodynamics of moving media, and formulated the basis of a mechanics appropriate to four-dimensional space with an indefinite squared interval. Minkowski's study represented the first elaboration of the principle of relativity by a mathematician in Germany.

Soon after its publication, the *Grundgleichungen* sustained restrained comment from Minkowski's former students Albert Einstein and Jakob Laub (1908a, 1908b). These authors rejected out of hand the four-dimensional apparatus of Minkowski's paper, the inclusion of which, they wrote, would have placed "rather great demands" on their readers (1908a: 532). No other reaction to Minkowski's work was published before the Cologne meeting.

By the fall of 1908, Minkowski had spoken publicly of his views on relativity on several occasions, but never outside of Göttingen. The annual meeting of the German Association was Minkowski's first opportunity to speak on relativity before an elite international audience of physicists, mathematicians, astronomers, chemists and engineers. At no other meeting could a scientist in Germany interact with other professionals working in disciplines outside of his own.

The organization of the various disciplinary sections of the annual meeting of the German Association fell to the corresponding professional societies (Forman 1967: 156). For example, the German Physical Society organized the physics section, and the German Society of Mathematicians managed the mathematics section. For the latter section, the theme of discussion was announced in late April by the society's president, Felix Klein. In a call for papers, Klein encouraged authors to submit works especially in the area of mechanics. Prior to the announcement, however, Klein must have already arranged at least one contribution in mechanics, since he added a teaser, promising an "expert aspect" of a recent investigation in this area.[14] It is tempting to identify this as a forward reference to Minkowski's lecture, a draft of which predates Klein's communication by a few days. The lecture was to be the first talk out of seven in the mathematics section, which doubled as a session of the German Society of Mathematicians.[15]

2. The Cologne lecture

The Göttingen archives contain four distinct manuscript drafts of Minkowski's Cologne lecture, none of which corresponds precisely to either of the two printed versions of the lecture in the original German.[16] Unless stipulated otherwise, we refer here to the longer essay published posthumously in both the *Physikalische*

[14] *Jahresbericht der deutschen Mathematiker-Vereinigung* **17** (1908): 61, dated 26 April 1908.

[15] Most of the lectures in the first section were published in volume 18 of the *Jahresbericht der deutschen Mathematiker-Vereinigung*. Shortly after the end of the First World War, the German Physical Society also held session at meetings of the German Association (see Forman 1967: 156).

[16] Niedersächsische Staats- und Universitätsbibliothek, Math. Arch. 60: 2 and 60: 4. An early draft is dated 24 April 1908 (60: 4, folder 1, p. 66); the other drafts are undated.

Zeitschrift and the *Jahresbericht der deutschen Mathematiker-Vereinigung* in early 1909.

From the outset of his lecture, Minkowski announced that he would reveal a radical change in the intuitions of space and time:

> Gentlemen! The conceptions of space and time which I would like to develop before you arise from the soil of experimental physics. Therein lies their strength. Their tendency is radical. From this hour on, space by itself and time by itself are to sink fully into shadows and only a kind of union of the two should yet preserve autonomy.
>
> First of all I would like to indicate how, [starting] from the mechanics accepted at present, one could arrive through purely mathematical considerations at changed ideas about space and time.[17] (Minkowski 1909: 75)

The evocation of experimental physics was significant in the first sentence of Minkowski's lecture, and it was deceptive. In what followed, Minkowski would refer to experimental physics only once, to invoke the null result of Albert A. Michelson's optical experiment to detect motion with respect to the luminiferous ether. Otherwise, Minkowski kept his promise of a "*rein mathematische*" exposé, devoid of experimental considerations. A purely theoretical presentation enabled Minkowski to finesse the recent well-known experimental results purporting to disconfirm relativity theory, obtained by Walter Kaufmann.[18]

Less illusory than the mention of experimental physics was Minkowski's announcement of a radical change in conceptions of space and time. That this revelation was local and immediate, is signaled by the phrase "from this hour on" [*von Stund' an*]. Here it was announced that a union of space and time was to be revealed, and for the first time. This was a rhetorical gesture (all of the results presented in the Cologne lecture had been published in the *Grundgleichungen*), but it was an effective one, because the phrase in question became emblematic of the theory of relativity in broader circles.

It may be noted from the outset that the claims Minkowski made for his theory fell into two categories. In one category were Minkowski's claims for scientific priority, which concerned the physical, mathematical and philosophical aspects of his theory of relativity. In what follows, we will concentrate on the second category of claims, which were *metatheoretical* in nature. The latter claims concerned the theory's type, not its constituent elements. Claims of the second sort, all having to do with the geometric nature of the theory, reinforced those of the first sort.

---

[17] "M. H.! Die Anschauungen über Raum und Zeit, die ich Ihnen entwickeln möchte, sind auf experimentell-physikalischem Boden erwachsen. Darin liegt ihre Stärke. Ihre Tendenz ist eine radikale. Von Stund' an sollen Raum für sich und Zeit für sich völlig zu Schatten herabsinken und nur noch eine Art Union der beiden soll Selbständigkeit bewahren. Ich möchte zunächst ausführen, wie man von der gegenwärtig angenommen Mechanik wohl durch eine rein mathematische Überlegung zu veränderten Ideen über Raum und Zeit kommen könnte."

[18] The empirical adequacy of the "Lorentz-Einstein" theory had been challenged by Walter Kaufmann in 1905, on the basis of his measurements of the magnetic deflection of cathode rays (see Miller 1981 and Hon 1995). Two days after Minkowski's lecture, Alfred Bucherer announced to the physical section the results of his deflection experiments, which contradicted those of Kaufmann and confirmed the expectations of the Lorentz-Einstein theory (Bucherer 1908). In the discussion of this lecture, Minkowski expressed joy in seeing the "monstrous" rigid electron hypothesis experimentally defeated in favor of the deformable electron of Lorentz's theory (see Bucherer 1908: 762).

The opening remarks provide an example: the allusion to changed ideas about space and time belongs to the first sort, while the claim of a purely mathematical development is of the second kind.

In order to demonstrate the difference between the old view of space and time and the new one, Minkowski distinguished two transformation groups with respect to which the laws of classical mechanics were covariant.[19] Considering first the same zero point in time and space for two systems in uniform translatory motion, he noted that the spatial axes x, y, z could undergo an arbitrary rotation about the origin. This corresponded to the invariance in classical mechanics of the sum of squares $x^2 + y^2 + z^2$, and was a fundamental characteristic of physical space, as Minkowski reminded his audience, that did not concern motion. Next, the second group was identified with the transformations:

$$x' = x + \alpha t,\ y' = y + \beta t,\ z' = z + \gamma t,\ t' = t.$$

Thus physical space, Minkowski pointed out, which one supposed to be at rest, could in fact be in uniform translatory motion; from physical phenomena no decision could be made concerning the state of rest (1909: 77).

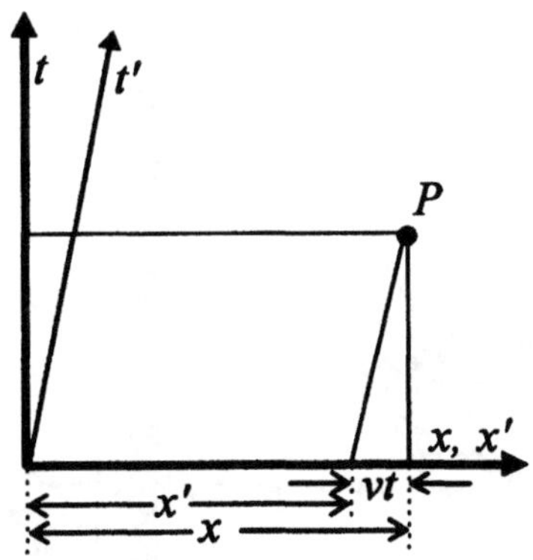

Figure 1. Classical displacement diagram.

After noting verbally the distinction between these two groups, Minkowski turned to the blackboard for a graphical demonstration. He drew a diagram to demonstrate that the above transformations allowed one to draw the time axis in any direction in the half-space $t > 0$. While no trace has been found of Minkowski's drawing, it may have resembled the one published later by Max Born and other expositors of the theory of relativity (see Figure 1).[20] This was the

[19] Minkowski introduced the use of *covariance* with respect to the Lorentz transformations in Minkowski 1908a: 473. In the Cologne lecture, the term *invariant* was employed in reference to both covariant and invariant expressions.

[20] Born 1920. A similar diagram appeared earlier in a work by Vito Volterra, who attributed it to a lecture given in Rome by Guido Castelnuovo (Volterra 1912: 22, fig. 5).

occasion for Minkowski to introduce a spate of neologisms (Minkowski 1909: 76–77): *Weltpunkt, Weltlinie* and *Weltachse*, as well as new definitions of the terms *Substanz* ['something perceptible'], and *Welt* [the manifold of all conceivable points x, y, z, t].

At this point, Minkowski raised the question of the relation between these two groups, drawing special attention to the characteristics of spatial orthogonality and an arbitrarily-directed temporal axis. In response, he introduced the hyperbolic equation:

$$c^2t^2 - x^2 - y^2 - z^2 = 1,$$

where c was an unspecified, positive-valued parameter (Minkowski 1909: 77). Suppressing two dimensions in y and z, he then showed how this unit hypersurface might be used to construct a group of transformations G_c, once the arbitrary displacements of the zero point were associated with rotations about the origin. The figure obtained was introduced on a transparent slide, showing two pairs of symmetric axes.[21]

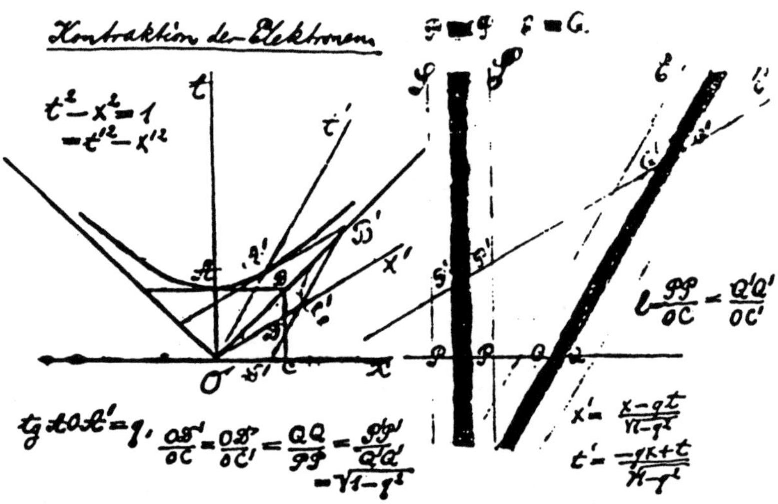

Figure 2. Minkowski's space-time and length-contraction diagrams.

Minkowski's hand-colored, transparent slide [10 × 15 cm], Niedersächsische Staats- und Universitätsbibliothek, Math. Archiv 60: 2, is reproduced here by courtesy of the Handschriftenabteilung.

Minkowski constructed the figure using the upper branch of the two-branched unit hyperbola $c^2t^2 - x^2 = 1$ to determine the parallelogram $OA'B'C'$, from which the x' and t' axes were established (see Figure 2, left, and the Appendix). The

[21] Similar figures appear in Minkowski 1909: 77.

relation between this diagram and the one corresponding to classical mechanics he pointed out directly: as the parameter c approached infinity,

> this special transformation becomes one in which the t' axis can have an arbitrary upward direction, and x' approaches ever closer to x.[22] (Minkowski 1909: 78)

In this way, the new space-time diagram collapsed into the old one, in a lovely graphic recovery of classical kinematics.[23]

The limit-relation between the group G_c and the group corresponding to classical mechanics (G_∞) called forth a comment on the history of the principle of relativity. Minkowski observed that in light of this limit-relation, and

> since G_c is mathematically more intelligible than G_∞, a mathematician would well have been able, in free imagination, to arrive at the idea that in the end, natural phenomena actually possess an invariance not with respect to the group G_∞, but rather to a group G_c, with a certain finite, but in ordinary units of measurement *extremely large* [value of] c. Such a premonition would have been an extraordinary triumph for pure mathematics.[24] (Minkowski 1909: 78)

To paraphrase, it was no more than a fluke of history that a nineteenth-century mathematician did not discover the role played by the group G_c in physics, given its greater mathematical intelligibility in comparison to the group G_∞. In other words, the theory of relativity was not a product of pure mathematics, although it could have been. Minkowski openly recognized the role—albeit a heuristic one—of experimental physics in the discovery of the principle of relativity. All hope was not lost for pure mathematics, however, as Minkowski continued:

> While mathematics displays only more staircase-wit here, it still has the satisfaction of realizing straight away, thanks to fortunate antecedents and the exercised acuity of its senses, the fundamental consequences of such a reformulation of our conception of nature.[25] (Minkowski 1909: 78)

Minkowski conceded that, in this instance, mathematics could only display wisdom after the fact, instead of a creative power of discovery. Again he stressed the mathematician's distinct advantage over members of other scientific disciplines in seizing the deep consequences of the new theoretical view.

---

[22] "jene spezielle Transformation in der Grenze sich in eine solche verwandelt, wobei die t'-Achse eine beliebige Richtung nach oben haben kann und x' immer genauer sich an x annähert."

[23] The elegance of Minkowski's presentation of relativistic kinematics with respect to classical kinematics was admired and appreciated by many, including Max Planck, who may have been in the audience. See Planck 1910b: 42.

[24] "Bei dieser Sachlage, und da G_c mathematisch verständlicher ist als G_∞, hätte wohl ein Mathematiker in freier Phantasie auf den Gedanken verfallen können, daß am Ende die Naturerscheinungen tatsächlich eine Invarianz nicht bei der Gruppe G_∞, sondern vielmehr bei einer Gruppe G_c mit bestimmtem endlichen, nur in den gewöhnlichen Maßeinheiten *äusßerst großen* c besitzen. Eine solche Ahnung wäre ein außerordentlicher Triumph der reinen Mathematik gewesen."

[25] "Nun, da die Mathematik hier nur mehr Treppenwitz bekundet, bleibt ihr doch die Genugtuung, daß sie dank ihren glücklichen Antezedenzien mit ihren in freier Fernsicht geschärften Sinnen die tiefgreifenden Konsequenzen einer solcher Ummodelung unserer Naturauffassung auf der Stelle zu erfassen vermag." We translate "Treppenwitz" literally as "staircase-wit," although the term was taken by Giuseppe Gianfranceschi and Guido Castelnuovo to mean that mathematics had not accomplished the first step: "Qui veramente la matematica non ha compiuro il primo passo ... " (see Minkowski 1909: 338).

2.1. Minkowski the mathematician

Minkowski's repetitive references to mathematicians and pure mathematics demand an explanation. Minkowski was a mathematician by training and profession. This fact is hardly obscure, but Minkowski's reasons for stressing his point may not be immediately obvious. Two suggestions may be made here.

In the first place, we believe that Minkowski and his contemporaries saw his work on relativity as an expansion of the disciplinary frontier of mathematics. Furthermore, this expansion was naturally regarded by some German physicists as imperialist, occurring at the expense of the nascent, growing sub-discipline of theoretical physics.[26] A desire to extend mathematical dominion over the newly-discovered region of relativistic physics would explain why Minkowski chose neither to describe his work as theoretical physics, nor to present himself as a theoretical (or mathematical) physicist.

Secondly, in relation to this, we want to suggest that Minkowski was aware of the confusion that his ideas were likely to engender in the minds of certain members of his audience. In effect, Minkowski's response to this expected confusion was to reassure his audience, by constantly reaffirming what they already knew to be true: he, Minkowski, was a mathematician.[27]

Minkowski's wide reputation and unquestioned authority in pure mathematics created a tension, which is manifest throughout his writings on relativity. As long as Minkowski signed his work as a mathematician, any theory he produced lacked the "authenticity" of a theory advanced by a theoretical physicist. No "guarantee" of physical relevance was attached to his work—on the contrary. With very few exceptions (the article on capillarity, for example), nothing Minkowski had published was relevant to physics.

Acutely aware of the cross-disciplinary tension created by his excursion into theoretical physics, Minkowski made two moves toward its alleviation. The first of these was to assert, at the outset of the lecture, that the basis of his theory was in experimental physics. The second was to display the physico-theoretical pedigree of the principle of relativity, aspects of which had been developed by the paragon of theoretical physicists, H. A. Lorentz, and by the lesser-known patent clerk and newly-named lecturer in theoretical physics in Bern, Albert Einstein.

Up to this point in his lecture, Minkowski had presented a new, real geometric interpretation of a certain transformation in x, y, z and t, which formed a group denoted by G_c. This group entertained a limit relation with the group under which the laws of classical mechanics were covariant. From this point on, until the end of the first section of his lecture, Minkowski presented what he, and soon

[26] The entry of mathematicians into the field of relativity was described by Einstein as an invasion, as Sommerfeld later recalled (1949: 102). On the emergence of theoretical physics in Germany, see Stichweh 1984; Jungnickel & McCormmach 1986; Olesko 1991. The term "disciplinary frontier" is borrowed from Rudolf Stichweh's writings.

[27] This is further suggested by the sociologist Erving Goffman's analysis of the presentation of self. Goffman noted that individuals present a different "face" to different audiences. The audience reserves the right to take the individual at his occupational face value, seeing in this a way to save time and emotional energy. According to Goffman, even if an individual were to try to break out of his occupational role, audiences would often prevent such action (see Goffman 1959: 57).

a great number of scientists, considered to be *his* theory.[28] What was this new theory? Once a system of reference x, y, z, t was determined from observation, in which natural phenomena agreed with definite laws, the system of reference could be changed arbitrarily without altering the form of these laws, provided that the transformation to the new system conformed to the group G_c. As Minkowski put it:

> The existence of the invariance of the laws of nature for the group G_c would now be understood as follows: from the entirety of natural phenomena we can derive, through successively enhanced approximations, an ever more precise frame of reference x, y, z and t, space and time, by means of which these phenomena can then be represented according to definite laws. This frame of reference, however, is by no means uniquely determined by the phenomena. *We can still arbitrarily change the frame of reference according to the transformations of the group termed* G_c *without changing the expression of the laws of nature.*[29] (Minkowski 1909: 78–79)

For anyone who might have objected that others had already pointed this out, Minkowski offered an interpretation of his theory on the space-time diagram.[30]

> We can, for example, also designate time [as] t', according to the figure described. However, in connection with this, space must then necessarily be defined by the manifold of three parameters x', y, z, on which physical laws would now be expressed by means of x', y, z, t' in exactly the same way as with x, y, z, t. Then from here on, we would no longer have *space* in the world, but endlessly many spaces; analogously, endlessly many planes exist in three-dimensional space. Three-dimensional geometry becomes a chapter of four-dimensional physics. You realize why I said at the outset: space and time are to sink into the shadows; only a world in and of itself endures.[31] (Minkowski 1909: 79)

The emphasis on space was no accident, as Minkowski presented the notion of "endlessly many spaces" as his personal contribution, in analogy to Einstein's concept of relative time. The grandiose announcement of the end of space and

[28] Examples of the identification of this passage as Minkowski's principle of relativity are found in several reports, such as Volkmann 1910: 148, and Wiechert 1915: 55.

[29] "Das Bestehen der Invarianz der Naturgesetze für die bezügliche Gruppe G_c würde nun so zu fassen sein: Man kann aus der Gesamtheit der Naturerscheinungen durch sukzessiv gesteigerte Approximationen immer genauer ein Bezugsystem x, y, z und t, Raum und Zeit, ableiten, mittels dessen diese Erscheinungen sich dann nach bestimmten Gesetzen darstellen. Dieses Bezugsystem ist dabei aber durch die Erscheinungen keineswegs eindeutig festgelegt. *Man kann das Bezugsystem noch entsprechend den Transformationen der genannten Gruppe* G_c *beliebig verändern, ohne daß der Ausdruck der Naturgesetze sich dabei verändert*."

[30] Neither Einstein, nor Lorentz, nor Poincaré attended the Cologne meeting, although in late February Einstein wrote to Johannes Stark of his intention to do so (Einstein *CP5*: doc. 88).

[31] "Z. B. kann man der beschriebenen Figur entsprechend auch t' Zeit benennen, muß dann aber im Zusammenhange damit notwendig den Raum durch die Mannigfaltigkeit der drei Parameter x', y, z definieren, wobei nun die physikalischen Gesetze mittels x', y, z, t' sich genau ebenso ausdrücken würden, wie mittels x, y, z, t. Hiernach würden wir dann in der Welt nicht mehr *den* Raum, sondern unendlich viele Räume haben, analog wie es im dreidimensionalen Raume unendlich viele Ebenen gibt. Die dreidimensionale Geometrie wird ein Kapitel der vierdimensionalen Physik. Sie erkennen, weshalb ich am Eingange sagte, Raum und Zeit sollen zu Schatten herabsinken und nur eine Welt an sich bestehen."

time served as a frame for the enunciation of Minkowski's principle of relativity.[32]

Rhetorical gestures such as this directed attention to Minkowski's theory; its acceptance by the scientific community, however, may be seen to depend largely upon the presence of two elements: empirical adequacy, claimed by Minkowski at the opening of the lecture, and the perception of an advantage over existing theories. Minkowski went on to address in turn the work of two of his predecessors, Lorentz and Einstein. Before discussing Minkowski's exposé of their work, however, we want to consider briefly the work of a third precursor, whose name was not mentioned at all in this lecture: Henri Poincaré.

2.2. Why did Minkowski not mention Poincaré?

Widely acknowledged at the turn of the century as the world's foremost mathematician, Henri Poincaré developed Lorentz's theory of electrons to a state formally equivalent to the theory published at the same time by Einstein.[33] Poincaré and Einstein both recognized that the Lorentz transformations (so named by Poincaré) form a group; Poincaré alone exploited this knowledge in the search for invariants.[34] Among Poincaré's insights relating to his introduction of a fourth imaginary coordinate in $t\sqrt{-1}$ (where $c = 1$), was the recognition of a Lorentz transformation as a rotation about the origin in four-dimensional space, and the invariance of the sum of squares in this space, which he described as a measure of distance (1906: 542). This analysis then formed the basis of his evaluation of the possibility of a Lorentz-covariant theory of gravitation.

It is unlikely that the omission of Poincaré's name was a simple oversight on Minkowski's part. The printed version of Minkowski's lecture, the corrected proofs of which were mailed only days before a fatal attack of appendicitis, was the result of careful attention in the months following the Cologne meeting.[35] This suggests that both the structure of the paper and the decision to include (or exclude) certain references were the result of deliberate choices on the part of the author.

A great admirer of Poincaré's science, Minkowski was familiar with his long paper on the dynamics of the electron, having previously cited it in the *Grundgleichungen*, in the appendix on gravitation. In an earlier, then-unpublished lecture to the Göttingen Mathematical Society on the principle of relativity, delivered on 5 November 1907, Minkowski went so far as to portray Poincaré as one of the four principal authors of the principle of relativity:

---

32 In Göttingen, Minkowski's lofty assertions were the target of student humor, as witnessed by a student parody of the course guide, see Galison 1979: 111, n. 69. Minkowski, whose lectures were said by Born (1959: 682) to be punctuated by witty remarks, undoubtedly found this amusing. His sharp sense of humor is also evident in the correspondence with Hilbert (see Rüdenberg & Zassenhaus 1973).

33 One sign of Poincaré's mathematical preeminence was the Bólyai Prize, awarded him by a unanimous jury in 1905. For studies of Poincaré's mathematical contributions to relativity theory see Cuvaj 1968 and Miller 1973. Poincaré's critique of fin-de-siècle electrodynamics is discussed in Darrigol 1995.

34 Poincaré proved that the Lorentz transformations form a group in a letter to Lorentz (reproduced in Miller 1980), and later pointed out to students the group nature of the parallel velocity transformations (see the notes by Henri Vergne of Poincaré's 1906/7 lectures, Poincaré 1906/7: 222).

35 On Minkowski's labors see Hilbert 1909a: xxix.

> Concerning the credit to be accorded to individual authors, stemming from the foundations of Lorentz's ideas, Einstein developed the principle of relativity more distinctly [and] at the same time applied it with particular success to the treatment of special problems in the optics of moving media, [and] ultimately [was] also the first to draw conclusions concerning the variability of mechanical mass in thermodynamic processes. A short while later, and no doubt independently of Einstein, Poincaré extended [the principle of relativity] in a more mathematical study to Lorentz electrons and their status in gravitation. Finally, Planck sought the basis of a dynamics grounded on the principle of relativity.[36] (Minkowski 1907b: 16–17)

Following their appearance in this short history of the principle of relativity, the theoretical physicists Lorentz, Einstein and Max Planck all made it into Minkowski's Cologne lecture, but the more mathematical Poincaré was left out.

At least one theoretical physicist felt Minkowski's exclusion of Poincaré in "Raum und Zeit" was unfair: Arnold Sommerfeld. In the notes he added to a 1913 reprint of this lecture, Sommerfeld attempted to right the wrong by making it clear that a Lorentz-covariant law of gravitation and the idea of a four-vector had both been proposed earlier by Poincaré.

Among the mathematicians following the developments of electron theory, many considered Poincaré as the founder of the new mechanics. For instance, the editor of *Acta Mathematica*, Gustav Mittag-Leffler, wrote to Poincaré on 7 July 1909 of Stockholm mathematician Ivar Fredholm's suggestion that Minkowski had given Poincaré's ideas a different expression:

> You undoubtedly know the pamphlet by Minkowski, "Raum und Zeit," published after his death, as well as the ideas of Einstein and Lorentz on the same question. Now, M. Fredholm tells me that you have touched upon similar ideas before the others, while expressing yourself in a less philosophical, more mathematical manner.[37] (Mittag-Leffler 1909)

It is unknown if Poincaré ever received this letter. Like Sommerfeld, Mittag-Leffler and Fredholm reacted to the omission of Poincaré's name from Minkowski's lecture.

The absence of Poincaré from Minkowski's speech was remarked by leading scientists, but what did Poincaré think of this omission? His first response, in any case, was silence. In the lecture Poincaré delivered in Göttingen on the new mechanics in April 1909, he did not see fit to mention the names of Minkowski and Einstein (Poincaré 1910a). Yet where his own engagement with the principle of relativity was concerned, Poincaré became more expansive. In Berlin the

[36] "Was das Verdienst der einzelnen Autoren angeht, so rühren die Grundlagen der Ideen von Lorentz her, Einstein hat das Prinzip der Relativität reinlicher herauspräpariert, zugleich es mit besonderem Erfolge zur Behandlung spezieller Probleme der Optik bewegter Medien angewandt, endlich auch zuerst die Folgerungen über Veränderlichkeit der mechanischen Masse bei thermodynamischen Vorgängen gezogen. Kurz danach und wohl unabhängig von Einstein hat Poincaré sich in mehr mathematischer Untersuchung über die Lorentzschen Elektronen und die Stellung der Gravitation zu ihnen verbreitet, endlich hat Planck einen Ansatz zu einer Dynamik auf Grund des Relativitätsprinzipes versucht."

[37] "Vous connaissez sans doute l'opuscule de Minkowski "Raum und Zeit," publié après sa mort ainsi que les idées de Einstein et Lorentz sur la même question. Maintenant M. Fredholm me dit que vous avez touché à des idées semblables avant les autres, mais en vous exprimant d'une manière moins philosophique et plus mathématique." It is a pleasure to acknowledge the assistance of Dr. K. Broms in providing me with a copy of this letter.

following year, for example, Poincaré dramatically announced that already back in 1874 (or 1875), while a student at the École polytechnique, he and a friend had experimentally confirmed the principle of relativity for optical phenomena (Poincaré 1910b: 104).[38] Less than five years after its discovery, the theory of relativity's prehistory was being revealed by Poincaré in a way that underlined its empirical foundations—in contradistinction to the Minkowskian version. If Poincaré expressed little enthusiasm for the new mechanics unleashed by the principle of relativity, and had doubts concerning its experimental underpinnings, he never disowned the principle.[39] In the spring of 1912, Poincaré came to acknowledge the wide acceptance of a formulation of physical laws in four-dimensional (Minkowski) space-time, at the expense of the Lorentz-Poincaré electron theory. His own preference remained with the latter alternative, which did not require an alteration of the concept of space (Poincaré 1912: 170).

In the absence of any clear indication why Minkowski left Poincaré out of his lecture, a speculation or two on his motivation may be entertained. If Minkowski had chosen to include some mention of Poincaré's work, his own contribution may have appeared derivative. Also, Poincaré's modification of Lorentz's theory of electrons constituted yet another example of the cooperative role played by the mathematician in the elaboration of physical theory.[40] Poincaré's "more mathematical" study of Lorentz's electron theory demonstrated the mathematician's dependence upon the insights of the theoretical physicist, and as such, it did little to establish the independence of the physical and mathematical paths to the Lorentz group. The metatheoretical goal of establishing the essentially mathematical nature of the principle of relativity was no doubt more easily attained by neglecting Poincaré's elaboration of this principle.

2.3. Lorentz and Einstein

Turning first to the work of Lorentz, Minkowski made another significant suppression. In the *Grundgleichungen*, Minkowski had adopted Poincaré's suggestion to give Lorentz's name to a group of transformations with respect to which Maxwell's equations were covariant (p. 473), but in the Cologne lecture, this convention was dropped. Not once did Minkowski mention the "Lorentz" transformations, he referred instead to transformations of the group designated G_c. The reason for

[38] The experiment was designed to test the validity of the principle of relativity for the phenomenon of double refraction. The telling of this school anecdote may also be connected to Mittag-Leffler's campaign to nominate Poincaré for the 1910 Nobel Prize for physics. Poincaré never mentioned the names of Einstein or Minkowski in print in relation to the theory of relativity, but during the course of this lecture, according to one witness, he mentioned Einstein's work in this area (see Moszkowski 1920: 15).

[39] In a lecture to the Saint Louis Congress in September 1904, Poincaré interpreted the "principe de relativité" with respect to Lorentz's theory of electrons, distinguishing this extended relativity principle from the one employed in classical mechanics (1904: 314).

[40] Willy Wien spelled out this role at the 1905 meeting of the German Society of Mathematicians in Meran. Wien suggested that "physics itself" required "more comprehensive cooperation" from mathematicians in order to solve its current problems, including those encountered in the theory of electrons (Wien 1906: 42; McCormmach 1976: xxix). While Poincaré's work in optics and electricity was well received, and his approach emulated by some German physicists (see Darrigol 1993: 223), mathematicians generally considered him their representative.

this suppression is unknown, but very probably is linked to Minkowski's discovery of a precursor to Lorentz in the employment of the transformations. In 1887, the Göttingen professor of mathematical physics, Woldemar Voigt, published his proof that a certain transformation in x, y, z and t (which was formally equivalent to the one used by Lorentz) did not alter the fundamental differential equation for a light wave propagating in the free ether with velocity c (Voigt 1887). For Minkowski, this was an essential application of the law's covariance with respect to the group G_c. Lorentz's insight he considered to be of a more general nature; Lorentz would have attributed this covariance to all of optics (Minkowski 1909; 80). By placing Voigt's transformations at the origins of the principle of relativity, Minkowski not only undercut Poincaré's attribution to Lorentz, he also emulated Hertz's epigram (Maxwell's theory is Maxwell's system of equations), whose underlying logic could only reinforce his own metatheoretical claims. In addition, he showed courtesy toward his colleague Voigt, who was not displeased by the gesture.[41]

Having dealt in this way with the origins of the group G_c, Minkowski went on to consider another Lorentzian insight: the contraction hypothesis. Using the space-time diagram, Minkowski showed how to interpret the hypothesis of longitudinal contraction of electrons in uniform translation (Figure 2, right). Reducing Lorentz's electron to one spatial dimension, Minkowski showed two bars of unequal width, corresponding to two electrons: one at rest with respect to an unprimed system and one moving with relative velocity v, but at rest with respect to the primed system. When the moving electron was viewed from the unprimed system, it would appear shorter than an electron at rest in the same system, by a factor $\sqrt{1 - v^2/c^2}$. Underlining the "fantastic" nature of the contraction hypothesis, obtained "purely as a gift from above," Minkowski asserted the complete equivalence between Lorentz's hypothesis and his new conception of space and time, while strongly suggesting that, by the latter, the former became "much more intelligible." In sum, Minkowski held that his theory offered a better understanding of the contraction hypothesis than did Lorentz's theory of electrons (1909: 80).[42]

In his discussion of Lorentz's electron theory, Minkowski was led to bring up the notion of local time, which was the occasion for him to mention Einstein.

> But the credit of first clearly recognizing that the time of one electron is just as good as that of the other, that is to say, that t and t' are to be treated identically, belongs to A. Einstein.[43] (Minkowski 1909: 81)

---

41 In response to Minkowski's attribution of the transformations to his 1887 paper, Voigt gently protested that he was concerned at that time with the elastic-solid ether theory of light, not the electromagnetic theory. At the same time, Voigt acknowledged that his paper contained some of the results later obtained from electromagnetic-field theory (see the discussion following Bucherer 1908: 762). Shortly afterwards, Lorentz generously conceded that the idea for the transformations might have come from Voigt (Lorentz 1916: 198, n. 1).

42 Lorentz's theory did not purport to explain the hypothetical contraction. Although he made no mention of this in the Cologne lecture, Minkowski pointed out in the *Grundgleichungen* that the (macroscopic) equations for moving dielectrics obtained from Lorentz's electron theory did not respect the principle of relativity (Minkowski 1908: 493).

43 "Jedoch scharf erkannt zu haben, daß die Zeit des einen Elektrons ebenso gut wie die des anderen ist, d.h. daß t und t' gleich zu behandeln sind, ist erst das Verdienst von A. Einstein."

This interpretation of Einstein's notion of time with respect to an electron was not one advanced by Einstein himself. We will return to it shortly; for now we observe only that Minkowski seemed to lend some importance to Einstein's contribution, because he went on to refer to him as having deposed the concept of time as one proceeding unequivocally from phenomena.[44]

2.4. Minkowski's distortion of Einstein's kinematics

At this point in his lecture, after having briefly reviewed the work of his forerunners, Minkowski was in a position to say just where they went wrong. Underlining the difference between his view and that of the theoretical physicists Lorentz and Einstein, Minkowski offered the following observation:

> Neither Einstein nor Lorentz rattled the concept of space, perhaps because in the above-mentioned special transformation, where the plane of $x't'$ coincides with the plane of xt, an interpretation is [made] possible by saying that the x-axis of space maintains its position.[45] (Minkowski 1909: 81–82)

This was the only overt justification offered by Minkowski in support of his claim to have surpassed the theories of Lorentz and Einstein. His rather tentative terminology [*eine Deutung möglich ist*] signaled uncertainty and perhaps discomfort in imputing such an interpretation to this pair. Also, given the novelty of Minkowski's geometric presentation of classical and relativistic kinematics, his audience may not have seen just what difference Minkowski was pointing to. Minkowski did not elaborate; but for those who doubted that a priority claim was in fact being made, he added immediately:

> Proceeding beyond the concept of space in a corresponding way is likely to be appraised as only another audacity of mathematical culture. Even so, following this additional step, indispensable to the correct understanding of the group G_c, the term *relativity postulate* for the requirement of invariance under the group G_c seems very feeble to me.[46] (Minkowski 1909: 82)

Where Einstein had deposed the concept of time (and time alone, by implication), Minkowski claimed in a like manner to have overthrown the concept of space, as Galison has justly noted (1979: 113). Furthermore, Minkowski went so far as to suggest that his "additional step" was essential to a "correct understanding" of what he had presented as the core of relativity: the group G_c. He further implied that the theoretical physicists Lorentz and Einstein, lacking a "mathematical culture," were one step short of the correct interpretation of the principle of relativity.

---

[44] "Damit war nun zunächst die Zeit als ein durch die Erscheinungen eindeutig festgelegter Begriff abgesetzt" (Minkowski 1909: 81).

[45] "An dem Begriffe des Raumes rüttelten weder Einstein noch Lorentz, vielleicht deshalb nicht, weil bei der genannten speziellen Transformation, wo die x', t'-Ebene sich mit der x, t-Ebene deckt, eine Deutung möglich ist, als sei die x-Achse des Raumes in ihrer Lage erhalten geblieben."

[46] "Über den Begriff des Raumes in entsprechender Weise hinwegzuschreiten, ist auch wohl nur als Verwegenheit mathematischer Kultur einzutaxieren. Nach diesem zum wahren Verständnis der Gruppe G_c jedoch unerläßlichen weiteren Schritt aber scheint mir das Wort *Relativitätspostulat* für die Forderung einer Invarianz bei der Gruppe G_c sehr matt."

Having disposed in this way of his precursors, Minkowski was authorized to invent a name for his contribution, which he called the postulate of the absolute world, or world-postulate for short (1909: 82). It was on this note that Minkowski closed his essay, trotting out the shadow metaphor one more time:

> The validity without exception of the world postulate is, so I would like to believe, the true core of an electromagnetic world picture; met by Lorentz, further revealed by Einstein, [it is] brought fully to light at last.[47] (Minkowski 1909: 88)

According to Minkowski, Einstein clarified the physical significance of Lorentz's theory, but did not grasp the true meaning and full implication of the principle of relativity. Minkowski marked his fidelity to the Göttingen electron-theoretical program, which was coextensive with the electromagnetic world picture. When Paul Ehrenfest asked Minkowski for a copy of the paper going by the title "On Einstein-Electrons," Minkowski replied that when used in reference to the *Grundgleichungen*, this title was "somewhat freely chosen." However, when applied to the planned sequels to the latter paper, he explained, this name would be "more correct."[48] Ehrenfest's nickname for the *Grundgleichungen* no doubt reminded Minkowski of a latent tendency among theoretical physicists to view his theory as a prolongation of Einstein's work, and may have motivated him to provide justification of his claim to have proceeded beyond the work of Lorentz and Einstein.

Did Minkowski offer a convincing argument for the superiority of his theory? The argument itself requires some clarification. According to Peter Galison's reconstruction (1979: 113), Minkowski "conjectures [that a] relativistically correct solution can be obtained" in one (spatial) dimension by rotating the temporal axis through a certain angle, leaving the x'-axis superimposed on the x-axis. Yet Minkowski did *not* suggest that this operation was either correct or incorrect. Rather, he claimed it was possible to interpret a previously-mentioned transformation in a way which was at odds with his own geometric interpretation. Proposed by Minkowski as Lorentz's and Einstein's view of space and time, such a reading was at the same time possible, and incompatible with Einstein's presentations of the principle of relativity.

The claim referred back to Minkowski's exposé of both classical and relativistic kinematics by means of space-time diagrams. As mentioned above, he had emphasized the fact that in classical mechanics the time axis may be assigned any direction with respect to the fixed spatial axes x, y, z, in the region $t > 0$. Minkowski's specification of the "special transformation" referred in all likelihood to

---

[47] "Die ausnahmslose Gültigkeit des Weltpostulates ist, so möchte ich glauben, der wahre Kern eines elektromagnetischen Weltbildes, der von Lorentz getroffen, von Einstein weiter herausgeschält, nachgerade vollends am Tage liegt."

[48] Minkowski to Paul Ehrenfest, 22 October 1908, Ehrenfest Papers, Museum Boerhaave, Leiden. Judging from the manuscripts in Minkowski's *Nachlaß* (Niedersächsische Staats- und Universitätsbibliothek, Math. Archiv 60: 1), he had made little progress on Einstein-electrons before an attack of appendicitis put an end to his life in January 1909, only ten weeks after writing to Ehrenfest. An electron-theoretical derivation of the basic electromagnetic equations for moving media appeared under Minkowski's name in 1910, but was actually written by Max Born (cf. Minkowski & Born 1910: 527).

the special Lorentz transformations, in which case Minkowski's further requirement of coincidence of the xt and $x't'$ planes was (trivially) satisfied; the term is encountered nowhere else in the text. By singling out the physicists' reliance on the special Lorentz transformation, Minkowski underlined his introduction of the inhomogeneous transformations, which accord no privilege to any single axis or origin.[49] He then proposed that Lorentz and Einstein *might* have interpreted the special Lorentz transformation as a rotation of the t'-axis alone, the x'-axis remaining fixed to the x-axis. Since Minkowski presented two geometric models of kinematics in his lecture, we will refer to them in evaluating his view of Lorentz's and Einstein's kinematics.

The first interpretation, and the most plausible one in the circumstances, refers to the representation of Galilean kinematics (see Figure 1). On a rectangular coordinate system in x and t, a t'-axis is drawn at an angle to the t-axis, and the x'-axis lies on the x-axis as required by Minkowski. Lorentz's electron theory held that in inertial systems the laws of physics were covariant with respect to a Galilean transformation, $x' = x - vt$.[50] In the $x't'$-system, the coordinates are oblique, and the relationship between t and t' is fixed by Lorentz's requirement of absolute simultaneity: $t' = t$. Where Poincaré and Einstein wrote the Lorentz transformation in one step, Lorentz used two, so that a Galilean transformation was combined with a second transformation containing the formula for local time.[51] The second transformation did not lend itself to graphical representation, and had no physical meaning for Lorentz, who understood the transformed values as auxiliary quantities. The first stage of the two-dimensional Lorentz transformation was identical to that of classical mechanics, and may be represented in the same way, by rotating the time axis while leaving the position of the space axis unchanged. When realized on a Galilean space-time diagram, and in the context of Lorentz's electron theory, Minkowski's description of the special Lorentz transformations seems quite natural. On the other hand, as a description of Einstein's kinematics it seems odd, because Einstein explicitly abandoned the use of the Galilean transformations in favor of the Lorentz transformations.[52]

Lorentz's theory of electrons provided for a constant propagation velocity of light *in vacuo*, when the velocity was measured in an inertial frame. However, this propagation velocity was not considered to be a universal invariant (as was maintained in the theories of both Einstein and Minkowski). In Lorentz's theory of electrons, retention of classical kinematics (with the adjoining notion of absolute simultaneity) meant that the velocity of light in a uniformly translating frame of

---

49 See Minkowski 1908a: § 5; 1909: 78.

50 The terminology of *Galilean transformations* was introduced by Philipp Frank (1908: 898) in his analysis of the *Grundgleichungen*.

51 Lorentz (1904) used the Galilean transformations separately from, and in conjunction with the following transformations (the notation is modified): $x' = \gamma x$, $y' = y$, $z' = z$, $t = t/\gamma - \gamma vx/c^2$, where $\gamma = 1/\sqrt{1 - v^2/c^2}$.

52 To suppose t equal to t', Einstein commented later, was to make an "arbitrary hypothesis" (1910: 26).

reference would in general depend on the frame's velocity with respect to the ether. Measurements of light velocity performed by observers in these frames, however, would always reveal the same value, due to compensating dilatory effects of motion on the tools of measurement (Lorentz 1916: 224–225).

The latter distinction enters into the second way by which Minkowski might have measured Einstein's kinematics. Referring now to a Minkowski diagram, two inertial systems S and S' may be represented, as in the left side of Figure 2. In system S, points in time and space are represented on general Cartesian axes, on which the units are chosen in such a way that the velocity of light *in vacuo* is equal to 1.[53] For an observer at rest in S, the system S' appears to be in uniform motion in a direction parallel to the x-axis with a sub-light velocity v, and the temporal axis ct' for the system S' is drawn at an angle to the axis ct. Einstein postulated that the velocity of light *in vacuo* was a universal constant, and asserted that units of length and time could be defined in the same way for all inertial systems (this definition will be discussed later, with respect to the concept of simultaneity). He showed that from the light postulate and a constraint on linearity, in accordance with his measurement conventions, it followed that light propagated with the same velocity in both systems. From the corresponding transformation equations, Einstein deduced the following equations for the surface of a light wave emitted from the origin of the space and time coordinates considered in the systems S (with coordinates x, y, z, t) and S' (coordinates designated ξ, η, ζ, τ):

$$x^2 + y^2 + z^2 = c^2t^2, \qquad \xi^2 + \eta^2 + \zeta^2 = c^2\tau^2.$$

Einstein initially presented this equivalence as proof that his two postulates were compatible; later he recognized that the Lorentz transformations followed from this equivalence and a requirement of symmetry (Einstein 1905: 901; 1907: 419). At the same time, he made no further comment on the geometric significance of this invariance and maintained at least a semantic distinction between kinematics and geometry.[54] Minkowski chose to fold one into the other, regarding $c^2t^2-x^2-y^2-z^2$ as a *geometric* invariant. Since y and z do not change in the case considered here, $c^2t^2-x^2$ is an invariant quantity when measured in an inertial system. Minkowski's space-time diagram is a model of the geometry based on this metric.

Following Minkowski's interpretation of Einstein's kinematics, the x'-axis (that which records the spatial distribution of events corresponding to $ct' = 0$) coincides with the x-axis. Recalling that the units of length and time for inertial systems were defined by Einstein in such a way that the quantity $c^2t^2 - x^2$ was invariant for any two points, the position of the x'-axis with respect to the x-axis depended only upon the relative velocity of S', manifest in the tilt angle of the ct'-axis with

53 This value of c itself implies the orthogonality of temporal and spatial axes in every inertial system, a feature which is not apparent on a Minkowski diagram. For his part, Einstein defined the units of length and time (ideal rods and clocks) in a coordinate-free manner.

54 On Einstein's reluctance to confound kinematics with geometry see his introduction of the terms "geometric shape" and "kinematic shape" to distinguish the forms of rigid bodies in a rest frame from those of rigid bodies in frames in uniform relative motion (Einstein 1907: 417, 1910: 28; Paty 1993: 170).

respect to the ct-axis (and vice-versa). Consequently, the requirement that the x'-axis coincide with the x-axis could not be met here, either, at least not without: (1) sacrificing one of Einstein's postulates, (2) abandoning Einstein's definition of time (and simultaneity), or (3) arbitrarily introducing an additional transformation in order to recover the special Lorentz transformation through composition.

Neither one of the first two options would have been considered natural or plausible to one familiar with Einstein's publications. As for the last option, since none of the properties of the Lorentz transformations are reflected geometrically, the operation is far from interpretative—it is pointless. It is also improbable that Minkowski would have attributed, even implicitly, the use of his space-time diagram to Lorentz or Einstein. For all these reasons, this reconstruction is far less plausible than the one considered previously.

If either of these two reconstructions reflects accurately what Minkowski had in mind, the upshot is an assertion that Lorentz and Einstein subscribed to a definition of space and time at variance with the one proposed by Einstein in 1905. Ascribing the first (Galilean) interpretation to Lorentz was unlikely to raise any eyebrows. The second interpretation is inconsistent with Einstein's presentation of relativistic kinematics. Furthermore, Minkowski imputed *one* interpretation [*eine Deutung*] to both Lorentz *and* Einstein.[55] Attentive to the distinction between Lorentz's theory of electrons and Einstein's theory of relativity, both Philipp Frank and Guido Castelnuovo rectified what they perceived to be Minkowski's error, as we will see later in detail for Castelnuovo.[56] On the other hand, Vito Volterra (1912: 23) and Lothar Heffter (1912: 4) adopted Minkowski's view of Einstein's kinematics, so it appears that no consensus was established on the cogency of Minkowski's argument in the pre-war period.

The confrontation of Einstein's articles of 1905 and 1907, both cited by Minkowski, with the interpretation charged to Einstein (and Lorentz) by Minkowski, offers matter for reflection. Indeed, the justification offered by Minkowski for his claim would seem to support the view, held by more than one historian, that Minkowski, to put it bluntly, did not understand Einstein's theory of relativity.[57]

2.5. Did Minkowski understand Einstein's concepts of relative time and simultaneity?

A detailed comparison of the theories of Einstein and Minkowski is called for, in order to evaluate Minkowski's understanding of Einsteinian relativity; here we review only the way in which Einstein's concepts of time and simultaneity were employed by both men up to 1908, concepts chosen for their bearing upon Minkowski's unique graphic representation of Lorentz's and Einstein's kinematics.

---

[55] A basis for this conflation was provided by Einstein in 1906, when he referred to the "*Theorie von Lorentz und Einstein*" (see the editorial note in Einstein *CP2*: 372).

[56] Frank 1910: 494; Castelnuovo 1911: 78. For later examples see Silberstein 1914: 134 and Born 1920: 170. Extreme discretion was exercised here, as none of these writers taxed Minkowski with error.

[57] Many historians have suggested that Minkowski never fully understood Einstein's theory of relativity, for example, Miller (1981: 241), Goldberg (1984: 193); Pyenson (1985: 130).

The relativity of simultaneity and clock synchronization via optical signals had been discussed by Poincaré as early as 1898, and several times thereafter (Poincaré 1898, 1904: 311). As mentioned above, Lorentz's theory of electrons did not admit the relativity of simultaneity; Lorentz himself used this concept to distinguish his theory from that of Einstein (Lorentz 1910: 1236).

Along with the postulation of the invariance of the velocity of light propagation in empty space and of the principle of relativity of the laws of physics for inertial frames of reference, Einstein's 1905 *Annalen* article began with a *definition* of simultaneity (1905: 891–893). He outlined a method for clock synchronization involving a pair of observers at rest, located at different points in space, denoted A and B, each with identical clocks. Noting that the time of an event at A may not be compared with the time of an event at B without some conventional definition of "time," Einstein proposed that time be defined in such a way that the delay for light traveling from A to B has the same duration as when light travels from B to A.

Einstein supposed that a light signal was emitted from A at time t_A, reflected at point B at time t_B, and observed at point A at time t'_A. The clocks at A and B were then synchronous, again by definition, if $t_B - t_A = t'_A - t_B$. After defining time and clock synchronicity, Einstein went on to postulate that the propagation velocity of light in empty space is a universal constant (1905: 894), such that

$$\frac{2\overline{AB}}{t'_A - t_A} = c.$$

Essentially the same presentation of time and simultaneity was given by Einstein in his 1908 review paper, except in this instance he chose to refer to one-way light propagation (1907: 416).

In summary, by the time of the Cologne lecture, Einstein had defined clock synchronicity using both round-trip and one-way light travel between points in an inertial frame. Furthermore, we know for certain that Minkowski was familiar with both of Einstein's papers. The formal equivalence of Einstein's kinematics with that of Minkowski is not an issue, since Minkowski adopted unequivocally the validity of the Lorentz transformations, and stated just as clearly that the constant appearing therein was the velocity of propagation of light in empty space. The issue is Minkowski's own knowledge of this equivalence. In what follows, we examine some old and new evidence concerning Minkowski's grasp of Einstein's time concept.

Insofar as meaning may be discerned from use, Minkowski's use of the concepts of time and of simultaneity was equivalent to that of Einstein. In the Cologne lecture, for example, Minkowski demonstrated the relativity of simultaneity, employing for this purpose his space-time diagram (1909: 83). A more detailed exposé of the concept—without the space-time diagram—had appeared in the *Grundgleichungen*. In the earlier paper, Minkowski examined the conditions under which the notion of simultaneity was well defined for a single frame of reference. His reasoning naturally supposed that the one-way light delay between two distinct points

A and *B* was equal to the ordinary distance *AB* divided by the velocity of light, exactly as Einstein had supposed. To conclude his discussion of the concept of time in the *Grundgleichungen*, Minkowski remarked by way of acknowledgment that Einstein had addressed the need to bring the nature of the Lorentz transformations physically closer (1908a: 487).

Notwithstanding Minkowski's demonstrated mastery of Einstein's concepts of time and of simultaneity, his understanding of Einstein's idea of time has been questioned. In particular, a phrase cited above from the Cologne lecture has attracted criticism, and is purported to be emblematic of Minkowski's unsure grasp of the difference between Lorentz's theory and Einstein's (Miller 1981: 241). In explaining how Einstein's notion of time was different from the "local time" employed by Lorentz in his theory of electrons, Minkowski recognized the progress made by his former student, for whom "the time of one electron is just as good as that of the other. ... " In his 1905 relativity paper, Einstein referred, not to the time of one electron, but to the time associated with the origin of a system of coordinates in uniform translation, instantaneously at rest with respect to the velocity of an electron moving in an electromagnetic field (1905: 917–918). Provided that such systems could be determined for different electrons, the time coordinates established in these systems would be related in Einstein's theory by a Lorentz transformation. In this sense, Minkowski's electronic interpretation of time was compatible with Einstein's application of his theory to electron dynamics.

Minkowski's interpretation of Einstein's time also reflects the conceptual change wrought in physics by his own notion of proper time (*Eigenzeit*). Near the end of 1907, Minkowski became aware of the need to introduce a coordinate-independent time parameter to his theory.[58] This recognition led him (in the appendix to the *Grundgleichungen*) to introduce proper time, which he presented as a generalization of Lorentz's local time (1908a: 515). From a formal perspective, proper time was closely related to Einstein's formula for time dilation.[59] Minkowski may have simply conflated proper time with time dilation, since the "time of one electron" that Minkowski found in Einstein's theory naturally referred in his view to the *time parameter along the world-line of an electron*, otherwise known as proper time. The introduction of proper time enabled Minkowski to develop the space-time formalism for Lorentz-covariant mechanics, which formed the basis for subsequent research in this area. In this way, proper time became firmly embedded in the Minkowskian view of world-lines in space-time, which Einstein also came to adopt several years later.[60]

[58] On Minkowski's discovery of proper time, see Walter 1996: 101.

[59] Minkowski's expression for proper time, $\int d\tau = \int dt\sqrt{1 - v^2/c^2}$, may be compared with Einstein's expression for time dilation, $\tau = t\sqrt{1 - v^2/c^2}$, although the contexts in which these formulae appeared were quite dissimilar (Einstein 1905: 904; Miller 1981: 271–272). The notation has been changed for ease of comparison.

[60] Einstein's research notes indicate that he adopted a Riemannian space-time metric as the basis of his theory of gravitation in the summer of 1912; see the transcriptions and editorial notes in Einstein *CP4*.

While the electronic interpretation of time has a clear relation to both Einstein's writings and Minkowski's proper time, the phrase "the time of one electron is just as good as that of the other" appears to belong to Lorentz. One of the drafts of the Cologne lecture features a discussion of the physical meaning of Lorentz's local time, which was not retained in the final version. Minkowski referred to a conversation with Lorentz during the mathematicians' congress in Rome, in early April 1908:

> For the uniformly moving electron, Lorentz had called the combination $t' = (-qx + t)/\sqrt{1 - q^2}$ the local time of the electron, and used this concept to understand the contraction hypothesis. Lorentz himself told me conversationally in Rome that it was to Einstein's credit to have recognized that *the time of one electron is just as good as that of the other*, i.e., that t and t' are equivalent. [Italics added][61] (Undated manuscript, Niedersächsische Staats- und Universitätsbibliothek, Math. Archiv 60:4, 11)

According to Minkowski's account, Lorentz employed the phrase in question to characterize Einstein's new concept of time. In fact, what Lorentz had called local time was not the above expression, but $t' = t/\gamma - \gamma vx/c^2$. When combined with a Galilean transformation, the latter expression is equivalent to the one Minkowski called Lorentz's local time. Minkowski must have recognized his mistake, because in the final, printed version of "*Raum und Zeit*" he rewrote his definition of local time and suppressed the attribution of the italicized phrase to Lorentz.

Based on the similarity of the treatment of simultaneity in the *Grundgleichungen* with that of Einstein's writings, Minkowski's acknowledgment of Einstein's contribution in this area, his extension via proper time of Einstein's relative time to the parameterization of world-lines, and the change he made to the definition of local time given in an earlier draft of the Cologne lecture, it appears that Minkowski understood Einstein's concepts of time and simultaneity. This means, of course, that Minkowski's graphic representation of Einstein's kinematics was uncharitable at best. Minkowski may have perceived the success of his own formulation of relativity to depend in some way upon a demonstration that his theory was not just an elaboration of Einstein's work. Likewise, some expedient was required in order for Minkowski to achieve the metatheoretical goal of demonstrating the superiority of pure mathematics over the intuitive methods of physicists; he found one in a space-time diagram.

3. Responses to the Cologne lecture

The diffusion of Minkowski's lecture was exceptional. A few months after the Cologne meeting, it appeared in three different periodicals, and as a booklet. By

[61] "Lorentz hatte für das gleichförmig bewegte Elektron die Verbindung $t' = (-qx + t)/\sqrt{1 - q^2}$ Ortszeit des Elektrons genannt, und zum Verständnis der Kontraktionshypothese diesen Begriff verwandt. Lorentz selbst sagte mir gesprächsweise in Rom, dass die Zeit des einen Elektrons ebensogut wie die des anderen ist, d.h. die Gleichwertigkeit zu t und t' erkannt zu haben, das Verdienst von Einstein ist." Minkowski's story was corroborated in part by his student Louis Kollros, who recalled overhearing Lorentz and Minkowski's conversation on relativity during a Sunday visit to the gardens of the Villa d'Este in Tivoli (Kollros 1956: 276).

the end of 1909, translations had appeared in Italian and French, the latter with the help of Max Born (Minkowski 1909: 517, n. 1). The response to these publications was phenomenal, and has yet to be adequately measured. In this direction, we first present some bibliometric data on research in non-gravitational relativity theory, then discuss a few individual responses to Minkowski's work.

In order to situate Minkowski's work in the publication history of the theory of relativity, we refer to our bibliometric analysis (Walter 1996). The temporal evolution in the number of articles published on non-gravitational relativity theory is shown in Figure 3, for West European-language journals worldwide from 1905 to 1915, along with the relative contribution of mathematicians, theoretical physicists, and non-theoretical physicists. These three groups accounted for nine out of ten papers published in this time period.

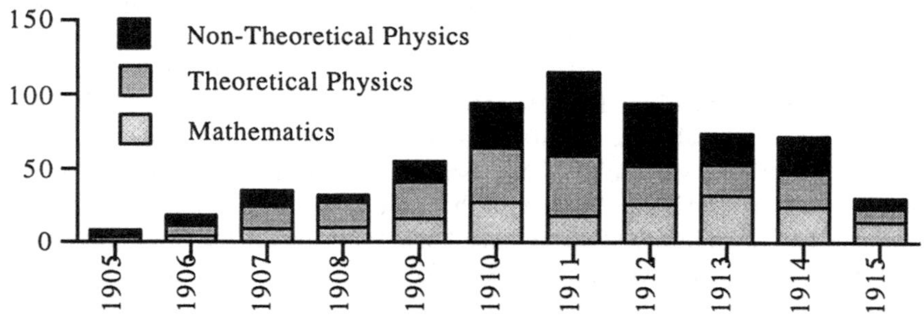

Figure 3. Papers on the non-gravitational theory of relativity.

The plot is based on 610 articles out of a total of 674 for all professions in the period from 1905 to 1915, inclusive. For details on sources and selection criteria, see chapter four of the author's Ph.D. dissertation (Walter 1996).

Starting in 1909, publication numbers increased rapidly until 1912, when the attention of theoretical physicists shifted to quantum theory and theories of gravitation. The annual publication total also declined then for non-theoretical physicists, but remained stable for mathematicians until the outbreak of war in 1914.

A comparison of the relative strength of disciplinary involvement with the theory of relativity can be made for a large group of contributors, if we categorize individuals according to the discipline they professed in the university. Factoring in the size of the teaching staff in German universities in 1911, and taking into consideration only research published by certified teaching personnel (more than half of all authors in 1911 Germany), we find the greatest penetration of relativity theory among theoretical physicists, with one out of four contributing at least one paper on this subject (Table 1, col. 5). Professors of mathematics and of non-theoretical physics largely outnumbered professors of theoretical physics in German universities, and consequently, the penetration of relativity theory in the former fields was significantly lower than the ratio for theoretical physics. The number of contributors for each of the three groups was roughly equivalent, yet theoretical physicists wrote three papers for every one published by their counterparts in mathematics or non-theoretical physics (Table 1, col. 4).

Discipline	Instructors	Relativists	Pubs.	Rel./Instr.
Theoretical Physics	23	6	21	26%
Non-Theoretical Physics	100	6	8	6%
Mathematics	86	5	7	6%

Table 1. Disciplinary penetration of relativity for university instructors in 1911 Germany.
The *relativist* category is taken here to include critics of the special theory of relativity; *physics* is taken to include applied physics. The number of teaching positions is compiled from Auerbach & Rothe 1911.

3.1. The physical reception of Minkowski's theory

The initial response by Einstein and Laub to the *Grundgleichungen*, we mentioned earlier, dismissed the four-dimensional approach, and criticized Minkowski's formula for ponderomotive force density. Others were more appreciative of Minkowski's formalism, including the co-editors of the *Annalen der Physik*, Max Planck and Willy Wien. According to Planck and Wien, Minkowski had put Einstein's theory in a very elegant mathematical form (Wien 1909a: 37; Planck 1910a: 110). In private, however, both men acknowledged a significant physical content to Minkowski's work; in a letter to Hilbert, Wien expressed hope that these ideas would be "thoroughly worked out" (Wien 1909b; Planck 1909). While Wien and Planck applauded Minkowski's mathematical reformulation of the theory of relativity, they clearly rejected his metatheoretical views, and since their public evaluation came to dominate physical opinion of Minkowski's theory, Minkowski's effort in the Cologne lecture to disengage his work from that of Einstein must be viewed as a failure, at least as far as most physicists were concerned.

Not all physicists agreed with Planck and Wien, however. The respected theorist Arnold Sommerfeld was the key exception to the rule of recognizing only Minkowski's formal accomplishment. A former student of Hurwitz and Hilbert, and an ex-protégé of Felix Klein, Sommerfeld taught mathematics in Göttingen before being called to the Aachen chair in mechanics. In 1906, on the basis of his publications on diffraction and on electron theory, and upon Lorentz's recommendation, he received a call to the chair in theoretical physics in Munich, where he was also to head a new institute.[62]

Sommerfeld was among the first to champion Minkowskian relativity for both its physical and mathematical insights. The enthusiasm he showed for Minkowski's theory contrasts with the skepticism with which he initially viewed Einstein's

[62] See Eckert & Pricha 1984; Jungnickel & McCormmach 1986: vol. 2, 274.

theory. The latter held little appeal for Sommerfeld, who preferred the Göttingen lecturer Max Abraham's rigid-sphere electron theory for its promise of a purely electromagnetic explanation of physical phenomena.[63] In Munich Sommerfeld's views began to change. The mathematical rigor of his papers on the rigid electron was subjected to harsh criticism by his former thesis advisor, now colleague, the professor of mathematics Ferdinand Lindemann. Vexed by these attacks, Sommerfeld finally suggested to Lindemann that the problems connected with time in electron theory were due not to its mathematical elaboration, but to its physical foundations (Sommerfeld 1907a: 281). Sommerfeld wrote a paper defending Einstein's theory against an objection raised by Wien (Sommerfeld 1907b), and in the summer of 1908, he exchanged correspondence with Minkowski concerning Einstein's formula for ponderomotive force, and Minkowski's description of the motion of a uniformly-accelerating electron (Minkowski 1908b).[64]

The nature of Sommerfeld's immediate reaction to Minkowski's lecture is unknown, although he was one of three members of the audience to respond during the discussion period, and the only physicist.[65] After the meeting, he wrote to Lorentz to congratulate him on the success of his theory, for Alfred H. Bucherer had presented results of Becquerel-ray deflection experiments that favored the "Lorentz-Einstein" deformable-electron theory over the rigid-electron theory (Sommerfeld 1908). In another letter to Lorentz, a little over a year later, Sommerfeld announced, "Now I, too, have adapted to the relative theory; in particular, Minkowski's systematic form and view facilitated my comprehension" (Sommerfeld 1910c).[66] Both Bucherer's experimental results and the Minkowskian theoretical view contributed to Sommerfeld's adjustment to the theory of relativity, but the latter was what he found most convincing.

In Sommerfeld's first publications on Minkowski's theory, he emphasized the geometric interpretation of the Lorentz transformations as a rotation in space-time; this was an aspect that also featured in lectures given in Munich during winter semester 1909/10.[67] He further enhanced the geometric view of relativity by deriving the velocity addition formula from spherical trigonometry with imaginary sides—a method that pointed the way to a reformulation of the theory of relativity in terms of hyperbolic trigonometry. Remarking that Einstein's formula "loses all strangeness" in the Minkowskian interpretation, Sommerfeld maintained that his only goal in presenting this derivation was to show that the space-time view was a

63 See the remarks made by Sommerfeld after a lecture by Planck (1906: 761).

64 In this letter, Minkowski extended an invitation to Sommerfeld to participate in a debate on electron theory to be held at the meeting of the Mathematical Society in Göttingen on the eighth of August.

65 Along with the mathematicians Eduard Study and Friedrich Engel. Only Study's remarks were recorded; see *Verhandlungen der Gesellschaft Deutscher Naturforscher und Ärzte* **80** (1909): vol. 2, 9.

66 "Ich bin jetzt auch zur Relativtheorie bekehrt; besonders die systematische Form und Auffassung Minkowski's hat mir das Verständnis erleichtert."

67 Sommerfeld (1909a); (1909b); lecture notes entitled "*Elektronentheorie*," Deutsches Museum, Sommerfeld *Nachlaß*; Archives for History of Quantum Physics, reel 22.

"useful guide" in special questions, in addition to facilitating development of the "relative theory" (Sommerfeld 1909a: 827, 829; Walter 1998).

Sommerfeld naturally considered Minkowski's view to be more geometric than Einstein's theory; he found also that Einstein and Minkowski differed on what appeared to be substantial questions of physics. The prime example of this difference concerned the correct expression for ponderomotive force density. The covariant expression employed by Minkowski was presented by Sommerfeld as "closer to the principle of relativity" than Einstein and Laub's formula (Sommerfeld 1909b: 815). Indeed, the latter formula was not Lorentz-covariant, but it had been proposed solely for a system at rest.[68]

Einstein appeared as a precursor to Minkowski in Sommerfeld's widely read publication on the theory of relativity in the *Annalen der Physik*. Offered in tribute to Minkowski, this work criticized "older theories" that employed the concept of absolute space, in what appears to be a response to Minkowski's self-presentation as genitor of a new notion of space. In Sommerfeld's view, Einstein's theory represented an intermediate step between Lorentz and Minkowski, who had rendered the work of both Lorentz and Einstein "irrelevant":

> The troublesome calculations through which Lorentz (1895 and 1904) and Einstein (1905) prove their validity independent of the coordinate system, and [for which they] had to establish the meaning of the transformed field vectors, become irrelevant in the system of the Minkowski "world."[69] (Sommerfeld 1910a: 224)

Sommerfeld depicted the technical difficulty inherent to Lorentz's and Einstein's theories as a thing of the past. Inasmuch as Minkowski appealed to mathematicians to study the theory of relativity in virtue of its essential mathematical nature, Sommerfeld encouraged physicists to take up Minkowski's theory in virtue of its new-found technical simplicity. The pair of *Annalen* publications delivered Minkowskian relativity in a form more palatable to physicists, by replacing the unfamiliar matrix calculus with a four-dimensional vector notation. Similar vectorial reformulations of Minkowski's work were published the same year by Max Abraham (1910) and Gilbert Newton Lewis (1910a, 1910b).

Apart from the change in notation, Sommerfeld's presentation was wholly consonant with Minkowski's reinterpretation of electron-theoretical results. He paraphrased, for example, Minkowski's remark to the effect that, far from being rendered obsolete by his theory, the results for retarded potentials from (pre-Einsteinian) electron-theoretical papers by Liénard, Wiechert and Schwarzschild "first reveal their inner nature in four dimensions, in full simplicity" (Sommer-

[68] Einstein later wrote to Laub that he had persuaded Sommerfeld of the correctness of their formula (27 August 1910; Einstein *CP5*: doc. 224). For a description of the physics involved, see the editorial note in Einstein *CP2*: 503. Debate on this question continued for several years, but by 1918, as Einstein candidly acknowledged to Walter Dällenbach, it had been known for a while that the formula he derived with Laub was wrong (Fölsing 1993: 276).

[69] "Die umständlichen Rechnungen, durch die Lorentz (1895 und 1904) und Einstein (1905) ihre vom Koordinatensystem unabhängige Gültigkeit erweisen und die Bedeutung der transformierten Feldvektoren feststellen mußten, werden also im System der Minkowskischen 'Welt' gegenstandslos."

feld 1909b: 813).[70] As mentioned above, Sommerfeld's reputation in theoretical physics had been established on the basis of his publications on the rigid-electron theory, which for years had formed the basis of the electromagnetic world picture. The rigid electron had now been repudiated empirically by Bucherer's results, but Minkowski felt it was still possible to pursue the electromagnetic world picture with Einstein-electrons, as we saw above.[71] Furthermore, this suggests that in supporting—unconditionally—Minkowski's view of relativity, Sommerfeld did not "burn his boats," as Thomas Kuhn has stated (1967: 141). Instead, Sommerfeld's active promotion and extension of Minkowski's theory is best understood as an *adaptation* of the framework of the electromagnetic world picture to the principle of relativity.[72]

An example of this adaptation may be seen in Sommerfeld's redescription of a primary feature of the electromagnetic world picture: the ether. For those scientists still attached to the concept of ether (or absolute space, in Sommerfeld's terminology), Sommerfeld proposed that they substitute Minkowski's notion of the absolute world, in which the "absolute substrate" of electrodynamics was now to be found (1910a: 189). In this way, Minkowski and Sommerfeld filled the conceptual void created by Einstein's brusque elimination of the ether.

Sommerfeld's mathematical background and close contacts with the Göttingen faculty distinguished him from other theoretical physicists, and enabled him to pass through the walls separating the mathematical and physical communities. In the direction of mathematics, Sommerfeld was a privileged interlocutor for Göttingen mathematicians. He shared their appreciation of the Lorentz transformation as a four-dimensional rotation; his derivation of the velocity addition theorem via spherical trigonometry stimulated dozens of publications by mathematicians in what became a mathematical sub-specialty: the non-Euclidean interpretation of relativity theory (Walter 1998). When David Hilbert needed an assistant in physics, he trusted Sommerfeld to find someone with the proper training.[73] Hilbert felt that Sommerfeld's view of theoretical physics could benefit research in Göttingen (including his own), and after Poincaré (1909), Lorentz (1910), and Michelson (1911), Sommerfeld received an invitation from the Wolfskehl Commission to give lectures on "recent questions in mathematical physics," in the summer of 1912.[74]

In the direction of physics, as we have mentioned, Sommerfeld rendered Min-

---

[70] "Enthüllen erst in vier Dimensionen ihr inneres Wesen voller Einfachheit" in a paraphrase of Minkowski 1909: 88. On this theme see also Sommerfeld 1910b: 249–250.

[71] Poincaré had shown that the stability of Lorentz's deformable electron required the introduction of a compensatory non-electromagnetic potential, producing what was later dubbed *Poincaré pressure*; for details, see Cuvaj 1968 and Miller 1973: 300.

[72] For an example of Sommerfeld's later fascination with the electromagnetic world picture, see Sommerfeld 1922: chap. 1, § 2.

[73] According to Reid 1970: 129, Sommerfeld sent his student P. P. Ewald to Hilbert in 1912.

[74] *Nachrichten von der Königlichen Gesellschaft der Wissenschaften zu Göttingen, geschäftliche Mitteilungen* (1910): 13, 117; (1913): 18; Born 1978: 147.

kowskian relativity comprehensible to physicists by introducing it in vector form. When chosen by the German Physical Society to deliver a report on the theory of relativity for the Karlsruhe meeting of the German Association in 1911, Sommerfeld announced that in the six years since Einstein's publication, the theory had become the "secure property of physics" (Sommerfeld 1911: 1057). His avowed enthusiasm for the theory, made manifest in publications, lectures and personal contacts, was essential in making this statement ring true.

3.2. Mathematicians and Minkowskian relativity

At the same time, there were many relativists who were convinced that the theory of relativity belonged to mathematics. Physicists typically rejected the Minkowskian view of the mathematical essence of the principle of relativity, but the message was heard in departments of mathematics around the world. Mathematicians were already familiar with the concepts and techniques from matrix calculus, hyperbolic geometry and group theory employed in Minkowski's theory, and were usually able to grasp its unified structure with ease. As Hermann Weyl recalled in retrospect, relativity theory seemed revolutionary to physicists, but it had a pattern of ideas which made a perfect fit with those already a part of mathematics (Weyl 1949: 541). Harry Bateman saw the the principle of relativity as unifying disparate branches of mathematics such as geometry, partial differential equations, generalized vector analysis, continuous groups of transformations, and differential and integral invariants (Bateman et al. 1911: 500). Mathematicians, from graduate students to full professors, some of whom had never made the least foray into physics, answered the call to study and develop the theory. According to our study (1996: chap. 4), between 1909 and 1915, sixty-five mathematicians wrote 151 articles on non-gravitational relativity theory, or one out of every four articles published in this domain. In 1913, mathematicians publishing articles worldwide on the theory of relativity (22 individuals) outnumbered their counterparts in both theoretical (16) and non-theoretical (15) physics.[75]

In addition to writing articles, some of these mathematicians introduced the theory of relativity to their research seminars, and taught its formal basis to an expanding student population eager to learn the "radical" theory of space-time. In Germany, according to the listings in the *Physikalische Zeitschrift*, out of thirty-nine regular course offerings on the theory of relativity up to 1915, eight were taught by mathematicians. This broad engagement with the theory of relativity ensured the institutional integration and intellectual propagation necessary to the survival of any research program.

While the impetus for mathematical engagement with the theory of relativity had several sources, the practical advantages offered by the Minkowskian space-time formalism were probably decisive for many 'relativist' mathematicians, who almost invariably employed this formalism in their work. Minkowskian mathematicians made significant contributions in relativistic kinematics and mechanics,

---

[75] These figures are based on primary articles only, excluding book reviews and abstracts; for details, see the author's Ph.D. dissertation (Walter 1996: chap. 4).

although their results were infrequently assimilated by physicists. A striking example of this failure to communicate was pointed out by Stachel (1995: 278), with respect to Émile Borel's 1913 discovery of Thomas precession.

Perhaps more significant to the history of relativity than any isolated mathematical discovery was the introduction of a set of techniques and ideas to the practice of relativity by Minkowskian mathematicians. In favor of this standpoint we recall Stachel's view (1989: 55) of the role of the rigidly-rotating disk problem in the history of general relativity, and Pais's conjecture (1982: 216) that Born's definition of the motion of a rigid body pointed the way to Einstein's adoption (in 1912) of a Riemannian metric in the *Entwurf* theory of gravitation and general relativity. These are particular cases of a larger phenomenon; non-Euclidean and nonstatic geometries were infused into the theory of relativity from late 1909 to early 1913, as a by-product of studies of accelerated motion in space-time by the Minkowskians Max Born, Gustav Herglotz, Theodor Kaluza, Émile Borel and others (Walter 1996: chap. 2).

The clarion call to mathematicians did not come from Minkowski alone. Felix Klein quickly recognized the great potential of Minkowski's approach, integrating Minkowski's application of matrix calculus to the equations of electrodynamics into his lectures on elementary mathematics (1908: 165). The executive committee of the German Society of Mathematicians, of which Klein was a member, chose geometric kinematics as one of the themes of the society's next annual meeting in Salzburg, but Klein did not wait until the fall to give his own view of this subject.[76] Developing his ideas before Göttingen mathematicians in April 1909, Klein pointed out that the new theory based on the Lorentz group (which he preferred to call "*Invariantentheorie*") could have come from pure mathematics (1910: 19). He felt that the new theory was anticipated by the ideas on geometry and groups that he had introduced in 1872, otherwise known as the Erlangen program (see Gray 1989: 229). The latter connection was not one made by Minkowski, yet it tended to anchor the theory of relativity ever more solidly in the history of late nineteenth century mathematics (for Klein's version see 1927: 28).

The subdued response of the physics elite towards Minkowskian relativity constrasts with the enthusiasm displayed by Göttingen mathematicians. Of course, Minkowski's sudden death just months after the Cologne meeting may have influenced early evaluations of his work. David Hilbert's appreciation of Minkowski's lecture, for example, was published as part of an obituary. In Hilbert's account appeared nothing but full agreement with the views expressed by Minkowski, including the assessment of the contributions of Lorentz and Einstein. A few years later, Hilbert portrayed Einstein's achievement as more fundamental than that of Minkowski, although this characterization appeared in a letter requesting financial support for visiting lecturers in theoretical physicists.[77]

---

[76] On the research themes chosen by the German Society of Mathematicians and Klein's role in promoting applied mathematics, see Tobies 1989: 229.

[77] Hilbert to Professor H. A. Krüss, undated typescript, Niedersächsische Staats- und Universitätsbibliothek, Hilbert *Nachlaß* 494. Hilbert gave Einstein credit for having drawn the "full logical conse-

The axiomatic look of the theory presented by Minkowski in the *Grundgleichungen* was perfectly in line with Hilbert's own aspirations for the mathematization of physics, which he had announced as number six in his famous list of worthy problems (Hilbert 1900; Rowe 1995; Corry 1997). In Hilbert's view, Minkowski's greatest positive result was not the discovery of the world postulate, but its application to the derivation of the basic electrodynamic equations for matter in motion (Minkowski *GA*: I, xxv). Hilbert did not publish on the non-gravitational theory of relativity, but like Einstein, he borrowed Minkowski's four-dimensional formalism for his work on the general theory of relativity in 1915 (Hilbert 1916).

In one sense, Minkowski's theory was the fruit of Hilbert's concerted efforts, first in bringing Minkowski to Göttingen from Zurich, then in creating jointly-led advanced seminars to enhance his friend's considerable knowledge and skills in geometry and mechanics, and to direct these toward the development of an axiomatically-based physics. The success of Minkowski's theory was also Hilbert's success and was, as David Rowe has remarked, a major triumph for the Göttingen mathematical community (Rowe 1995: 24). In 1909, on the occasion of Klein's sixtieth birthday, and in the presence of Henri Poincaré, David Hilbert offered his thoughts on the outlook for mathematics:

> What a joy to be a mathematician today, when mathematics is seen sprouting up everywhere and blossoming, when it is shown ever more to advantage in application in the natural sciences as well as in the philosophical direction, and stands to reconquer its former central position.[78] (Hilbert 1909b)

Minkowski's theory of relativity was no doubt a prime example for Hilbert of the reconquest of physics by mathematicians.

So far we have encountered the responses to Minkowski's work by his Göttingen colleagues, who of course had a privileged acquaintance with his approach to electrodynamics. In this respect, most mathematicians were in a position closer to that of our third and final illustration of mathematical responses to the Cologne lecture, from Guido Castelnuovo. This case, however, is chosen primarily for its bearing on Minkowski's interpretation of Einsteinian kinematics, and should not be taken as definitive of mathematical opinion of his work outside of Göttingen.

Castelnuovo was a leading figure in algebraic geometry, a professor of mathematics at the University of Rome and president of the Italian Mathematical Society. In an article published in *Scientia*, he reviewed the notions of space and time according to Minkowski, closely following the thematic progression of the Cologne lecture. With an important difference, however: when Castelnuovo came to discuss the difference between classical and relativistic space-time, he credited the latter to Einstein instead of Minkowski. What is more, where Minkowski main-

---

quence" of the Einstein addition theorem, while the "definitive mathematical expression of Einstein's idea" was left to Minkowski. See also Pyenson 1985: 192.

[78] "Lust ist er heute, Mathematiker zu sein, wo allerwegen die Math. emporspriesst und die emporgesprossene erblickt, wo in ihrer Anwendung auf Naturwissenschaft wie andererseits in der Richtung nach der Philosophie hin die Math. immer mehr zur Geltung kommt und ihre ehemalige zentrale Stellung zurückzuerobern ein Begriff steht." For a full translation of Hilbert's address, differing slightly from my own, see Rowe 1986: 76.

tained that Einstein did *not* modify the classical notion of space, Castelnuovo insisted upon the contrary:

> The statement that the velocity of light is always equal to 1 for any observer is equivalent to the statement that a change in the temporal axis also brings a change to the spatial axes. ...[79] (Castelnuovo 1911: 78)

In light of our earlier reconstruction of Minkowski's argument, it would seem that Castelnuovo denied the possibility of the interpretation imputed to Einstein by Minkowski, in which a rotation of the temporal axis left the spatial axis unchanged; in Castelnuovo's view, Einstein's theory required that the temporal and spatial axes rotate together. From a disciplinary standpoint, it is remarkable that Castelnuovo claimed to be giving an authentic account of *Minkowski's view* of Einstein's kinematics.

Since Castelnuovo apparently contested, and effectively silenced the reasoning given by Minkowski to differentiate his theory from that of Einstein, he might have gone on to assert the equivalence of the two theories. Instead, he affirmed one of Minkowski's metatheoretical claims. Following his exposé of classical and Einsteinian kinematics, Castelnuovo reiterated that in the latter, a rotation of the temporal axis is necessarily accompanied by a rotation of the spatial axes. He continued:

> In truth, this change could be perceived solely by [an observer moving with the speed of light]. Yet if our senses were sufficiently acute, certain differences in the details of the presentation of phenomena would not escape us.[80] (Castelnuovo 1911: 78)

Despite his destruction of the basis to Minkowski's priority claim, Castelnuovo acknowledged the cogency of his geometric approach, while recognizing the change in the concept of space brought about by Einsteinian relativity. The perception of the aforementioned rotation of the spatial axes concomitant with a rotation of the temporal axis required either the adoption of Minkowski's point of view, or the results of experimental physics. Of course, this was a paraphrase of Minkowski; we saw earlier how he conceded that the results of experimental physics had led to the discovery of the principle of relativity, and argued that pure mathematics could have done as well without Michelson's experiment. For Castelnuovo, the acceptance of Minkowski's metatheoretical view of the mathematical essence of the principle of relativity apparently did not conflict with a rejection of his theoretical claim on a new view of space.

---

79 "Affermare che la velocità della luce vale sempre 1, qualunque sia l'osservatore, equivale ad asserire che il cambiamento nell'asse del tempo porta pure un cambiamento nell'asse dello spazio."

80 "Il cambiamento a dir vero sarebbe solo percepito dal demone di Minkowski. Ma di qualche differenza nelle particolarità dei fenomeni dovremmo accorgerci noi pure, quando i nostri sensi fossero abbastanza delicati." The artifice of a demon—recalling Maxwell's demon—was attributed to Minkowski by Castelnuovo earlier in his article, and connected to H. G. Wells' writings. According to Castelnuovo, Minkowski "immagina uno spirito superiore al nostro, il quale concepisca il tempo come une quarta dimensione dello spazio, e possa seguire l'eroe di un noto romanzo di Wells nel suo viaggio meraviglioso attraverso ai secoli" (Castelnuovo 1911: 76).

4. Concluding remarks

Minkowski's semi-popular Cologne lecture was an audacious attempt, seconded by Göttingen mathematicians and their allies, to change the way scientists understood the principle of relativity. Henceforth, this principle lent itself to a geometric conception, in terms of the intersections of world-lines in space-time. Considered as a sales pitch to mathematicians, Minkowski's speech appears to have been very effective, in light of the substantial post-1909 increase in mathematical familiarity with the theory of relativity. Minkowski's lecture was also instrumental in attracting the attention of physicists to the principle of relativity. The Göttingen theorists Walter Ritz, Max Born and Max Abraham were the first to recommend Minkowski's formalism, and following Sommerfeld's intervention, the space-time theory seduced Max von Laue and eventually even Paul Ehrenfest, both of whom had strong ties to Göttingen.

For a mathematician of Minkowski's stature there was little glory to be had in dotting the *i*'s on the theory discovered by a mathematically unsophisticated, unknown, unchaired youngster. In choosing to publish his space-time theory, Minkowski put his personal reputation at stake, along with that of his university, whose identification with the effort to develop the electromagnetic world picture was well established. As a professor of mathematics in Göttingen, Minkowski engaged the reputation of German mathematics, if not that of mathematics in general. From both a personal and a disciplinary point of view, it was essential for Minkowski to show his work to be different from that of Lorentz and Einstein. At the same time, the continuity of his theory with those advanced by the theoretical physicists was required in order to overcome his lack of authority in physics. This tension led Minkowski to assimilate Einstein's kinematics with those of Lorentz's electron theory, contrary to his understanding of the difference between these two theories. Minkowski was ultimately unable to detach his theory from that of Einstein, because even if he convinced some mathematicians that his work stood alone, the space-time theory came to be understood by most German physicists as a purely formal development of Einstein's theory.

Einstein, too, seemed to share this view. It is well known that after unifying geometry and physics on electrodynamic foundations, Minkowski's theory of space-time was instrumental to the geometrization of the gravitational field. In one of Einstein's first presentations of the general theory of relativity, he wrote with some understatement that his discovery had been "greatly facilitated" by the form given to the special theory of relativity by Minkowski (Einstein 1916: 769).

The pronounced disciplinary character of this episode in the history of relativity is undoubtedly linked to institutional changes in physics and mathematics in the decades preceding the discovery of the theory of relativity. For some mathematicians, the dawn of the twentieth century was a time of conquest, or rather reconquest, of terrain occupied by specialists in theoretical physics in the latter part of the nineteenth century. In time, with the growing influence of this new sub-discipline, candidates for mathematical chairs were evaluated by theoretical physicists, and chairs of mathematics and mathematical physics were converted to chairs in theoretical physics. After a decade of vacancy, Minkowski's chair in Zurich, for example, was accorded to Einstein.[81] It seems that a critical shift took place in this period, as a new sense emerged for the role of mathematics in the construction of physical theories, which was reinforced by Einstein's discovery of the field equations of general relativity. Mathematicians followed this movement closely, as Tullio Levi-Civita, Hermann Weyl, Élie Cartan, Jan Schouten and L. P. Eisenhart, among others, revived the tradition of seeking in the theories of physics new directions for their research.

[81] Robert Gnehm to Einstein, 8 December 1911 (Einstein *CP5*: doc. 317).

Appendix. **Minkowski's space-time diagram and the Lorentz transformations**

The relation between the Minkowski space-time diagram and the special Lorentz transformations is presented in many treatises on special relativity. One way of recovering the transformations from the diagram, recalling a method outlined by Max Laue (1911: 47), proceeds as follows.

A two-dimensional Minkowski space-time diagram represents general Cartesian systems with common origins, whereby we constrain the search to linear, homogeneous transformations. For convenience, we let $\ell = ct$ and $\beta = v/c$. These conditions determine the form of the desired transformations:

$$x = \nu\ell' + \rho x' \qquad \text{and} \qquad \ell = \lambda\ell' + \mu x'.$$

On a Minkowski diagram (where the units are selected so that $c = 1$) we draw the invariant curves $\ell^2 - x^2 = \ell'^2 - x'^2 = \pm 1$ (see Figure 4).

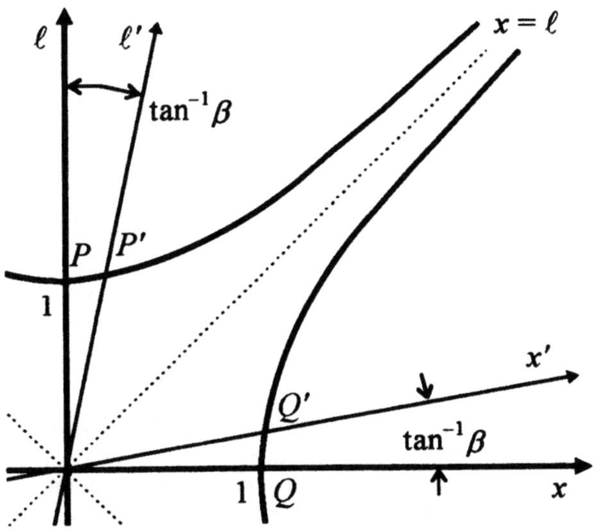

Figure 4. Minkowski diagram of systems S and S'.

Next, we mark two points in the coordinate system $S(x, \ell)$, $P = (0, 1)$ and $Q = (1, 0)$, located at the intersections of the ℓ-axis and x-axis with these hyperbolae. Another system S' translates uniformly at velocity $v = c\beta$ with respect to S, such that the origin of S' appears to move according to the expression $x = \beta\ell$. This line is taken to be the ℓ'-axis. From the expression for the hyperbolae, it is evident that the x'-axis and the ℓ'-axis are mutually symmetric, and form the same angle $\tan^{-1}\beta$ with the x-axis and the ℓ-axis, respectively. The two points in S are denoted here as $P' = (0, 1)$ and $Q' = (1, 0)$ and marked accordingly, at the intersections of the hyperbolae with the respective axes. The ℓ'-axis, $x = \beta\ell$, intersects the hyperbola $\ell^2 - x^2 = 1$ at P'. Using this data, we solve for the coefficients ν and λ:

$$\nu = \frac{\beta}{\sqrt{1-\beta^2}} \quad \text{and} \quad \lambda = \frac{1}{\sqrt{1-\beta^2}}.$$

Applying the same reasoning to the x'-axis ($x = \ell\,\beta$), we solve for the coefficients ρ and μ, evaluating the expressions for x and ℓ at the intersection of the x'-axis with the hyperbola $\ell^2 - x^2 = -1$, at the point labeled Q', and we find

$$\rho = \frac{1}{\sqrt{1-\beta^2}} \quad \text{and} \quad \mu = \frac{\beta}{\sqrt{1-\beta^2}}.$$

Substituting these coefficients into the original expressions for x and ℓ, we obtain the following transformations:

$$x = \frac{x' + \beta\ell'}{\sqrt{1-\beta^2}} \quad \text{and} \quad \ell = \frac{\ell' + \beta x'}{\sqrt{1-\beta^2}}.$$

The old form of the special Lorentz transformations is recovered by substituting $\ell = ct$ and $\beta = v/c$,

$$x = \frac{x' + vt'}{\sqrt{1-v^2/c^2}} \quad \text{and} \quad t = \frac{t' + vx'/c^2}{\sqrt{1-v^2/c^2}}.$$

Invoking the property of symmetry, the transformations for x' and t' may be calculated in the same fashion as above, by starting with S' instead of S.

Acknowledgements

For their critiques of preliminary versions of this paper, my warmest thanks go to Olivier Darrigol, Peter Galison, Christian Houzel, Arthur Miller, Michel Paty, Jim Ritter and John Stachel. The themes of this paper were presented in seminars at the University of Paris 7, at University College London, and at the 1995 HGR congress; I am grateful to their participants and organizers for stimulating discussions. Financial support was provided by a fellowship from the French Ministry of Research and Higher Education, and archival research was made possible by travel grants from the American Institute of Physics and the University of Paris 7.

References

Abraham, Max. (1910). "Sull'elettrodinamica di Minkowski." *Rendiconti del Circolo Matematico di Palermo* **30**: 33–46.

Auerbach, Felix & Rothe, Rudolf, eds. (1911). "Verzeichnis der Hochschullehrer." *Taschenbuch für Mathematiker und Physiker* **2**: 535–544.

BATEMAN, Harry et al. (1911). "Mathematics and Physics at the British Association, 1911." *Nature* **87**: 498–502.

BORN, Max. (1906). *Untersuchungen über die Stabilität der elastischen Linie in Ebene und Raum, unter verschiedenen Grenzbedingungen.* Göttingen: Dieterichsche Univ.-Buchdruckerei.

——— (1920). *Die Relativitätstheorie Einsteins und ihre physikalischen Grundlagen, gemeinverständlich dargestellt.* Berlin: Springer.

——— (1959). "Erinnerungen an Hermann Minkowski zur 50. Wiederkehr seines Todestages." *Die Naturwissenschaften* **46**: 501–505. Reprinted in Born 1963: vol. 2, 678–680.

——— (1962). *Einstein's Theory of Relativity.* New York: Dover.

——— (1963). *Ausgewählte Abhandlungen.* Göttingen: Vandenhoeck & Ruprecht.

——— (1968). *My Life and My Views.* New York: Scribner's.

——— (1978). *My Life: Recollections of a Nobel Laureate.* New York: Scribner's.

BUCHERER, Alfred Heinrich. (1908). "Messungen an Becquerelstrahlen. Die experimentelle Bestätigung der Lorentz-Einsteinschen Theorie." *Physikalische Zeitschrift* **9**: 755–762.

CASTELNUOVO, Guido. (1911). "Il principio di relatività e fenomeni ottica." *Scientia (Rivista di Scienza)* **9**: 64–86.

CORRY, Leo. (1997). "Hermann Minkowski and the Postulate of Relativity." *Archive for History of Exact Sciences* **51**: 273–314.

CUVAJ, Camillo. (1968). "Henri Poincaré's Mathematical Contributions to Relativity and the Poincaré Stresses." *American Journal of Physics* **36**: 1102–1113.

DARRIGOL, Olivier. (1993). "The Electrodynamic Revolution in Germany as Documented by Early German Expositions of 'Maxwell's Theory'." *Archive for History of Exact Sciences* **45**: 189–280.

——— (1995). "Henri Poincaré's Criticism of Fin de Siècle Electrodynamics." *Studies in History and Philosophy of Modern Physics* **26**: 1–44.

ECKERT, Michael & PRICHA, Willibald. (1984). "Boltzmann, Sommerfeld und die Berufungen auf die Lehrstühl für theoretische Physik in Wien und München 1890–1917." *Mitteilungen der Österreichischen Gesellschaft für Geschichte der Naturwissenschaften* **4**: 101–119.

EINSTEIN, Albert. (1905). "Zur Elektrodynamik bewegter Körper." *Annalen der Physik* **17**: 891–921 [= EINSTEIN *CP2*: doc. 23].

——— (1907). "Relativitätsprinzip und die aus demselben gezogenen Folgerungen," *Jahrbuch der Radioaktivität und Elektronik* **4**: 411–462 [= Einstein *CP2*: doc. 47].

——— (1910). "Le Principe de Relativité et ses conséquences dans la physique moderne." *Archives des Sciences physiques et naturelles* **29**: 5–28 [= EINSTEIN *CP3*: doc. 2].

——— (1916). "Die Grundlage der allgemeinen Relativitätstheorie." *Annalen der Physik* **49**: 769–822 [= EINSTEIN *CP6*: doc. 30].

——— (*CP2*). *The Collected Papers of Albert Einstein.* Vol. 2: *The Swiss Years: Writings, 1900–1909.* J. Stachel, D. C. Cassidy, J. Renn, & R. Schulmann, eds. Princeton: Princeton University Press (1989).

——— (*CP4*). *The Collected Papers of Albert Einstein.* Vol. 4: *The Swiss Years: Writings, 1912–1914.* M. J. Klein, A. J. Kox, J. Renn & R. Schulmann, eds. Princeton: Princeton University Press (1995).

——— (*CP5*). *The Collected Papers of Albert Einstein.* Vol. 5: *The Swiss Years: Correspondence, 1902–1914.* M. J. Klein, A. J. Kox, & R. Schulmann, eds. Princeton: Princeton University Press (1993).

——— (*CP6*). *The Collected Papers of Albert Einstein*. Vol. 6. *The Berlin Years: Writings, 1914–1917*. A. J. Kox, M. J. Klein, & R. Schulmann, eds. Princeton: Princeton University Press (1996).

Einstein, Albert & Laub, Jakob. (1908a). "Über die elektromagnetischen Grundgleichungen für bewegte Körper." *Annalen der Physik* **26**: 532–540 [= Einstein *CP2*: doc. 51].

——— (1908b). "Über die im elektromagnetischen Felde auf ruhende Körper ausgeübten ponderomotorischen Kräfte." *Annalen der Physik* **26**: 541–550 [= Einstein *CP2*: doc. 52].

von Ferber, Christian. (1956). *Die Entwicklung des Lehrkörpers in der deutschen Hochschulen 1864–1954*. Göttingen: Vandenhoeck & Ruprecht.

Fölsing, Albrecht. (1993). *Albert Einstein: Eine Biographie*. Frankfurt am Main: Suhrkamp.

Forman, Paul. (1967). "The Environment and Practice of Atomic Physics in Weimar Germany: a Study in the History of Science." Ph.D. dissertation, University of California, Berkeley.

Frank, Philipp. (1908). "Das Relativitätsprinzip der Mechanik und die Gleichungen für die elektromagnetischen Vorgänge in bewegten Körpern." *Annalen der Physik* **27**: 897–902.

——— (1910). "Das Relativitätsprinzip und die Darstellung der physikalischen Erscheinungen im vierdimensionalen Raum." *Zeitschrift für physikalische Chimie* **74**: 466–495.

Galison, Peter. (1979). "Minkowski's Spacetime: From Visual Thinking to the Absolute World." *Historical Studies in the Physical Sciences* **10**: 85–121.

Goffman, Erving. (1959). *The Presentation of Self in Everyday Life*. New York: Penguin.

Goldberg, Stanley. (1984). *Understanding Relativity*. Boston & Basel: Birkhäuser.

Gray, Jeremy J. (1989). *Ideas of Space*. 2d ed. Oxford: Oxford University Press.

Heffter, Lothar. (1912). "Zur Einführung der vierdimensionalen Welt Minkowskis." *Jahresbericht der deutschen Mathematiker-Vereinigung* **21**: 1–8.

Hilbert, David. (1900). "Mathematische Probleme." *Nachrichten von der Königlichen Gesellschaft der Wissenschaften zu Göttingen. Mathematisch-Physikalische Klasse*: 253–297.

——— (1909a). "Hermann Minkowski: Gedächtnisrede." *Nachrichten von der Königlichen Gesellschaft der Wissenschaften zu Göttingen, Mathematisch-Physikalische Klasse*: 72–101 [Reprinted: *Mathematische Annalen* **68** (1910): 445–471; Minkowski *GA*: I, v–xxxi].

——— (1909b). "An Klein zu seinem 60sten Geburts-Tage, 25 April 1909." In Hilbert *Nachlaß* 575, Niedersächsische Staats- und Universitätsbibliothek.

——— (1916). "Die Grundlagen der Physik (Erste Mitteilung)." *Nachrichten von der Königlichen Gesellschaft der Wissenschaften zu Göttingen. Mathematisch-Physikalische Klasse*: 395–407.

Hirosige, Tetu. (1968). "Theory of Relativity and the Ether." *Japanese Studies in the History of Science* **7**: 37–53.

——— (1976). "The Ether Problem, the Mechanistic Worldview, and the Origins of the Theory of Relativity." *Historical Studies in the Physical Sciences* **7**: 3–82.

Hon, Giora. (1995). "The Case of Kaufmann's Experiment and its Varied Reception." In *Scientific Practice: Theories and Stories of Doing Physics*. Jed Z. Buchwald, ed. 170–223. Chicago: University of Chicago Press.

Illy, József. (1981). "Revolutions in a Revolution." *Studies in History and Philosophy of Science* **12**: 173–210.

JUNGNICKEL, Christa & MCCORMMACH, Russell. (1986). *Intellectual Mastery of Nature: Theoretical Physics from Ohm to Einstein*. Chicago: University of Chicago Press.

KLEIN, Felix. (1908). *Elementarmathematik vom höheren Standpunkt aus*. Vol. 1: *Arithmetik, Algebra, Analysis. Vorlesungen gehalten im Wintersemester 1907–08*. Leipzig and Berlin: Teubner.

——— (1910). "Über die geometrischen Grundlagen der Lorentzgruppe." *Jahresbericht der deutschen Mathematiker-Vereinigung* **19**: 281–300. [Reprinted: *Physikalische Zeitschrift* **12** (1911): 17–27].

——— (1926–27). *Vorlesungen über die Entwicklung der Mathematik im 19. Jahrhundert*, 2 vols. Berlin: Springer.

KOLLROS, Louis. (1956). "Albert Einstein en Suisse: Souvenirs." *Helvetica Physica Acta. Supplementum* **4**: 271–281.

KUHN, Thomas S. et al. (1967). *Sources for History of Quantum Physics: An Inventory and Report*. Philadelphia: American Philosophical Society.

VON LAUE, Max. (1911). *Das Relativitätsprinzip. Die Wissenschaft* **38**. Braunschweig: Vieweg.

LEWIS, Gilbert Newton. (1910a). "On Four-Dimensional Vector Analysis and its Application in Electrical Theory." *Proceedings of the American Academy of Arts and Science* **46**: 165–181.

——— (1910b). "Über vierdimensionale Vektoranalysis und deren Anwendung auf die Elektrizitätstheorie." *Jahrbuch der Radioaktivität und Elektronik* **7**: 329–347.

LORENTZ, Hendrik Antoon. (1904). "Electromagnetic Phenomena in a System Moving with any Velocity Less than that of Light." *Proceedings of the Section of Sciences. Koninklijke Akademie van Wetenschappen te Amsterdam* **6**: 809–831.

——— (1910). "Alte und neue Fragen der Physik." *Physikalische Zeitschrift* **11**: 1234–1257.

——— (1916). *The Theory of Electrons and its Application to the Phenomena of Light and Radiant Heat*. 2d ed. Leipzig & Berlin: Teubner.

MCCORMMACH, Russell. (1976). "Editor's Forward." *Historical Studies in the Physical Sciences* **7**: xi–xxxv.

MILLER, Arthur I. (1973). "A Study of Henri Poincaré's 'Sur la dynamique de l'électron'." *Archive for History of Exact Sciences* **10**: 207–328.

——— (1980). "On Some Other Approaches to Electrodynamics in 1905." In *Some Strangeness in the Proportion: A Centennial Symposium to Celebrate the Achievements of Albert Einstein*. Harry Woolf, ed. 66–91. Reading: Addison-Wesley.

——— (1981). *Albert Einstein's Special Theory of Relativity: Emergence (1905) and Early Interpretation (1905–1911)*. Reading: Addison-Wesley.

MINKOWSKI, Hermann. (1906). "Kapillarität." In *Encyklopädie der mathematischen Wissenschaften*. Vol. 5: *Physik*. A. Sommerfeld, ed. 558–613. Leipzig & Berlin: Teubner. [= MINKOWSKI *GA*: II, 298–351].

——— (1907a). *Diophantische Approximationen. Eine Einführung in die Zahlentheorie*. Leipzig & Berlin: Teubner.

——— (1907b). "Das Relativitätsprinzip." Typescript. Math. Archiv 60: 3. Niedersächsische Staats- und Universitätsbibliothek.

——— (1908a). "Die Grundgleichungen für die electromagnetischen Vorgänge in bewegten Körpern." *Nachrichten von der Königlichen Gesellschaft der Wissenschaften zu Göttingen. Mathematisch-Physikalische Klasse*: 53–111. [= MINKOWSKI *GA*: II, 352–404].

——— (1908b). [Hermann Minkowski to Arnold Sommerfeld]. 21 July 1908. Sommerfeld *Nachlaß*. Deutsches Museum, Munich.

——— (1909). "Raum und Zeit." *Jahresbericht der deutschen Mathematiker-Vereinigung* **18**: 75–88 = *Physikalische Zeitschrift* **10**: 104–111. [= MINKOWSKI *GA*: II, 431–446].

——— (*GA*). *Gesammelte Abhandlungen*. 2 vols. D. Hilbert, ed. Leipzig & Berlin: Teubner (1911).

MINKOWSKI, Hermann & BORN, Max. (1910). "Eine Ableitung der Grundgleichungen für die elektromagnetischen Vorgänge in bewegten Körpern. Aus dem Nachlaß von Hermann Minkowski bearbeitet von Max Born in Göttingen." *Mathematische Annalen* **68**: 526–550. [= MINKOWSKI *GA*: II, 405–430].

MITTAG-LEFFLER, Gustav. (1909). [Gustav Mittag-Leffler to Henri Poincaré]. 7 July 1909. Mittag-Leffler Institute, Djursholm.

MOSZKOWSKI, Alexander. (1920). *Einstein: Einblicke in seine Gedankenwelt. Gemeinverständliche Betrachtungen über die Relativitätstheorie und ein neues Weltsystem.* Berlin & Hamburg: Fontane.

OLESKO, Kathryn M. (1991). *Physics as a Calling: Discipline and Practice in the Königsberg Seminar for Physics*. Ithaca: Cornell University Press.

PAIS, Abraham. (1982). *"Subtle is the Lord ... "—The Science and the Life of Albert Einstein*. Oxford: Oxford University Press.

PATY, Michel. (1993). *Einstein philosophe*. Paris: Presses Universitaires de France.

PAULI, Wolfgang. (1958). *The Theory of Relativity*. Oxford: Pergamon.

PLANCK, Max. (1906). "Die Kaufmannschen Messungen der Ablenkbarkeit der β-Strahlen in ihrer Bedeutung für die Dynamik der Elektronen." *Physikalische Zeitschrift* **7**: 753–761. [= PLANCK *PAV*: II, 121–135].

——— (1909). [Max Planck to Wilhelm Wien]. 30 November 1909. Wien *Nachlaß* 38, Staatsbibliothek Preußischer Kulturbesitz.

——— (1910a). *Acht Vorlesungen über theoretische Physik*. Leipzig: Hirzel.

——— (1910b). "Die Stellung der neueren Physik zur mechanischen Naturanschauung." *Physikalische Zeitschrift* **11**: 922–932. [= PLANCK *PAV*: III, 30–46].

——— (*PAV*). *Physikalische Abhandlungen und Vorträge*. 3 vols. Verband Deutscher Physikalischer Gesellschaften and Max-Planck-Gesellschaft zur Förderung der Wissenschaften, eds. Braunschweig: Vieweg (1958).

POINCARÉ, Henri. (1898). "La Mesure du temps." *Revue de Métaphysique et de Morale* **6**: 1–13.

——— (1904). "L'État actuel et l'avenir de la physique mathématique." *Bulletin des Sciences Mathématiques* **28**: 302–324.

——— (1906). "Sur la dynamique de l'électron." *Rendiconti del Circolo Matematico di Palermo* **21**: 129–176. [Reprinted: *Œuvres de Henri Poincaré*. Vol. 9: 494—550. G. Petiau, ed. Paris: Gauthier-Villars (1954)].

——— (1906/7). Personal course notes by Henri Vergne. Published as "Les Limites de la loi de Newton." *Bulletin Astronomique* **17** (1953): 121–269.

——— (1910a). "La Mécanique nouvelle." In *Sechs Vorträge über ausgewählte Gegenstände aus der reinen Mathematik und mathematischen Physik*. 51–58. Leipzig & Berlin: Teubner.

——— (1910b). "Die neue Mechanik." *Himmel und Erde* **23**: 97–116.

——— (1912). "L'Espace et le Temps." *Scientia (Rivista di Scienza)* **12**: 159–170.

PYENSON, Lewis. (1985). *The Young Einstein: The Advent of Relativity*. Bristol: Adam Hilger.

——— (1987). "The Relativity Revolution in Germany." In *The Comparative Reception of Relativity*. Thomas F. Glick, ed. 59–111. Dordrecht: Reidel.

REID, Constance. (1970). *Hilbert*. Berlin: Springer.

ROWE, David E. (1986). "David Hilbert on Poincaré, Klein, and the World of Mathematics." *Mathematical Intelligencer* **8**: 75–77.

——— (1995). "The Hilbert Problems and the Mathematics of a New Century." *Preprint-Reihe des Fachbereichs Mathematik* **1**, Johannes Gutenberg-Universität, Mainz.

RÜDENBERG, Lily & ZASSENHAUS, Hans, eds. (1973). *Hermann Minkowski. Briefe an David Hilbert*. Berlin: Springer.

SEELIG, Carl. (1956). *Albert Einstein: A Documentary Biography*. M. Savill, trans. London: Staples.

SILBERSTEIN, Ludwik. (1914). *The Theory of Relativity*. London: Macmillan.

SOMMERFELD, Arnold. (1907a). "Zur Diskussion über die Elektronentheorie." *Sitzungsberichte der Königlichen Bayerischen Akademie der Wissenschaften, Mathematisch-Physikalische Klasse* **37**: 281.

——— (1907b). "Ein Einwand gegen die Relativtheorie der Elektrodynamik und seine Beseitigung." *Physikalische Zeitschrift* **8**: 841. [= SOMMERFELD *GS*: II, 183–184].

——— (1908). [Arnold Sommerfeld to H. A. Lorentz]. 16 November 1908. Lorentz Papers. Rijksarchief in Noord-Holland te Haarlem.

——— (1909a). "Über die Zusammensetzung der Geschwindigkeiten in der Relativtheorie," *Physikalische Zeitschrift* **10**: 826–829. [= SOMMERFELD *GS*: II, 185–188].

——— (1909b). [Review of MINKOWSKI 1908 and 1909]. *Beiblätter zu den Annalen der Physik* **33**: 809–817.

——— (1910a). "Zur Relativitätstheorie I: Vierdimensionale Vektoralgebra." *Annalen der Physik* **32**: 749–776. [= SOMMERFELD *GS*: II, 189–216].

——— (1910b). "Zur Relativitätstheorie II: Vierdimensionale Vektoranalysis." *Annalen der Physik* **33**: 649–689. [= SOMMERFELD *GS*: II, 217–257].

——— (1910c). [Arnold Sommerfeld to H. A. Lorentz]. 9 January 19[10]. Lorentz Papers 74: 4. Rijksarchief in Noord-Holland te Haarlem.

——— (1911). "Das Plancksche Wirkungsquantum und seine allgemeine Bedeutung für die Molekularphysik." *Physikalische Zeitschrift* **12**: 1057–1069. [= SOMMERFELD *GS*: III, 1–19].

——— (1913). "Anmerkungen zu Minkowski, Raum und Zeit." In *Das Relativitätsprinzip: Eine Sammlung von Abhandlungen*. Otto Blumenthal, ed. 69–73. Leipzig: Teubner.

——— (1922). *Atombau und Spektrallinien*. 3rd ed. Braunschweig: Vieweg.

——— (1949). "To Albert Einstein's Seventieth Birthday." In *Albert Einstein: Philosopher-Scientist*. Paul A. Schilpp, ed. 99–105. Evanston (IL): The Library of Living Philosophers.

——— (*GS*). *Gesammelte Schriften*. 4 vols. Fritz Sauter, ed. Braunschweig: Vieweg (1968).

STACHEL, John (1989). "The Rigidly Rotating Disk as the 'Missing Link' in the History of General Relativity." In *Einstein and the History of General Relativity* (Einstein Studies, vol. 1). Don Howard and John Stachel, eds. 48–62. Boston: Birkhäuser.

——— (1995). "History of Relativity." In *Twentieth Century Physics*. Vol. 1: 249–356. Laurie M. Brown et al., eds. New York: American Institute of Physics Press.

STICHWEH, Rudolf. (1984). *Zur Entstehung des modernen Systems wissenschaftlicher Disziplinen*. Frankfurt am Main: Suhrkamp.

TOBIES, Renate. (1989). "On the Contribution of Mathematical Societies to Promoting Applications of Mathematics in Germany." In *The History of Modern Mathematics*. Vol. 2: 223–248. David Rowe and John McCleary, eds. Boston: Academic Press.

VOIGT, Woldemar. (1887). "Über das Doppler'sche Princip." *Nachrichten von der Königlichen Gesellschaft der Wissenshaften und der Georg-August-Universität zu Göttingen*: 41–51. [Reprinted with additions: *Physikalische Zeitschrift* **16** (1915) 381–386].

Volkmann, Paul. (1910). *Erkenntnistheoretische Grundzüge der Naturwissenschaften und ihre Beziehungen zum Geistesleben der Gegenwart. Wissenschaft und Hypothese* **9**. Leipzig & Berlin: Teubner.

Volterra, Vito. (1912). *Lectures Delivered at the Celebration of the 20th Anniversary of the Foundation of Clark University*. Worcester: Clark University.

Walter, Scott. (1996). "Hermann Minkowski et la mathématisation de la théorie de la relativité restreinte, 1905–1915." Ph.D. dissertation, University of Paris 7.

——— (1998). "The Non-Euclidean Style of Minkowskian Relativity." In *The Symbolic Universe*. Jeremy J. Gray, ed. Oxford: Oxford University Press.

Weyl, Hermann. (1949). "Relativity Theory as a Stimulus in Mathematical Research." *Proceedings of the American Philosophical Society* **93**: 535–541.

Wiechert, Emil. (1915). "Die Mechanik im Rahmen der allgemeinen Physik." In *Die Kultur der Gegenwart*. Teil 3, Abt. 3, Bd. 1: *Physik*. Emil Warburg, ed. 1–78. Leipzig & Berlin: Teubner.

Wien, Wilhelm. (1906). "Über die partiellen Differentialgleichungen der Physik." *Jahresbericht der deutschen Mathematiker-Vereinigung* **15**: 42–51.

——— (1909a). "Über die Wandlung des Raum- und Zeitbegriffs in der Physik." *Sitzungsberichte der physikalisch-medicinischen Gesellschaft zu Würzburg*: 29–39.

——— (1909b). Wilhelm Wien to David Hilbert. 15 April 1909. Hilbert *Nachlaß* 436, Niedersächsische Staats- und Universitätsbibliothek.

Heuristics and Mathematical Representation in Einstein's Search for a Gravitational Field Equation

Jürgen Renn and Tilman Sauer

Einstein first dealt with the problem of a relativistic theory of gravitation in 1907 (Einstein 1907). He was then confronted with the task of revising the classical Newtonian theory of gravitation in the light of the relativity theory of 1905 since he had to write a review paper that would have to cover the implications of this theory for various areas of physics. Indeed, contrary to the field theory of electrodynamic interactions, Newton's theory of gravitation implies an instantaneous action at a distance, incompatible with the requirement of special relativity that the propagation speed of physical interactions is limited by the speed of light. Hence the revision of Newtonian gravitation theory entered Einstein's intellectual horizon as a necessity of the day, imposed by the need to integrate new results with the traditional body of knowledge.

The advent of the theory of special relativity had sharpened the conflict, but an incompatibility of classical mechanics and electrodynamics had widely been recognized by physicists before. Hence not only Einstein but also several of his contemporaries addressed the problem of formulating a field theory of gravitation that was to be in agreement with the principles suggested by the theory of the electrodynamic field, and most importantly with the new kinematics of relativity theory. In addressing this problem, however, Einstein quickly decided to take a step different from that of his contemporaries. Instead of formulating a special relativistic law of gravitational interaction, he searched for a generalization of

This article is based on joint work with Michel Janssen, John Norton, and John Stachel. Preliminary versions of this paper were presented at the Fourth International Conference on the History of General Relativity held in Berlin, 31 July–3 August 1995, and at the Boston Colloquium for Philosophy of Science "Einstein in Berlin: The First Ten Years," Boston, 3–4 March 1997. A larger joint work is in preparation.

H. Goenner, J. Renn, J. Ritter, T. Sauer (eds.)
The Expanding Worlds of General Relativity
(Einstein Studies, volume 7) pp. 87–125

the relativity principle and for a new theory of gravitation that closely associated gravitation with inertia. The peculiarity of Einstein's approach is embodied in the two heuristic principles that guided his search from the beginning and that will be designated here throughout as the *equivalence principle* and the *principle of generalized relativity*, although these principles did not bear any specific names in Einstein's first paper on the subject. In his 1907 paper the first principle was formulated as the assumption of physical indistinguishability between a homogeneous gravitational field and a uniformly accelerated frame of reference. The second principle was adumbrated by the question:

> Is it conceivable that the principle of relativity also applies to systems that are accelerated relative to each other?[1] (Einstein 1907: 454).

These principles do not express features of classical or special relativistic physics as it was then understood but rather requirements to be satisfied only by the new theory of gravitation. And indeed, they are since the time of the completion of the theory of general relativity in 1915, with its generally covariant field equation, often claimed to capture the most essential aspects of this theory, although this claim has not remained undisputed (Norton 1993).

If accepted, this claim seems to imply, however, that precisely what later turned out to be the essential and most innovative conceptual developments brought about by general relativity were already anticipated by Einstein long before he found the field equations of his new theory. Is general relativity hence the result of a conceptual leap which belongs to a context of discovery that can only be further elucidated by studying the psychological roots of Einstein's creativity but will ultimately remain inexplicable? The history of the discovery of the field equations of general relativity seems to present a strong argument in favor of this conjecture. In fact, Einstein formulated his heuristic guidelines at the very beginning of his search for a new theory of gravitation and he stubbornly held on to them for roughly eight years, in spite of the considerable difficulties he found in implementing his original ideas. But which exactly were the difficulties that Einstein encountered in formulating the field equations of general relativity?

At first glance there seem to be two alternatives: If Einstein's difficulties were predominantly of a technical nature, due either to the complexity of the mathematical language to be learned and adapted to the problem at hand or to some false working hypotheses about the mathematical representation to be eliminated along the way, then the conceptual innovation brought about by general relativity would indeed be contained *in nuce* in Einstein's starting point. The discovery of the field equations of general relativity could then be recounted as a comedy of errors, with a brilliant beginning, some deviations, and a happy conclusion, yet without explanation of the crucial steps of conceptual development involved. The role of the formalism for this discovery would then be merely that of a medium in which preconceived physical conceptions are expressed more or less appropriately.

If Einstein's difficulties were, however, themselves of a conceptual nature then it

[1] "Ist es denkbar, daß das Prinzip der Relativität auch für Systeme gilt, welche relativ zueinander beschleunigt sind?"

rather seems that his eventual success must be interpreted as a case of serendipity, as a lucky finding that cannot be accounted for by the quality of his heuristic starting point. The role of the formalism for Einstein's discovery of the field equations would then be rather that of an ill-mastered language expressing a meaning different from the intended one. In this version, conceptual development would hence be accounted for by a lucky misunderstanding. In other words, these two extreme alternatives both fail to provide any rational account of the conceptual development from classical and special relativistic physics to the physics of general relativity. They also fail to take into account any non-trivial interaction between heuristics and representation. In the present paper we will analyze this interaction more closely and propose, on the basis of this analysis, a third, alternate account of conceptual development in the case of general relativity.

Fortunately, Einstein's research in the crucial years 1907 to 1915 is not only documented by his publication of preliminary versions of general relativity but also by research notebooks, as well as by his contemporary correspondence. In order to analyze the relationship between heuristics and formalism in the emergence of general relativity we will make use of preliminary results of an extensive study of these documents in the context of a cooperative research project involving Michel Janssen, John Norton, John Stachel, as well as the authors.[2]

1. Einstein's "Zurich Notebook" and the genesis of general relativity

The task of reconstructing Einstein's discovery of the theory of general relativity has challenged historians of science for a long time.[3] Major progress in this endeavour is due to John Stachel's and John Norton's path-breaking investigations.[4]

In the course of preparing the editorial project of the *Collected Papers of Albert Einstein* John Stachel first realized the significance of the so-called Zurich Notebook (Einstein *CP4*: doc. 10) for the reconstruction of the genesis of the theory of general relativity. In 1984, John Norton published a comprehensive reconstruction of Einstein's discovery process (Norton 1984). In this paper, he was able to correct some widespread prejudices about Einstein's path towards general relativity by an analysis of some key pages of the Zurich Notebook. But even after this pioneering work large parts of the Notebook remained obscure and poorly understood. In the course of a research project involving Peter Damerow, Werner Heinrich, Michel Janssen, John Norton, John Stachel, as well as the authors[5] the Zurich Notebook

2 While the reconstruction of the Zurich Notebook on which we base our analysis is a joint achievement of this project, the authors alone bear the responsibility for the views expressed in this paper.

3 A (certainly incomplete) list of the older secondary literature includes Hoffmann 1972; Lanczos 1972; Mehra 1974; Earman & Glymour 1978a, 1978b; Vizgin & Smorodinskii 1979; Pais 1982. More recent literature may be found in the volumes of the "Einstein Studies" series, Howard & Stachel 1989; Eisenstaedt & Kox 1992; Earman, Janssen & Norton 1993.

4 See Stachel 1980, 1982, 1989 and Norton 1984.

5 The research project was begun under the direction of Peter Damerow and Jürgen Renn with the Working Group Albert Einstein, funded by the Senate of Berlin and affiliated with the Center for Development and Socialization headed by Wolfgang Edelstein at the Max Planck Institute for Human

could now be fully reconstructed and comprehensively analyzed. It turned out that this research manuscript allows for a detailed understanding of Einstein's working procedures and the heuristic principles guiding them. The analysis presented in the following is based on the results of this collaboration.

The Zurich Notebook originally comprised 96 pages, the cover bearing the title "Relativität" in Einstein's hand. Eighty-four pages of this Notebook contain calculations or short notes on various problems of physics, mainly on gravitation theory. Due to the character of a research notebook, most of the calculations are extremely sketchy, display false starts, and come with almost no explanatory text. The dating of these entries to the critical period between summer 1912 and spring 1913, as well as the identification of their chronological sequence, must hence rely on the mathematical notation used, and, most importantly, on a detailed reconstruction of their meaning.

Let us briefly review the principal steps of Einstein's discovery of the theory of general relativity and locate the Zurich Notebook in this context. As mentioned above, in 1907 when Einstein was still an employee at the patent office in Bern he had already laid out crucial elements of the heuristics he would follow in the years to come, in particular the equivalence principle and the principle of generalized relativity. As early as 1907 he also considered three possible physical consequences of the equivalence principle, the gravitational red-shift, the bending of light in a gravitational field (Einstein 1907: sect. V), and the perihelion advance of Mercury.[6] He inferred from the equivalence principle that the speed of light must be variable, in contrast to one of the fundamental principles of the special theory of relativity of 1905. In spite of this rapid and impressive initial progress, however, he did not yet begin to work out a theory of gravitation based on a generalized principle of relativity. In fact, when the history of Einstein's work on his theory in the following years is judged from hindsight, it may appear as a sequence of missed opportunities, characterized by some reluctance to react to work on gravitation by his colleagues and in particular to that work which actually paved the way for further development. Not only did Einstein fail to follow up on his own first successful steps towards a relativistic theory of gravitation, he also failed for some time to adopt Minkowski's reformulation of special relativity in terms of a four-dimensional space-time manifold, a crucial instrument for the further development of a relativistic theory of gravitation. A pivotal element of Minkowski's four-dimensional representation of special relativity was the invariance under linear,

---

Development and Education in Berlin. It was continued under the direction of Jürgen Renn as part of the project of studies of the integration and disintegration of knowledge in modern science at the Max Planck Institute for the History of Science in Berlin.

6 See Einstein to Conrad Habicht, 24 December 1907: "At the moment I am working on a relativistic analysis of the law of gravitation by means of which I hope to explain the still unexplained secular changes in the perihelion of Mercury." ("Jetzt bin ich mit einer ebenfalls relativitätstheoretischen Betrachtung über das Gravitationsgesetz beschäftigt, mit der ich die noch unerklärten Aenderungen der Perihellänge des Merkur zu erklären hoffe.") Einstein *CP5*: 82. Although Einstein did not say *how* he tried to calculate the perihelion advance it seems rather unlikely that he would not have based his consideration on the equivalence principle expounded shortly before in the 1907 review paper.

orthogonal transformations of the quantity

$$x^2 + y^2 + z^2 - (ct)^2,$$

formed from Cartesian coordinates x, y, z, t, with c denoting the speed of light (Minkowski 1908). This quantity, when taken infinitesimally as

$$ds^2 = dx^2 + dy^2 + dz^2 - (cdt)^2, \tag{1}$$

in fact later turned out to be suitable not only for representing the 'flat' space-time of special relativity but also the 'curved' space-times of general relativity.

Only in 1911, now a professor in Prague, did Einstein come back to the problem of gravitation, when it occurred to him that a physical consequence of the equivalence principle, the bending of light in a gravitational field, might be observable for light rays passing near the sun (Einstein 1911). The publication of this remarkable physical prediction drew the attention of other scientists to Einstein's approach, with important consequences for the further development. In January 1912, it was Max Abraham who published a theory of gravitation (Abraham 1912a) with a gravitational field equation, formulated in terms of Minkowski's four-dimensional space-time formulation, which yielded Einstein's basic relation between the variable velocity of light and the gravitational potential as a special case.

Einstein responded with the publication of a different theory of gravitation (Einstein 1912a), based on the equivalence principle, but still not formulated in a four-dimensional formulation and restricted to the special case of static fields. He also attacked Abraham's theory because of what Einstein saw as the incompatibility between Abraham's simultaneous use of both Minkowki's formalism and Einstein's variable speed of light. To this, Abraham in turn responded, in an extremely brief note of correction (Abraham 1912b),[7] with the suggestion that his use of Minkowski's geometry is to be understood as being restricted to an infinitesimally small local region of the four-dimensional space-time, with the line element (1) (and hence the geometry of space-time) in general being dependent on the gravitational field through the coefficient $c^2 = c^2(x, y, z)$ in front of dt^2. It was thus Abraham who effectively introduced the line element of a non-flat space-time into gravitation theory.[8]

Although this insight would shortly become the basis of Einstein's further search for the gravitational field equation, it was at this point taken up neither by Abraham himself nor by Einstein—another missed opportunity. Only after further elaborating his theory for the special case of static fields, and after familiarizing himself with Minkowski's formalism more carefully, did Einstein, at some point in the summer of 1912 (Stachel 1980), eventually adopt a 10-component metric tensor

---

[7] Abraham's note suggests that Einstein expressed his criticism not only in print but also in private correspondence. Unfortunately, it seems that no such correspondence has been preserved.

[8] For a discussion of the historical evidence for this period, see Pais 1982: sect. 12b; see also Stachel 1980 and Maltese & Orlando 1995.

$g_{\mu\nu}$, describing a variable line-element in four dimensions,

$$ds^2 = \sum_{\mu,\nu=1}^{4} g_{\mu\nu}\, dx^\mu dx^\nu,$$

as the representation of the gravitational potential, and hence as the basis for his further search for a relativistic field theory of gravitation.

Earlier, probably in the spring of 1912, Einstein had conceived of yet another possibility for gaining observational support for a new theory of gravitation—gravitational lensing, at the time so distant from observational possibilities that Einstein did not even publish his prediction of this effect (Renn, Sauer & Stachel 1997).

In August 1912 Einstein left Prague, where he had stayed for one and a half years, for Zurich. There he remained, as a professor of the Eidgenössische Technische Hochschule (ETH), until he left for Berlin in the spring of 1914. During his time in Zurich he collaborated with his former ETH classmate, now professor of mathematics at the ETH, Marcel Grossmann, in his attempts to find a gravitational field equation for the metric tensor. Grossmann was particularly helpful in making the literature on invariant theory and on the absolute differential calculus accessible to Einstein. He also suggested to Einstein some mathematically plausible candidates for a gravitational field equation.

It has long been known that during this period of time, Einstein missed, so at least it appears from hindsight, another important opportunity. In fact, he had come close to the later field equation of general relativity by considering the Ricci tensor,

$$R_{\mu\nu} = \frac{\partial \Gamma^\lambda_{\mu\nu}}{\partial x^\lambda} - \frac{\partial \Gamma^\lambda_{\mu\lambda}}{\partial x^\nu} - \Gamma^\kappa_{\mu\lambda}\Gamma^\lambda_{\nu\kappa} + \Gamma^\kappa_{\mu\nu}\Gamma^\lambda_{\kappa\lambda},$$

given to him by Grossmann as a possible candidate for the left-hand side of such a field equation, whose right-hand side is given by the energy-momentum tensor $T_{\mu\nu}$.[9] Numerous speculations have been ventured for why he needed another two and a half years before he took up this promising line of attack once more.[10] The detailed examination of the Zurich Notebook has meanwhile increased the list of missed opportunities. John Norton found in his first pioneering analysis of the Notebook that Einstein not only considered the Ricci tensor in 1912, about three years before he came back to it in November 1915, but that Grossmann showed him also a way of deriving another candidate for the left-hand side of the field equation closely related to the Ricci tensor. Indeed, one may subtract from

---

[9] Here

$$\Gamma^\lambda_{\mu\nu} = \frac{1}{2} g^{\lambda\kappa} \left(\frac{\partial g_{\mu\kappa}}{\partial x^\nu} + \frac{\partial g_{\kappa\nu}}{\partial x^\mu} - \frac{\partial g_{\mu\nu}}{\partial x^\kappa} \right)$$

denotes the Christoffel symbols of the second kind. Summation over repeated indices is implied throughout this paper albeit in some equations below we will write out redundant summation signs in order to be closer to the formulas as written in the Zurich Notebook.

[10] See the references cited in (Norton 1984: nn. 8, 9).

the Ricci tensor a part which transforms tensorially under a restricted group of coordinate transformations and take the remaining part,

$$N_{\mu\nu} = \frac{\partial \Gamma^{\lambda}_{\mu\nu}}{\partial x^{\lambda}} - \Gamma^{\kappa}_{\mu\lambda}\Gamma^{\lambda}_{\nu\kappa},$$

as a new candidate for the gravitational field equation.

This 'tensor'[11] was also reconsidered by Einstein in November 1915 (Einstein 1915a), and for this reason will in the following simply be called the "November tensor."

The analysis of the Zurich Notebook carried out in the context of our joint research project has revealed another surprising missed opportunity. At the end of 1912 or in the beginning of 1913 Einstein even happened to consider, in linearized form, the final version of the field equation of general relativity. Adding a trace term ($R \equiv g^{\mu\nu} R_{\mu\nu}$) to the Ricci tensor, he actually considered what is now called the Einstein tensor,

$$E_{\mu\nu} = R_{\mu\nu} - \frac{1}{2} g_{\mu\nu} R,$$

as the left-hand side of a field equation, and discarded it as well. But in spite of pursuing these promising candidates, the Notebook instead ends with a short derivation of the left-hand side of the curious *Entwurf* field equations,

$$\begin{aligned} \frac{1}{\sqrt{-g}} \frac{\partial}{\partial x^{\alpha}} \left(\sqrt{-g}\, g^{\alpha\beta} \frac{\partial g_{\mu\nu}}{\partial x^{\beta}} \right) - g^{\alpha\beta} g^{\tau\rho} \frac{\partial g_{\mu\tau}}{\partial x^{\alpha}} \frac{\partial g_{\nu\rho}}{\partial x^{\beta}} \\ - \frac{1}{2} \frac{\partial g_{\tau\rho}}{\partial x^{\mu}} \frac{\partial g^{\tau\rho}}{\partial x^{\nu}} + \frac{1}{4} g_{\mu\nu} g^{\alpha\beta} \frac{\partial g_{\tau\rho}}{\partial x^{\alpha}} \frac{\partial g^{\tau\rho}}{\partial x^{\beta}}, \end{aligned} \tag{2}$$

so called because Einstein and Grossmann published field equations with this differential expression in a paper entitled "Entwurf einer verallgemeinerten Relativitätstheorie und einer Theorie der Gravitation" in the spring of 1913 (Einstein & Grossmann 1913).[12]

The *Entwurf* field equations are represented by a complex mathematical differential operator which is covariant only under some restricted class of coordinate transformations, and for this reason make no sense from the point of view of modern general relativity. Clearly, the *Entwurf* equations are covariant at least under linear transformations but the precise transformational properties were unknown to Einstein and Grossmann when they published their paper and hence also the

---

[11] For the restricted group of unimodular coordinate transformations which leave $g \equiv \det(g_{\mu\nu})$ invariant, the quantity $\Gamma^{\lambda}_{\mu\lambda} = \partial(\ln g)/\partial x^{\mu}$ transforms as a vector, and hence its covariant derivative $\Gamma^{\lambda}_{\mu\lambda;\nu} = \partial\Gamma^{\lambda}_{\mu\lambda}/\partial x^{\nu} - \Gamma^{\kappa}_{\mu\nu}\Gamma^{\lambda}_{\kappa\lambda}$ transforms as a tensor. This is just the quantity by which the Ricci tensor differs from the November tensor $N_{\mu\nu}$ which therefore transforms also covariantly under unimodular coordinate transformations.

[12] The article was published jointly by Einstein and Grossmann according to the title page. In our context it may be noteworthy, however, that the paper itself was divided into a 'physical' part for which Einstein alone took the responsibility, and a 'mathematical' part which was written by Grossmann. For an historical discussion of the *Entwurf* theory, see Norton 1984.

extent to which their field equation represented a realization of the principle of generalized relativity.

Until the fall of 1915 Einstein continued to elaborate on and refine the *Entwurf* theory, at first together with his friends Marcel Grossmann and Michele Besso, then, for the most part, on his own. He explored the transformational properties of its field equation and developed arguments that were intended to explain its failure to be generally covariant, i.e., not to realize the generalized principle to the utmost extent.[13] Einstein also developed new ways of deriving the *Entwurf* field equation from fundamental assumptions and probed the consequences of the theory, in particular—albeit again without success—the possibility of explaining on its basis the anomalous perihelion advance of Mercury.[14] He also studied the relation of the *Entwurf* theory to alternative relativistic theories of gravitation, in particular to that formulated by Gunnar Nordstrøm.[15] Nordstrøm's theory was, however, as little successful as Einstein's in explaining the perihelion shift of Mercury, the only direct empirical consequence not explained by Newtonian gravitation theory that was then accessible to observational control.

Although Einstein knew of the failure of the *Entwurf* theory to pass the test of explaining the Mercury anomaly since the middle of 1913, he nevertheless continued to hold on to this theory for more than two years. When in September 1915 he definitely found out that its field equation is not satisfied by the gravitational field of a uniformly rotating system, spoiling temporary hopes that this might be, after all, the case, he at first also missed the opportunity of giving up this theory.[16] In the fall of 1915 Einstein also had to realize that an earlier, mathematical derivation of its field equation was flawed. But even the discovery of this flaw did not imply the immediate demise of the *Entwurf* field equation as Einstein was quick to add additional assumptions to make its derivation work.[17]

It may thus almost come as a surprise that, in November 1915, Einstein did give up the *Entwurf* theory, after all. It was at this point that he once more turned to the discarded candidate field equations considered in the Zurich Notebook. But he still did not take up the correct field equation involving the Einstein tensor, the field equation whose linearized version he had considered three years earlier when working with the Zurich Notebook. Instead, he returned to the non-generally covariant candidate which he had also abandoned earlier and which we have called the "November tensor" (Einstein 1915a). For some days, he was convinced he had solved all the problems which he had found earlier with this tensor. He then

---

13 The most prominent of these is the so-called "hole argument." See Stachel 1980 and Norton 1987 for a discussion of this argument.

14 For an account of Einstein's computation of the perihelion shift on the basis of the *Entwurf* theory, see Earman & Janssen 1993 and Einstein *CP4*: 344–359.

15 For a discussion of Nordstrøm's theory, see Norton 1992.

16 For an account of Einstein's discovery of the incompatibility of the *Entwurf* equations with Minkowski space-time in rotating Cartesian coordinates (in itself a sequence of missed opportunites), see Janssen 1997.

17 See Einstein to H. A. Lorentz, 12 December 1915 (EA 16–442).

abandoned the November tensor once again but still did not return to the correct Einstein tensor. He rather examined the Ricci tensor first (Einstein 1915b), only to give it up as well, also for a second time, after about a week. Only then did he at last come back to the Einstein tensor and he presented the correct field equations based on it to the Prussian Academy on 25 November 1915 (Einstein 1915d).

In order to analyze Einstein's willingness to give up generally covariant field equations and his motives behind these rather erratic looking leaps from one candidate field equation to another, and in order to understand the reasons for what appear to be so many missed opportunities on his part, one has to take a closer look at his heuristic principles and their effect on his research. More than any of Einstein's published papers, the Zurich Notebook provides an ideal source for such an analysis.

2. Elements of Einstein's heuristics

As pointed out above, the early summer of 1912 represents a crucial turning point in the history of Einstein's discovery of the gravitational field equation because it is then that he realized the significance of the metric tensor and the general line element for a generalized theory of gravitation. Following this insight Einstein started to study (again) the mathematics of Gaussian surface theory, and in the course of a more systematic survey of the relevant mathematical literature, undertaken in the collaboration with Grossmann, became acquainted with Beltrami invariants and the "absolute differential calculus" of Ricci and Levi-Civita.[18] The mathematical formalism provided by the theory of differential invariants and tensor calculus for Einstein presented an as yet unfamiliar network of deductive possibilities and constraints which had to be explored and endowed with physical meaning.

Before illustrating in Section 3 the interactive process of representation and interpretation, of deductivity and heuristics, by analyzing an episode of this process documented in the Zurich Notebook, we briefly review the building blocks out of which Einstein wanted to construct his new theory of gravitation and discuss the crucial principles guiding his heuristics.

2.1. Metric tensor, special relativity, and theory of static fields

Einstein assembled the building blocks for his new theory of gravitation on an early page (39L, see Figure 1) of the Zurich Notebook: the metric tensor, Minkowski's four-dimensional reformulation of special relativity, and the left-hand side of the gravitational equation of his second theory of static gravitational fields.

The general line element with a metric tensor $g_{\lambda\mu}$,

$$ds^2 = \sum g_{\lambda\mu}\, dx^\lambda dx^\mu,$$

for the first time altogether appears on this early page of the Zurich Notebook.

---

[18] For a historical survey of the relevant mathematics, including a discussion of Einstein's pertinent mathematical background, see Reich 1994.

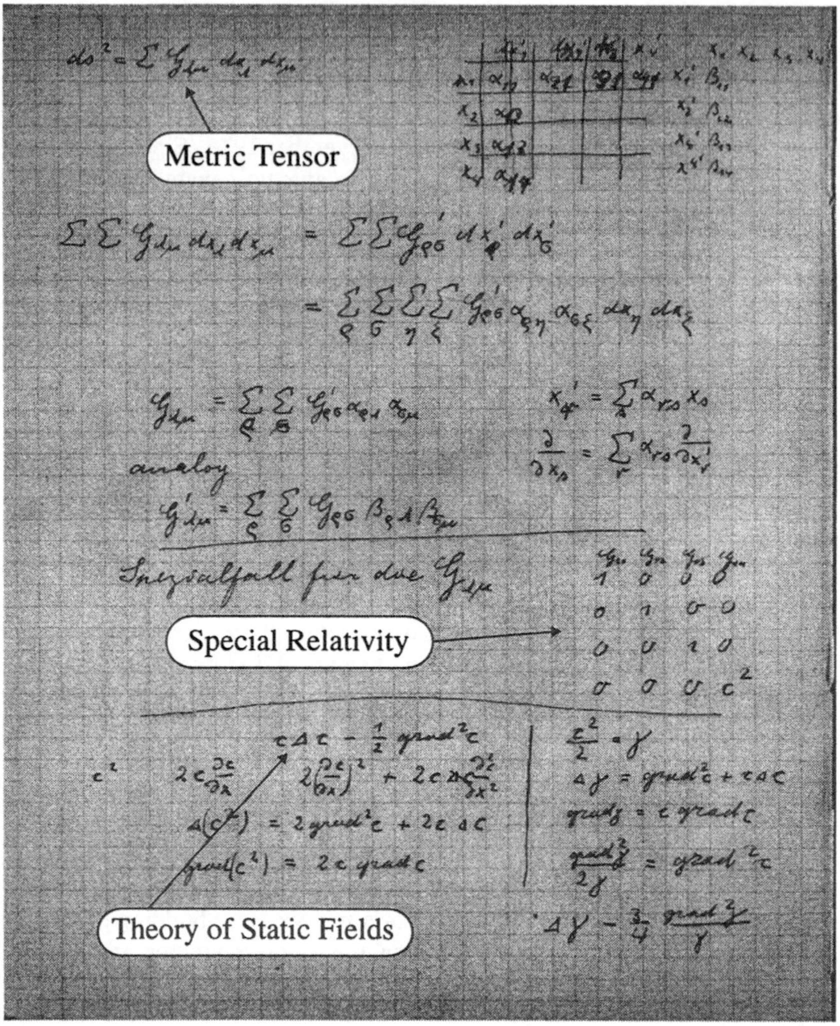

Figure 1. Page 39L of the Zurich Notebook. This early page of the Notebook contains Einstein's basic elements for a relativistic theory of gravitation.

The identification of p. 39L as the earliest page in the Notebook dealing with gravitation is corroborated both by Einstein's initial use of the majuscule G for denoting the metric tensor (in the remainder of the manuscript he uses the usual minuscule g) and by the elementary character of the calculations. Here in fact Einstein was still checking the tensorial character and the transformational properties of the metric tensor by explicit coordinate transformations. In later parts of the manuscript his techniques became much more sophisticated.

The special form of a diagonal metric,

$$\begin{pmatrix} 1 & 0 & 0 & 0 \\ 0 & 1 & 0 & 0 \\ 0 & 0 & 1 & 0 \\ 0 & 0 & 0 & c^2 \end{pmatrix}, \tag{3}$$

with the g_{44}-component given by the square of the velocity of light refers to the four-dimensional reformulation of special relativity with an imaginary time coordinate. It represented the special case to which tentative generalizations had to reduce in the absence of gravitational fields. Einstein expected that his theory of static gravitational fields with the (spatially variable) velocity of light representing the gravitational potential could also be embedded into a generalized tensorial theory by means of this particular form of a metric, as becomes clear from the third and last building block displayed on this page.

The differential expression for the velocity of light c written on the bottom of the page can be identified as the left-hand side of the "field equation" for static gravitation,

$$c\Delta c - \frac{1}{2}(\operatorname{grad} c)^2 = kc^2\sigma, \tag{4}$$

which Einstein had published in March 1912 (Einstein 1912b). Equation (4) was obtained as a nonlinear generalization of the classical scalar Poisson equation with c representing the potential of the static gravitational field, k being a constant, and σ denoting the field-generating mass and (non-gravitational) energy density.

Einstein's intention for the transformation $c^2/2 = \gamma$, performed on the bottom of p. 39L, obviously was to put the differential equation (4) into a form that allowed its interpretation as one particular component of a 10-component tensorial field equation for the metric tensor.

This calculation therefore closely pertains to the central question governing all calculations of the Zurich Notebook dealing with the problem of gravitation: *What is the appropriate differential expression $\Gamma_{\mu\nu}$ which is formed from the metric tensor and its first and second derivatives and which enters a field equation of the form*

$$\Gamma_{\mu\nu} = \kappa T_{\mu\nu}, \tag{5}$$

with the stress-energy tensor $T_{\mu\nu}$ of matter as the source term on the right-hand side?

2.2. The heuristic requirements

Finding an answer to this question had two different aspects. On the one hand, Einstein had to find suitable candidates for the differential expression acting on and involving the metric tensor. On the other hand, he had to check, for each such candidate, whether it actually satisfied all his heuristic requirements.

Einstein's heuristic criteria were, of course, variable to some extent and were indeed modified in the course of and adapted to the explorative experiences accumulated in the research process. As we will illustrate below, the interplay of these general heuristic ideas with the deductive structure materialized in the mathematical representation governed the calculations contained in the Zurich Notebook and make their sequence understandable. The reconstruction of these calculations, together with an analysis of Einstein's publications and his contemporary correspondence, allows the identification of four distinct heuristic requirements which had to be checked for each of the candidate gravitational field equations.

2.2.1. *The Requirement of Equivalence*

Already in his review of 1907, Einstein formulated the assumption of complete physical equivalence between a uniformly accelerated reference frame and a constant homogeneous gravitational field. He asserted:

> The heuristic value of this assumption rests on the fact that it permits the replacement of a homogeneous gravitational field by a uniformly accelerated reference system, the latter case being to some extent accessible to theoretical treatment.[19] (Einstein 1907: 454).

Einstein's assumption was, however, much more than a means useful to make gravitational fields accessible to theoretical treatment. It was motivated by the requirement of keeping inertial and gravitational mass equal even in a relativistic gravitation theory, a requirement that was by no means self-evident in the contemporary discussion. In classical mechanics, the proportionality of inertial and gravitational mass, which guarantees Galileo's principle of the equality of accelerations in free fall, is merely an empirical statement. Einstein's approach to the new theory of gravitation, on the other hand, is characterized by transforming this empirical statement into a heuristic principle of fundamental importance for this new theory.[20]

Whereas the equivalence between a uniformly accelerated reference frame and a homogeneous static gravitational field forms the essence of the equivalence principle,[21] Einstein also attempted to extend this relation to other accelerated frames. Many years later, he once gave the following general characterization of the equivalence principle:

> The equivalence principle asserts that the qualities of physical space, as they present themselves from an accelerated coordinate system, represent a special case of the gravitational field. (Einstein to A. Rehtz, 12 July 1953, EA 27–134)

The paradigmatic case for such an extension of the equivalence principle to other accelerated frames was the "rotating bucket," i.e., a uniformly rotating system of reference. The rotating bucket had been used by Newton and much later by Mach in order to discuss the question of whether inertia is a property of absolute space or, as Mach suggested, caused by the interaction of masses. If the acceleration

---

19 "Der heuristische Wert der Annahme liegt darin, daß sie ein homogenes Gravitationsfeld durch ein gleichförmig beschleunigtes Bezugssystem zu ersetzen gestattet, welch letzterer Fall bis zu einem gewissen Grade der theoretischen Behandlung zugänglich ist."

20 For a discussion of the peculiarity of Einstein's approach in its historical context, see Renn 1994.

21 For a historical discussion of the equivalence principle, see Norton 1985.

field of such a rotating frame of reference could be interpreted as a gravitational field, then rotation could be conceived as a state of rest, as Einstein once put it.[22] In other words, locally, a rotating frame of reference could be considered, according to Einstein, as being equivalent to an inertial frame equipped with a special gravitational field somehow caused by the presence of masses. In this way, the equivalence between acceleration fields and special cases of the gravitational field could be exploited in order to realize Mach's idea of interpreting inertial effects as being due to the interaction of masses. As a consequence, the inertial frames of reference of classical mechanics and special relativity no longer had to be considered as privileged frames of reference, fundamentally distinct from accelerated frames of reference. For Einstein this idea became the core of his search for a generalization of the relativity principle.

2.2.2. *The Requirement of Generalized Relativity*

Einstein attempted to generalize the principle of relativity by requiring that the covariance group of his new theory of gravitation be larger than the group of Lorentz transformations of special relativity. In his understanding, this requirement was optimally satisfied if the field equation of the new theory could be shown to possess the mathematical property of general covariance.

Einstein saw a generalized principle of relativity as guaranteeing the satisfaction of the equivalence principle as well. In fact, according to the equivalence principle, an arbitrarily accelerated frame of reference in Minkowski space-time can precisely then be considered as being physically equivalent to an inertial frame if a gravitational field can be introduced which accounts for the inertial effects in the accelerated frame. If now gravito-inertial fields are described by a generally covariant field equation, this must (at least locally) always be possible, since the metric tensor describing the inertial effects in the arbitrarily accelerated frame is then obviously a solution of this field equation and hence represents, in Einstein's understanding, a particular instance of a gravitational field. But in the period under consideration, the generalized relativity principle was not yet explicitly distinguished from what Einstein later introduced as "Mach's principle."[23] It was only the later development that necessitated this distinction; Einstein initially hoped that accounting for inertial effects in terms of gravitational fields would automatically provide an explanation of these effects by the interaction of masses.

2.2.3. *The Requirement of Correspondence*

Einstein further required that the new theory would describe, under certain limiting conditions, the gravitational effects familiar from Newtonian physics. For this

[22] "You see that I am still far from being able to conceive of rotation as rest!" ("Du siehst, dass ich noch weit davon entfernt bin, die Drehung als Ruhe aufassen zu können!") Einstein to M. Besso, 26 March 1912 (Einstein *CP5*: 436).

[23] The term "Mach's principle" was introduced in Einstein 1918a in the context of a controversy with E. Kretschmann on whether or not the general covariance of the field equations implied a generalization of the relativity principle of classical mechanics and special relativity or would only be a mathematical property. In response to this argument, Einstein explicitly distinguished between the principle of relativity and Mach's principle. For further historical discussion of Mach's principle, see Hoefer 1994; Renn 1994; and Barbour & Pfister 1995.

reason, he expected that the unknown gravitational field equation for the metric tensor would reduce to the Poisson equation for the scalar gravitational potential of the classical theory and that, under the same limiting conditions, the equation of motion of his new theory would yield Newton's second law with the force derived from this classical potential. Finally, he assumed that this Newtonian limit can be obtained from the full field equation via an intermediate step characterized by weak static fields leading to a linearized field equation for the metric tensor. Specifically, Einstein expected that the 'Newtonian limit' should be obtained via a metric of the form (3) so that a spatially variable g_{44}-component would guarantee the link to his earlier theory of static gravitational fields (Stachel 1980, 1982, 1989).

We subsume these various demands under what we call Einstein's *correspondence principle* of general relativity. The realization of this principle was a crucial condition for conveying physical meaning to the various mathematical constructs he elaborated since only in this way could they be brought into contact with the empirical knowledge embodied in Newtonian gravitation theory.

2.2.4. *The Requirement of Conservation*

When Einstein began his systematic search for a field equation in 1912, the development of special relativistic dynamics, including the four-dimensional formulation of electrodynamics and of continuum mechanics,[24] offered formulations of basic physical laws which suggested a plausible generalization to a generally relativistic setting. In particular, special relativistic dynamics displayed a generalizable model for the formulation of the conservation of energy and momentum centered upon a four-dimensional stress-energy tensor. By the fall of 1912, Einstein had indeed found a plausible, generally covariant equation involving this tensor, which he interpreted as representing both a generalization of the special-relativistic formulation of the conservation of energy and momentum as well as of the Newtonian law of motion for continuous matter in a gravitational field. It was therefore natural to use this equation as a touchstone for the gravitational field equation to be found; we call this requirement the *conservation principle*. In the Zurich Notebook this equation appears in the following form, valid for a symmetric stress-energy tensor $T^{\mu\nu}$

$$\frac{\partial}{\partial x^{\nu}}(\sqrt{-g}\, g_{\mu\lambda}T^{\mu\nu}) - \frac{1}{2}\sqrt{-g}\frac{\partial g_{\mu\nu}}{\partial x^{\lambda}}T^{\mu\nu} = 0. \tag{6}$$

In modern terms, this equation corresponds to the vanishing of the covariant divergence of the stress energy tensor of matter, $T^{\nu}_{\lambda;\nu} = 0$.

The principle of conservation of energy also motivated the search for an expression which could be interpreted as a (not necessarily generally covariant) stress-energy 'tensor' of the gravitational field, allowing one to rewrite the second term of the above equation as the coordinate divergence of this tensor. The conservation of energy and momentum would then be expressed by the vanishing of the coordinate divergence of the sum of the energy-momentum tensors of matter and field.

---

[24] For a contemporary review, see von Laue 1911.

These four heuristic requirements played a dominant role in Einstein's search for a relativistic theory of gravitation, acting in an oscillating manner either as starting points or as touchstones for tentative field equations. As it turned out, they actually overdetermined the problem and Einstein in his investigation of concrete field equations was forced to weigh the different requirements against each other. Attempting to make ends meet he was faced with the necessity of weakening or modifying one or more of them.

2.3. EINSTEIN'S DOUBLE STRATEGY

In the course of Einstein's work on the problem of gravitation two distinct and complementary strategies emerged for the construction of suitable candidates for the differential operator entering the left-hand side of a tentative gravitational field equation. These two strategies take the above-introduced heuristic requirements in a complementary manner either as points of departure or as touchstones. The strategies and the role of the heuristic requirements in following these strategies can be identified most distinctly in the reconstruction of the calculations of the Zurich Notebook.

A 'mathematical' strategy started from the requirement of the generalized relativity principle, The ground for pursuing this strategy had to be prepared by scanning the mathematical literature for suitable differential expressions with a well-known covariance group. The entries in the Notebook indeed reveal Einstein's increasing familiarity with the relevant mathematics of invariant theory and of Riemann calculus. Tentative field equations with candidate differential operators thus obtained from the mathematical literature then of course had to be tested for the principle of conservation of energy. It also had to be checked whether the Newtonian limit could be realized in the manner expected by Einstein.

The complementary 'physical' strategy started from the well-known limiting case of special relativity and the apparently also well established and firmly founded special theory of static gravitational fields. Along this strategy Einstein sought to construct physically plausible generalizations whose specialization to the Newtonian limit was obvious. Here again conservation of energy was an independent heuristic requirement which had to be checked for each candidate field equation. But most importantly, the covariance group of the differential expressions constructed along this strategy was unknown from the beginning. It hence remained to be investigated to what extent the principle of generalized relativity had actually been realized.

The identification of these two complementary strategies turned out to be the key for understanding many of Einstein's considerations and calculations documented in the Zurich Notebook, as well as for understanding many of his considerations from the period between 1913 and 1915. In the reconstruction of the genesis of the theory of general relativity the identification of these strategies allows in particular a detailed understanding of the complicated process of expressing physical concepts by means of a mathematical framework with its own logical and deductive structure and, vice versa, of endowing the objects of the mathematical representation with a meaning in the conceptual framework of physics.

3. A case study: discovery and rejection of the Einstein tensor in the Zurich Notebook

In this Section we illustrate the interaction between heuristics and representation with an episode of calculations documented in the Zurich Notebook. The episode culminates in Einstein's discovery and rejection of the Einstein tensor in late 1912 or early 1913. The reconstruction is based on the comprehensive analysis of the Notebook undertaken by our group.[25]

In the episode presented in the following, Einstein first investigated a plausible generalization of the Laplacian operator along the 'physical' strategy but was then distracted from this path and driven toward the 'mathematical' strategy when Grossmann showed him the Riemann tensor. Investigating differential expressions derived from the Riemann tensor along the 'mathematical' strategy, Einstein came first to consider the Ricci tensor, the very same object he would consider as the left-hand side of a gravitational field equation three years later in his second memoir of November 1915 (Norton 1984). In the Notebook, Einstein then turned, guided by his heuristic principles and the conflicts generated by their mathematical implementation, to the gravitation tensor of the final theory, the Einstein tensor—if only in linearized approximation. The same heuristic principles that led him to the Einstein tensor then forced him, however, to give it up. Eventually, he returned again to the physical strategy.

3.1. The heuristic ideas meet the mathematical representation

3.1.1. *The core operator—the ideal starting point for the physical strategy*

Both directions of Einstein's heuristic double strategy had their natural starting points: the correspondence principle in the case of the physical strategy, the relativity principle in the case of the mathematical strategy. Each of these alternative starting points required a concretization in the mathematical representation. Along the 'physical' strategy, an obvious way of generalizing the Laplacian operator of the classical Poisson equation for the Newtonian gravitational field was to apply the Laplace-Beltrami operator directly to the components of the metric:

$$K_{\mu\nu} \equiv \sum_{\alpha\beta} \frac{1}{\sqrt{-g}} \frac{\partial}{\partial x^{\alpha}} \left(\sqrt{-g}\, g^{\alpha\beta} \frac{\partial g^{\mu\nu}}{\partial x^{\beta}} \right). \tag{7}$$

We call this object the *core operator* since it is the starting point and the core of many of Einstein's considerations in the years 1912–1915.[26] Obviously, the core operator reduces to the d'Alembertian operator for weak fields, and it reduces to the Laplacian operator for the special case of the static metric (3). Hence it satisfies

---

[25] The discovery that Einstein actually considered the (linearized) Einstein tensor in the Notebook, while unknown in the literature, has meanwhile become the basis for the annotation of the relevant pages of the Notebook in Einstein *CP4*; it has been summarized in Castagnetti, Damerow et al. 1994 and Renn & Sauer 1996.

[26] Strictly speaking, the expression as written in the text is not an operator since it contains the metric components on which operators act.

the heuristic Requirement of Correspondence, i.e., it allows taking the Newtonian limit in the expected manner.

For the special case in which the determinant of the metric is constant and the "Hertz-condition":[27]

$$\sum \frac{\partial g^{\mu\nu}}{\partial x^{\mu}} = 0, \tag{8}$$

holds,[28] the core operator reduces to the simple form

$$g^{\alpha\beta} \frac{\partial^2 g^{\mu\nu}}{\partial x^{\alpha} \partial x^{\beta}}. \tag{9}$$

While it was clear that the core operator satisfies the correspondence principle, Einstein had to check its compatibility with his other heuristic principles. Several pages of the Zurich Notebook can be reconstructed as documenting just such checks.

On page 13R of the Notebook (Figure 2), e.g., Einstein started an investigation of the compatibility of this operator with his heuristic condition of conservation of energy. Specifically, he checked the consequences of energy-momentum conservation (6) for a tentative field equation which had expression (9) on its left-hand side by substituting it for the energy-momentum tensor. Such a substitution amounts to a consistency check between field equation and energy-momentum conservation. This consistency check led him to consider the expression[29]

$$\sum_{\alpha\beta\mu\nu} \frac{\partial}{\partial x^{\mu}} \left(g_{m\nu} g^{\alpha\beta} \frac{\partial^2 g^{\mu\nu}}{\partial x^{\alpha} \partial x^{\beta}} \right) - \frac{1}{2} \sum_{\alpha\beta\mu\nu} \frac{\partial g_{\mu\nu}}{\partial x^{m}} g^{\alpha\beta} \frac{\partial^2 g^{\mu\nu}}{\partial x^{\alpha} \partial x^{\beta}}. \tag{10}$$

Considered as an equation representing energy-momentum conservation (i.e., set equal to 0), this expression should not introduce any new conditions on the gravitational field in addition to the field equation, at least if the core operator is a plausible candidate for the left-hand side of the field equation. Indeed, in the correct field equation of general relativity no such new conditions are introduced by the requirement of energy-momentum conservation due to the contracted Bianchi identities. Furthermore, it should be possible to rewrite the above expression in terms of a stress-energy tensor of the gravitational field.

In the Notebook, Einstein observed, however, that the above expression would normally contain derivatives of third order, unless condition (8) holds. In other words, he found that energy-momentum conservation amounts to the introduction

[27] This condition is mentioned in a letter by Einstein to Paul Hertz on 22 August 1915 (EA 12–203). The letter and its relevance for the history of general relativity are discussed in extenso in Howard & Norton 1993.

[28] Both conditions together imply the condition $\partial/\partial x^{\mu}(\sqrt{-g}\, g^{\mu\nu}) = 0$ which is equivalent to the harmonic condition (12), to be discussed below.

[29] In the Notebook, contravariant components of the metric are denoted by a Greek letter, $\gamma_{\mu\nu}$, rather than by the modern 'upstairs' indices, $g^{\mu\nu}$. Also, indices to coordinates and coordinate differentials were always written 'downstairs'.

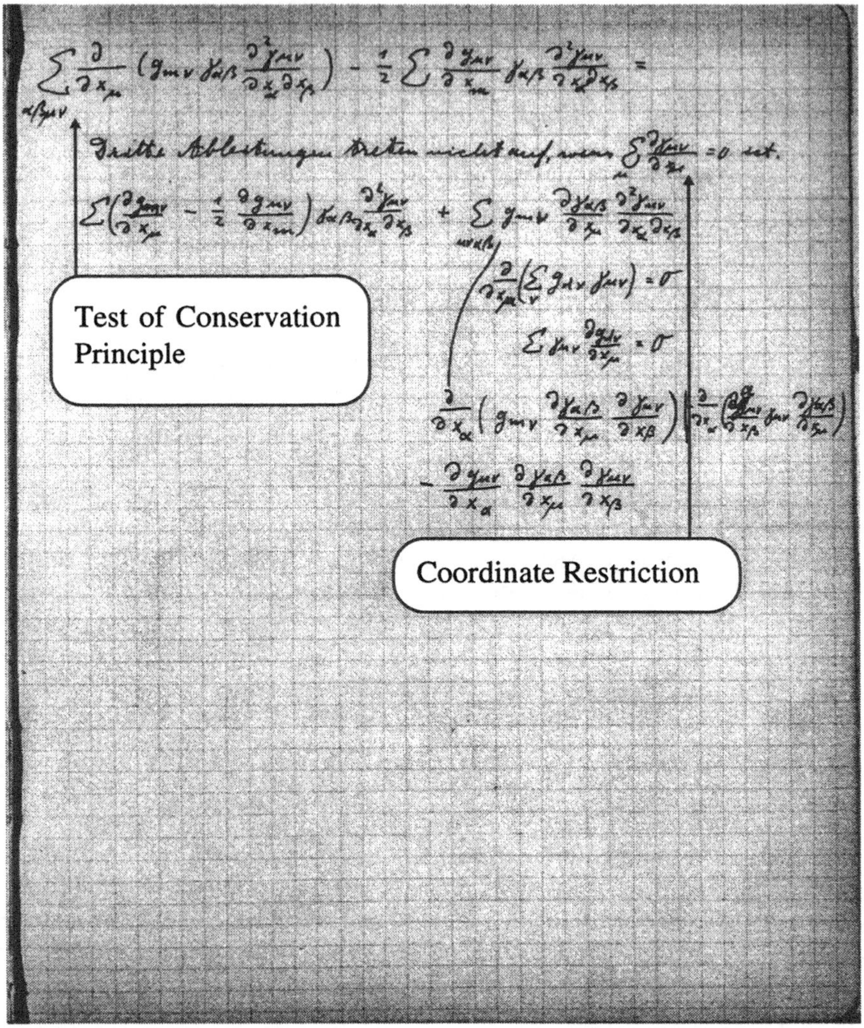

Figure 2. Page 13R of the Zurich Notebook. Here Einstein checked the compatibility of the *core operator* with his heuristic requirement of energy-momentum conservation.

Reproduced with permission of the Albert Einstein Archives, The Hebrew University of Jerusalem, Israel.

of additional requirements. After an attempt at further simplifying expression (10) under the assumption of Equation (8), Einstein broke off. At this point, he apparently found no obvious way of interpreting Equation (10) and of showing the compatibility of the core operator with energy-momentum conservation. Probably for this reason he did not pursue this line of thought any further.

3.1.2. *The Riemann tensor—the ideal starting point for the mathematical strategy*

Following the mathematical strategy, it is the Riemann tensor which could be taken as a natural starting point because it opens up the perspective of finding generally covariant field equations by employing mathematically well understood objects and techniques. Einstein became aware of the Riemann tensor through the help of his mathematician friend Marcel Grossmann. In October 1912, Einstein wrote to Arnold Sommerfeld:

> I am now working exclusively on the gravitation problem and I believe that I can overcome all difficulties with the help of a mathematician friend of mine here.[30] (Einstein *CP5*: 505).

This letter is the only external evidence for dating a page of the Notebook which has Grossmann's name on it next to the definition of what is called a "tensor of fourth manifold,"[31]

$$(i\kappa, lm) = \frac{1}{2}\left(\frac{\partial^2 g_{im}}{\partial x^\kappa \partial x^l} + \frac{\partial^2 g_{\kappa l}}{\partial x^i \partial x^m} - \frac{\partial^2 g_{il}}{\partial x^\kappa \partial x^m} - \frac{\partial^2 g_{\kappa m}}{\partial x^i \partial x^l}\right) + \sum_{\rho\sigma} g^{\rho\sigma}\left(\begin{bmatrix} im \\ \sigma \end{bmatrix}\begin{bmatrix} \kappa l \\ \rho \end{bmatrix} - \begin{bmatrix} il \\ \sigma \end{bmatrix}\begin{bmatrix} \kappa m \\ \rho \end{bmatrix}\right),$$

readily recognized to be the fully covariant Riemann tensor (Figure 3).[32]

With such a generally covariant object at hand, the 'mathematical' strategy must have appeared quite promising since the heuristic Requirement of Generalized Relativity could be fulfilled to the utmost extent. But as such the Riemann tensor was, of course, not a suitable object. In order to obtain a field equation of the form (5), Einstein had to extract a two-index object from the Riemann tensor, in particular since the energy-momentum tensor on the right-hand side has only two indices. Indeed, on p. 14L, Einstein contracts the Riemann tensor once,

$$\sum g^{\kappa l}(i\kappa, lm),$$

in order to obtain such a two-index object, the covariant Ricci tensor, which might be taken as a candidate for the left-hand side of a tentative field equation.

As with the core operator, Einstein now had to examine this new candidate against the list of his heuristic requirements. For objects constructed along the 'mathematical' strategy the requirement of the correspondence principle was the first and most important one on the list. A short consideration, however, made Einstein realize that for weak fields the Ricci tensor does not reduce to the required form of the d'Alembertian acting on the weak field components. In addition to a core operator term of the form (9), the Ricci tensor contains three other terms with

---

30 "Ich beschäftige mich jetzt ausschliesslich mit dem Gravitationsproblem und glaube nun mit Hilfe eines befreundeten Mathematikers aller Schwierigkeiten Herr zu werden."

31 "Tensor vierter Mannigfaltigkeit," see Figure 3.

32 $\begin{bmatrix} im \\ \sigma \end{bmatrix} \equiv \frac{1}{2}\left(\frac{\partial g_{i\sigma}}{\partial x_m} + \frac{\partial g_{\sigma m}}{\partial x_i} - \frac{\partial g_{im}}{\partial x_\sigma}\right)$ denotes the Christoffel symbol of the first kind.

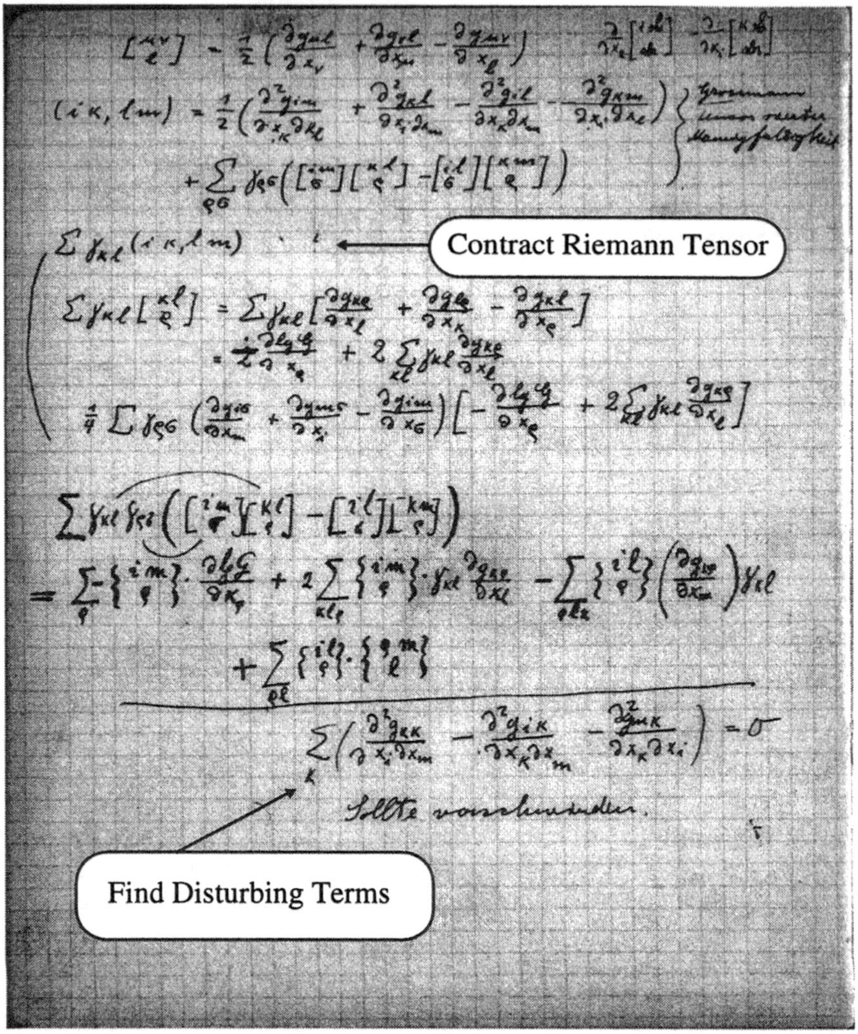

Figure 3. Page 14L of the Zurich Notebook. On this page Einstein started his investigation of the Riemann tensor following the 'mathematical' strategy.

Reproduced with permission of the Albert Einstein Archives, The Hebrew University of Jerusalem, Israel.

second derivatives of the metric. These terms obstruct a smooth transition to the Newtonian limit for weak, static fields in the manner expected by Einstein. In the Notebook, Einstein set these terms equal to zero,

$$\sum_{\kappa}\left(\frac{\partial^2 g_{\kappa\kappa}}{\partial x^i \partial x^m} - \frac{\partial^2 g_{i\kappa}}{\partial x^\kappa \partial x^m} - \frac{\partial^2 g_{m\kappa}}{\partial x^\kappa \partial x^i}\right) = 0, \tag{11}$$

with the comment: "should vanish."[33]

The upshot of this first investigation was that the Ricci tensor violated Einstein's heuristic Requirement of Correspondence.

3.2. The mathematical representation generates conflicts and novelties

3.2.1. *Coordinate restrictions as a novelty suggested by the conflict between the Requirements of Correspondence and Relativity*

We have seen that the consideration of the Ricci tensor revealed a conflict between the Requirements of Generalized Relativity and of Correspondence. This conflict did, however, not force Einstein to give up the Riemann tensor as his starting point, as the formalism still provided more paths, the exploration of which made good sense from the perspective of Einstein's heuristics.

But the mathematical representation did not only suggest alternative objects that could be tested against Einstein's heuristics, it also offered possibilities for modifying the heuristics, e.g., by restricting or reinterpreting one of the heuristic requirements. One obvious candidate for such a modification was the requirement of general relativity because it represented a 'maximalist' goal of Einstein's search.

The expression of the conflict between correspondence and relativity principles by the 'disturbing terms' (11) suggested, in particular, that these terms could be brought to vanish by imposing an additional restrictive condition. On p. 19L, Einstein in fact restricted the group of admissible coordinate systems to those which satisfy the following condition (see Figure 4)

$$\sum_{k,l} g^{kl}\begin{bmatrix} kl \\ i \end{bmatrix} = 0. \tag{12}$$

Coordinates satisfying this condition at the time were called "isothermal"[34] and are called 'harmonic' coordinates today. For these coordinates the Ricci tensor reduces to the form

$$\sum_{k,l} g^{kl}\frac{\partial^2 g_{im}}{\partial x^k \partial x^l} + \Delta_{im}, \tag{13}$$

where we have defined Δ_{im} by

$$\Delta_{im} \equiv -\frac{1}{2}\frac{\partial g^{kl}}{\partial x^m}\frac{\partial g_{kl}}{\partial x^i} + \frac{\partial g^{kl}}{\partial x^m}\frac{\partial g_{il}}{\partial x^k} + \frac{\partial g^{kl}}{\partial x^i}\frac{\partial g_{mk}}{\partial x^l} - g^{\rho\sigma}g^{kl}\frac{\partial g_{i\rho}}{\partial x^l}\frac{\partial g_{m\sigma}}{\partial x^k} + g^{\rho\sigma}g^{kl}\frac{\partial g_{il}}{\partial x^\rho}\frac{\partial g_{m\sigma}}{\partial x^k},$$

a term that contains only products of first derivative terms, which do not affect the Newtonian limit in Einstein's understanding of weak and static fields.

---

33 "Sollte verschwinden," see Figure 3.

34 See Bianchi 1910: § 36–37.

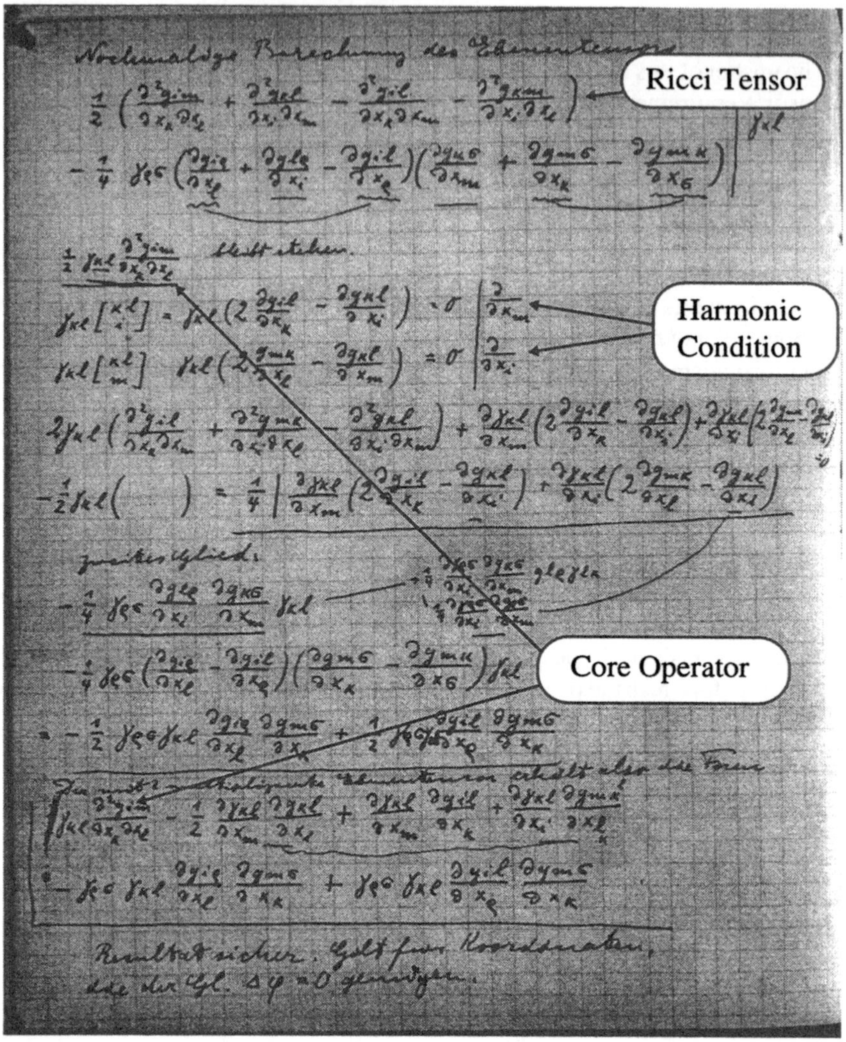

Figure 4. Page 19L of the Zurich Notebook. Einstein succeeded in reducing the Ricci tensor using harmonic coordinates.

Reproduced with permission of the Albert Einstein Archives, The Hebrew University of Jerusalem, Israel.

It is unclear how exactly Einstein understood the additional condition, whether indeed as a restriction of the relativity principle as we suggest here or as a coordinate condition in the modern sense.[35] Nowadays coordinate conditions such as

---

[35] Looking only at p. 19L, it is quite possible that Einstein understood the harmonic coordinate condition appearing on this page in the modern sense. This is the way the appearance of Equation (12) was interpreted in Norton 1984. But a comprehensive reconstruction of the Notebook has revealed hints that the interpretation given in the text may be in better agreement with other parts of the Notebook.

Equation (12) are invoked if a generally covariant equation is applied to a physical situation which suggests the specialization of the coordinate system. The obvious example is Newtonian gravitation theory. In whichever way Einstein saw the restrictive condition (12), it is clear that the mathematical concretization of Einstein's heuristic requirements created a conflict which in turn led to a new insight, the possibility of avoiding this conflict by adding coordinate restrictions to a generally covariant field equation.

The reduced object (13) must have looked promising to Einstein. The only term containing second derivative terms was a term of the form of the core operator

$$\gamma_{kl}\frac{\partial^2 g_{im}}{\partial x_k \partial x_l}$$

and the other terms could be neglected for weak fields. The reduced object (13) was consequently in agreement with the heuristic Requirement of Correspondence and one might proceed to check the corresponding field equation against the other heuristic requirements. Since the Ricci tensor could only be brought into the form (13) for coordinates satisfying condition (12), however, it also remained to be checked whether this additional condition was compatible with his other heuristic requirements.

3.2.2. *A conflict between the Requirements of Correspondence and Conservation*

Einstein now considered the case of weak fields $g_{\mu\nu} = \eta_{\mu\nu} + h_{\mu\nu}$, where the Minkowski metric with imaginary time coordinate is given by $\eta_{\mu\nu} = \mathrm{diag}(1, 1, 1, 1)$. In this case the object (13) reduces to the core operator (9), which further simplifies to the d'Alembertian operator acting on the components $h_{\mu\nu}$. In linearized approximation one therefore obtains the following field equation with an energy-momentum tensor $\rho_0 u^\mu u^\nu$ for incoherent matter ("dust") as source term

$$\sum_\kappa \frac{\partial^2 h_{\mu\nu}}{\partial x^{\kappa 2}} = \kappa\rho_0 u_\mu u_\nu, \tag{14}$$

and this linearized field equation was written down explicitly on p. 19R of the Notebook (Figure 5).

Einstein first convinced himself that this equation satisfied the principle of conservation since, for the approximation at hand, it allowed the derivation of an energy-momentum 'tensor' of the gravitational field.[36]

However, there was another problem which immediately arose from the consideration of energy-momentum conservation. In linearized approximation, energy-momentum conservation is expressed by the vanishing of the simple coordinate

In particular, Einstein repeatedly performed coordinate transformations of such coordinate conditions (e.g., p. 22L), which does not make sense from a modern point of view. This question will be discussed in detail elsewhere.

36 "The energy and momentum theorem holds to the relevant approximation." ("Energie- und Impulssatz gilt mit der in Betr[acht] kommenden Annäherung," see Figure 5.

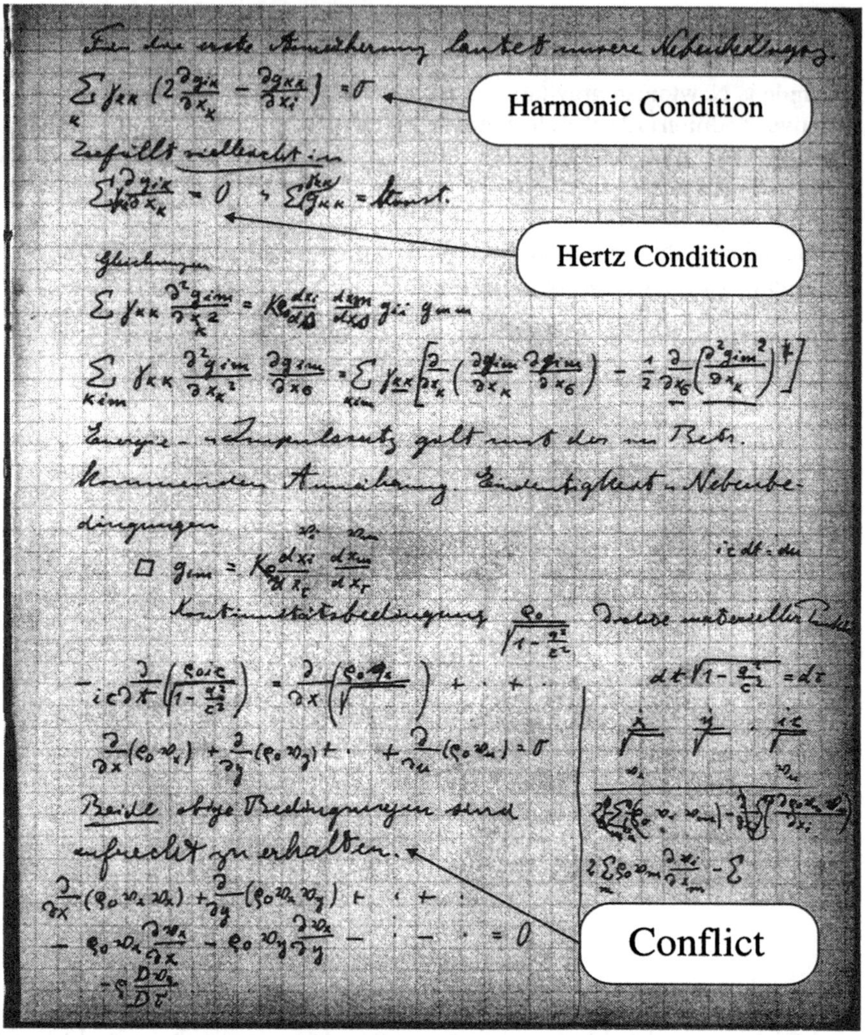

Figure 5. Page 19R of the Zurich Notebook. Einstein discovered a conflict between the harmonic coordinate restriction, necessary to satisfy the Requirement of Correspondence, and the "Hertz condition," necessary to satisfy the requirement of energy and momentum conservation.

divergence of the energy-momentum tensor of matter. Together with the linearized field equation (14) this condition immediately implies the linearized Hertz condition,

$$\sum_{\nu} \frac{\partial h_{\mu\nu}}{\partial x^{\nu}} = 0, \tag{15}$$

as an additional restriction on the coordinates. Taking into account his two heuristic Requirements of Correspondence and Conservation, Einstein at this point thus had to deal with two restrictive conditions on the coordinates. Comparing condition (15) with the "isothermal" or "harmonic" condition (12), which in linearized approximation reads

$$\sum_{\nu} \left(\frac{\partial h_{\kappa\lambda}}{\partial x^{\kappa}} - \frac{1}{2} \frac{\partial h_{\kappa\kappa}}{\partial x^{\lambda}} \right) = 0, \tag{16}$$

one recognizes that the simultaneous stipulation of the two conditions (15) and (16) further implies the condition

$$\sum_{\nu} \frac{\partial h_{\kappa\kappa}}{\partial x^{\lambda}} = 0. \tag{17}$$

This latter condition, however, was unacceptable for two reasons. First, it implied together with the linearized field equation (14) the constancy of the trace of the energy-momentum tensor of matter. This consequence is obviously violated even by the simple case of dust. Second, Equation (17) implies that for weak fields the tensorial field equation does not reduce to a simple equation for the g_{44} component representing the scalar gravitational potential in the Newtonian limit. Instead, one has to take into account also non-trivial spatial g_{ii}-components at the same level of weak field approximation. This last consequence means that the Newtonian limit could not be attained in the expected way in the special case of weak, static fields, via the metric (3) for Einstein's theory of static gravitation.

In summary, Einstein had, on the one hand, succeeded at this point in extracting from the Ricci tensor a linearized field equation that was compatible with his heuristic Requirement of Correspondence. But his difficulty was that he had only been able to do so by imposing the harmonic condition (12) resp. (16). Energy-momentum conservation on the other hand implied the restrictive condition (15) and the simultaneous stipulation of these two conditions implied the untenable restriction (17).

3.2.3. *The Einstein tensor as a novelty suggested by the conflict between two heuristic requirements*

This dilemma of two incompatible restrictions triggered a further exploration of the formalism. On the next page of the Notebook Einstein considered a modification of the linearized field equation by which he hoped to avoid this dilemma (see Figure 6). This modification effectively made him consider the linearized Einstein tensor as a possible candidate for the left-hand side of the field equation.

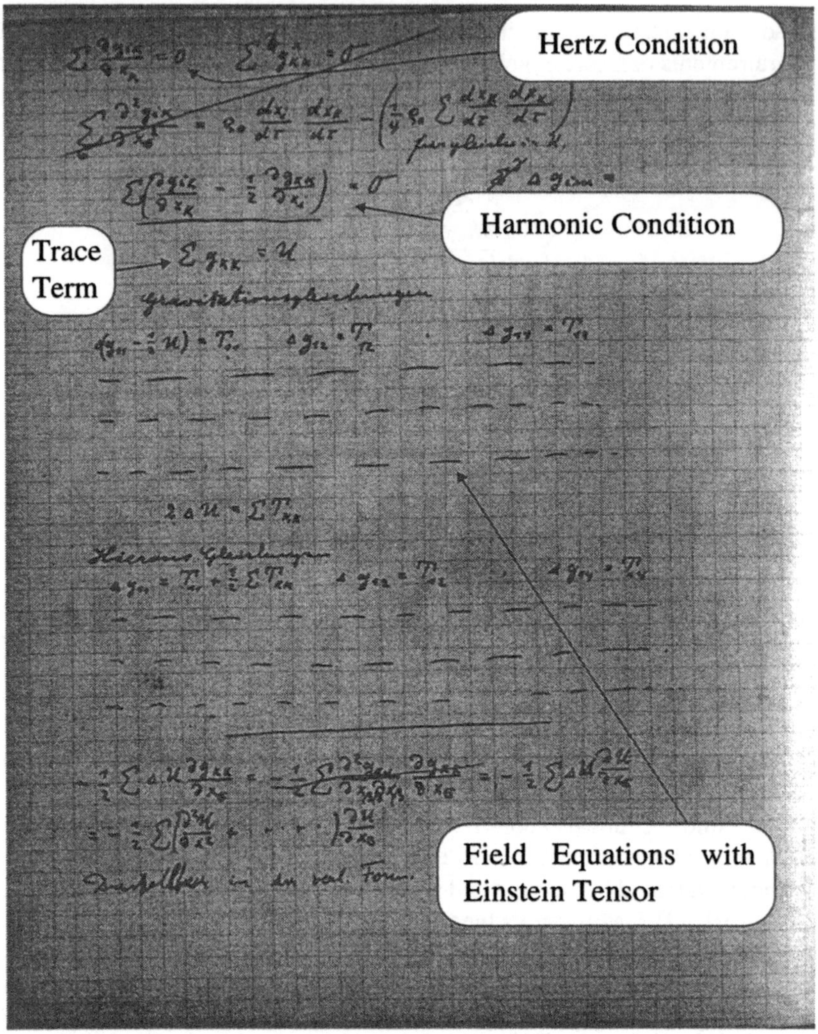

Figure 6. Page 20L of the Zurich Notebook. Einstein tried to solve the conflict between the harmonic coordinate restriction and the Hertz restriction by an ad hoc modification of the field equations. Einstein here effectively wrote down the field equations of the final theory in linear approximation.

Looking at the linearized field equation (14), he noticed that the first of the two above-mentioned consequences, i.e., that the trace of $T_{\mu\nu}$ had to be a constant, could be fixed by adding a trace term to the source term on the right-hand side of the field equation,[37]

$$\sum_{\sigma} \frac{\partial^2 h_{i\kappa}}{\partial x^{\sigma 2}} = K\rho_0 u_i u_\kappa - K\frac{1}{4}\delta_{i\kappa}\rho_0 \sum_k u_k u_k. \tag{18}$$

He must then have remembered that his current considerations were made under the general assumption, formulated on the top of p. 19R, that the harmonic restriction (16) "*perhaps* splits up"[38] into the two stronger conditions (15) and (17). Now that this train of thought had led him into severe difficulties Einstein reconsidered the harmonic coordinate restriction. This reconsideration probably suggested the idea of adding a trace term $\sum h_{\kappa\kappa} = U$ to the left-hand side of the linearized field equation (15) resulting in the following field equation[39]

$$\sum_{\sigma} \frac{\partial^2}{\partial x^{\sigma 2}} \left\{ h_{i\kappa} - \frac{1}{2}\delta_{i\kappa} U \right\} = T_{i\kappa}. \tag{19}$$

Such an additional trace term has the consequence that energy-momentum conservation, expressed in terms of the divergence of $T_{\mu\nu}$, no longer implies the Hertz restriction (15) but rather the harmonic restriction (16). Hence the coordinate restriction arising from energy-momentum conservation does not impose an additional restriction but coincides with the restrictive harmonic condition.

With this modification, Einstein had now arrived at a new candidate field equation by considering what we now identify as the linearized Einstein tensor. The new field equation, generated by a further elaboration of the mathematical representation triggered by a conflict between two heuristic principles, now had to be tested again with respect to Einstein's heuristic requirements.

Before Einstein turned to a consideration of his other heuristic principles, he added a multiple of the traced equation and thus transformed the equation to an alternative form which has the trace term on the same side as the source term. This might have been done in order to check whether his new modification was equivalent to his previous adding of a trace term. Einstein at this point made an algebraic error which, however, was inconsequential. He formed the trace of the gravitational field equation (19) erroneously obtaining $2\partial^2 U/\partial x^{\sigma 2} = \sum T_{\kappa\kappa}$ instead of $-\partial^2 U/\partial x^{\sigma 2} = \sum T_{\kappa\kappa}$. Adding 1/2 of the traced equation to the field equation (19) he obtained (again erroneously)

$$\sum_{\sigma} \frac{\partial^2 h_{i\kappa}}{\partial x^{\sigma 2}} = T_{i\kappa} + \frac{1}{2}\delta_{i\kappa} T, \tag{20}$$

---

[37] In the Notebook, Einstein explicitly noted that the trace term was to be subtracted only "in the case of the same i and κ" ("für gleiche i u[nd] κ," see Figure 6) instead of using the compact notation of the 'Kronecker delta' $\delta_{i\kappa}$ as is done in our transcription. In the manuscript these lines were immediately deleted, possibly because Einstein at some point realized that such a modification had to address the second of the above-mentioned problems.

[38] "Zerfällt *vielleicht* in," see Figure 6.

[39] In the Notebook, Einstein explicitly put down the 11-, 12-, and 14-component of this equation, expressing the d'Alembertian $\sum_\sigma \frac{\partial^2}{\partial x^{\sigma 2}}$ by a triangle Δ, and indicated the other components by horizontal dashes.

whereas the version which would correctly be equivalent to (19) would have $-1/2$ instead of $+1/2$. The algebraic error was again inconsequential because in any case equation (20) was different from Equation (18). This was an expected result since the field equation (18) was obtained by requiring that the trace of the source term vanish on the basis of the assumption that the linearized harmonic condition (16) splits up into the two stronger conditions (15) and (17). The field equation (20) was obtained by reverting to the harmonic condition (16) and requiring that the vanishing of the coordinate divergence of $T_{i\kappa}$ was consistent with it.

3.2.4. *The conflict between the Einstein tensor and the Requirement of Correspondence*

As with the simple linearized field equation (14), the first thing to check for the modified field equation (19) was to see whether one could still satisfy the requirement of energy conservation by deriving a gravitational energy-momentum tensor. By a brief consideration Einstein found that also with the additional trace term one could find such a gravitational energy tensor and he noted that it would be "representable in the required form."[40]

Adding the trace term to the linearized field equation either in the form of Equation (18) or in the form of (19) did not solve, however, the second of the two problems mentioned above, i.e., the incompatibility with the correspondence principle. In either case, the trace term implied that the metric (3) for static fields could not be a solution of the field equation and was hence incompatible with it. The trace term precluded that the tensorial multi-component field equation reduces to a simple one-component Poisson equation for the g_{44}-component of the metric representing the Newtonian gravitational potential. Hence, the Newtonian limit could not be realized in the way expected by Einstein and this was probably the reason why he deleted Equation (18) in the Notebook. In view of this consequence, he either had to reconsider his correspondence principle or he had to give up the whole approach of reducing the Ricci tensor using the harmonic condition (12). On the next pages of the Notebook (20R–21R), he indeed reconsidered this requirement and in particular his assumption that the metric of static gravitational fields has to be of the form (3). But he apparently found his arguments in favor of the correspondence principle strong enough to reject the Einstein tensor.[41] The same heuristic principles that had led Einstein in 1912–1913 to consider the final field equation of general relativity, thus also forced him to reject it.

In spite of the problems with the Ricci and the Einstein tensors, as the Notebook documents, Einstein continued for some time to search for a suitable gravitation tensor on the basis of the Riemann tensor (22L–25R). A couple of pages later, however, Einstein eventually gave up the idea, after having explored all the possibilities that the formalism appeared to offer. The 'mathematical strategy' of

---

40 "Darstellbar in der verl[angten] Form," see Figure 6.

41 One such argument was explicitly formulated as late as 1915 in a letter to Freundlich (EA 11-208). Here Einstein reports that he examined the question whether matter at rest produces metric components other than g_{44}. Apparently assuming a weak-field condition of the form (14), he concluded that this cannot be the case. A more detailed discussion of these arguments will be presented elsewhere.

starting from a mathematically well defined object with well-known covariance group had definitely failed to yield a field equation that was compatible also with the more physically motivated heuristic Requirements of Correspondence and Conservation.

3.3. THE *ENTWURF* OPERATOR—A METASTABLE STATE IN THE INTERACTION OF HEURISTICS AND MATHEMATICAL REPRESENTATION

After the failure of the 'mathematical' strategy, Einstein, at the end of the Notebook (26L ff), returned to the 'physical' strategy. He now started from the core operator (7) which, as we have discussed, reduces to the Laplacian operator without any additional constraints. For this operator, the satisfaction of the correspondence principle represented no difficulty. In order to also comply with the conservation principle, Einstein then introduced higher-order correction terms to this core operator, constructed in just such a way that the divergence equation (6) could be satisfied. The conservation principle was now employed in a constructive manner rather than only being used as a criterion for accepting or rejecting candidate field equations. The procedure yielded expression (2) for the left-hand side of the field equation (5), and the problem for the core operator which we discussed earlier could thus be resolved from the beginning. In March 1913, Einstein and Grossmann published this expression as the left-hand side of a gravitational field equation in their joint *Entwurf* article, Einstein & Grossmann 1913. The field equation of this theory is compatible with all of Einstein's heuristic requirements—save the relativity principle. In fact, the transformational behavior of the *Entwurf* field equation was not exactly known and left as an open problem when Einstein and Grossmann published their paper.[42] But since Einstein anyhow was not sure to what extent this principle could be realized, his search for a new theory of gravitation had thus reached a state in which it made sense to elaborate the consequences of this theory instead of continuing to produce alternative versions. Historically, this state of the theory proved to be metastable, bound to decay after a couple of years.

4. The interplay between heuristics and deductivity

Among the missed opportunities in Einstein's long and winding path towards general relativity, his rejection of gravitational field equations based on the Einstein tensor certainly is the gravest. With the field equations of his final theory already before his eyes—if only in linear approximation—he could happily, or so at least it may appear, have achieved his most brilliant scientific achievement already in 1912 or 1913. The fact that Einstein missed this opportunity seems to present a strong case against the interpretation of his final achievement as a case of serendipity,

[42] "But we do not know whether there exists a general transformation group under which the equations are covariant. The question of the existence of such a group ... is the most important one which follows up on the considerations expounded here." ("Wir wissen aber nicht, ob es eine allgemeine Transformationsgruppe gibt, der gegenüber die Gleichungen kovariant sind. Die Frage nach der Existenz einer derartigen Gruppe ... ist die wichtigste, welche sich an die hier gegebenen Ausführungen anknüpft.") (Einstein & Grossmann 1913: 18)

which would have had Einstein find the field equation by mere luck. If in late 1915 Einstein had accidentally come back to general covariance and, equally by accident, to the Einstein tensor, had found it acceptable and had published it, he might well have done so two years earlier.

But the alternative interpretation, according to which Einstein's return to general covariance in 1915 might be due to a last minute *conceptual* breakthrough by which he redirected his *heuristics* along a more successful line of attack has other problems. The 1915 paper itself (Einstein 1915d), as well as the other papers of that period, do not bear evidence of such a radical change. Also the commentaries available from Einstein's correspondence, or from the papers he published shortly afterwards, do not indicate a fundamental revision of his heuristic principles. The equivalence principle is still understood as being included in the generalized principle of relativity which Einstein believed to be satisfied precisely because of the general covariance of the new field equation. The conservation principle is, even in Einstein's conclusive 1915 paper, still considered as an additional requirement to be imposed on the field equation rather than as being implied by it via the contracted Bianchi identity. The fundamental features of Einstein's understanding of the correspondence principle had not changed either. He still did not have a mathematical framework in which the dynamics and the space-time structure of classical gravitation theory could be obtained from general relativity by a consistent limiting procedure.[43]

Nevertheless, some years later, in late November 1915, Einstein formulated the 'final' field equation of general relativity (Einstein 1915d). How then did Einstein succeed in formulating the field equation of general relativity if not by a conceptual breakthrough? First of all, from a modern perspective, his heuristics did cover the mathematical requirements that uniquely define the Einstein tensor. If one looks for a generally covariant, second-rank tensor that is linear in the second derivatives of the metric components, contains no higher derivatives, vanishes in flat space-time and satisfies the Bianchi identity, then no other alternative is available. Whatever else was implied by Einstein's heuristics, these mathematical requirements are in fact suggested by his generalized principle of relativity, by his correspondence principle, and by his conservation principle. Einstein's problem was that these requirements did not exhaust his heuristics and that their consequences were not fully compatible with it. In the face of such conflicts, he was hence forced to weigh the various components of his heuristics against each other, and he had to be prepared to reduce his ambitious goals, for instance by restricting the demand for general covariance or by weakening his criteria for the satisfaction of energy-momentum conservation. But his judgement about the proper equilibrium between his conflicting heuristic components depended on the state of elaboration of their deductive consequences. This is why the balance between the heuristic components turned out to favor different candidates in the course of Einstein's first examination of the Ricci, the November and the Einstein tensor in the years 1912–1913 than in the course of his second examination of these tensors

---

[43] For a modern discussion of this limiting procedure, see Ehlers 1981.

in 1915. The interplay between heuristics and deductivity is the crucial intellectual process that prepared the ground for the discovery of the field equation of general relativity.

Let us recapitulate. At the time of his considerations documented by the Zurich Notebook, it was his heuristic Requirement of Correspondence, together with the requirement of conservation, that made Einstein reject the Einstein tensor. Einstein's understanding of the Newtonian limit was that for weak and static fields the 10-component tensorial field equations would reduce to the Poisson equation for the time-time component of the metric, representing the single, scalar potential of Newtonian gravitation. Similarly, this metric component would have to govern the equations of motion for a material point moving in a Newtonian gravitational field. But the trace term introduced on p. 20L of the Notebook (see Equation (19) above) inevitably implied that non-trivial spatial components had to be taken into account on the same level of approximation as that which yielded the time-time component as the Newtonian gravitational potential, and this violation of his heuristic requirements made it inacceptable for Einstein.

It was thus Einstein's heuristic requirements that prevented an early breakthrough in 1912 or 1913. On the other hand, as we have seen, it was the very same heuristic requirements that had led to the introduction of the trace term in the first place since this term had only been considered in an attempt to solve a conflict between the Requirements of Correspondence and Conservation. In hindsight, it must be observed, however, that the motivation for adding a trace term to the Ricci tensor in 1912 was markedly different from the justification of the Einstein tensor in the final theory. In the Zurich Notebook, Einstein had to resolve the conflict between the harmonic coordinate condition on the one hand, necessary to satisfy the Requirement of Correspondence, and the Hertz condition on the other hand, necessary to satisfy the requirement of energy conservation. He had tried to resolve this conflict by adding a trace term to the field equation which would turn the coordinate restriction stemming from the requirement of conservation from the Hertz condition to the harmonic condition as well. Thus it was the trace term appearing in the harmonic condition that was the basis for the trace term in the field equation. From a modern perspective, however, the trace term in the harmonic coordinate condition has nothing to do with the fact that the Einstein tensor satisfies the contracted Bianchi identity. It was thus a contingent feature of the mathematical representation of his heuristic ideas that led Einstein to consider the correct field equation in linearized approximation in 1912. When Einstein stumbled, so to say, upon this field equation in an early stage of his exploration of the deductive structure implied by the mathematical representation, and found problems in its physical interpretation, he chose to give up the mathematical strategy rather than his more physically motivated heuristic requirements.

In the years following the publication of the *Entwurf* field equations, Einstein had explored many consequences of this theory. In the course of this exploration, he had become familiar with more aspects of the mathematical representation, and in particular began to employ a mathematical representation, the variational formulation, which allowed him to draw far-reaching conclusions about the transformational properties of the theory. The use of variational techniques for some time helped to stabilize the conceptual framework of the *Entwurf* theory, in particular through the identification of the covariance group of the field equations and its interpretation in terms of the hole argument. Eventually, however, properties of the theory were elaborated to a point where they led to manifestly contradictory consequences in various instances. As we have mentioned, already in 1913 Einstein and Besso had found that the *Entwurf* equations did not account for the anomalous advance of the perihelion of Mercury. In an epistolary controversy with Levi-Civita in the spring of 1915, Einstein had to realize that an alleged proof of the covariance properties of the *Entwurf* equations was faulty.[44] He then found that the equations would not hold for Minkowski space-time in rotating Cartesian coordinates. Finally, he had to realize that an allegedly unique derivation of the field equations was faulty.[45] Turning to a reconsideration of alternatives to the *Entwurf* equations, Einstein was now able to examine the candidates he had previously considered in the Zurich Notebook with a new level of knowledge about deductive consequences and equipped with powerful tools such as the calculus of variations. In the course of this reconsideration, he first returned to the November tensor, then to the Ricci tensor, and only then published the final field equations with the Einstein tensor.

But what precisely turned the balance in 1915 in favor of the Einstein tensor, in spite of the remaining conflicts with Einstein's heuristic principles and in spite of the availability of alternatives? The solution came through a technical loophole, which until November 1915 had remained unnoticed, but which then opened up the way if not to a final resolution but at least to a circumvention of the most serious objection against the Einstein tensor. As we have seen, this objection resulted from the correspondence principle and consisted in the observation that the weak field equation resulting from the Einstein tensor involves a metric with more than one variable component. Such a weak field equation can hence not simply be reduced to the classical Poisson equation for one scalar potential.

Surprisingly, a closer inspection of the Newtonian limit of the equation of motion and, in particular, its interpretation as an independent postulate of the theory—contrary to the modern understanding—opened up a way to avoid this dilemma. The geodesic equation reduces in fact, under the mathematical conditions which correspond to classical physics (weak static fields, low velocities, appropriate coordinates), to the Newtonian equation of motion. Under these circumstances, only one component of the metric tensor actually matters, while the other components of the metric can be neglected. As Einstein explicitly noted some years later in his Princeton lectures:

> We see ... that even in first approximation the structure of the graviational field is fundamentally different from the one according to Newton's theory; this is a consequence of the tensorial, non-scalar character of the gravita-

[44] For a discussion of the controversy between Einstein and Levi-Civita, see Cattani & De Maria 1989.

[45] See Einstein to Sommerfeld, 28 November 1915, published in Hermann 1968: 32–36.

> tional potential. The fact that this had not been noticed before is due to the fact that exclusively the component g_{44} enters the equations of motion in first approximation.[46] (Einstein 1956: 89).

In other words, here was the loophole through which Einstein could escape from the argument that weak static fields have to be described by a metric tensor with only a single variable component, corresponding to the Newtonian gravitational potential. If other variable components could exist without affecting the equation of motion in the Newtonian limit, this anomaly became tolerable and now shifted the balance in favor of the Einstein tensor. The fact that the metric associated with this peculiar way of attaining the Newtonian limit also explained the perihelium shift of Mercury protected it against the criticism of being just a dubious technical trick and stabilized Einstein's network of conclusions well beyond what, at that point, could have been achieved by its still fragile internal coherency.

5. General relativity—an accidental discovery?

The triumph of November 1915 was hence not the victory of new concepts over old ones but just the temporary stabilization of a complex network made up of still largely traditional concepts, of Einstein's original heuristic arguments with only slight adjustments, and of unforeseen results on the level of the mathematical representation of the new theory. If this interpretation is indeed correct, then the conflicts between the original heuristics and the deductive consequences of the new theory could not have been settled yet. Einstein's continued search for a realization of his original ideas after 1915, beyond the field equation of general relativity, provides in fact strong support in favor of this interpretation (Renn 1994). He recognized, for instance, that his strategy of implementing Machian ideas in the new theory via the requirement of general covariance did not work, since general relativity allows solutions in which inertial effects are present even without being caused by masses. As a reaction to this unexpected difficulty, Einstein did not simply give up his original goal as part of a context of discovery that had been superseded by his results. He rather strengthened his heuristic requirements by now demanding that the metric tensor be "completely" determined by the masses of the bodies that act as a source of the gravitational field. While he had previously simply assumed that such a determination would be an automatic consequence of his heuristic program, in 1918 he felt compelled to introduce this requirement as an additional condition, complementing his original heuristics (Einstein 1918a). He gave it the name "Mach's principle" and made clear that he had hitherto included this requirement in his understanding of the generalized principle of relativity.

That Einstein's reinforcement of his original heuristics was not a matter of philosophical dispute over what had already been achieved becomes clear from

[46] "Man sieht . . . , daß auch in erster Näherung die Struktur des Gravitationsfeldes von derjenigen gemäß NEWTONS Theorie prinzipiell abweicht; es liegt dies daran, daß das Gravitationspotential tensoriellen und nicht skalaren Charakter hat. Daß sich dies nicht längst bemerkbar gemacht hat, kommt davon, daß in die Bewegungsgleichung des Massenpunktes in erster Näherung ausschließlich die Komponente g_{44} eingeht."

the drastic consequences he was still prepared to draw from it. In 1917 he was ready to modify the field equation of general relativity by the introduction of the cosmological term, with the intention of confining the solutions of his theory precisely to those that satisfy his demand for a Machian explanation of inertia (Einstein 1917). But even when it turned out that the modified theory also had non-Machian solutions, Einstein continued to believe that

> the General Theory of Relativity only forms a satisfactory system if, according to it, the physical qualities of space are *completely* determined by matter alone.[47] (Einstein 1918b: 271; emphasis in the original)

He now hoped that, even if other solutions would be mathematically possible, at least nature would have a preference for a static cosmological model compatible with his Machian understanding of inertia. Only in early 1931 was Einstein forced to recognize that this hope was not borne out by the astronomical data in favor of an expanding universe.[48]

Did Einstein, after this apparently definite failure of his heuristic program, now finally restrict himself to the technical and conceptual exploration of general relativity as he had established it in 1915? Surprisingly not, because he had other reasons to remain unsatisfied with this theory, reasons that were apparently in flat contradiction with his Machian heuristics. He pursued, at least since 1919, an alternative strategy to modify or further develop general relativity in such a way that space is not conceived as a field effect ultimately caused by matter, but so that, vice versa, matter is rather being constructed in terms of a universal field representing the physical qualities of space.[49] Einstein's hopes to reach a unified field theory that would integrate gravitation theory with electromagnetism and also explain the quantum structure of matter remained as unfulfilled as his Machian program. But if we take into account these more ambitious goals as a context for Einstein's work on gravitation, it seems that the formulation of the field equation of general relativity in 1915 was, after all, nothing more than an accidental discovery in the framework of a research project that was guided by concepts, heuristic principles, and expectations which cannot be reconciled with those associated with general relativity as we understand it today.

6. Einstein's vision redeemed

The opposition between Einstein's goals of explaining space in terms of matter or matter in terms of space resembles somewhat today's alternative between the program to quantize gravitation in the framework of quantum field theory and the program to revise the concepts of space and time as they are used in quantum field theory according to general relativity. The striking fact, however, that one

[47] "Bildet die allgemeine Relativitätstheorie nämlich nur dann ein befriedigendes System, wenn nach ihr die physikalischen Qualitäten des Raumes allein durch die Materie *vollständig* bestimmt werden."

[48] See Einstein 1931; for further discussion, see Renn 1994: 49–50.

[49] For a discussion of Einstein's investigation of modified field equations in the context of his program of finding a unified field theory, see Jim Ritter's forthcoming article.

and the same man could entertain such contrary options raises the question of whether there may not be some deeper level to Einstein's heuristics that might also explain its contribution to his success with general relativity, without reducing this achievement to a pure case of serendipity. In fact, on closer inspection, Einstein's Machian program, as well as the apparently opposite goal of a unified field theory, turn out both to be the result of his striving for a conceptual unity of physics that would overcome the dualism between particle and field physics. That this fundamental problem occupied his mind is not new. More interesting is Einstein's particular approach in addressing this issue, which, in our view, accounts to a large extent for the success of his heuristics.

Instead of relying on the concepts of either mechanics or electrodynamics, Einstein took the knowledge accumulated in both branches of physics and the structures organizing them as equally fundamental. Contrary to Lorentz's electrodynamics, for instance, Einstein's special theory of relativity treated the relativity principle of mechanics as just as foundational as the laws of the propagation of light established by electrodynamics, at the price of a revision of the concepts of classical physics.[50] Similarly, instead of attempting to resolve Mach's paradox of the privileged role of inertial frames in the context of a revised version of classical mechanics, as did several followers of Mach, Einstein addressed this problem in the context of a field theory of gravitation in which inertial forces could be understood as an aspect of a unified gravito-inertial field. And finally, instead of constraining the understanding of this gravito-inertial field by the chronogeometry of special relativity, which to most of his contemporaries had quickly acquired a universal, almost a priori status for physics, Einstein realized that chronogeometry and gravito-inertial structure are equally universal for physics.

Mathematical tools such as the affine connection for describing the gravito-inertial structure were still lacking when Einstein began his work. But even in the absence of such tools, Einstein's uncommon openness to philosophical questions had helped him to recognize the universality of the gravito-inertial structure where others just saw one more force to be subsumed in a special relativistic treatment. In spite of the mathematical and conceptual difficulties that he encountered, his heuristic principles guided a reconciliation between gravito-inertial structure and chronogeometry which excluded any prior geometry, thus effectively determining the characteristic features of general relativity, even as we understand it today. In other words, Einstein's heuristics was so successful not because it anticipated the conceptual novelties of general relativity but because it molded the knowledge accumulated in the different branches of classical physics in such a way that its integration into one coherent framework became possible. This makes Einstein's philosophical openness and his integrative outlook on the foundations of physics into an historical lesson from which even today's unifying ventures may profit.

When studying this lesson it is, however, important to keep in mind the very specific historical circumstances under which a single scientist could make such an integrative achievement. Just as the development of Einstein's *political* thinking

---

50 For a further discussion of Einstein's perspective on Lorentz' electrodynamics, see Renn 1993.

was dependent on its *social* context (Scheideler & Goenner 1997), Einstein's integrative capabilities in the *conceptual* realm were dependent on historically specific features of his *intellectual* biography.[51]

And already the contemporary attempt to exploit Einstein's integrative capabilities in the *conceptual* realm for an *organization* of science—aimed at fostering an integration of physics and chemistry—unmistakably failed (Castagnetti & Goenner 1997).

References

ABRAHAM, Max. (1912a). "Zur Theorie der Gravitation." *Physikalische Zeitschrift* **13**: 1–4.

——— (1912b). "Berichtigung." *Physikalische Zeitschrift* **13**: 176.

BARBOUR, Julian & PFISTER, H., eds. (1995). *Mach's Principle. From Newton's Bucket to Quantum Gravity* (Einstein Studies, vol. 6). Boston: Birkhäuser.

BIANCHI, Luigi. (1910). *Vorlesungen über Differentialgeometrie.* Leipzig & Berlin: Teubner.

CASTAGNETTI, Giuseppe & GOENNER, Hubert. (1997). "Directing a Kaiser Wilhelm Institute: Einstein, Organizer of Science?" In G. Castagnetti, H. Goenner, J. Renn, T. Sauer & B. Scheideler. "Foundations in Disarray: Essays on Einstein's Science and Politics in the Berlin Years." 55–80. (Papers presented at the Boston Colloquium for Philosophy of Science, March 1997). Berlin: Max-Planck-Institut für Wissenschaftsgeschichte, Preprint 63.

CASTAGNETTI, Giuseppe, DAMEROW, Peter, HEINRICH, W., RENN, Jürgen & SAUER, Tilman. (1994). "Wissenschaft zwischen Grundlagenkrise und Politik: Einstein in Berlin." Arbeitsbericht der Arbeitsstelle Albert Einstein 1991–1993, Max-Planck-Institut für Bildungsforschung.

CATTANI, Carlo & DE MARIA, Michelangelo. (1989). "The 1915 Epistolary Controversy between Einstein and Tullio Levi-Civita." In HOWARD & STACHEL 1989: 175–200.

EARMAN, John & GLYMOUR, Clark. (1978a). "Lost in the Tensors: Einstein's Struggles with Covariance Principles, 1912–1916." *Studies in History and Philosophy of Science* **9**: 251–278.

——— (1978b). "Einstein and Hilbert: Two Months in the History of General Relativity." *Archive for History of Exact Sciences* **19**: 291–308.

EARMAN, John & JANSSEN, Michel. (1993). "Einstein's Explanation of the Motion of Mercury's Perihelion." In EARMAN, JANSSEN & NORTON 1993: 129–172.

EARMAN, John, JANSSEN, Michel & NORTON, John D., eds. (1993). *The Attraction of Gravitation. New Studies in the History of General Relativity* (Einstein Studies, vol. 5). Boston: Birkhäuser.

EHLERS, Jürgen. (1993). "Über den Newtonschen Grenzwert der Einsteinschen Gravitationstheorie." In *Grundlagenprobleme der modernen Physik.* J. Nitsch et al, eds. 65–84. Mannheim: Bibliographisches Institut.

EINSTEIN, Albert. (1907). "Über das Relativitätsprinzip und die aus demselben gezogenen Folgerungen." *Jahrbuch der Radioaktivität und Elektronik* **4**: 411–462 [= EINSTEIN *CP2*: doc. 47].

——— (1911). "Über den Einfluß der Schwerkraft auf die Ausbreitung des Lichtes." *Annalen der Physik* **35**: 898–908 [= EINSTEIN *CP3*: doc. 23].

[51] A thorough discussion of this dependence would go beyond the scope of this article. For a brief discussion of some aspects of this issue, see Renn & Schulman 1992: introduction.

——— (1912a). "Lichtgeschwindigkeit und Statik des Gravitationsfeldes." *Annalen der Physik* **38**: 355–369 [= Einstein *CP4*: doc. 3].

——— (1912b). "Zur Theorie des statischen Gravitationsfeldes." *Annalen der Physik* **38**: 443–458 [= Einstein *CP4*: doc. 4].

——— (1915a). "Zur allgemeinen Relativitätstheorie." *Sitzungsberichte der Königlichen Preussischen Akademie der Wissenschaften.* (Berlin) (1915): 778–786 [= Einstein *CP6*: doc. 21].

——— (1915b). "Zur allgemeinen Relativitätstheorie (Nachtrag)." *Sitzungsberichte der Königlichen Preussischen Akademie der Wissenschaften.* (Berlin) (1915): 799–801 [= Einstein *CP6*: doc. 22].

——— (1915c). "Die Feldgleichungen der Gravitation." *Sitzungsberichte der Königlichen Preussischen Akademie der Wissenschaften.* (Berlin) (1915): 844–847 [= Einstein *CP6*: doc. 25].

——— (1917). "Kosmologische Betrachtungen zur allgemeinen Relativitätstheorie." *Sitzungsberichte der Königlichen Preussischen Akademie der Wissenschaften.* (Berlin) (1917) 142–152 [= Einstein *CP6*: doc. 43].

——— (1918a). "Prinzipielles zur allgemeinen Relativitätstheorie." *Annalen der Physik* **55**: 241–244.

——— (1918b). "Kritisches zu einer von Hrn. De Sitter gegebenen Lösung der Gravitationsgleichungen." *Sitzungsberichte der Königlichen Preussischen Akademie der Wissenschaften* (Berlin) (1918): 270–272.

——— (1931). "Zum kosmologischen Problem der allgemeinen Relativitätstheorie." *Sitzungsberichte der Preussischen Akademie der Wissenschaften.* (Berlin). (1931): 235–237.

——— (1956). *Grundzüge der Relativitätstheorie.* Braunschweig: Vieweg.

——— (*CP2*). *The Collected Papers of Albert Einstein.* Vol. 2: *The Swiss Years: Writings, 1900–1909.* J. Stachel, D. C. Cassidy, J. Renn, & R. Schulmann, eds. Princeton: Princeton University Press (1989).

——— (*CP3*). *The Collected Papers of Albert Einstein.* Vol. 3. *The Swiss Years: Writings, 1909–1911.* M. J. Klein, A. J. Kox, J. Renn & R. Schulmann, eds. Princeton: Princeton University Press (1993).

——— (*CP4*). *The Collected Papers of Albert Einstein.* Vol. 4. *The Swiss Years: Writings, 1912–1914.* M. J. Klein, A. J. Kox,, J. Renn & R. Schulmann, eds. Princeton: Princeton University Press (1995).

——— (*CP5*). *The Collected Papers of Albert Einstein.* Vol. 5. *The Swiss Years: Correspondence, 1902–1914.* M. J. Klein, A. J. Kox & R. Schulmann, eds. Princeton: Princeton University Press (1993).

——— (*CP6*). *The Collected Papers of Albert Einstein.* Vol. 6. *The Berlin Years: Writings, 1914–1917.* A. J. Kox, M. J. Klein, & R. Schulmann, eds. Princeton: Princeton University Press (1996).

Einstein, Albert & Grossmann, Marcel. (1913). *Entwurf einer verallgemeinerten Relativitätstheorie und einer Theorie der Gravitation.* Leipzig: Teubner. [Reprinted with added "Bemerkungen," *Zeitschrift für Mathematik und Physik* **62** (1914): 225–261 = Einstein *CP4*: doc. 13].

Eisenstaedt, Jean & Kox, Anne J., eds. (1992). *Studies in the History of General Relativity* (Einstein Studies, vol. 3). Boston: Birkhäuser.

Hermann, Armin, ed. (1968). *Albert Einstein–Arnold Sommerfeld. Briefwechsel.* Basel & Stuttgart: Schwabe.

Hoefer, C. (1994). "Einstein and Mach's Principle." *Studies in History and Philosophy of Science* **25**: 287–335.

HOFFMANN, Banesh. (1972). "Einstein and Tensors." *Tensor* **26**: 157–162.

HOWARD, Don & NORTON, John D. (1993). "Out of the Labyrinth? Einstein, Hertz, and the Göttingen Answer to the Hole Argument." In EARMAN, JANSSEN & NORTON 1993: 30–62.

HOWARD, Don & STACHEL, John, eds. (1989). *Einstein and the History of General Relativity* (Einstein Studies, vol. 1). Boston: Birkhäuser.

JANSSEN, Michel (1997). "Rotation as the Nemesis of the Einstein's *Entwurf* Theory." Paper presented at the Fourth International Conference on the History of General Relativity, Berlin 1995. [This volume].

LANCZOS, C. (1972). "Einstein's Path From Special to General Relativity." In *General Relativity: Papers in Honour of J. L. Synge*. L. O'Raifertaigh, ed. 5–19. Oxford: Clarendon Press.

VON LAUE, Max. (1911). *Das Relativitätsprinzip*. Braunschweig: Vieweg.

MALTESE, G. & ORLANDO, L. (1995). "The Definition of Rigidity in the Special Theory of Relativity and the Genesis of the General Theory of Relativity." *Studies in History and Philosophy of Modern Physics* **26**: 263–306.

MEHRA, J. (1974). *Einstein, Hilbert, and the Theory of Gravitation. Historical Origins of General Relativity Theory*. Dordrecht & Boston: D. Reidel.

MINKOWSKI, Hermann. (1908). "Die Grundgleichungen für die electromagnetischen Vorgänge in bewegten Körpern." *Nachrichten von der Königlichen Gesellschaft der Wissenschaften zu Göttingen. Mathematisch-Physikalische Klasse*: 53–111. [Reprinted: H. Minkowski. *Gesammelte Abhandlungen*. 2 vols. D. Hilbert, ed. Leipzig & Berlin: Teubner (1911): II, 352–404].

NORTON, John D. (1984). "How Einstein Found His Field Equations, 1912–1915." *Historical Studies in the Physical Sciences* **14**: 253–316. [Reprinted in HOWARD & STACHEL 1989: 101–159].

——— (1985). "What Was Einstein's Principle of Equivalence?" *Studies in History and Philosophy of Science* **16**: 203–246. [Reprinted in HOWARD & STACHEL 1989: 5–47].

——— (1987). "Einstein, the Hole Argument and the Reality of Space." In *Measurement, Realism and Objectivity*. J. Forge, ed. 153–188. Dordrecht: D. Reidel.

——— (1992). "Einstein, Nordstrøm and the Early Demise of Scalar, Lorentz-Covariant Theories of Gravitation." *Archive for History of Exact Sciences* **45**: 17–94.

——— (1993). "General Covariance and the Foundations of General Relativity: Eight Decades of Dispute." *Reports on Progress in Physics* **56**: 791–858.

PAIS, Abraham. (1982). *'Subtle is the Lord ... ' The Science and the Life of Albert Einstein*. Oxford: Oxford University Press.

REICH, Karin. (1994). *Die Entwicklung des Tensorkalküls* (Science Networks, vol. 11). Basel: Birkhäuser.

RENN Jürgen. (1993). "Einstein as a Disciple of Galileo: A Comparative Study of Concept Development in Physics." *Science in Context* **6**: 311–341.

——— (1994). "The Third Way to General Relativity." Max Planck Institute for the History of Science Preprint Nr. 9.

RENN Jürgen & SAUER, Tilman. (1996). "Einstein's Züricher Notizbuch." *Physikalische Blätter* **52**: 865–872.

RENN Jürgen, SAUER, Tilman & STACHEL, John. (1997). "The Origin of Gravitational Lensing: A Postscript to Einstein's 1936 *Science* Paper." *Science* **275**: 184–186.

RENN Jürgen & SCHULMANN, Robert, eds. (1992). *Albert Einstein–Mileva Maric. The Love Letters*. Princeton: Princeton University Press.

SCHEIDELER, B. & GOENNER, Hubert. (1997). "Albert Einstein in Politics—A Comparative Approach." In G. Castagnetti, H. Goenner, J. Renn, T. Sauer & B. Scheideler. "Foundations in Disarray: Essays on Einstein's Science and Politics in the Berlin Years."

1–28. (Papers presented at the Boston Colloquium for Philosophy of Science, March 1997). Max-Planck-Institut für Wisenschaftsgeschichte, Preprint 63.

STACHEL, John. (1980). "Einstein's Search for General Covariance." Paper delivered at the Ninth International Conference on General Relativity and Gravitation, Jena, Germany (DDR), 17 July 1980. [Revised version in HOWARD & STACHEL 1989: 63–100].

——— (1982). "The Genesis of General Relativity." In *Einstein Symposion Berlin aus Anlaß der 100. Wiederkehr seines Geburtstages*. H. Nelkowski, H. Poser, R. Schrader and R. Seiler, eds. 428–442. Berlin: Springer.

——— (1989). "The Rigidly Rotating Disk As the 'Missing Link' in the History of General Relativity." In HOWARD & STACHEL 1989: 48–62.

VIZGIN, Vladimir P. & SMORODINSKIY, Y. A. (1979). "From the Equivalence Principle to the Equations of Gravitation." *Soviet Physics–Uspekhi* **22**: 489–513.

Rotation as the Nemesis of Einstein's *Entwurf* Theory

Michel Janssen

SHORTLY AFTER THE PUBLICATION OF THE FIELD EQUATIONS of general relativity in November 1915, Einstein explained in letters to Sommerfeld and Lorentz why he had grown dissatisfied with his earlier non-generally covariant field equations, known in the historical literature, after the title of the paper in which they were first published (Einstein & Grossmann 1913), as the *Entwurf* field equations. Einstein gave three reasons in these letters. One of them was that the metric field of a Minkowski space-time in a coordinate system rotating with respect to some inertial Lorentz frame is not a solution of the *Entwurf* field equations. In this paper, I want to tell the story of how Einstein discovered this problem with rotation in his *Entwurf* theory, which, for the purposes of this paper, can be thought of simply as general relativity with its generally covariant field equations replaced by the *Entwurf* field equations. I was able to reconstruct this story in considerable detail, drawing on correspondence and research notes preserved in the Einstein Archive, notably a letter from Einstein to Erwin Freundlich of September 1915 and some calculations of 1913 in the so-called "Einstein-Besso manuscript."[1] The story suggests that it was the problem with rotation rather than the other two problems mentioned in the letters to Sommerfeld and Lorentz that dealt the decisive blow to the *Entwurf* theory.

[1] A substantial part of my reconstruction can already be found, in a very condensed form, in the annotation for the relevant pages of the Einstein-Besso manuscript in Einstein *CP4*: doc. 14, pp. [41–42]. The letter to Freundlich and other correspondence from the period 1915–1917 that I drew on for this paper appear in Einstein *CP8*. I wrote this paper in the context of a larger project of the Max-Planck-Institut für Wissenschaftsgeschichte which aims at giving the most detailed reconstruction yet of Einstein's path to general relativity. My paper does not necessarily reflect the views of the other members of the group working on this project. See Renn & Sauer 1996 for a preliminary report on the group's findings.

H. Goenner, J. Renn, J. Ritter, T. Sauer (eds.)
The Expanding Worlds of General Relativity
(Einstein Studies, volume 7) pp. 127–157

It turns out that in 1913 Einstein had already gone through the exact same calculation that would reveal the problem with rotation to him in 1915. In 1913, he missed the problem, partly because of an elementary mistake in reading off the components of the metric tensor from the relevant line element, partly because he was too willing to assume he must have made an error somewhere when his calculations did not bear out his expectations. Einstein's conclusion at the end of his calculation in 1913 was simply an innocuous "correct" [*stimmt*]. After redoing the calculation in 1915 without the errors of 1913, he wrote to Freundlich: "*This is a blatant contradiction*" [*Dies ist ein flagranter Widerspruch*; Einstein's emphasis].

1. Einstein on why he abandoned the *Entwurf* theory

Before I introduce any other source material, I want to quote the two well-known passages from Einstein's correspondence in which he stated his reasons for abandoning the *Entwurf* theory. On 28 November 1915, shortly after publishing the final version of the field equations of general relativity (Einstein 1915c), he wrote to Sommerfeld:

> I realized that my previous gravitational field equations were completely untenable! This was indicated by the following points:
> 1) I proved that the gravitational field on a uniformly rotating system does not satisfy the field equations.
> 2) The motion of the perihelion of Mercury came out as 18″ instead of 45″ per century.
> 3) The covariance argument in my paper of last year does not give the Hamiltonian function H. When suitably generalized, it allows an arbitrary H. It followed that covariance relative to "adapted" coordinate systems was a wild goose chase.[2] (Einstein *CP8*: doc. 153; translation from Stachel 1980: 85)

About a month later, on 1 January 1916, he wrote to Lorentz:

> The gradually dawning knowledge of the incorrectness of the old gravitational field equations gave me a rotten time last autumn. I had already found earlier that the perihelion motion of Mercury was too small. In addition, I found that the equations were not covariant for substitutions which corresponded to a uniform rotation of the (new) reference system. Finally I found that my approach of last year to the determination of Lagrange's function H of the gravitational field was illusory throughout, since it could be easily modified so that one needed to apply no limiting condition at all

---

[2] "Ich erkannte nämlich, dass meine bisherigen Feldgleichungen der Gravitation gänzlich haltlos waren! Dafür ergaben sich folgende Anhaltspunkte:
1) Ich bewies, dass das Gravitationsfeld auf einem gleichförmig rotierenden System den Feldgleichungen nicht genügt.
2) Die Bewegung des Merkur-Perihels ergab sich zu statt pro Jahrhundert.
3) Die Kovarianzbetrachtung in meiner Arbeit vom letzten Jahre liefert die Hamilton-Funktion H nicht. Sie lässt, wenn sie sachgemäss verallgemeinert wird, ein beliebiges H zu. Daraus ergab sich, dass die Kovarianz bezüglich 'angepasster' Koordinatensysteme ein Schlag ins Wasser war." [facsimile in Hermann 1968: 34–35]

> to H, so that it could have been chosen quite freely.[3] (Einstein *CP8*: doc. 177; translation from Norton 1984: 138)

The three reasons for abandoning the *Entwurf* theory given in these letters have been discussed by Norton (1984: 138–141). From the Einstein-Besso manuscript, discovered a few years later, we now know that Einstein collaborated with his friend Michele Besso in June 1913 to calculate the perihelion motion of Mercury predicted by the *Entwurf* theory (see Earman & Janssen 1993: 135–136, and the editorial note on the Einstein-Besso manuscript in Einstein *CP4*: 344–359). The Einstein-Besso manuscript also sheds important new light on the problem with rotation, as I will show in this paper.

Notice the difference between the two statements of the problem with rotation that I quoted above. To Sommerfeld, Einstein wrote: "I proved that the gravitational field on a uniformly rotating system does not satisfy the field equations." To Lorentz, he wrote: "I found that the equations were not covariant for substitutions which corresponded to a uniform rotation of the (new) reference system." As one would expect, and as we will see below, exactly what it was that Einstein had discovered is captured more accurately by the earlier statement than by the later one.

A year later, in a letter to De Sitter of 23 January 1917, Einstein once again described the problem, using a formulation close to the one in the letter to Sommerfeld. In this letter, Einstein identified two passages in his lengthy exposition of the *Entwurf* theory from October 1914 (Einstein 1914) where his argument was flawed. The first passage is § 12 on the so-called hole argument (see Norton 1984: 128–132; Stachel 1980: 71–81), the second occurs in § 14, the section leading up to the fallacious argument for the uniqueness of the Lagrangian for the gravitational field mentioned in the letters to Sommerfeld and Lorentz above.[4] Einstein then makes the following comment:

> I noticed my mistakes at that point when I calculated directly that the field equations I held then are not satisfied for a system rotating in a Galilean space.[5] (Einstein *CP8*: doc. 290)

Notice that Einstein now explicitly talks about a "Galilean space," by which he means an inertial frame in Minkowski space-time.[6] As we will see below, the direct

---

[3] "Die allmählich aufdämmernde Erkenntnis von der Unrichtigkeit der alten Gravitations-Feldgleichungen hat mir letzten Herbst böse Zeiten bereitet. Ich hatte schon früher gefunden, dass die Perihelbewegung des Merkur sich zu klein ergab. Dazu fand ich, dass die Gleichungen nicht kovariant waren für Substitutionen, die einer gleichförmigen Rotation des (neuen) Bezugssystems entsprachen. Endlich fand ich, dass meine letztes Jahr aufgestellte Betrachtung zur Bestimmung der Lagrange'schen Funktion H des Gravitationsfeldes durchaus illusorisch war, indem sie leicht so modifiziert werden konnte, dass man H überhaupt keiner einschränkenden Bedingung zu unterwerfen brauchte, sodass es ganz frei gewählt werden konnte."

[4] The second "printing error" is discussed briefly in Norton 1984: 155, n. 36, in the context of the problem of the non-uniqueness of the Lagrangian for the gravitational field. I want to emphasize that these are two different problems.

[5] "Ich merkte meine damaligen Irrtümer daran, dass ich direkt ausrechnete dass meine damaligen Feldgleichungen für ein in einem Galileischen Raum rotierendes System nicht erfüllt waren."

[6] In a letter to De Sitter of 15 July 1916, Einstein explained his usage of this phrase: "It is not just

calculation Einstein is referring to is not quite as straightforward as his comment to De Sitter suggests.

2. What exactly was the problem with rotation that Einstein discovered?

The identification of the problem with rotation that I gave in the introduction is obtained by combining Einstein's statements to Sommerfeld and De Sitter: the metric field of a Minkowski space-time in a coordinate system rotating uniformly with respect to some inertial Lorentz frame is not a solution of the *Entwurf* field equations. Since the Minkowski metric in its usual diagonal form, i.e., in some Lorentz frame, *is* a solution of the equations, it follows that the *Entwurf* field equations can not be invariant under transformations to a new coordinate system rotating uniformly with respect to the original one. The latter assertion is the most natural reading of Einstein's statement of the problem with rotation in the letter to Lorentz.[7]

These two forms of the problem with rotation are not equivalent. If they were, it would be the case that field equations are invariant under rotation whenever the Minkowski metric is a solution of those equations both in a Lorentz frame and in a rotating frame. This is obviously not true. After all, for field equations to be invariant under rotation it has to be the case that starting from an *arbitrary* solution in an *arbitrary* coordinate system, and transforming to a new coordinate system rotating with respect to the old one, one still has a solution. This clearly is much stronger than saying that starting from a *particular* solution in a *particular* coordinate system, such as a Minkowski metric in a Lorentz frame, and transforming to a rotating frame, one still has a solution. In other words, the condition that the Minkowski metric in rotating coordinates be a solution of the *Entwurf* field equations is a necessary but not a sufficient condition for these equations to be invariant under rotation.

The story that will emerge in this paper strongly suggests that throughout the reign of the *Entwurf* theory, Einstein believed that the Minkowski metric in rotating coordinates is a solution to the *Entwurf* field equations. However, there is also strong evidence to suggest that there was a period in late-1913/early-1914 in which Einstein believed that the *Entwurf* field equations were invariant under linear transformations only.[8] The elementary analysis of the logical situation that we

---

the 'space' which is 'Galilean,' but the 'space' together with the reference frame which makes the $g_{\mu\nu}$ constants" ["'Galileisch' ist ... nicht nur der 'Raum' sondern der 'Raum' zusammen mit dem Bezugssystem, welches die $g_{\mu\nu}$ zu Konstanten macht."] (Einstein *CP8*: doc. 235)

[7] It is also the way in which Einstein made it clear in the paper in which he retracted the *Entwurf* field equations and replaced them with equations constructed out of the Ricci tensor (Einstein 1915a) that these new field equations do not face the problem with rotation. At the end of the paper, Einstein simply points out that transformations to rotating coordinate systems are unimodular and therefore belong to the class of "allowed" [*erlaubten*] transformations, i.e., transformations under which the equations are invariant (ibid., p. 786).

[8] For detailed references to the relevant passages in Einstein's writing and correspondence in this

have given above shows that these two beliefs need not be incompatible with one another.[9]

3. The context of the letter to Freundlich in which the problem with rotation is reported

The calculation that showed Einstein that the metric field of a Minkowski space-time in a rotating coordinate system is not a solution of the *Entwurf* field equations is described in a letter to Freundlich.[10] The letter is dated only as "30.IX," but it is clear that the year must be 1915. This can be inferred from the beginning of the letter:

> Dear Freundlich, I am happy to write to Naumann, and I will do so within the next few days. Early tomorrow morning I am seeing Planck, with whom I will also talk about it. At this point, I am writing to you concerning a scientific matter which electrifies me greatly.[11] (Einstein *CP8*: doc. 123)

Before I turn to this electrifying scientific matter, I want to examine briefly the background to the first two sentences, which are important for the dating of this letter and the dating of another document that will play a role in my story. Material documenting the episode referred to in these sentences can be found in Kirsten & Treder 1979. Additional documents plus a detailed discussion can be found in Hentschel 1992, 1994. In the following brief account, I rely heavily on Hentschel's work.

Freundlich studied mathematics and astronomy in Göttingen. He obtained his doctorate in 1910 with a dissertation on analytic functions. His thesis supervisor, Felix Klein, arranged a position for him at the observatory in Babelsberg near Berlin, as assistant to Karl Hermann Struve, the director of the institute. The arrangement was unhappy from the beginning. Struve was an old-fashioned bread-and-butter astronomer, interested only in high-precision measurements of stellar positions, whereas Freundlich was more interested in theoretical astrophysical issues.

---

period, see Einstein *CP4*: 582, n. 4; 622, n. 28. See also a letter from Einstein to Ernst Mach of December 1913 (Einstein *CP5*: doc. 495).

9 This way of accounting for what seems to be a tension in Einstein's position in this period is not as artificial as it may appear to be at first sight. The notion of transformations between different coordinate systems for particular solutions of the *Entwurf* field equations would come to play an important role in Einstein's work on the *Entwurf* theory in 1914. In Einstein & Grossmann 1914: 220–225 and Einstein 1914: 1067–1074 the covariance properties of the *Entwurf* field equations are discussed in terms of "justified" [*berechtigte*] coordinate transformations, which are coordinate transformations between so-called "adapted" [*angepasste*] coordinate systems, where "adapted" means "adapted to the metric field." So, justified coordinate transformations are transformations to new coordinates given not only the old coordinates but also a particular metric field in the old coordinates. This notion of justified coordinate transformations may also be related to what Einstein, in a letter to Lorentz of 14 August 1913 (Einstein *CP5*: doc. 467), called "dependent transformations" [*unselbständige Transformationen*], i.e., coordinate transformations represented by a matrix that depends on the metric field.

10 I am indebted to John Norton for drawing my attention to this letter.

11 "Lieber Freundlich, Ich will Naumann gerne schreiben, und zwar schon in den nächsten Tagen. Morgen früh sehe ich Planck, mit dem ich auch darüber spreche. Ich schreibe Ihnen jetzt in einer wissenschaftlichen Angelegenheit, die mich ungeheur elektrisiert."

In August 1911, Freundlich received a letter from the astronomer Pollak about the possibility of astronomical tests of the gravitational theory developed by Einstein, Pollak's new physics colleague in Prague. Not surprisingly, given his interests, Freundlich immediately contacted Einstein and began to spend more and more time working on tests of Einstein's theories, at the expense of his normal duties at the observatory. This caused considerable friction between Freundlich and Struve.

After several unsuccessful attempts of Freundlich to verify the prediction of the bending of light,[12] Freundlich turned to the other effect predicted by the theory, the gravitational red shift (Hentschel 1992: chap. 5, pp. 38–50; 1994: 155–161). He noticed that stars showing a red shift seemed to outnumber stars showing a blue shift. Tacitly assuming a static universe (so that we only have the ordinary Doppler shift and the gravitational red shift and no cosmological red shift), he tried to see whether the balance could be restored by taking into account the gravitational red shift effect predicted by Einstein. In 1915, he published the first results of these endeavors (Freundlich 1915b), claiming that the results supported Einstein's prediction. Unfortunately, Hugo von Seeliger, astronomer in Munich, discovered an elementary mistake in Freundlich's paper. Freundlich had annoyed Seeliger earlier in 1915 with a paper (Freundlich 1915a) purporting to show that the latter's proposal for explaining the anomalous motion of the perihelion of Mercury does not work. Seeliger did not waste this opportunity to get even with Freundlich. Rather tellingly, he wrote to Struve and not to Freundlich himself to point out the error in the paper.[13] Freundlich had to promise Struve to publish a correction to his paper.[14] He would eventually publish this correction (Freundlich 1915/16), but in doing so he would ignite Seeliger even more. Not only did he fail to acknowledge Seeliger for discovering the error in the original paper, he also made a rather clumsy attempt to save his conclusion that the data supported Einstein's theory by arbitrarily adjusting some parameters. Seeliger thereupon sent more angry letters to Struve,[15] and published a devastating critique of Freundlich's work (Seeliger 1916).[16]

It is clear that something needed to be done about the growing tension between Freundlich and Struve. Freundlich wanted to be relieved of his normal duties at the observatory so that he could spend all his time testing Einstein's theories. Einstein agreed to help him achieve this goal. Freundlich was under the impression

---

[12] Not only did he analyze photographic plates of past solar eclipses, he also led the ill-fated eclipse expedition to the Crimea at the outbreak of World War I (Hentschel 1992: 30–37; 1994: 149–154).

[13] Seeliger to Struve, 27 June 1915 (see Hentschel 1992: 42; 1994: 159).

[14] Freundlich to Struve, 7 August 1915 (quoted and discussed in Hentschel 1992: 42–43; 1994: 159).

[15] Seeliger to Struve, 12 and 26 January 1916. The first of these two letters is published in Kirsten & Treder 1979: 171–172 and is discussed by Hentschel (1992: 44–45; 1994: 161–162). The second is partly quoted and discussed in Hentschel 1992: 54.

[16] Hentschel (1992: 45; 1994: 162) quotes another letter from Seeliger to Struve of 15 August 1915, showing quite clearly that Seeliger's overzealous pursuit of Freundlich was motivated in part by his skepticism about Einstein's new gravitational theory.

that Otto Naumann, the Ministerial Director at the Prussian Ministry of Education [Kultusministerium], who was responsible for the observatory, was amenable to the idea of creating a special position for him. Einstein promised to write to Naumann. This is what lies behind the opening sentence of his letter to Freundlich of 30 September. The reference to Naumann makes it clear that the year has to be 1915.

We need to take a closer look at the exchange with Naumann. The Einstein Archive contains an undated draft of a letter to Naumann written on a sheet of paper that also has some equations written on it which are related to the problem with rotation. The document is briefly discussed in Norton 1984: 139. Norton writes that the draft is of a letter that "can be dated by content to late November or perhaps early December 1915," probably because of the similarity between this draft and an actual letter to Naumann of 7 December 1915. I will argue that it is much more likely that the draft was written in early October.[17]

From the way the draft is written around the calculation, it is clear that the calculation was there before the draft. Presumably, the calculation was done *before* the letter to Freundlich, for the upshot of the calculation is reported in the letter. It also seems clear that the calculation was done *after* the recent visit by Freundlich referred to in the draft. Otherwise, Einstein would presumably have told Freundlich about the matter on that occasion rather than explaining it to him in a letter. So, most likely, the calculation dates from shortly before 30 September 1915. This dating of the calculation seems to be at odds with dating the draft to late November or early December 1915.

The draft begins:

> Letter to Naumann.
> The other day, Dr. Freundlich from the observatory N[eubabelsberg][18] visited me. He told me that, on the occasion of an official meeting, you indicated the possibility that he would be freed from his duties as assistant at the observatory for a few years, without losing his salary as assistant. I was very pleased to hear this news and so was my colleague Planck, who recently encouraged me to ask you in writing not to drop this liberating idea.[19] (Einstein *CP8*: doc. 124)

Einstein then explains how the time-consuming and routine nature of the measurements Freundlich is supposed to perform at the observatory keep him from pursuing more important matters. He praises what Freundlich has been able to accomplish despite these difficult circumstances and intimates what valuable contributions are to be expected from him if only he is freed from his current duties.

---

[17] I am grateful to Giuseppe Castagnetti for helpful discussions in reconstructing Einstein's exchange with Naumann.

[18] In an actual letter to Naumann of 7 December 1915, Einstein refers to the "Sternwarte in Neubabelsberg" (Kirsten & Treder 1979: I, 168; Einstein *CP8*, doc. 124).

[19] "Brief an Naumann. Letzter Tage war Herr Dr. Freundlich von der Sternwarte N bei mir. Er erzählte mir, dass Sie ihm bei Gelegenheit einer dienstlichen Besprechung andeutungsweise von der Möglichkeit gesprochen haben, dass er für einige Jahre von seinen Verpflichtungen als Assistent der Sternwarte entbunden werden könnte, ohne dass ihm sein Gehalt als Assistent entzogen werden würde. Ich war über diese Mitteilung hoch erfreut und ebenso Kollege Planck, der mich neulich dazu ermunterte, Sie brieflich zu bitten diese erlösende Idee nicht fallen zu lassen."

It seems that Einstein did not actually send this letter.[20] He did pay Naumann a visit, however, during which he made essentially the same points. This can be gathered from another letter to Freundlich, dated only as "Berlin, Mittwoch," but probably written in November 1915, in which Einstein gave a report of such a visit, immediately after it took place.[21] Among other things, Einstein informed Freundlich in this letter that he asked Naumann to contact Planck in this matter. Early December 1915, Naumann actually talked to Planck. This can be inferred from the letter Einstein finally did send to Naumann on 7 December 1915, which begins:

> My colleague Planck just told me on the telephone that he has talked to you about the Freundlich issue. It gives me great pleasure to see that you have not lost sight of the issue on behalf of which I recently took the liberty of visiting you.[22] (Kirsten & Treder 1979: I, 167; Einstein *CP8*: doc. 160)

There are at least three clear indications that the draft that I quoted above is *not* the draft of this letter of 7 December. First, in the letter Einstein assumes that Naumann is familiar with the "Freundlich issue," whereas Einstein carefully explains the issue in the draft. Second, Planck is mentioned both in the draft and the letter, but in the draft there is no mention of any contact between Planck and Naumann concerning the Freundlich issue. Third, and most convincingly, in the draft Einstein does not refer to his own visit with Naumann. It seems safe to conclude that the draft is of a letter that either was never sent or has not survived, and that it was written shortly after Einstein's letter to Freundlich of 30 September 1915 in which he promised to write to Naumann "over the next few days." That places the draft close enough in time to the calculation related to the problem of rotation for it to be plausible that draft and calculation appear on the same sheet of paper.

Although this is of no importance for the dating of documents bearing on the problem with rotation, the reader may be curious to hear the rest of the story about the "Freundlich issue." In his letter to Naumann, Einstein gave a fairly detailed outline of the important work that could be done by Freundlich if a special position were created for him. Naumann passed a copy of this letter on to Struve, asking the latter to accept Einstein's proposal.[23] Predictably upset that Freundlich and Einstein had gone to Naumann behind his back, Struve sent a lengthy reply to Naumann in which he strongly advised against creating a special position for Freundlich.[24] About a month later, Struve would receive the angry letters from

---

[20] Castagnetti has pointed out to me that, if he had, it would almost certainly have survived in the archives of the Kultusministerium.

[21] The letter begins: "I have just come from Naumann" ("Ich komme soeben von dem Naumann"). (Einstein *CP8*: doc. 151)

[22] "Soeben erfuhr ich telephonisch von Kollegen Planck, dass er mit Ihnen über die Angelegenheit Freundlich gesprochen habe. Es gereicht mir zur grossen Freude zu sehen, dass Sie die Angelegenheit, um deretwegen ich Sie jüngst aufzusuchen mir erlaubte, nicht aus den Augen verloren haben."

[23] Naumann to Struve, 16 December 1915 (Kirsten & Treder 1979: I, 166–168).

[24] Struve to Naumann, 20 December 1915 (Kirsten & Treder 1979: I, 169–171). See also the excerpt of an internal report by Struve dated "later than 10 December 1915" (Kirsten & Treder 1979: I, 68–69).

Seeliger I already mentioned, in which Seeliger expressed his dismay at the way Freundlich had handled the criticism of his paper on gravitational red shift. At this time, even Einstein was beginning to voice reservations about Freundlich.[25] However, he remained loyal enough to use his power as the director of the new Kaiser-Wilhelm Institut für physikalische Forschung in late 1917 to make sure that the first full-time position of the institute went to Freundlich (Hentschel 1992: 55; 1994: 166).

4. The letter to Freundlich

I now return to the problem with rotation and to the scientific part of Einstein's letter to Freundlich of 30 September 1915. After the sentences quoted at the beginning of Section 3, the letter continues:

> In fact, I have hit upon a logical contradiction of a quantitative nature in the theory of gravitation which proves to me that there must be a computational incorrectness somewhere in my construction. Imagine an infinitesimally slowly rotating coordinate system (with velocity of rotation ω). In this system, as can easily be shown through a simple transformation, the gravitational field is given by the $g_{\mu\nu}$-system
>
> $$\begin{pmatrix} -1 & 0 & 0 & \omega y \\ 0 & -1 & 0 & -\omega x \\ 0 & 0 & -1 & 0 \\ \omega y & -\omega x & 0 & 1 \end{pmatrix}$$
>
> With the help of the equations, I can now calculate the next approximation (terms proportional to ω^2) and from the last gravitational field equation I find
>
> $$g_{44} = 1 - \frac{3}{4}\omega^2(x^2 + y^2),$$
>
> whereas direct transformation from the Galilean case leads to
>
> $$g_{44} = 1 - \omega^2(x^2 + y^2).$$
>
> *This is a blatant contradiction.* I do not doubt, therefore, that the theory of the perihelion motion also suffers from the same mistake. Either the equations are numerically incorrect (number coefficients) or I apply the equations in a fundamentally wrong way. I do not believe I am able to find the mistake myself, for in this matter my mind is too set in a deep rut. More likely, I have to rely on some fellow human being with unspoiled brain matter finding the mistake. Don't forget to spend some time on the matter when you can. Best wishes, your A. Einstein.[26] (Einstein *CP8*: doc. 123)

---

[25] Hentschel (1992: 47–48, 54; 1994: 165) cites: Einstein to Schwarzschild, 9 January 1916 (Einstein *CP8*: doc. 181); Einstein to Sommerfeld, 2 February 1916 (Hermann 1968: 38–39; Einstein *CP8*: doc. 186); Einstein to Struve, 13 February 1916 (Kirsten & Treder 1979: I, 173; Einstein *CP8*: doc. 190).

[26] "Ich bin nämlich in der Gravitationstheorie auf einen logischen Widerspruch quantitativer Art gestossen, der mir beweist, dass in meinem Gebäude irgendwo eine rechnerische Unrichtigkeit stecken muss. Denken Sie ein unendlich langsam rotierendes Koordinatensystem (Rotationsgeschwindigkeit

First, I want to go over the "simple transformation" that leads to the "$g_{\mu\nu}$-system" given in the letter. An error in this simple transformation was the main culprit concealing the problem with rotation in 1913.

In some arbitrary Lorentz frame with coordinates $x'_\mu = (x', y', z', t')$ [$\mu = 1, 2, 3, 4$], the metric field of Minkowski space-time is given by the diagonal matrix $g'_{\mu\nu} = \text{diag}(-1, -1, -1, c^2)$, where c is the velocity of light in vacuo. The corresponding line element is

$$ds^2 = -dx'^2 - dy'^2 - dz'^2 + c^2\, dt'^2.$$

Consider the coordinate transformation

$$\begin{aligned} x &= x' \cos \epsilon t' + y' \sin \epsilon t' \\ y &= -x' \sin \epsilon t' + y' \cos \epsilon t' \\ z &= z' \\ t &= t', \end{aligned} \tag{1}$$

i.e., a transformation from the x'_μ-frame to a new x_μ-frame rotating with respect to the old x'_μ-frame with angular velocity ϵ (counterclockwise if ϵ is positive and clockwise if ϵ is negative). When the differentials of the x'_μ-coordinates are expressed in terms of the differentials of the x_μ-coordinates with the help of the inverse of Equation (1), the Minkowski line element can be written as:

$$\begin{aligned} ds^2 = -dx^2 - dy^2 - dz^2 + 2\epsilon y\, dx\, dt - 2\epsilon x\, dy\, dt \\ + c^2 \left(1 - \frac{\epsilon^2 r^2}{c^2}\right) dt^2, \end{aligned} \tag{2}$$

where $r^2 \equiv x^2 + y^2$.[27] From this line element, the components $g_{\mu\nu}$ of the metric tensor in the rotating x_μ-frame can be read off. For instance, the 14- and 41-components (i.e., the xt- and tx-components) are equal to ϵy; the 24- and 42-components are equal to $-\epsilon x$.

ω). In diesem ist, wie sich durch einfache Transformation leicht zeigen lässt, das Gravitationsfeld durch das $g_{\mu\nu}$-System ... gegeben. Ich kann nun mittelst der Gleichungen die nächste Approximation berechnen (Glieder proportional ω^2) und finde aus der letzten Feldgleichung der Gravitation ... während die unmittelbare Transformation aus dem galileischen Fall ergibt ...

Dies ist ein flagranter Widerspruch. Ich zweifle daher nicht daran, dass auch die Theorie der Perihelbewegung an der gleichen Fehler krankt. Entweder sind die Gleichungen schon numerisch unrichtig (Zahlenkoeffizienten) oder ich wende die Gleichungen prinzipiell falsch an. Ich glaube nicht, dass ich selbst imstande bin, den Fehler zu finden, da mein Geist in dieser Sache zu ausgefahrene Geleise hat. Ich muss mich vielmehr darauf verlassen, dass ein Nebenmensch mit unverdorbener Gehirnmasse den Fehler findet. Versäumen Sie nicht, wenn Sie Zeit haben, sich mit dem Gegenstande zu beschäftigen. Mit bestem Gruss, Ihr A. Einstein."

27 The inverse of the transformation in Equation (1) is given by:

$$x' = x \cos \epsilon t - y \sin \epsilon t, \qquad y' = x \sin \epsilon t + y \cos \epsilon t, \qquad z' = z, \qquad t' = t,$$

It follows that

$$\begin{aligned} dx' &= dx \cos \epsilon t - dy \sin \epsilon t - \epsilon(x \sin \epsilon t + y \cos \epsilon t)dt \\ &= dx \cos \epsilon t - dy \sin \epsilon t - \epsilon y' dt \\ dy' &= dx \sin \epsilon t + dy \cos \epsilon t - \epsilon(x \cos \epsilon t - y \sin \epsilon t)dt \\ &= dx \sin \epsilon t - dy \cos \epsilon t + \epsilon x' dt. \end{aligned}$$

5. The query in the "Scratch Notebook"

The calculation I just outlined can be found on the last page of the so-called "Scratch Notebook" (Einstein *CP3*: appendix A, p. [66]).[28] This particular entry, I think, dates from around 1913. For a facsimile reproduction, see Figure 1.

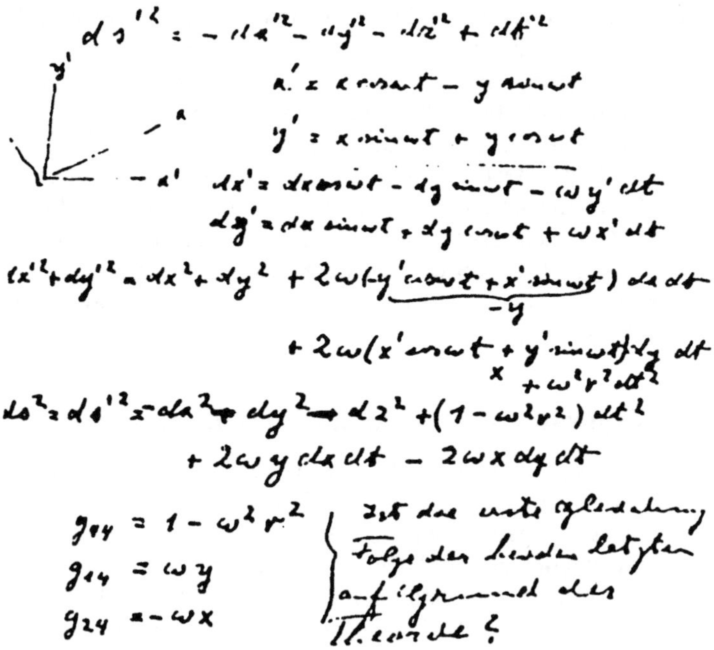

Figure 1. Facsimile of calculation on p. [66] of Einstein's "Scratch Notebook."

Reproduced with permission of the Albert Einstein Archives, The Hebrew University of Jerusalem, Israel, and Princeton University Press.

Squaring these expressions, one finds

$$dx'^2 = dx^2 \cos^2 \epsilon t + dy^2 \sin^2 \epsilon t + \epsilon^2 y'^2 dt^2$$
$$- 2 \cos \epsilon t \sin \epsilon t \, dx \, dy - 2\epsilon y' \cos \epsilon t \, dx \, dt + 2\epsilon y' \sin \epsilon t \, dy \, dt$$
$$dy'^2 = dx^2 \sin^2 \epsilon t + dy^2 \cos^2 \epsilon t + \epsilon^2 x'^2 dt^2$$
$$+ 2 \sin \epsilon t \cos \epsilon t \, dx \, dy + 2\epsilon x' \sin \epsilon t \, dx \, dt + 2\epsilon x' \cos \epsilon t \, dy \, dt$$

Adding these expressions to $-dz^2 + c^2 dt^2$, using Equation (1) to write

$$-2\epsilon y' \cos \epsilon t \, dx \, dt + 2\epsilon x' \sin \epsilon t \, dx \, dt = -2\epsilon \, y \, dx \, dt$$
$$2\epsilon y' \cos \epsilon t \, dy \, dt + 2\epsilon x' \sin \epsilon t \, dy \, dt = 2\epsilon \, x \, dy \, dt,$$

one arrives at Equation (2).

[28] I am indebted to Jürgen Renn for drawing my attention to this passage.

At the end of the calculation—using coordinates such that $c = 1$ and writing ω instead of ϵ[29] —Einstein writes down three components of the Minkowski metric in rotating coordinates:

$$g_{44} = 1 - \omega^2 r^2, \quad g_{14} = \omega\, y, \quad g_{24} = \omega\, x,$$

and then raises the question: "Is the first equation a consequence of the two last ones on the basis of the theory?" [*Ist die erste Gleichung Folge der beiden letzten auf Grund der Theorie?*]. For the *Entwurf* theory, the answer to this question is no. As Einstein writes to Freundlich, given the metric to first order in ω, the 44-component of the *Entwurf* field equations gives $g_{44} = 1 - (3/4)\,\omega^2\, r^2$ in second order approximation instead of $g_{44} = 1 - \omega^2\, r^2$. The calculation on the draft of the letter to Naumann mentioned in Section 3 is a cryptic version of the derivation of this result. As I mentioned in the Introduction, Einstein had done this same calculation in 1913. It is part of the Einstein-Besso manuscript. At that point, Einstein managed to convince himself that the answer to the question raised in the "Scratch Notebook" was yes.

6. The problem with rotation and the Einstein-Besso manuscript

The calculation of g_{44} on the basis of the field equations can be reconstructed with the help of the Einstein-Besso manuscript (Einstein *CP4*: doc. 14). The relevant part of this manuscript dates from June 1913, possibly with some later additions during the remainder of that year.[30] The iterative approximation procedure Einstein is referring to in his letter to Freundlich for determining terms of second order in ω given the metric up to first order in ω is the same as the approximation procedure he and Besso used to find the lowest order contribution to the metric field of the sun that gives rise to a perihelion precession of planetary orbits (Einstein *CP4*: doc. 14, [pp. [1–6]). The approximation procedure is discussed in detail in Earman & Janssen 1993: 142–143 and in section II of the editorial note on the Einstein-Besso manuscript in Einstein *CP4*: 346–349.

The connection to the perihelion problem immediately explains Einstein's remark in the letter to Freundlich that he does not doubt "that the theory of the perihelion motion also suffers from the same mistake." It probably also accounts

[29] The convention I used in Section 4 of using primed coordinates for the old (inertial) frame and unprimed coordinates for the new (rotating) frame is in accordance with Einstein's notation on this page of the "Scratch Notebook."

[30] For a detailed discussion of the dating of the manuscript, see section III of the editorial note on the Einstein-Besso manuscript in Einstein *CP4*: 356–359. For the purposes of this paper, the following brief remarks suffice. The manuscript was found in the Besso Nachlass. An extremely plausible explanation of how it ended up there is that this is the manuscript Einstein referred to when he wrote to Besso in a letter of January 1914 "Here you finally have your manuscript bundle" [*Hier enthältst Du endlich Dein Manuskriptbündel*] (Einstein *CP5*: doc. 499). If the "manuscript bundle" was indeed the Einstein-Besso manuscript, Einstein's contributions to the manuscript must all date from 1913. Independently of these considerations, it is clear that Einstein's contributions must be earlier than late October 1914, since by that time he had changed the notation used throughout the manuscript (see Norton 1984: 154, n. 32).

for why Einstein turned to Freundlich for suggestions as to what might be the origin of the "blatant contradiction" he had found in his theory.[31] Freundlich had recently published his critique of Seeliger's solution of the perihelion problem (Freundlich 1915a), perhaps even at Einstein's instigation, and the two men in all likelihood discussed the problem in the context of the *Entwurf* theory.[32]

On pp. [41–42] of the manuscript, one finds the actual application of the approximation procedure mentioned in the letter to Freundlich, i.e., finding the ω^2-term in the 44-component of a metric which to first order in ω is just the Minkowski metric in rotating coordinates. As I indicated above, the conclusion of the calculation was that the 44-component found through an application of this approximation procedure is precisely the 44-component found by directly transforming the diagonal Minkowski metric to a rotating coordinate system.

7. Reconstruction of the calculation which in 1915 revealed the problem with rotation

I will give a self-contained presentation of the calculation needed to answer Einstein's query in the "Scratch Notebook." I essentially follow the notation of the *Entwurf* paper (Einstein & Grossmann 1913).[33] In this notation, all indices are written 'downstairs.' Covariant components are indicated by Latin letters (e.g., $g_{\mu\nu}$), contravariant components by Greek ones (e.g., $\gamma_{\mu\nu}$).

The basic *Ansatz* for the approximation procedure Einstein used is that the metric field can be written as a power series in some small expansion parameter. It will be convenient to introduce a special notation to distinguish the different terms in such expansions, even though Einstein did not do so. Whenever I write "(n)" ($n = 0, 1, 2 \ldots$) above some quantity, I mean the nth-order term in the power series expansion of that quantity. With the help of this notation the power series expansions of the covariant and contravariant components of the metric can be written as

$$g_{\mu\nu} = \overset{(0)}{g}_{\mu\nu} + \overset{(1)}{g}_{\mu\nu} + \overset{(2)}{g}_{\mu\nu} + \ldots$$

$$\gamma_{\mu\nu} = \overset{(0)}{\gamma}_{\mu\nu} + \overset{(1)}{\gamma}_{\mu\nu} + \overset{(2)}{\gamma}_{\mu\nu} + \ldots.$$

For small angular velocities ϵ, the metric field $g_{\mu\nu}$ that can be read off from the

---

[31] It is interesting to note that Einstein apparently did not turn to Besso. There is also no mention of this problem in a letter to Lorentz of 12 October 1915, in which Einstein first reported that, contrary to what he had claimed in Einstein 1914, the Lagrangian for the gravitational field is not unique (see Sections 1 and 10 for further discussion).

[32] See Einstein to Freundlich, 19 March 1915 (Einstein *CP8*: doc. 63). The "problem of the planets" [*Planetenproblem*] mentioned in this letter is a reference, I think, to the problem of accounting for the Mercury anomaly (cf. Norton 1984: 138). Freundlich is also acknowledged in the famous perihelion paper of November 1915 for providing Einstein with Newcomb's values for the perihelion advance of several planets (Einstein 1915b: 839).

[33] Unlike Einstein in 1913, I will not write any summation signs. In the following, summation over repeated indices is always understood.

line element (2) has this form:

$$g_{\mu\nu} = \begin{pmatrix} -1 & 0 & 0 & \epsilon y \\ 0 & -1 & 0 & -\epsilon x \\ 0 & 0 & -1 & 0 \\ \epsilon y & -\epsilon x & 0 & c^2\left(1 - \dfrac{\epsilon^2 r^2}{c^2}\right) \end{pmatrix} \tag{3}$$

The same is true for the contravariant components $\gamma_{\mu\nu}$ of this metric field. Inverting the matrix (3), we find for $\gamma_{\mu\nu}$:

$$\gamma_{\mu\nu} = \begin{pmatrix} -1 + \dfrac{\epsilon^2 y^2}{c^2} & -\dfrac{\epsilon^2 xy}{c^2} & 0 & \dfrac{\epsilon y}{c^2} \\ -\dfrac{\epsilon^2 xy}{c^2} & -1 + \dfrac{\epsilon^2 x^2}{c^2} & 0 & -\dfrac{\epsilon x}{c^2} \\ 0 & 0 & -1 & 0 \\ \dfrac{\epsilon y}{c^2} & -\dfrac{\epsilon x}{c^2} & 0 & \dfrac{1}{c^2} \end{pmatrix} \tag{4}$$

From Equations (3) and (4), one can read off that the terms of zeroth order in ϵ in the expansions of $g_{\mu\nu}$ and $\gamma_{\mu\nu}$ are

$$\overset{(0)}{g}_{\mu\nu} = \mathrm{diag}(-1, -1, -1, c^2), \quad \overset{(0)}{\gamma}_{\mu\nu} = \mathrm{diag}\left(-1, -1, -1, \frac{1}{c^2}\right), \tag{5}$$

and that the only non-vanishing terms of first order in ϵ are

$$\begin{aligned} \overset{(1)}{g}_{14} = \overset{(1)}{g}_{41} &= \epsilon y & \overset{(1)}{\gamma}_{14} = \overset{(1)}{\gamma}_{41} &= \frac{\epsilon y}{c^2} \\ \overset{(1)}{g}_{24} = \overset{(1)}{g}_{42} &= -\epsilon x & \overset{(1)}{\gamma}_{24} = \overset{(1)}{\gamma}_{42} &= -\frac{\epsilon x}{c^2} \end{aligned} \tag{6}$$

Equations (5)–(6) can be looked upon as giving the starting terms of a power series expansion of some metric field with components $g_{\mu\nu}$ and $\gamma_{\mu\nu}$. As we will see below, in this first order approximation, this metric field is a solution of the *Entwurf* field equations. The field equations up to second order in ϵ plus the assumption that the metric is stationary (so that $\partial g_{\mu\nu}/\partial x_4 = \partial \gamma_{\mu\nu}/\partial x_4 = 0$) can then be used to determine the second order terms in the power expansion of $g_{\mu\nu}$ and $\gamma_{\mu\nu}$. The task before us is to check whether these second order terms are identical to those that can be read off from Equations (3) and (4), in particular, whether the second order contribution to g_{44} is equal to $-\epsilon^2 r^2$.

In the Einstein-Besso manuscript, Einstein used the *Entwurf* field equations in their "covariant" form to determine $\overset{(2)}{g}_{44}$, and in their "contravariant" form to determine $\overset{(2)}{\gamma}_{44}$. So, we need to consider both forms of the equations. As a matter of convenience and in accordance with Einstein's usage of the terms, I will continue to use the designations "contravariant" and "covariant"—without quotation marks from now on—even though the equations are invariant only under a limited class of coordinate transformations and even though there is an extra minus sign in the relation between the contravariant and covariant form of one of their ingredients.

In their contravariant form, the *Entwurf* field equations are (Einstein & Grossmann 1913: 17, eq. 18)

$$\Delta_{\mu\nu}(\gamma) = \kappa(\Theta_{\mu\nu} + \vartheta_{\mu\nu}), \tag{7}$$

where $\Theta_{\mu\nu}$ is the contravariant stress-energy tensor for matter (p. 11), and where both $\Delta_{\mu\nu}(\gamma)$ and $\vartheta_{\mu\nu}$ are functions of the metric tensor and its derivatives. $\Delta_{\mu\nu}(\gamma)$ is given by (p. 16, eq. 15):

$$\Delta_{\mu\nu}(\gamma) = \frac{1}{\sqrt{-g}}\frac{\partial}{\partial x_\alpha}\left(\gamma_{\alpha\beta}\sqrt{-g}\,\frac{\partial \gamma_{\mu\nu}}{\partial x_\beta}\right) - \gamma_{\alpha\beta}\, g_{\tau\rho}\frac{\partial \gamma_{\mu\tau}}{\partial x_\alpha}\frac{\partial \gamma_{\nu\rho}}{\partial x_\beta},$$

and $\vartheta_{\mu\nu}$, the "contravariant stress-energy tensor of the gravitational field" (p. 16), is given by (p. 15, eq. 13):

$$-2\kappa\,\vartheta_{\mu\nu} = \gamma_{\alpha\mu}\,\gamma_{\beta\nu}\frac{\partial g_{\tau\rho}}{\partial x_\alpha}\frac{\partial \gamma_{\tau\rho}}{\partial x_\beta} - \frac{1}{2}\gamma_{\mu\nu}\,\gamma_{\alpha\beta}\frac{\partial g_{\tau\rho}}{\partial x_\alpha}\frac{\partial \gamma_{\tau\rho}}{\partial x_\beta}.$$

In their covariant form, the *Entwurf* field equations are (p. 17, eq. 21):

$$-D_{\mu\nu}(g) = \kappa(T_{\mu\nu} + t_{\mu\nu}), \tag{8}$$

where $T_{\mu\nu}$ is the covariant stress-energy tensor for matter, and where $D_{\mu\nu}(g)$ and $t_{\mu\nu}$ are given by expressions very similar to the ones for $\Delta_{\mu\nu}(\gamma)$ and $\vartheta_{\mu\nu}$. $D_{\mu\nu}(g)$ is given by (p. 16, eq. 16):

$$D_{\mu\nu}(\gamma) = \frac{1}{\sqrt{-g}}\frac{\partial}{\partial x_\alpha}\left(\gamma_{\alpha\beta}\sqrt{-g}\,\frac{\partial g_{\mu\nu}}{\partial x_\beta}\right) - \gamma_{\alpha\beta}\,\gamma_{\tau\rho}\frac{\partial g_{\mu\tau}}{\partial x_\alpha}\frac{\partial g_{\nu\rho}}{\partial x_\beta}, \tag{9}$$

and $t_{\mu\nu}$, the covariant stress-energy tensor of the gravitational field, is given by (p. 16, eq. 14)

$$-2\kappa\, t_{\mu\nu} = \frac{\partial g_{\tau\rho}}{\partial x_\mu}\frac{\partial \gamma_{\tau\rho}}{\partial x_\nu} - \frac{1}{2} g_{\mu\nu}\,\gamma_{\alpha\beta}\frac{\partial g_{\tau\rho}}{\partial x_\alpha}\frac{\partial \gamma_{\tau\rho}}{\partial x_\beta}. \tag{10}$$

One easily verifies that

$$t_{\mu\nu} = g_{\mu\kappa} g_{\nu\lambda}\,\vartheta_{\kappa\lambda}, \qquad D_{\mu\nu}(g) = -g_{\mu\kappa} g_{\nu\lambda}\,\Delta_{\kappa\lambda}(g),$$

from which it follows that Equations (7) and (8) are equivalent. Notice that, contrary to what is suggested by the notation, $D_{\mu\nu}(g)$ is *minus* the covariant version of $\Delta_{\mu\nu}(g)$.

To first order in ϵ, the metric field described in Equations (3) and (4) is a solution of the vacuum *Entwurf* field equations. This can be seen as follows. The derivatives of the components of the metric up to first order are either zero or ϵ. It follows that second order derivatives of the metric in this approximation all vanish and that terms quadratic in derivatives of the metric are of order ϵ^2 and therefore negligible in this approximation. Since all terms in the vacuum *Entwurf* field equations involve either second order derivatives or products of two first order derivatives, it follows that the metric field indeed satisfies the equations in this approximation.

Given the metric field up to first order, the second order contributions to the vacuum field equations in their covariant and contravariant forms can be used to determine $\overset{(2)}{g}_{\mu\nu}$ and $\overset{(2)}{\gamma}_{\mu\nu}$ respectively. Following Einstein, I will focus on the 44-components.

Consider the 44-component of the vacuum *Entwurf* field equations in their covariant form:

$$\overset{(2)}{D}_{44}(g) + \kappa\, \overset{(2)}{t}_{44} = 0. \tag{11}$$

The second-order contribution to $D_{44}(g)$ is given by:[34]

$$\overset{(2)}{D}_{44}(g) = -\Delta \overset{(2)}{g}_{44} - 2\epsilon^2. \tag{12}$$

[34] To find the second-order contribution to $D_{44}(g)$, it is convenient to write the 44-component of Equation (9) as

$$D_{44}(g) = \frac{1}{\sqrt{-g}} \frac{\partial}{\partial x_\alpha}\left(\gamma_{\alpha\beta}\sqrt{-g}\right) \frac{\partial g_{44}}{\partial x_\beta} + \gamma_{\alpha\beta} \frac{\partial^2 g_{44}}{\partial x_\alpha \partial x_\beta} - \gamma_{\alpha\beta}\, \gamma_{\tau\rho} \frac{\partial g_{4\tau}}{\partial x_\alpha} \frac{\partial g_{4\rho}}{\partial x_\beta}.$$

Since $\overset{(1)}{g}_{44} = 0$, the first term does not give any second-order contributions. The second term gives

$$\overset{(0)}{\gamma}_{\alpha\beta} \frac{\partial^2 \overset{(2)}{g}_{44}}{\partial x_\alpha \partial x_\beta} = -\Delta \overset{(2)}{g}_{44}$$

where Δ is the Laplacian. The third term gives (identical) contributions for the index combinations ($\alpha = \beta = 1$, $\tau = \rho = 2$) and ($\alpha = \beta = 2$, $\tau = \rho = 1$). Consider, for instance, the former combination:

$$-\overset{(0)}{\gamma}_{11}\, \overset{(0)}{\gamma}_{22} \frac{\partial \overset{(1)}{g}_{42}}{\partial x_1} \frac{\partial \overset{(1)}{g}_{42}}{\partial x_1} = -\epsilon^2.$$

where I used Equations (5)–(6). Adding these contributions from the second and the third terms, one arrives at Equation (12).

The second order contribution to $\kappa\, t_{44}$ is given by:[35]

$$\kappa \overset{(2)}{t_{44}} = -\epsilon^2. \tag{13}$$

Inserting Equations (12)–(13) into Equation (11), one arrives at

$$-\Delta \overset{(2)}{g_{44}} - 2\epsilon^2 - \epsilon^2 = 0. \tag{14}$$

or, equivalently,

$$\Delta \overset{(2)}{g_{44}} = -3\epsilon^2. \tag{15}$$

An analogous argument can be made in terms of the contravariant form of the vacuum *Entwurf* field equations. The second order contribution to the 44-component of these equations is

$$\overset{(2)}{\Delta_{44}}(\gamma) - \kappa \overset{(2)}{\vartheta_{44}} = 0. \tag{16}$$

Through calculations almost identical to the ones spelled out above, one finds that

$$\overset{(2)}{\Delta_{44}}(\gamma) = -\Delta \overset{(2)}{\gamma_{44}} - \frac{2\epsilon^2}{c^4}.$$

and that

$$\kappa \overset{(2)}{\vartheta_{44}} = -\frac{\epsilon^2}{c^4}.$$

Inserting these last two equations into Equation (16), one arrives at

$$-\Delta \overset{(2)}{\gamma_{44}} - \frac{2\epsilon^2}{c^4} + \frac{\epsilon^2}{c^4} = 0. \tag{17}$$

---

[35] To find $\kappa\, t_{44}$ one needs to evaluate (see Eq. (10))

$$\kappa \overset{(2)}{t_{44}} = -\frac{1}{2}\frac{\partial \overset{(1)}{g_{\tau\rho}}}{\partial x_4}\frac{\partial \overset{(1)}{\gamma_{\tau\rho}}}{\partial x_4} + \frac{1}{4}\overset{(0)}{g_{44}}\,\overset{(0)}{\gamma_{\alpha\beta}}\frac{\partial \overset{(1)}{g_{\tau\rho}}}{\partial x_\alpha}\frac{\partial \overset{(1)}{\gamma_{\tau\rho}}}{\partial x_\beta}.$$

Recall that the metric is stationary so that $\partial \overset{(1)}{g_{\tau\rho}}/\partial x_4 = \partial \overset{(1)}{\gamma_{\tau\rho}}/\partial x_4 = 0$, which means that the first term in the expression above vanishes. The second term gives (identical) contributions for the index combinations

$$\begin{aligned} \tau &= 1, \quad \rho = 4, \quad \alpha = \beta = 2 \\ \tau &= 2, \quad \rho = 4, \quad \alpha = \beta = 1 \\ \tau &= 4, \quad \rho = 1, \quad \alpha = \beta = 2 \\ \tau &= 4, \quad \rho = 2, \quad \alpha = \beta = 1 \end{aligned}$$

Consider, for example, the first of these index combinations:

$$\frac{1}{4}\overset{(0)}{g_{44}}\,\overset{(0)}{\gamma_{22}}\frac{\partial \overset{(1)}{g_{14}}}{\partial x_2}\frac{\partial \overset{(1)}{\gamma_{14}}}{\partial x_2} = -\frac{1}{4}c^2\frac{\epsilon^2}{c^2},$$

where I used Equations (5)–(6). In this way, one arrives at Equation (13).

or, equivalently,

$$\Delta \overset{(2)}{\gamma}_{44} = -\frac{\epsilon^2}{c^4}. \tag{18}$$

Now, assume that $\overset{(2)}{\gamma}_{44}$ and $\overset{(2)}{g}_{44}$ are of the form

$$\overset{(2)}{g}_{44} = C_1\,\epsilon^2 r^2, \quad \overset{(2)}{\gamma}_{44} = C_2\,\frac{\epsilon^2 r^2}{c^4}, \tag{19}$$

where C_1 and C_2 are constants. Recall that $r^2 \equiv x^2 + y^2$, so that $\Delta\, r^2 = 4$. Inserting (19) into Equations (15) and (18), we find

$$4C_1 = -3, \quad 4C_2 = -1.$$

Inserting these values for C_1 and C_2 back into (19), we find

$$\overset{(2)}{g}_{44} = -\frac{3}{4}\,\epsilon^2 r^2, \quad \overset{(2)}{\gamma}_{44} = -\frac{1}{4}\,\frac{\epsilon^2 r^2}{c^4}, \tag{20}$$

which differ from the expressions one reads off from Equations (3) and (4), i.e.,

$$\overset{(2)}{g}_{44} = -\epsilon^2 r^2, \quad \overset{(2)}{\gamma}_{44} = 0. \tag{21}$$

Here we have the "blatant contradiction" Einstein reports in the letter to Freundlich I quoted. Starting from the Minkowski metric in rotating coordinates to first order in the angular velocity, the *Entwurf* field equations give a second order contribution to the 44-component of this metric field that differs from the corresponding term in the Minkowski metric in rotating coordinates by a factor 3/4.[36]

8. How Einstein missed the problem with rotation in the Einstein-Besso manuscript in 1913

How did Einstein miss this factor 3/4 when he did this same calculation in 1913? To answer this question, we need to examine Einstein's calculations on pp. [41–42] of the Einstein-Besso manuscript. For a facsimile reproduction of the relevant portions of these pages, see Figures 2 and 3.[37] These calculations in the Einstein-Besso manuscript closely match the calculations presented above (Equations (11)–(20)). On p. [41], the *Entwurf* field equations are used in their covariant form to compute the second order contribution to the covariant 44-component of a metric

[36] The 44-components are not the only components of $\overset{(2)}{g}_{\mu\nu}$ and $\overset{(2)}{\gamma}_{\mu\nu}$ that deviate from the ϵ^2-terms one reads off from Equations (3) and (4). Through calculations similar to the ones in Equations (11)–(18), one finds, for instance, that the 11-component of $\overset{(2)}{D}_{\mu\nu}(g) + \kappa\,\overset{(2)}{t}_{\mu\nu} = 0$ gives $\Delta \overset{(2)}{g}_{11} = \epsilon^2/c^2$. So, $\overset{(2)}{g}_{11}$ contains an ϵ^2-term, contrary to what one reads off from Equation (3).

[37] For a facsimile reproduction of the complete pages, see Einstein *CP4*: 670–671; for the annotated transcription, see pp. 442–449.

which to first order is the Minkowski metric in rotating coordinates (except for some errors). On p. [42], the equations are used in their contravariant form to compute the second order contribution to the contravariant 44-component of this metric, which is then used to compute the covariant 44-component once again.

The crucial error in these calculations occurs in the very first step. For the components of the metric to first order in the angular velocity—α in Einstein's notation on these pages—Einstein writes (in the upper right corners of p. [41] and p. [42]; see Figures 2 and 3):

$$g_{14} = -2\alpha\, y, \quad \gamma_{14} = -2\alpha\, y$$

$$g_{24} = 2\alpha\, x, \quad \gamma_{24} = 2\alpha\, x$$

Compare these expressions with the ones in Equation (6) for $\epsilon = \alpha$. First, the equations differ by a minus sign. This does not matter. It only means that the metric Einstein considers is for a coordinate system rotating clockwise rather than counterclockwise. Second, there is a factor $1/c^2$ missing in Einstein's expressions for γ_{14} and γ_{24}. In the actual calculations, however, these factors are taken into account. Third, and most important, the expressions above all contain an extra factor 2. This error is not corrected in the actual calculations.

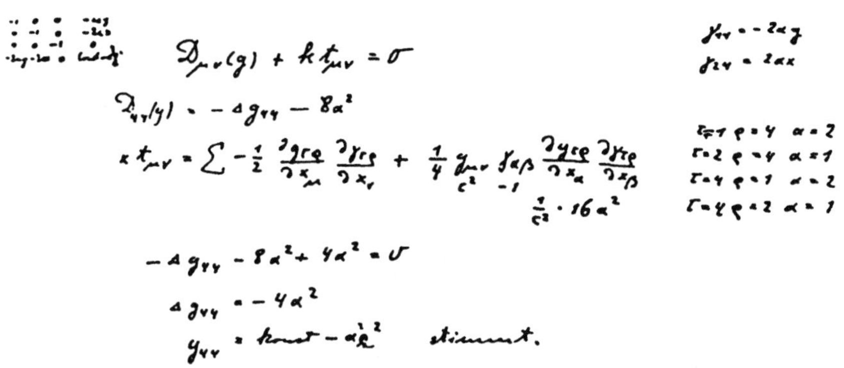

Figure 2. Facsimile of part of p. [41] of the Einstein-Besso manuscript. The matrix in the upper left corner is in Besso's hand.

Reproduced with permission of the Albert Einstein Archives, The Hebrew University of Jerusalem, Israel, and Princeton University Press.

Presumably, the error occurred in reading off the components of the metric from the line element in eq. (2). Setting the term $2\epsilon y\, dx\, dt$ for instance, equal to $g_{14}\, dx^1\, dx^4$ instead of $g_{14}\, dx^1\, dx^4 + g_{41}\, dx^4\, dx^1$, one arrives at the erroneous identification of g_{14} given by Einstein. Amazing as it may seem that Einstein would make such a trivial mistake, I see no other plausible reconstruction of how he ended up with the extra factors of 2. It would also not be the only time Einstein made this mistake. It occurs in two different places in the so-called "Zurich Notebook" from late 1912/early 1913 (Einstein *CP4*: doc. 10, p. [8] and p. [24]).[38]

Inserting $\epsilon = -2\alpha$ into Equations (12)–(13) and suppressing the notation I introduced to distinguish between different terms in the power series expansion of the metric field, one arrives at those equations for $D_{44}(g)$ and $\kappa\, t_{44}$ found by Einstein at the top of p. [41] (see Figure 2):[39]

$$D_{44}(g) = -\Delta g_{44} - 8\alpha^2, \quad \kappa\, t_{44} = -4\alpha^2.$$

Einstein loses the minus sign on the right hand side of the expression for $\kappa\, t_{44}$ when he inserts these results into the field equation $D_{44}(g) + \kappa\, t_{44} = 0$:

$$-\Delta g_{44} - 8\alpha^2 + 4\alpha^2 = 0.$$

In this way, he arrives at:

$$\Delta g_{44} = -4\alpha^2.$$

This is just the equation Einstein hoped to find, for one immediately sees that $g_{44} = c^2 - \alpha^2 r^2$, the 44-component of the Minkowski metric in rotating coordinates, is a solution of this equation. This probably accounts for the sign error just mentioned. Without this sign error, he would have found $\Delta\, g_{44} = -12\alpha^2$, in which case $g_{44} = c^2 - \alpha^2 r^2$ would not be a solution. Immediately below the equation for g_{44}, Einstein does actually write down a solution that has exactly the form of the 44-component of the Minkowski metric in rotating coordinates and writes next to it: "correct" [*stimmt*].

Einstein came very close to discovering his error on the very next page, p. [42], of the Einstein-Besso manuscript (see Figure 3). On this page, we find the analogue of the argument on p. [41] in terms of the contravariant field equations. Inserting $\epsilon = -2a$ into Equations (17)–(18), while again suppressing the additional notation I introduced, one arrives at:

$$-\Delta\gamma_{44} - \frac{8\alpha^2}{c^4} + \frac{4\alpha^2}{c^4} = 0,$$

and

$$\Delta\gamma_{44} = -\frac{4\alpha^2}{c^4}.$$

the first two equations on p. [42] (see Figure 3). Assuming that the second order contribution to γ_{44} is of the form $\beta\, \rho^2$, where $\rho \equiv x^2 + y^2$ and β is some constant,[40]

[38] In terms of the numbers assigned to the pages of the notebook when they were first catalogued, p. [8] is the right hand side of page 42 ('42R') and p. [24] is the right hand side of page 12 ('12R').

[39] For the significance of the index combinations next to the expression for $\kappa\, t_{\mu\nu}$ in Figure 2, see the calculation in note 35.

[40] Instead of writing $\overset{(2)}{\gamma}_{44} = C\,\epsilon^2 r^2$ (see Eq. (19)), Einstein writes $\overset{x}{\gamma}_{44} = \beta\,\rho^2$ (see Figure 3). The usage of the superscript 'x' to indicate a second-order term in a power series expansion of the metric can also be found elsewhere in the Einstein-Besso manuscript (see p. [3]; Einstein *CP4*: 632). In solving $\Delta\gamma_{44} = -4\alpha^2/c^4$ for γ_{44}, Einstein at first only assumes that γ_{44} is cylindrically symmetric and rewrites $\Delta\,\gamma_{44}(\rho)$ in cylindrical coordinates. It is only at that point that he makes the further assumption that $\overset{x}{\gamma}_{44}$ is proportional to ρ^2.

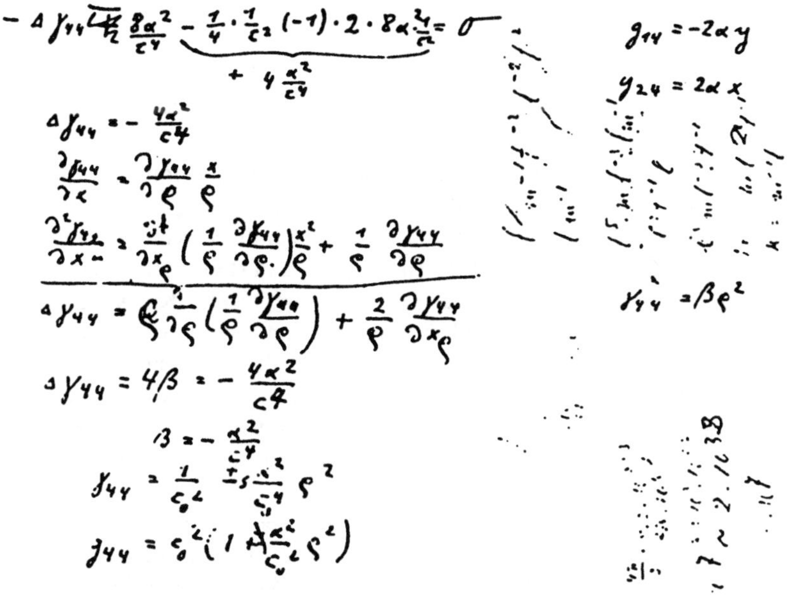

Figure 3. Facsimile of part of p. [42] of the Einstein-Besso manuscript. The material written sideways is in Besso's hand and is not related to Einstein's calculation.

Einstein finds that $\beta = -\alpha^2/c^4$. So, for γ_{44} up to second order in α we find

$$\gamma_{44} = \frac{1}{c^2} - \frac{\alpha^2 r^2}{c^4}.$$

Einstein subsequently added a factor 3 to the α^2-term (see Figure 3).[41] Presumably, this was to make the expression for γ_{44} compatible with the expression for g_{44} he had found on p. [41] and which had the desired form of the 44-component of the Minkowski metric in rotating coordinates. Because of the relation $g_{\mu\rho}\,\gamma_{\rho\nu} = \delta_{\mu\nu}$, where $\delta_{\mu\nu}$ is the Kronecker delta, the components of the metric field computed here must satisfy $g_{4\rho}\,\gamma_{\rho 4} = 1$ to order α^2. Given the spurious factors 2 in Einstein's identification of g_{41}, g_{42}, γ_{14}, and γ_{24}, this means that the numerical coefficient of the α^2-terms in g_{44} and γ_{44} must add up to -4.[42]

Inserting Einstein's expressions for g_{41}, g_{42}, γ_{14}, and γ_{24}, along with $g_{44} = c^2(1 - C_1\alpha^2\rho^2/c^2)$ and $\gamma_{44} = 1/c^2(1 - C_2\alpha^2\rho^2/c^2)$ where C_1 and C_2 are

---

41 The notations 'c' and 'c_0' are used interchangeably in the manuscript.

42 Without these extra factors, they should add up to -1, as is actually the case for the numerical coefficients of the ϵ^2-terms in g_{44} and γ_{44} in both Equations (20) and (21).

constants, into $g_{4\rho}\,\gamma_{\rho 4}$, we find

$$\frac{4\alpha^2 y^2}{c^2}+\frac{4\alpha^2 x^2}{c^2}+1-(C_1+C_2)\frac{\alpha^2\rho^2}{c^2}+\mathcal{O}(\alpha^4).$$

Since this should be equal to 1 to second order in α, it must be the case that $C_1+C_2=4$. Since Einstein wanted $C_1=1$, he needed $C_2=3$.

Such considerations are probably behind Einstein's corrections to the final expressions for g_{44} and γ_{44} on p. [42]. The sequence of steps seems to have been the following. Einstein found that $\gamma_{44} = 1/c^2 - \alpha^2\rho^2/c^4$. Knowing that $g_{4\rho}\,\gamma_{\rho 4}$ has to be 1 to order α^2, he realized that in that case g_{44} would be equal to $c^2(1-3\,\alpha^2\rho^2/c^2)$. At first, he accepted this result. Then, however, he decided that he must have made a mistake in the calculation of γ_{44}, crossed out the factor 3 in the equation for g_{44}, and put it in the expression for γ_{44} instead. What remains puzzling on this account is that it apparently did not bother Einstein that γ_{44} did not have the right form (see Eq. (4)).[43]

Whether or not one accepts the details of this scenario, one can, I think, at least say the following. Einstein's calculation of the α^2-term in γ_{44} on p. [42] gives a result that is incompatible with the result of his calculation of the α^2-term in g_{44} on p. [41]. The occurrence of the factors 3 in the expressions for g_{44} and γ_{44} on p. [42] strongly suggest that Einstein realized the two results were incompatible. Yet, this did not prompt him to carefully redo both calculations to see whether his conclusion that the g_{44} is just the 44-component of the Minkowski metric in rotating coordinates is actually warranted.

What conclusion can be drawn from the close examination of this calculation in the Einstein-Besso manuscript? I would say the following. Einstein was so strongly convinced that his approximation procedure would reproduce the 44-component of the Minkowski metric in rotating coordinates that, when his calculations did not bear out this expectation, he assumed he must have made an error somewhere. I identified two such instances: the sign error on p. [41] that gave him the desired equation $\Delta g_{44} = -4\,\alpha^2$ instead of $\Delta g_{44} = -12\,\alpha^2$, and the factor 3 that was moved from g_{44} to γ_{44} on p. [42]. The erroneous factors of 2 in the expressions for g_{41}, g_{42}, γ_{14}, and γ_{24} that Einstein started from do not fall in this category. In fact, it is hard to see how Einstein could have failed to recognize that the calculation does *not* give the result he expected, had he started from the correct expressions for these components.

---

[43] In the Einstein-Besso manuscript, only a few components of the metric of Minkowski space-time in rotating coordinates are explicitly given. In the Zurich notebook, however, the full metric for the $2+1$-dimensional case is given both in covariant and in contravariant form (Einstein *CP4*: doc. 10, p. [8] or '42R').

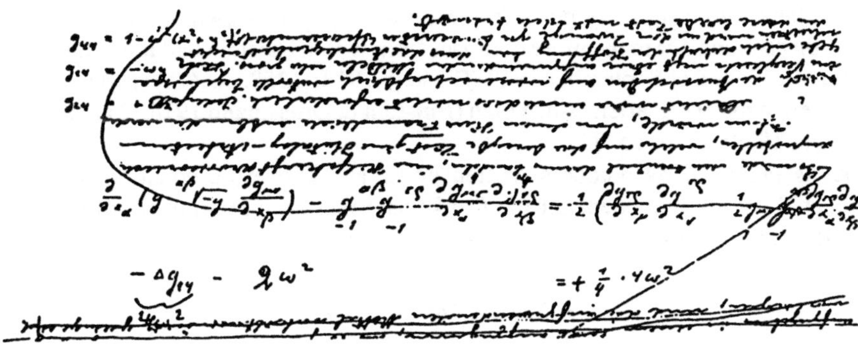

Figure 4. Facsimile of a calculation on the draft of a letter to Naumann.

9. The calculation on the Naumann draft

The Einstein Archive also contains a document with a version of this calculation without any of the mistakes made in the Einstein-Besso manuscript. This document is the draft of a letter to Naumann that I mentioned in Section 3. I argued that the draft was written shortly after Einstein's letter to Freundlich from 30 September 1915, while the calculation on the same page was done shortly before that letter.

As in the Einstein-Besso manuscript, the calculation is given in a rather cryptic form, but after everything that has been said so far it is easy to follow. First, the three components of a Minkowski metric in rotating coordinates from Einstein's query in the "Scratch Notebook" are given, this time without any spurious factors of 2:[44]

$$g_{44} = 1 - \omega^2(x^2 + y^2), \quad g_{14} = -\omega\, y, \quad g_{24} = \omega\, x.$$

Then an equation is given in which one recognizes the vacuum *Entwurf* field equations in the form $D_{\mu\nu}(g) = -\kappa\, t_{\mu\nu}$ (cf. Equations (8)–(10)):

$$\frac{\partial}{\partial x_\alpha}\left(g^{\alpha\beta}\sqrt{-g}\,\frac{\partial g_{\mu\nu}}{\partial x_\beta}\right) - g^{\alpha\beta}\, g^{\tau\rho}\,\frac{\partial g_{\mu\tau}}{\partial x_\alpha}\,\frac{\partial g_{\nu\rho}}{\partial x_\beta}$$
$$= \frac{1}{2}\left(\frac{\partial g_{\tau\rho}}{\partial x_\mu}\,\frac{\partial g^{\tau\rho}}{\partial x_\nu} - \frac{1}{2} g_{\mu\nu} g^{\alpha\beta}\,\frac{\partial g_{\tau\rho}}{\partial x_\alpha}\,\frac{\partial g^{\tau\rho}}{\partial x_\beta}\right).$$

Only a factor $1/\sqrt{-g}$ in front of the first term on the left hand side is missing. Notice that the contravariant components of the metric are written as $g^{\mu\nu}$ rather than as $\gamma_{\mu\nu}$, a clear indication that this calculation and the calculation in the Einstein-Besso manuscript date from different periods in the genesis of general relativity. Notice also, however, that the field equations are given in the form in

[44] Implicitly, units are chosen such that $c = 1$. The x_μ-frame rotates clockwise.

which they were originally given in the *Entwurf* paper—down to the labeling of the indices—and not in the more compact form which Einstein had introduced in a lecture in Vienna of September 1913 (Einstein 1913: 1258, eq. (7b)) and which he had used in his publications since then (Einstein & Grossmann 1914: 217, eq. (II); Einstein 1914: 1077, eq. 81).[45]

As is indicated by numbers written under and above the equation, Einstein evaluated its 44-component for the Minkowski metric in rotating coordinates up to order ω^2. The result is written on the next line:

$$-\Delta g_{44} - 2\,\omega^2 = \frac{1}{4}\,4\,\omega^2.$$

This is equivalent to Equation (14) above (with $\omega = -\epsilon$). The result of applying minus the Laplacian to $g_{44} = 1 - \omega^2(x^2 + y^2)$, i.e., $4\,\omega^2$, is then written under the term $-\Delta g_{44}$. This shows that the 44-component of the Minkowski metric in rotating coordinates is not a solution of the equation above. What is a solution is $g_{44} = 1 - (3/4)\,\omega^2(x^2 + y^2)$, the result Einstein reported to Freundlich.

10. The problem with rotation and the demise of the *Entwurf* theory

The result reported in the letter to Freundlich must have come as a shock to Einstein. The requirement that the Minkowski metric in rotating coordinates be a solution of any acceptable gravitational field equations had played an important role in Einstein's search for such equations. If this requirement was not satisfied, it would not be possible to interpret the inertial forces in a rotating frame as gravitational forces.[46] As our examination of the calculations on pp. [41–42] of the Einstein-Besso manuscript in Section 8 suggests, Einstein had been under the impression ever since 1913 that the *Entwurf* field equations do satisfy this important requirement. In 1914, this belief had been reinforced when Einstein convinced himself—again erroneously—that the Minkowski metric in rotating coordinates satisfies the condition for adapted coordinates, another necessary condition for this metric field to be a solution of the *Entwurf* field equations.[47] Now, in September

---

[45] It looks as if Einstein first wrote $\gamma_{\alpha\beta}$ instead of $g^{\alpha\beta}$ in the last term on the right hand side. Einstein may actually have copied the left and right hand sides of the equation from the *Entwurf* paper. Given that he had done the calculation in this form before, one can understand that he went back to the earlier form of his field equations.

[46] For the reasoning behind this claim, see Einstein 1914: 1031–1032 and 1067–1068.

[47] This condition is $B_\sigma = 0$, where B_σ is defined as

$$B_\sigma = \frac{\partial}{\partial x_\nu}\frac{\partial}{\partial x_\alpha}\left(\sqrt{-g}\,\gamma_{\alpha\beta}\,g_{\sigma\mu}\frac{\partial \gamma_{\mu\nu}}{\partial x_\beta}\right)$$

(see Einstein & Grossmann 1914: 218). In March 1914, Einstein wrote to Besso: "I have been able to prove through a *simple* calculation *that the gravitational field equations hold in every system that is adapted to this condition*. It follows that there is a wide variety of transformations corresponding to accelerations which transform the equations back into themselves (also rotation, for instance)." ["Ich habe beweisen können durch *einfache* Rechnung, *dass die Gleichungen der Gravitation für*

1915, his calculations strongly suggested that the Minkowski metric in rotating coordinates is not a solution after all. One can readily understand why Einstein calls the result a "blatant contradiction" in his letter to Freundlich.

Still, Einstein was not ready to give up the *Entwurf* theory yet. The impression conveyed by his diagnosis of the problem in the letter to Freundlich is one of attempted damage control. Einstein writes: "Either the equations are numerically incorrect (number coefficients) or I apply the equations in a fundamentally wrong way."

The latter alternative, I take it, refers to the approximation procedure. If there is something wrong with the approximation procedure, the *Entwurf* equations themselves might be off the hook. Presumably, Einstein did not find anything wrong on this score. He would use the same approximation procedure a little over a month later to compute the perihelion advance predicted by the successor to the *Entwurf* theory.

It can not have been long before the *Entwurf* field equations themselves came under increased scrutiny. At this point, one would expect Einstein to re-examine carefully the arguments leading to the choice of these equations. It should not come as a surprise, therefore, that, just around this time, Einstein finally discovered the flaw in the covariance and variational considerations that he had used a year earlier to argue for the uniqueness of the *Entwurf* field equations (Einstein 1914: 1074–1076).[48] Einstein must have discovered this problem shortly before 12 October 1915, when he reported it in a letter to Lorentz (Einstein *CP8*: doc. 129). He still did not give up the *Entwurf* field equations in that letter. He retreated to a more physical derivation of the equations.

At that point, it would seem Einstein no longer had a way out of the problem with rotation. It could not be blamed on the approximation procedure and it could not be resolved through some minor adjustments of the *Entwurf* field equations either. Shortly afterwards, Einstein abandoned the whole line of reasoning leading to the *Entwurf* equations, and returned to a very different line of reasoning that he had explored but rejected three years earlier in the Zurich notebook (see Norton 1984: 142–143).

---

jedes Bezugssystem gelten, welches dieser Bedingung angepasst ist. Hieraus geht hervor, dass es Beschleunigungstransformationen mannigfaltigster Art gibt, welche die Gleichungen in sich selbst transformieren (z.B. auch Rotation."] (Speziali 1972: 53; Einstein *CP5*: doc. 514). See also a letter from Einstein to Joseph Petzoldt of April 1914 (Einstein *CP8*: doc. 5). The letter to Besso suggests that Einstein checked whether $B_\sigma = 0$ for a Minkowski metric in rotating coordinates and that he convinced himself that this is indeed the case. In fact, when Equations (3) and (4) are inserted into the expression for $B_\sigma = 0$, a tedious but straightforward calculation shows that $B_\sigma = (-4\epsilon^4 x/c^4, -4\epsilon^4 y/c^4, 0, 0)$ for this metric field.

[48] Commenting on this argument, Norton (1984: 137) writes: "Einstein's satisfaction with his new treatment of the field equations was short-lived. He soon found that the last step in his derivation was incorrect." In fact, it took Einstein about a year to find the problem. It may be worth emphasizing in this context that Einstein's debate with Levi-Civita in early 1915 about the variational techniques Einstein employed in his 1914 paper did not touch upon the argument for fixing the Lagrangian in that paper, even though the exchange was apparently triggered by a letter from Abraham to Levi-Civita in which Abraham complained about the arbitrariness of Einstein's choice of the Lagrangian (see Cattani & De Maria 1989: 183–185).

There is an interesting remark by Einstein in this context in an undated letter to his friend Heinrich Zangger, which from context can be dated to 15 October 1915.[49] Einstein writes: "I wrote a supplementary paper to my paper of last year on general relativity."[50] "supplementary paper" [*ergänzende Arbeit*]? The first published paper on general relativity after 15 October 1915 is Einstein 1915a, the first of his four November communications to the Prussian Academy. This is the paper in which Einstein renounced the *Entwurf* field equations. Although he retained the rest of the formalism of the *Entwurf* theory, one would not expect Einstein to refer to this paper as a supplement to his earlier work. Given, that the letter to Zangger was apparently written three days after the letter to Lorentz in which Einstein first explained the problem with the uniqueness argument for the *Entwurf* field equations and given, that it is this problem which Einstein used in the introduction of Einstein 1915a to explain why he had lost all confidence in these equations, I propose the following scenario. When Einstein discovered the problem with his uniqueness argument, he originally planned to publish a paper in which the flawed derivation of the *Entwurf* field equations of Einstein 1914 would be replaced by the alternative derivation he had outlined in his letter to Lorentz. This is the "supplementary paper" Einstein mentioned to Zangger. Shortly afterwards, Einstein realized that the *Entwurf* field equations could not be salvaged this way, and this paper turned into the first November communication.[51]

The admittedly speculative reconstruction given in this Section of how Einstein came to abandon the *Entwurf* theory fits nicely with the chronology suggested by Einstein's letter to Lorentz of 1 January 1916, quoted in Section 1. According to this letter, Einstein "had already found earlier" that the theory cannot fully account for the anomalous perihelion advance of Mercury (cf. the Einstein-Besso manuscript), he "in addition" found the problem with rotation (cf. the letter to Freundlich of 30 September 1915), and he "finally" found the fallacy in his 1914 argument for the uniqueness of the Lagrangian (cf. the letter to Lorentz of 12 October 1915).

The reconstruction given here also suggests that, of the three problems Einstein listed in his letters to Sommerfeld and Lorentz, the problem with rotation was the decisive one. Einstein had known since June 1913 that the *Entwurf* theory could only account for part of the anomalous advance of Mercury's perihelion, yet had continued to work on the theory without even mentioning the problem in print. The problem with the uniqueness argument was probably a more serious blow, but,

---

[49] I am grateful to A. J. Kox and Robert Schulmann for drawing my attention to this passage.

[50] "Ich schrieb eine ergänzende Arbeit zu meiner letztjährigen Untersuchung über die allgemeine Relativität." (Einstein *CP8*: doc. 130)

[51] As Tilman Sauer pointed out to me, there is another possible candidate for the "supplementary paper," viz. Einstein 1916. Einstein already referred to the results published in this paper on 23 September 1915, in a letter to Lorentz (Einstein *CP8*: doc. 122). I still like to think that the "supplementary paper" is an early version of Einstein 1915a, since it provides an explanation for what otherwise would be a bit of an anomaly in my account, viz. the fact that in the introduction of Einstein 1915a the renouncement of the *Entwurf* equations is motivated by the failure of the uniqueness argument for the equations and not by the problem with rotation, which is only referred to obliquely at the end of the paper (see note 7 above).

as he wrote to Lorentz, he still had an alternative derivation of the equations. In the end, it seems to me, the problem with rotation was the nemesis of the *Entwurf* theory.

11. Epilogue: Einstein looking back on the problem with rotation in 1916

In a letter of 31 July 1916, Einstein wrote to Besso:

> The field of a rotating ring in the vicinity of the axis can be found as follows. In *first* approximation, the field is obtained easily by direct integration of the field equations. The second approximation is obtained from the vacuum field equations as the next approximation. The first approximation gives the Coriolis forces, the second the centrifugal forces.[52] (Speziali 1972: 77; Einstein *CP8*: doc. 245)

The calculation Einstein outlines here for Besso is very reminiscent of the calculations of the Einstein-Besso manuscript, the Naumann draft and the Freundlich letter that I analyzed in this paper. There is one important difference. This time the starting point of the iterative approximation procedure is not the Minkowski metric in rotating coordinates to first order, a vacuum solution of the field equations, but the first-order solution for the case of a rotating ring. Through his Machian hopes of attributing inertial forces to the gravitational effect of distant masses, these two solutions were closely related to one another for Einstein. It is no coincidence, I think, that on pp. [36–37] of the Einstein-Besso manuscript, one finds a calculation showing that to first order in the angular velocity, the field inside a rotating spherical shell has the same form as the metric field of a Minkowski space-time in rotating coordinates.[53] This Machian connection between the first-order Minkowski metric in rotating coordinates and the first-order metric field of a rotating shell may also explain Einstein's somewhat peculiar way of checking, through his iterative approximation procedure, whether the Minkowski metric in rotating coordinates is a solution of the *Entwurf* field equations. Presumably, what Einstein had in mind when he started from the first-order Minkowski metric in rotating coordinates, was that this metric field could be interpreted as the field of a rotating shell or a rotating ring with appropriately chosen mass and diameter.

If one replaces "the metric field of a rotating ring" by "the metric field of Minkowski space-time in rotating coordinates," the last sentence of the quotation above becomes particularly easy to understand.

In a Cartesian coordinate system (x, y, z) rotating in a counterclockwise direction around the z-axis with angular velocity ω with respect to an inertial frame,

---

[52] "Das Feld des rotierenden Ringes in der Nähe der Achse ist so zu finden. Das Feld in erster Näherung ergibt sich leicht durch unmittelbare Integration der Feldgleichungen. Die zweite Näherung ergibt sich aus den Vakuumfeldgleichungen als nächste Näherung. Die erste Näherung liefert die Korioliskräfte, die zweite die Zentrifugalkräfte."

[53] For a detailed analysis of this calculation, see Einstein *CP4*: 432–435.

the Newtonian equations of motion for a free particle are:[54]

$$\ddot{x} - 2\omega\,\dot{y} - \omega^2 x = 0, \quad \ddot{y} + 2\omega\,\dot{x} - \omega^2 y = 0, \quad \ddot{z} = 0, \tag{22}$$

(where dots indicate time derivatives), or, in vector form:

$$\ddot{\vec{x}} + 2\vec{\omega} \times \dot{\vec{x}} + \vec{\omega} \times (\vec{\omega} \times \vec{x})$$

where $\vec{\omega} \equiv (0, 0, \omega)$. The second term on the left-hand side is the Coriolis force, the third term is the centrifugal force. In general relativity (as well as in the *Entwurf* theory), the analogue of these equations is the geodesic equation:

$$\frac{d^2x^\alpha}{ds^2} + \left\{ \begin{matrix} \alpha \\ \rho\sigma \end{matrix} \right\} \frac{dx^\rho}{ds}\frac{dx^\sigma}{ds} = 0,$$

in a rotating frame in Minkowski space-time. Contracting with $g_{\mu\nu}$ and inserting

$$\left\{ \begin{matrix} \alpha \\ \rho\sigma \end{matrix} \right\} = \frac{1}{2} g^{\alpha\beta} \left(g_{\beta\rho,\sigma} + g_{\beta\sigma,\rho} - g_{\rho\sigma,\beta} \right)$$

for the Christoffel symbols, one can rewrite the geodesic equation as

$$g_{\mu\alpha}\frac{d^2x^\alpha}{ds^2} + \left(g_{\mu\rho,\sigma} - \frac{1}{2} g_{\rho\sigma,\mu} \right) \frac{dx^\rho}{ds}\frac{dx^\sigma}{ds} = 0.$$

For slowly moving particles, the proper time s/c can be replaced by the coordinate time t, and the previous equation becomes

$$g_{\mu\alpha}\ddot{x}^\alpha + \left(g_{\mu\rho,\sigma} - \frac{1}{2} g_{\rho\sigma,\mu} \right) \dot{x}^\rho\,\dot{x}^\sigma = 0.$$

For the metric of the form of a Minkowski metric in rotating coordinates (see Eq. (3)), the previous equation gives

$$\begin{aligned} \mu = 1: &\qquad -\ddot{x}^1 + g_{14,2}\dot{x}^2 - g_{24,1}\dot{x}^2 - \frac{1}{2} g_{44,1} = 0 \\ \mu = 2: &\qquad -\ddot{x}^2 + g_{24,1}\dot{x}^1 - g_{14,2}\dot{x}^1 - \frac{1}{2} g_{44,2} = 0 \\ \mu = 3: &\qquad \ddot{x}^3 = 0. \end{aligned} \tag{23}$$

Inserting Equation (3) for $g_{\mu\nu}$ into these equations, and writing ω instead of ϵ, one finds

$$-\ddot{x} + 2\omega\,\dot{y} + \omega^2 x = 0, \quad -\ddot{y} - 2\omega\,\dot{x} + \omega^2 y = 0, \quad \ddot{z} = 0. \tag{24}$$

---

[54] For an elementary derivation of these equations in Einstein's hand, see his "Lecture Notes for Introductory Course on Mechanics at the University of Zurich, Winter Semester 1909–1910" (Einstein *CP3*: doc. 1, p. [51]).

which is just Equation (22) from Newtonian mechanics. Eqs. (23)–(24) show that the components of $g_{\mu\nu}$ proportional to ω (i.e., g_{14} and g_{24}) give rise to the Coriolis force and that the component proportional to ω^2 (i.e., g_{44}) gives rise to the centrifugal force.[55]

It is very interesting to see what Einstein goes on to say in his letter to Besso about the determination of the centrifugal forces in the second approximation (in other words, the ω^2-term in g_{44}) given the Coriolis forces of the first approximation (in other words, the ω-terms of g_{14} and g_{24}). These sentences read like a direct commentary on the problem with rotation in the *Entwurf* theory. They also convey a strong sense of relief on Einstein's part for having solved the problem. It therefore seems appropriate to close my paper by quoting them. Immediately following the sentences quoted above, Einstein writes:[56]

> That the latter come out correctly is obvious given the general covariance of the equations, so that it is of no further interest whatsoever to actually do the calculation. This is of interest only if one does not know whether rotation-transformations are among the "allowed" ones, i.e., if one is not clear about the transformation properties of the equations, a stage which, thank God, has definitively been surpassed.[57] (Speziali 1972: 77; Einstein *CP8*: doc. 245)

Acknowledgements

This paper was born in Pittsburgh in 1992 when I showed my reconstruction of pp. [41–42] of the Einstein-Besso manuscript to John Norton, who then pulled the Freundlich letter and the Naumann draft discussed in this paper out of his files. Jürgen Renn then recalled the query in the "Scratch Notebook." I am grateful to both of them for drawing my attention to these documents and for their subsequent support. Further research for this paper was supported by the Max-Planck-Institut für Wissenschaftsgeschichte in Berlin, where I spent the summers of 1994 and 1995. I want to thank the institute and its staff for their support and hospitality. I also want to express my gratitude to the Hebrew University of Jerusalem for granting me permission to quote from unpublished material in the Albert Einstein Archives, and to both Hebrew University and Princeton University Press for granting me permission to reproduce several excerpts of material in the Albert Einstein Archives in facsimile.

[55] The $i4$-components of the Minkowski metric in rotating coordinates are explicitly called "the 'Coriolis field'" [*Das "Coriolis-Feld"*] in a letter from Einstein to Besso of 31 October 1916 (Speziali 1972: 86; Einstein *CP8*: doc. 270)

[56] In his Princeton lectures of 1921, Einstein made a similar comment, this time carefully distinguishing between the Minkowski metric in rotating coordinates and the metric field of a rotating shell. In a footnote to a brief discussion of gravitational effects inside a rotating shell, Einstein writes: "Daß die Zentrifugalwirkung mit der Existenz des Koriolisfeldes unzertrennlich verbunden sein muß, erkennt man auch ohne Rechnung an dem Spezialfall des relativ zu einem Inertialsystem gleichförmig rotierenden Koordinatensystems, welcher Fall unseren allgemein kovarianten Gleichungen natürlich genügen muß" (Einstein 1922: 66, footnote). The English translation of this footnote is: "That the centrifugal action must be inseparably connected with the experience of the Coriolis field may be recognized, even without calculation, in the special case of a co-ordinate system rotating uniformly relatively to an inertial system; our general co-variant equations naturally must apply to such a case" (Einstein 1956: 102, footnote).

[57] "Dass letztere richtig heraus kommen, ist bei der allgemeinen Kovarianz der Gleichungen selbstverständlich, sodass ein wirkliches Durchrechnen keinerlei Interesse mehr hat. Dies Interesse ist nur dann vorhanden, wenn man nicht weiss ob Rotations-transformationen zu den 'erlaubten' gehören, d.h. wenn man sich über die Transformationseigenschaften der Gleichungen nicht im Klaren ist, welches Stadium gottlob endgültig überwunden ist."

Finally, I want to thank Giuseppe Castagnetti, Rita Fountain, A. J. Kox, Annette Pringle, Tilman Sauer, Robert Schulmann and John Stachel for helpful comments, discussions, and references.

References

CATTANI, Carlo & DE MARIA, Michelangelo. (1989). "The 1915 Epistolary Controversy Between Einstein and Tullio Levi-Civita." In HOWARD & STACHEL 1989: 175–200.

EARMAN, John & JANSSEN, Michel. (1993). "Einstein's Explanation of the Motion of Mercury's Perihelion." In *The Attraction of Gravitation: New Studies in the History of General Relativity* (Einstein Studies, vol. 5). J. Earman, M. Janssen and J. D. Norton, eds. 129–172. Boston: Birkhäuser.

EINSTEIN, Albert. (1913). "Zum gegenwärtigen Stande des Gravitationsproblem." *Physikalische Zeitschrift* **14**: 1249–1266 [= EINSTEIN *CP4*: doc. 17].

——— (1914). "Die formale Grundlage der allgemeinen Relativitätstheorie." *Sitzungsberichte der Königlichen Preussischen Akademie der Wissenschaften* (Berlin): 1030–1085 [= EINSTEIN *CP6*: doc. 9].

——— (1915a). "Zur allgemeinen Relativitätstheorie." *Sitzungsberichte der Königlichen Preussischen Akademie der Wissenschaften* (Berlin): 778-786 [= EINSTEIN *CP6*: doc. 21].

——— (1915b). "Erklärung der Perihelbewegung des Merkur aus der allgemeinen Relativitätstheorie." *Sitzungsberichte der Königlichen Preussischen Akademie der Wissenschaften* (Berlin): 831–839 [= EINSTEIN *CP6*: doc. 24].

——— (1915c). "Die Feldgleichungen der Gravitation." *Sitzungsberichte der Königlichen Preussischen Akademie der Wissenschaften* (Berlin): 844–847 [= EINSTEIN *CP6*: doc. 25].

——— (1916). "Eine neue formale Deutung der Maxwellschen Feldgleichungen der Elektrodynamik." *Sitzungsberichte der Königlichen Preussischen Akademie der Wissenschaften* (Berlin): 184–188 [= EINSTEIN *CP6*: doc. 27].

——— (1922). *Vier Vorlesungen über Relativitätstheorie.* Braunschweig: Friedr. Vieweg & Sohn.

——— (1956). *The Meaning of Relativity.* 5th ed. Princeton: Princeton University Press.

——— (*CP3*). *The Collected Papers of Albert Einstein.* Vol. 3. *The Swiss Years: Writings, 1909–1911.* M. J. Klein, A. J. Kox, J. Renn & R. Schulmann, eds. Princeton: Princeton University Press (1993).

——— (*CP4*). *The Collected Papers of Albert Einstein.* Vol. 4. *The Swiss Years: Writings, 1912–1914.* M. J. Klein, A. J. Kox, J. Renn & R. Schulmann, eds. Princeton: Princeton University Press (1995).

——— (*CP5*). *The Collected Papers of Albert Einstein.* Vol. 5. *The Swiss Years: Correspondence, 1902–1914.* M. J. Klein, A. J. Kox & R. Schulmann, eds. Princeton: Princeton University Press (1993).

——— (*CP6*). *The Collected Papers of Albert Einstein.* Vol. 6. *The Berlin Years: Writings, 1914–1917.* A. J. Kox, M. J. Klein, & R. Schulmann, eds. Princeton: Princeton University Press (1996).

——— (*CP8*). *The Collected Papers of Albert Einstein.* Vol. 8. *The Berlin Years: Correspondence, 1914–1918.* R. Schulmann, A. J. Kox, M. Janssen and J. Illy, eds. Princeton: Princeton University Press, (1998).

EINSTEIN, Albert & GROSSMANN, Marcel. (1913). *Entwurf einer verallgemeinerten Relativitätstheorie und einer Theorie der Gravitation.* Leipzig: Teubner. [Reprinted with added "Bemerkungen," *Zeitschrift für Mathematik und Physik* **62** (1914): 225–261 = EINSTEIN *CP4*: docs. 13 and 26].

——— (1914). "Kovarianzeigenschaften der Feldgleichungen der auf die verallgemeinerte Relativitätstheorie gegründeten Gravitationstheorie." *Zeitschrift für Mathematik und Physik* **63**: 215–225 [= Einstein *CP6*: doc. 2].

Freundlich, Erwin Finlay. (1915a). "Über die Erklärung der Anomalien im Planeten-System durch die Gravitationswirkung interplanetarer Massen." *Astronomische Nachrichten* **201** (Nr. 4803): 49–56.

——— (1915b). "Über die Gravitationsverschiebung der Spektrallinien bei Fixsternen." *Physikalische Zeitschrift* **16**: 115–117.

——— (1915/16). "Über die Gravitationsverschiebung der Spektrallinien bei Fixsternen." *Astronomische Nachrichten* **202** (Nr. 4826): 17–24.

Hentschel, Klaus. (1992). *Der Einstein Turm: Erwin F. Freundlich und die Relativitätstheorie; Ansätze zu einer "dichten Beschreibung" von institutionellen, biographischen, und theoriengeschichtlichen Aspekten.* Heidelberg, Berlin & New York: Spektrum Akademie. [English translation: *The Einstein Tower. An Intertexture of Dynamical Construction, Relativity Theory, and Astronomy*. Stanford (CA): Stanford University Press (1997).

——— (1994). "Erwin Finlay Freundlich and Testing Einstein's Theory of Relativity." *Archive for the History of Exact Sciences* **47**: 143–201.

Hermann, Armin, ed. (1968). *Albert Einstein–Arnold Sommerfeld. Briefwechsel.* Basel & Stuttgart: Schwabe.

Howard, Don & Stachel, John, eds. (1989). *Einstein and the History of General Relativity* (Einstein Studies, vol. 1). Boston: Birkhäuser.

Kirsten, Christa & Treder, Hans-Jürgen. (1979). *Albert Einstein in Berlin, 1913–1933.* Berlin: Akademie-Verlag.

Norton, John D. (1984). "How Einstein Found His Field Equations: 1912–1915." *Historical Studies in the Physical Sciences* **14**: 253–316. [Reprinted in Howard & Stachel 1989: 101–159. Page references are to this reprint].

Renn, Jürgen & Sauer, Tilman. (1996). "Einsteins Züricher Notizbuch. Die Entdeckung der Feldgleichungen der Gravitation im Jahre 1912." *Physikalische Blätter* **52**: 865–872.

von Seeliger, Hugo. (1916). "Über die Gravitationswirkung auf die Spektrallinien." *Astronomische Nachrichten* **202** (Nr. 4829): 83–86.

Speziali, Pierre, ed. (1972). *Albert Einstein–Michele Besso. Correspondance.* Paris: Hermann.

Stachel, John. (1980). "Einstein's Search for General Covariance, 1912–1915." Paper delivered at the Ninth International Conference on General Relativity and Gravitation, Jena, Germany (DDR), 17 July 1980. [Quotations and page numbers from revised version in Howard & Stachel 1989: 63–100.

Part II

Relativity at Work

Einstein, Relativity and Gravitation Research in Vienna before 1938

Peter Havas

MUCH OF THE EARLY WORK stimulated by the creation of the special theory of relativity in 1905 and of the general theory in 1915 by Einstein was carried out by Einstein himself, wherever he was, and by scientists in Germany as well as in Holland and somewhat later also in England; the best known early researchers are Laue, Weyl, De Sitter, and Eddington (Pauli 1921, Thirring 1922). It is little known that significant research in both areas was also carried out in Vienna, the capital of the Austro-Hungarian Empire until 1918 and then of the Austrian Republic until the *Anschluss*, the incorporation of Austria into Hitler's Germany, in March 1938. Similarly, while a search for a theory of gravitation to replace Newton's had been done in many countries since the nineteenth century (Havas 1991) and continued even after the early successes of the general theory of relativity (Kienle 1924), an important part of this now only historically significant work was carried out by Viennese scientists.

The main reason that so much of the Viennese contributions to relativity and gravitation is overlooked is that while they were important, no fundamental results were obtained that became part of the standard literature; while some of the researchers are very well known, their fame rests on work outside relativity (and incidentally carried out mainly outside Austria), like that of Erwin Schrödinger in Quantum Mechanics and of Philipp Frank in Philosophy of Science. Also, the major contribution of Viennese physicists to relativity was pedagogical, principally in numerous review articles in then standard *Handbücher*, which obviously have long been superseded by more up-to-date ones, but also by many surveys on more elementary levels. Furthermore while the *Anschluss* into a country which had suppressed relativity and had driven Einstein and many other relativists into exile would have killed off work in relativity in Vienna in any case, such work had essentially ceased years before, for reasons to be discussed later.

H. Goenner, J. Renn, J. Ritter, T. Sauer (eds.)
The Expanding Worlds of General Relativity
(Einstein Studies, volume 7) pp. 161–206

Figure 1. The Austro-Hungarian Empire 1867–1918.
German-speaking areas are indicated by cross-hatching.

To understand the central role of Vienna and of its university for theoretical research before 1918 several historical facts must be kept in mind:[1] Vienna was then the capital of a large multinational and multilingual empire; while since 1867 it was nominally a *Doppelmonarchie* with equality of the Austrian and Hungarian parts, in effect it was politically and culturally dominated by German-speaking Austria proper. A small part of Hungary and some significant parts of the other regions of the empire also had German-speaking populations, as shown in Figure 1, but its only German university of significance outside Austria proper was that of Prague,[2] the capital of Bohemia, which contained the German-speaking Sudetenland. It, just like the universities within Austria proper, was subordinated to and financially dependent on the Austrian Ministry of Education in Vienna. Within Austria proper, all universities outside Vienna were small provincial ones. The intellectual dominance of Vienna was such that almost all German-speaking

1 For recent histories of Austria see e.g. Kann 1974, Stadler 1971; a detailed history of the first republic is provided by Gulick 1948 and a brief discussion in the context of the natural sciences in that period is given by Broda 1981b. Broda 1979b outlines the historical neglect of the natural sciences from the time of the empire to the present.

2 In 1910, in the Austrian part of the empire, 35.6% of the population were German-speaking (27.6% in Moravia); in the Hungarian part there were 9.8% (Hantsch 1953). In the academic year 1913/1914, in the Austrian part there were 13 German universities, 6 Czech, 5 Polish, and one Italian commercial school. The only German universities of significance for physics outside Vienna were those of Prague and Graz. Although noteworthy work on relativity was done at the Polish Jagellonian University of Cracow (Šredniawa 1985, 1987, 1991a, 1991b, 1994), apparently there was essentially no interaction with the physicists of the Austrian German universities.

men desiring an academic career (no such chance then existed for women) longed for positions in Vienna and aimed at least at publishing lengthy treatises in the *Sitzungsberichte* of the Viennese Academy of Sciences (which, incidentally, are not widely available outside central-European libraries, one of the reasons for the neglect of Viennese contributions).

Figure 2. The Austrian Republic 1918–1938.

At the end of World War I, the empire collapsed and broke up into a large number of nation-states. Austria was reduced to a small part of its previous size as shown in Figure 2, losing even some contiguous German-speaking areas in the 1919 treaty of St. Germain; the German University of Prague continued to exist, but lost its ties to Vienna.[3] Vienna was reduced to the capital of a country of barely three times its own population, bereft of most of its previous industrial and agricultural base, its economy in shambles. It had barely recovered from a super-inflation when the Great Depression struck. The Austrian universities, as always state-owned and state-run, were permanently under-financed, and for most scientists academic careers within Austria were pipe dreams.

The way to such careers involved a number of steps after the doctorate, which was a must: 1. (neither necessary nor sufficient, but helpful) to become an *Assistent* to one of the professors holding a chair after obtaining the doctorate or the corresponding engineering degree, 2. (almost necessary, but not sufficient) to obtain the *venia legendi* at a university, the right to give lectures (unpaid) at that university and carry the title *Privatdozent*, through *Habilitation*; the submission of a lengthy original paper, the *Habilitationsschrift*, to the appropriate faculty and its acceptance after an oral examination, 3. (neither necessary nor sufficient, but helpful) to receive a *Lehrauftrag*, i.e., being assigned to teach a course (and being paid for it), 4. (again necessary, but not sufficient) to be considered a candidate

[3] At the Munich conference of September 1938 England and France, without the consent of Czechoslovakia, permitted Hitler to occupy the Sudetenland and incorporate it into the German Reich; in the spring of 1939, he occupied all of Czechoslovakia and established a "Protektorat" and an "independent" Slovak state. At the end of World War II, in retaliation for the barbaric treatment of the Czechs during the occupation, the newly established Czechoslovak government expelled almost all Sudeten Germans regardless of their political background; thousands were killed, the others were deported to Germany and Austria. Many details are given in Schechtman 1962, others have been reported only recently. The German University in Prague became Czech.

for a professorship once an appropriate vacancy arose (the creation of a new chair was almost unheard of); the lower rank was *ausserordentlicher* (a.o.) Professor, the higher one *ordentlicher* (o.) or *Ordinarius*, which also involved administrative duties. The faculty then submitted a list of the top three candidates ranked *primo*, *secundo* and *tertio loco* to the Austrian Minister of Education, who made the final decision. This decision was influenced by political and religious considerations as well as, all too frequently, by any existence of a family tie of a candidate to a current or previous holder of a chair.

These considerations were influenced by the fact that, except for the first two years after the collapse of the Habsburg empire, the Social Democratic Party, although representing almost half the electorate, was not represented in the government and thus the Ministry of Education, which controled the universities, both financially and through the power of appointments, was in the hands of a conservative, Catholic party, headed for many years by Ignaz Seipel, a prelate. Vienna, on the other hand, was dominated by the Social Democrats, who held a two-thirds majority, were in charge of elementary education, and had instituted a wide-ranging reform of the educational system. The Viennese intellectual elite was mostly of the left, with a significant number of Jews (who comprised more than ten percent of the population), and functioned in a comparatively areligious and tolerant atmosphere. None of this was reflected in the composition and spirit of the university faculties.

While step 2 described above allowed one to get a foot in the door at a particular university, this might well be the end of the line; it did not put food on the table and one either had to have independent means, be able to obtain a fellowship (frequently from abroad) or take a nonacademic position, most often as a high school teacher. Such a position frequently was too time-consuming to allow further research and thus ended any hope for further advancement. Therefore many promising young scientists chose to seek their fortune abroad, many years before any political, religious, or 'racial' persecution might have forced some of them to do so. Some went to Germany, others to England or the United States.

In this article, I shall concentrate on contributions in relativity and on related research on space and time, as well as on various aspects of gravitation by scientists who worked in Vienna or were at least educated there or otherwise closely associated with it. Also included are those by Einstein which resulted from visits and lectures in Vienna, from friendship or at least correspondence, and published debates with scientists there, as well as his occasional intervention in Austrian controversies, often nonscientific ones. Furthermore, in 1911/1912 he was a professor at the German University in Prague (which, incidentally, automatically made him an Austrian citizen for that period), and thus necessarily became involved in various Austrian academic problems.

Before embarking on a detailed discussion of the Viennese contributions, it should be noted that the experimental investigations of gravitation of most importance for relativity were carried out in neighboring Hungary by Roland Eötvös (1848–1918). His measurements establishing the universal proportionality of inertial and gravitational mass with extreme precision were carried out over three decades, starting in the 1880's; the first results were published in 1890 and the last paper appeared posthumously in 1922. But he also contributed many other important studies of gravitation. Except for one paper in French all his articles were published in German-language periodicals (a few also simultaneously in Hungarian) and thus were readily accessible to the Viennese scientific community. They were later collected in Eötvös 1953 together with a brief biography.

All publications related to our topic I am aware of are included, and the more important ones are discussed briefly, avoiding technical details as far as possible; familiarity with the ideas and results of the early papers of Einstein laying the foundations of the special and the general theory of relativity is assumed. Many physicists and their contributions to relativity deserve a more detailed treatment, which I hope to provide elsewhere, and many publications deserve to be read even today.

1. The relevant academic community of Vienna

To facilitate keeping track of the large number of scientists associated with the university or the Technische Hochschule ('Technical University,' in the following referred to as TH) of Vienna who either worked in areas of relativity or gravitation research or played a significant role in the education of these scientists, an alphabetical list of these scientists is given in Table 1. Unless designated by EP (experimental physicist), M (mathematician) or A (astronomer), they were all theoretical physicists (although Ernst Mach occupied the chair of Philosophy and Johann Sahulka that of electrical engineering). The dates given are those of birth and death and of the year of obtaining the Dr. phil. (equivalent to the Ph.D.), the *Promotion*, as well as of the Habilitation, if known, and of professorship (both only if in Vienna; roman type refers to the university, italic type to the TH). It is remarkable that with the exception of that of Rothe, who had a Dr. techn. awarded by the TH, and of Smekal, whose degree was earned at the University of Graz, every single doctor's degree was awarded by the University of Vienna, which on the one hand shows the high quality of the education there, but on the other, more importantly, the extreme reluctance of the Viennese establishment to offering a university chair to anyone not their own.

The most important theoretical physicist and influential educator in Vienna in the prerelativistic period had been Ludwig Boltzmann, who committed suicide in 1906. Another dominant influence was that of Ernst Mach, who had been professor of physics at the German University in Prague before occupying the chair of philosophy in Vienna in 1895. He had to relinquish that position in 1901 for reasons of health and lived in retirement in Vienna until 1913, when he moved to Bavaria. He remained active until his death there in 1916. (For Mach's and Boltzmann's influence on Einstein see Broda 1979c, 1981a.) Boltzmann's successor was the very gifted Friedrich Hasenöhrl, whose work anticipated some results of the special theory of relativity; he volunteered for front line duty in World War I and was killed in 1915. The most important mathematician at the university was Wilhelm Wirtinger, appointed in 1903. Some of the more important professors of

Name		Vital Dates	Year of Dr. phil. *Promotion*	Year of *Habilitation* in Vienna	Year of professorship in Vienna
Bauer, Hans		1891–1953	1916	*1925*	*1937–1953*
				1931	1940–1953
Beck, Guido		1903–1988	1925		
Behacker, Max		1885–1915	1912		
Bernheimer, Walter E.	A	1892–1937	1922	1928	
Boltzmann, Ludwig Eduard		1844–1906	1866	1867	1873–1876
					1894–1900
					1902–1906
Duschek, Adalbert Ludwig	M	1895–1957	1921	*1925*	*1936–1938*
				1930	*1945–1957*
Ehrenfest, Paul		1880–1933	1904		
Ehrenhaft, Felix		1879–1952	1903	1905	1920–1938
					1947
Flamm, Ludwig		1885–1964	1909	*1916*	*1919-1956*
Frank, Philipp		1884–1966	1906	1910	
Geiringer, Hilda	M	1893–1973	1917		
Haas, Arthur		1884–1941	1906	1912	1923–1936
Hahn, Hans	M	1879–1934	1902	1905	1921–1934
Halpern, Otto		1899–19 ?	1922		
Hasenöhrl, Friedrich		1874–1915	1897	1900	*1905–1907*
					1907–1915
Hepperger, Josef	A	1855–1928	1879		1901– ?
Kohl, Emil Viktor		1862–1924	1890	1903	1923–1924
Kottler, Friedrich		1887–1965	1912	1919	1927–1938
					1956– ?
Lampa, Anton	EP	1868–1938	1893		1904–1909
					1921–1934
Lense, Josef	M	1890–1985	1914	1921	
Mach, Ernst		1838–1916	1860	1861	1895–1901
Mayer, Walther	M	1887–1948	1912	1926	1931–1938
Meyer, Stefan	EP	1872–1949	1896	?	1908–1938
					1945–1947
Oppenheim, Samuel	A	1857–1928	1884	?	1911– ?
Rothe, Hermann	M	1882–1923	1909	1910	*1913–1923*
Sahulka, Johann	EP	1857– ?	1881	?	*1903–* ?
Schrödinger, Erwin		1887–1961	1910	1914	1956–1958
Schweidler, Egon	EP	1873–1948	1895	1899	1906–1911
					1926–1939
Smekal, Adolf Gustav		1895–1959	1917	1920	1927–1928
				1921	
Thirring, Hans		1888–1976	1911	1919	1921–1938
					1945–1958
Weitzenböck, Roland	M	1885–1955	1910		
Wirtinger, Wilhelm	M	1865–1945	1887	1890	1903–1935

Table 1. The relevant Viennese academic community.

The data is taken from various editions of *J. C. Poggendorffs biographisch-literarisches Handwörterbuch* and from published obituaries.

astronomy were Josef Hepperger, Samuel Oppenheim, and later Kasimir Graff, and of experimental physics were Anton Lampa (Professor at the German University in Prague from 1909 to 1919 between professorships in Vienna), Felix Ehrenhaft (baptized to obtain a professorship, dismissed in 1938 for being 'Jewish'), Stefan

Meyer (also dismissed as Jewish in 1933) and Egon Schweidler.[4]

A new generation of theoretical physicists and mathematicians, all with doctorates obtained between 1902 and 1917, included (in chronological order of *Promotion*): Hans Hahn, Paul Ehrenfest, Philipp Frank, Arthur Haas, Ludwig Flamm, Hermann Rothe, Erwin Schrödinger, Roland Weitzenböck, Hans Thirring, Max Behacker, Friedrich Kottler, Walther Mayer, Josef Lense, Hans Bauer, and Adolf Smekal. Behacker was killed in World War I (Frank 1915). Ehrenfest and Mayer, being Jewish, saw no chance of an academic career and left Austria; Mayer returned in 1926 with the help of Einstein. Weitzenböck became a professional officer before returning to academic life after the war (a professorship in Graz and later in Prague and then in Amsterdam). Frank became the successor of Einstein at the German University of Prague in 1913. Haas's unusual career will be discussed in detail in Section 3.2; he became a professor in Leipzig in 1913 and in Vienna in 1923. Flamm (who married Boltzmann's youngest daughter in 1920, which prompted Einstein to say that "one can now see clearly who among the new blood of the Viennese physicists is most lovingly concerned with Boltzmann's works," Thirring 1965) became professor at the TH in 1919 and spent his entire career there, as did the mathematician Rothe.

Although Schrödinger was considered for an a.o. professorship in Vienna in 1920, he preferred to accept a succession of appointments as *Ordinarius* in Germany and Switzerland, not believing that he had a chance to become one in Austria because of his lack of proper connections (Schrödinger 1985, Moore 1989). He later became Professor in Berlin; he left there after Hitler's rise to power and stayed in Oxford for three years. While there, he was awarded the Nobel prize in 1933. Nevertheless, in 1936 Austria only offered him a chair in Graz, which he occupied until the *Anschluss* of Austria in 1938. Thirring took the position in Vienna Schrödinger had declined and held the chair of theoretical physics until 1938, when he was fired because of his well-known pacifism. He stayed in Vienna and regained his chair in 1945.

Kottler, who had written to Einstein in 1920 that he despaired of an academic career in Vienna (EA 14-326), did become a.o. Professor there in 1923. Lense went to Munich in 1927. Bauer was a Dozent at the TH for 22 years, teaching there regularly before finally receiving a professorship; in all that time he had to make a living teaching at a local high school. Smekal briefly held a professorship in Vienna before moving permanently to Germany. Of the next group (Adalbert Duschek, Hilda Geiringer, Otto Halpern, Guido Beck, Paul Lazarsfeld) only Duschek's academic career was in Vienna; Halpern appears to have attempted

[4] Here and in the following, 'Jewish' does not necessarily refer to a person's religion. In pre-Hitler Austria, it was mainly religion which counted in official life (though not among *deutschnational* students), and some 'Jewish' university professors, such as Ehrenhaft, had agreed to be baptized to be 'acceptable' for appointment. But during the Hitler regime, one's fate depended on whether one fitted the definition of 'Jew' under the Nuremberg laws, i.e., on having three or four grandparents who were Jewish by religion; the person's own religion, unless Jewish, was irrelevant. Similarly, an 'Aryan' [one who had to carry an *Arierpass* (Aryan passport)] was anybody who could prove that he had four grandparents who were not Jewish by religion. These definitions were adopted in Nuremberg because the Nazis, in spite of all their efforts, failed to find a 'racial' definition of 'Jewish'.

Habilitation, but was forced to leave, mainly due to Smekal's anti-Semitism and personal antagonism (Beck 1967). He had a distinguished academic career abroad, as did Beck and Geiringer. None of the Austrian or ex-Austrian relativists active after World War II owed their research interests to any intellectual influence of the interwar academic life of Vienna anymore.

2. Special relativity: forerunners, foundations, applications, and generalizations

2.1. Forerunners

The outstanding younger Viennese theoretical physicist at the beginning of the twentieth century was Fritz Hasenöhrl. His main research interest was in thermodynamics. As many others at that time, he was concerned with the negative results of the Michelson-Morley experiment which was designed to show the effect of the motion of the earth through an ether assumed to be at absolute rest. In a lengthy study he showed that it was possible to design a cyclic process involving a body in motion which would lead to a violation of the second law of thermodynamics unless one introduced "a new hypothesis" and concluded:[5]

> Such a hypothesis, which also resolves this contradiction, is ... the assumption of Lorentz and Fitzgerald. Of course it is impossible to prove that this assumption is the only one which is able to restore agreement with the second law; but since it is also able to explain the result of the Michelson-Morley experiment, it deserves to be noted in preference to all other possible ones. (Hasenöhrl 1904a: 470)

In an early paper (Hasenöhrl 1902) he derived Lorentz's electromagnetic field equations for moving bodies under Lorentz's assumption of an ether at rest, but without assuming that these bodies were composed of ions. He also studied the problem of electromagnetic radiation in cavities in slow motion in a series of papers (Hasenöhrl 1904b–d, 1905, 1907b). In the second of these he concluded that "all bodies must have an apparent mass that depends on the temperature, which must be added to the mass in the usual sense" and there and in the fourth of these he showed that the radiation acted as if it had mass, and that "the ratio of this apparent mass to the energy of the cavity at rest equals $8/3c^2$." In the 1905 paper, based on an objection in a letter to him by Abraham, he corrected an error in calculation and obtained a value of $4/3c^2$. This, of course, was before the creation of the special theory of relativity and the recognition of the general relation between mass and energy by Einstein.[6] Hasenöhrl 1907a is the text of a lecture on the ether given by him to a non-specialist audience; while it notes all the difficulties associated with this concept and of the puzzle posed by the Michelson-Morley results, there is no

---

[5] All documents referred to in the present article are in German, unless otherwise noted or apparent from the title. All translations are my own.

[6] The factor $4/3$ is the same which bedeviled not only Abraham's theory of the electromagnetic mass of the electron, but mystified even many relativists up to the present, although it had been rectified already in Fermi 1922.

indication that he was even aware of Einstein's 1905 results which disposed of the notion of an ether altogether, and no mention is made of Lorentz and Fitzgerald's work which he obviously had been familiar with since at least 1904, only of unspecified hypotheses "not sufficiently founded" to "remove the contradictions completely". However, he later seems to have fully accepted Einstein's work. A detailed study of the thermodynamics of moving systems (Hasenöhrl 1907–08) as well as his earlier papers were the main basis of his review article (Hasenöhrl 1909) of the relation between inertia and energy. There he stressed that he wanted to deduce this relation insofar as it does *not* (his emphasis) follow from the principle of relativity, and to establish the complete identity of the results. While he mainly based the discussion on his own previous (slow motion) calculations, full credit was given to the exact results (Mosengeil 1906, 1907) obtained by Max Planck's by then deceased student. He also included a detailed discussion of the special theory of relativity in his many-semester course in theoretical physics in the 1912 summer semester; a full set of notes was taken by Hans Thirring, by then his Assistent, and is preserved in his *Nachlaß* (W35-197 and 198).

The negative results of Michelson's experiments prompted several theoretical investigations. Egon Oppolzer (1902) showed that astronomical observations severely restricted an explanation based on the earth's dragging along of the ether. Emil Kohl (1906, 1907, 1909) questioned the usual interpretation of the experiments by providing a detailed discussion of the propagation of electromagnetic waves in dielectrics which he thought to allow an explanation of the null effect.

2.2. Special relativity

Objections against the special theory of relativity were raised by the young Paul Ehrenfest (1907, 1909, 1912), who later became Einstein's best friend and was considered by him to be the closest in spirit; their relation is discussed in detail in M. Klein 1970: chap. 12, and the papers quoted in chap. 7. The objections in the first paper were answered in Einstein 1907. The second one thought to have established what became known as "Ehrenfest's paradox," a velocity greater than light in the rotation of a rigid body. Einstein responded in Einstein 1911 and showed that it was based on a misunderstanding of the Lorentz contraction.[7] Ehrenfest 1910 discussed a problem raised by a hypothesized dispersion of the ether.

The main contributors to the special theory of relativity were Philipp Frank and Friedrich Kottler, both then Privatdozenten at the university. Frank was a student of Hasenöhrl and received his doctorate in 1906, the year after Einstein's *annus mirabilis*. He immediately embraced the theory of relativity fully, as well as its four-dimensional formulation given by Hermann Minkowski (1908). In his first paper (Frank 1908a) he first of all introduced the concept (and, incidentally, coined the name) of the Galilei group in complete analogy to the Lorentz group, an analogy which the famous mathematician Felix Klein apparently recognized only much later (F. Klein 1927: 53–58). Frank then showed that if one transformed

---

[7] For a later analysis see Cavalleri 1968.

Maxwell's electromagnetic equations for a body at rest by the Galilei group rather than the Lorentz one as Minkowski had done, one would be led to Heinrich Hertz's equations for a moving body. In Frank 1908b he showed that whereas Minkowski's results did not fully agree with Lorentz's, the latter's derivation from his theory of electrons (Lorentz 1904) could be modified to agree with Minkowski's for unmagnetizable bodies. In an important, very detailed, paper based on his Habilitationsschrift (Frank 1909) he continued the investigations of his first paper with one on Newtonian mechanics, partly elaborating on an earlier study (Schütz 1897) which had shown that the basic equations of Newtonian point mechanics can be derived from Galilei invariance and the law of conservation of energy alone. His main result was that special relativistic mechanics can be established in complete analogy to Newtonian mechanics by keeping all definitions and assumptions, only replacing Galilei by Lorentz transformations. He then prepared a pedagogical article (Frank 1910, 1911b) in which he discussed the four-dimensional formulation of mechanics as well as of relativistic optics in an elementary fashion at some length. In Frank 1911a he showed that the Lorentz transformations are the only linear ones that leave Maxwell's equations invariant. In Frank 1911c, 1911d and 1912 he showed that the formulae of special relativistic dynamics can be obtained without recourse to the Lorentz transformation by a generalization of an axiom formulated in Duhem 1892: 304 ("neuvième convention"), requiring that if a mass point having a velocity u is given an additional velocity perpendicular to u, the additional work should be equal to that needed to bring the mass point from rest to the final velocity v.

Frank then collaborated with the mathematician Hermann Rothe (who had obtained his Habilitation the same year as Frank) on three papers. In the first one (Frank & Rothe 1910a) they considered transformations of space-time (using, apparently for the first time, the German unhyphenated term *Raumzeit*) coordinates that they required to form a one-parameter group, which should be projective and linear homogeneous. Requiring a generalized mechanics satisfying certain axioms, they were led to an infinity of possibilities and then studied the additional axioms needed to arrive at the special relativistic one. In two further papers (Frank & Rothe 1910b, 1911) they required that transformations involving the velocity as parameter should form a group and that any length contraction should not depend on the sign of the velocity. They showed that this would either lead to the Lorentz group or to two possible groups with zero contraction, the Galilei group and one not previously considered, which they called the Doppler group. Rothe did no further work in relativity; a textbook on tensor analysis was published posthumously (Rothe 1924) with an introduction by Weitzenböck.

In 1912 Einstein, then at the German University in Prague, was offered a chair at the ETH Zurich, which he accepted. A report by a committee concerning a successor to his chair (Einstein, Lampa & Pick 1912) suggested as candidates Frank, Ehrenfest, and Emil Kohl (an older man, a Privatdozent in Vienna since 1903, who had similarly been the third candidate when Einstein was chosen for the chair the previous year), in that order. It gave an extensive laudatory description of Frank's work in relativity and appears to have been written by Einstein, since

Pick was a mathematician and there is no discussion of any mathematical aspects in the report and Lampa (an experimentalist, who had left Vienna for Prague in 1909 and had also been on the committee recommending Einstein) had a somewhat limited understanding of relativity or exchange of ideas with Einstein, as he himself stated somewhat regretfully in a 1920 letter to Einstein (EA 15-054, published in part in Blackmore & Hentschel 1985: 81). Ehrenfest took himself out of the running, because he was *konfessionslos* ('without religious affiliation,' as Einstein too had been before coming to Prague); at the time a condition of appointment was such an affiliation and he, unlike Einstein, refused to return to Judaism (M. Klein 1970: 178).[8] Kohl apparently was not seriously considered, but obtained an a. o. professorship in Vienna a decade later. Frank received the appointment in Prague in September 1912 and stayed there until 1936, but remained in close contact with the Viennese physicists as well as later with the philosophers of the Vienna Circle, and therefore a discussion of his work on relativity (including aspects of philosophy of science, see Sec. 4) in Prague is included in this article.

In Frank 1917, completed in Vienna, a method developed by Sommerfeld and Runge to derive the theorems of geometrical optics by means of vector calculus was extended to the optics in moving bodies, including a derivation of Fresnel's results for aberration. Most of Frank's subsequent work in physics was pedagogical. Frank 1920a is a textbook of the theory of relativity, which was announced, but apparently was not published. Similarly Frank 1930a, an important nonstandard review of relativistic mechanics, based to a considerable extent on his earlier papers, was almost totally ignored, as was his review of celestial mechanics (Frank 1930b) in the same *Handbuch* of mechanics. Frank 1922 discusses the foundations of the special theory; Frank 1924a and 1924b are contributions to a standard *Handwörterbuch*. Frank also became the co-editor of new, entirely reworked, editions of the old Riemann-Weber *Handbuch* (originally based on Riemann's lectures) together with Richard Mises (Frank & Mises 1930–35). He was responsible for the second volume, which contained the articles on physics, but rather surprisingly did not include any relativity. By the time this volume appeared, Hitler had come to power in Germany, and several of its authors as well as Mises had emigrated. Frank, who was also Jewish, left for the United States shortly thereafter; there he no longer did any research in relativity.

---

[8] While the German University accepted Jews on its faculty already during the Monarchy (and Einstein disclaimed having been exposed to any prejudice, Illy 1979), there still existed a 'Gentleman's Agreement' even in the Czechoslovak Republic that if it was a Jew's turn to serve as *Rektor* he would refuse to do so (Beck 1967). The committee's report also noted that it was restricting its choices to Austrians and that all candidates were "of German nationality." While this nowadays would be considered as a code word for 'not Jewish,' this cannot be the case here, since two members of the committee and at least one of the candidates were Jewish. (More than half of the part of the population of Prague declaring itself German was Jewish. Georg Pick, like many thousands of them, was deported by the Nazis to the Terezin [Theresienstadt] Ghetto and perished there, Frank 1947, Pini 1974). It seems rather to have been meant to assure the faculty that nobody of Slavic origin would 'pollute' the German University, most likely at the suggestion of Lampa; he was co-publisher of the militantly pro-German journal *Deutsche Arbeit* and in 1919 he was the only professor at the German University who voluntarily resigned his chair, apparently because he found it impossible to accept Czechoslovak citizenship, and returned to Vienna without assurance of a position there (Kleinert 1976).

Although Anton Lampa was an experimentalist and had shown little interest in relativity while a colleague of Einstein in Prague, he appears to have become a late convert in Vienna. Just after retiring he published a theoretical paper fully based on Einstein's work (Lampa 1934) in which he studied the question of the measurement of the length of a moving rod, correctly taking account of the fact that, for a stationary observer at a point off the line of motion, in general the light observed is emitted at different times from the two ends of the rod. While probably most relativists were aware of the problem, this paper anticipated the published discussion of the visual appearance of moving bodies by several decades.

Another physicist working both in theory and experiment interested in special relativity was Adolf Smekal (for an outline of his career in Vienna giving details of some of the post-World War I problems of physics in academia see Flamm 1960). Working at the time in Schweidler's Radiuminstitut, he was apparently the first physicist to discuss possible relativistic effects in nuclear physics in several papers (Smekal 1920a, 1920b, 1920c, 1921).

In Prague, Einstein had become close to Georg Pick, who suggested to him that for the further development of his ideas to generalize the theory of relativity the appropriate tool might be Ricci and Levi-Civita's absolute differential calculus (Frank 1947). In Vienna, Friedrich Kottler had already used this calculus in his dissertation, completed and published in 1912 (Kottler 1912), in studying the world lines in Minkowski space. He also used it to extend the Lorentz group to motions with constant rectilinear or rotational accelerations; he applied the results to the model of a rigid electron in such motions (Kottler 1914a, 1914b) and extended the study in Kottler 1921b. His further work will be discussed in the next Section.

After World War I little original research on the special theory was carried out. A brief discussion of the foundations of the special theory, suggesting an alternate approach leading to the same kinematics, was given by Karl Wessely (1920–21). Bernheimer 1926 is a discussion of the ballistic theory of light propagation in view of astronomical observations. Thirring published a large number of short papers, sometimes rejoinders to objections, but containing little original work, on various topics in the special theory (Thirring 1921 c–f, 1924, 1925a–d, 1926a–b). He also wrote a short popular and widely read book on the theory (Thirring 1921b), which was praised by Einstein (EA 23-058), went through three editions, and was translated into English, French (by Einstein's friend Solovine), Swedish, and Japanese (Zimmel & Kerber 1992), an early review (Thirring 1922), and two articles in the *Handbuch der Physik* on the foundations of field theory and on electrodynamics and the theory of relativity (Thirring 1927, 1929a), for which he also edited some volumes. The *Handbuch* also contains a survey article on relativistic mechanics by Otto Halpern, written while he was an Assistent in Ehrenhaft's institute (Halpern 1927). He had started out as an experimentalist and had then switched to work in quantum theory; while the article was very competently written, he had never worked in relativity and this remained his only publication in the field. Thus his selection, probably at the suggestion of Thirring, is rather surprising, given the large number of physicists in Vienna and elsewhere who had done research in this area.

Early pedagogical expositions were provided by Viktor Lang (1911), a retired professor at the university, and by Ludwig Flamm (1914a, 1914b), Space and time were discussed in a popular article by Ernst Mach (1910) and a wide-ranging non-technical discussion of four-dimensional space was given in a book by Roland Weitzenböck (1929). Arthur Haas wrote a textbook of theoretical physics based on lectures given at the universities of Leipzig and Vienna and continuously updated (Haas 1919, reedited in 1923 and 1930) that appears to be the first general text that included a detailed discussion of the theory of relativity. He also published several brief notes linking relativity and quantum theory (Haas 1927a, 1927b, 1927c, 1928); this subject was also considered in Bauer 1928b. An introduction to the special theory for chemists was provided by Emil Abel, then a Dozent and later Professor for physical chemistry at the TH (Abel 1918). In addition, several Viennese physicists contributed numerous abstracts of articles and books on relativity to *Physikalische Berichte*, as well as a number of reviews of books on relativity in various journals. Of particular interest is Thirring 1923, a lengthy review of a booklet by the violently anti-relativistic Philipp Lenard (the future author of the infamous *Deutsche Physik*).

3. General relativity and gravitation

3.1. Gravitational research other than in general relativity

The earliest investigations relevant to this article were carried out by astronomers. Josef Hepperger (1888) studied the question of the velocity of propagation of gravitation, as did Samuel Oppenheim (1917) later. Oppenheim (1903) studied the validity of Newton's law of gravitation in a high school publication and surveyed problems in astronomy in a book (Oppenheim 1911) before obtaining his appointment in Vienna, and later contributed review articles (Oppenheim 1919, 1920a, 1923) to the *Encyklopädie der mathematischen Wissenschaften*, for which he also served as the editor of the volume on astronomy until has death, as well as a more popular review (Oppenheim 1920b). Bernheimer 1928 is a review article on the structure of the universe in a volume of the *Handbuch der Physik* edited by Thirring. Hepperger 1922 briefly discusses the deflection of light in gravitational fields.

A serious attempt at a (scalar) field theory of gravitation fitting in his general continuum theory was provided by Gustav Jaumann (1911 and 1912). While it accounted for the advance of the perihelion of Mercury by proper adjustment of arbitrary constants, it was only Galilei invariant and described the propagation of gravitation in analogy to thermal diffusion.

Jaumann was professor at the German Technische Hochschule in Brünn (Moravia, now Brno, Czech Republic). For a decade he had been associated with Mach in Prague, whose philosophical attitude, including his anti-atomism, he shared, and was joint author of an elementary text (Mach & Jaumann 1891), but in an 1894 testamentary memorandum he was disowned by Mach, who insisted that in case of his death Jaumann should not take over his institute (Thiele 1978: 229). He was author of a respectable body of work and forever hoped for an appointment in Vienna. In 1910 he was one of the three candidates for the chair of theoretical physics of the German university in Prague. Einstein had been listed first on the basis of his achievements; however, the *secundo loco* Jaumann was approached first by the Ministry of Education, probably because, unlike Einstein, he was Austrian. But he responded that if Einstein was proposed as first choice because of the belief that he had greater achievements to his credit, then he did not want to have anything to do with a university that "chased after modernity and did not appreciate true merit" (Frank 1947).

A study of free fall and of the motion of the planets in Nordstrøm's theory of gravitation was published by Max Behacker (1913). Another theory of gravitation, not based on any field concepts but on atomic collision processes, was developed by Johann Sahulka (1907) and given an elementary exposition in Sahulka 1909. Sahulka was professor of electrical engineering at the Vienna TH; although his attempt was somewhat primitive and might appear strange in the present era dominated by field theory, a similar, but more sophisticated and Lorentz-invariant approach was considered much later by Synge (1934).

3.2. General relativity

An important impetus for the development of gravitation research in Vienna was provided in 1913, when the annual meeting of the German society of natural scientists and physicians was held there, with many important theorists in attendance, and Einstein gave a lecture on the current state of the problem of gravitation (Einstein 1913), describing the theories of Abraham and Nordstrøm as well as his present version of the general theory of relativity (the *Entwurf* by him and Marcel Grossmann). In the course of his report he mentioned that Friedrich Kottler had already solved the problem of the effects of gravitation on electromagnetic phenomena by replacing the equations of the special theory of relativity by those following from the use of the absolute differential calculus (Kottler 1912), and asked him to stand up, since he had not met him (Frank 1947). Of course Kottler had not attempted to solve the reverse problem of the effect of the electromagnetic field on the gravitational one, however.

During the lengthy and very interesting discussion, Gustav Mie mentioned that Abraham had been the first to develop "somewhat reasonable" equations of gravitation; Jaumann, true to his nature, took offense again and published a rejoinder claiming priority (Jaumann 1914). A question by Reissner was misunderstood by Einstein, who later published a revised answer (Einstein 1914).

Continuing his previous work on accelerated reference frames, Kottler (1914b) studied such systems in free fall performing hyperbolic motion. Errata to this paper (Kottler 1916a) appeared two years later with apologies for the delay "due to service in the field." In August 1914 World War I had started and all younger physicists were drafted into or volunteered for the Austrian army, except for Flamm, apparently exempted for medical reasons, and Thirring, who as the result of a skiing injury was initially deferred (Moore 1989); however, he volunteered in 1915, but obtained a leave to pursue war research after a few months (Zimmel & Kerber 1992, Flamm 1958). Behacker, who had spent some time in Zurich to study with Einstein and had just been offered a position as Assistent by Frank, was killed (Frank 1915), as was Hasenöhrl. According to a brief curriculum vitae he sent to Einstein in 1938 (EA 14-329), Kottler, at the time Assistent to the experimentalist Lecher, served in the army from 28 July 1914 to 21 November 1918. Schrödinger served even somewhat longer, others for shorter periods.

Erwin Schrödinger had attended Einstein's lecture and learned about Einstein's latest work on general relativity in the field (Schrödinger 1985). When he was transferred to Vienna in 1917, he was able to exchange ideas with Thirring and Flamm; Kottler too was transferred there around that time to be attached to a technical military committee (letter to Einstein, EA 14-318). Continuing his work on accelerated reference systems, he deduced that motions he called "acceleration-relative," which may not be noticed by the comoving observer, form a one-parameter conformal group (Kottler 1916b). In a further paper (Kottler 1916c) which he considered provisional since he could not continue the investigation due to military service, he criticized Einstein's treatment of the principle of equivalence and claimed that Einstein had abandoned it. Einstein, while considering the criticism "noteworthy, because this colleague has really penetrated into the spirit of the theory" of general relativity, disagreed and stated that the conflict was due to a different understanding of the principle of equivalence and that he was sure that Kottler would realize that general covariance necessitated the abandonment of Euclidean geometry (Einstein 1916a).

Though still in the army, Kottler completed a lengthy paper on the foundations of Einstein's theory (Kottler 1918), in which he claimed an alternative derivation of Einstein's field equations. He based it on a modification of Newton's law of inertia, but required a restriction on allowed reference systems as well as to static fields. He sent a copy of the manuscript (submitted on 14 March) to Einstein in Berlin, who found it "hard to understand" and commented critically on various points in a letter which is not preserved, but to which Kottler responded at length on 30 March (EA 14-318). No further communication on this subject is preserved, and nothing further on it was published. He later contributed an excellent survey of the relativistic theory of gravitation as a continuation of Oppenheim 1920a to the *Encyklopädie* (Kottler 1921a), but wrote to Einstein that he was leaving a detailed exposition of general relativity to Pauli (EA 14-320). Subsequently he wrote two important articles on the role (or lack of it) of the metric in Newton's and Maxwell's theories (Kottler 1922a, 1922b) and a historical study on the principle of relativity (Kottler 1924).

An early contribution to general relativity was an investigation by Flamm (1916) of the geometry of light rays in the (just published) exterior and interior Schwarzschild solution. The mathematician Josef Lense, whose dissertation and first few publications actually were in astronomy, contributed a discussion of Newton's law in non-Euclidean spaces (Lense 1917) and a study of relativistic effects on the motion of moons (Lense 1918). Motivated by 'Mach's principle,' Thirring (1918a and 1921a) investigated the effect of distant rotating masses (represented by a spherical shell) on the motion of a mass point, and showed that it produced effects analog to centrifugal and Coriolis forces. In a famous joint paper (Lense

& Thirring 1918) the shell was replaced by a solid sphere and the effects on the motion of planets and moons was calculated. Thirring (1918b) noted the formal analogy of the differential equations of Maxwell's theory and the first-order approximation of Einstein's, a result frequently rediscovered. Thirring 1926c briefly reviewed new experimental results relevant for special and for general relativity.

Still during the war, Bauer (1918) and Schrödinger (1918) published critical remarks on Einstein's choice of the (nontensorial) components of the energy and momentum of the gravitational field. Schrödinger's paper was criticized by Einstein (1918a). Another paper on this subject was published by Wolfgang Pauli, a godchild of Mach, two months out of high school (Pauli 1919a), but already completely familiar with the special and general theory, which he had studied on his own. The next year he wrote two papers on Weyl's theory of gravitation (Pauli 1919b, 1919c). By then he had moved to Munich to study with Sommerfeld, who at first wanted only his assistance in writing a survey of the theory for the *Encyklopädie der mathematischen Wissenschaften*, but was so impressed by his expertise that he entrusted him with its sole authorship (Pauli 1921). A brief paper on a cosmological solution of the field equations given by Einstein was reinterpreted by Schrödinger (1918b); it was commented on critically by Einstein (1918b). He returned to cosmological problems years later, after he had left Austria (see Rüger 1988, Urbantke 1992), mainly inspired by Eddington's ideas. He also suggested later that the concepts of geometry are not applicable in the small (Schrödinger 1934).

Hans Bauer discussed introducing Einstein's principle of general relativity into high school instruction (Bauer 1920). He also gave lectures on general relativity at the Verein deutscher Mathematiker und Physiker at the university and expanded them into a book (Bauer 1922) which was intended to form a bridge between the many popular expositions and the highly technical ones. His research on Einstein's equations resulted in a very lengthy, regrettably almost totally overlooked, study of spherically symmetric fluid solutions with a linear equation of state (Bauer 1918b, 1918c) and later in a similar one of the general static spherically symmetric solution (Bauer 1928a).

Arthur Haas, a multiply gifted scientist and prolific writer, was equally interested in physics and in its history. In Haas 1910 he had been the first to apply quantum theory to the atom, working with the Thomson model, and to obtain a relation between the Rydberg constant, the electronic charge and mass, and Planck's constant. It differed from Bohr's 1913 result by a factor of 8 (easily overlooked because of the 1910 uncertainties in the values of the constants). The negative, even mocking, reception of his ideas by Hasenöhrl and other Viennese physicists left him very discouraged (Hermann 1965) and from then on his principal efforts went into his historical and pedagogical work. His Habilitationsschrift in 1908 was in the area of the history of physics, which was unprecedented; he was therefore asked to supplement it with a part in physics alone, and thus its acceptance was delayed. But the same year he had met and impressed Karl Sudhoff, Professor of history of medicine in Leipzig, who invited him in 1911 to accept a professorship there. It was unpaid, however, and therefore this tentative offer was combined with that of taking over as editor of the next edition of *Poggendorffs Biographisch-Literarisches Handwörterbuch* (covering the years 1904–1913). Haas delayed an answer until his Viennese Habilitation was approved in 1912 and Leipzig made the offer official the next year. In the meantime he had given a course in Vienna in 1912/1913 on history of physics and also had lectured frequently at the Volksbildungshaus Urania, one of several institutions offering lectures and courses for the general population; these lectures, including one on cosmology, were published in Haas 1913. In accepting the Leipzig offer he had to resign from the faculty of the University of Vienna, although he continued his course unofficially during vacation periods (Hermann 1965). In Leipzig he at first lectured on the history of physics in the nineteenth century.

At the outbreak of the war Haas was drafted, but in 1917 he was released "due to the importance of the *Poggendorff* undertaking for the war." He resumed lecturing in Leipzig, soon concentrating entirely on theoretical physics. The list of courses for the summer semester 1918 announced one on modern problems of theoretical physics, including relativity, which appears to be the first one in the physics department of any German or Austrian university that discussed the general theory apart from Einstein's own lectures. His courses formed the basis of the textbook already mentioned, which he considered to be his life's work, and which contained the first extended discussion of both the special and the general theory in a general textbook. Lectures on cosmology at the University of Vienna formed the basis of another book (Haas 1934); its first half was elementary, the second half, devoted to the relativistic theory, was more technical. Brief original contributions were notes on the possible connection of the gravitational constant to the constants of the theory of electricity (Haas 1918) and on the mass density of the universe (Haas 1930).

Since he still received no pay in Leipzig, Haas asked for and received leaves for the winter semester 1919/1920 and for the academic years 1920/1921 and 1921/1922 (always for one semester only) to stay in Vienna, where he still had some means, and where he continued to work on the *Poggendorff* until the expiration of the contract in 1921. He renewed his *venia legendi* that year and received an appointment as a.o. Professor in 1923. Apparently his leave from Leipzig was extended indefinitely, although no documentation has been located. He finally asked the Ministry of Education there to be relieved of his post (letter to the *Dekanat* of the Faculty of Philosophy of 3 January 1927), and the application was promptly granted by the Ministry (letter to Haas of 17 January).

The economic collapse and hyperinflation had cost Haas (as well as Kottler, EA 14-329) his inherited fortune, and therefore in 1925 he had been forced to accept a position as a minor official of the Viennese Academy. He also continued to lecture at the Urania; five lectures given there in 1919/1920, the fourth of which dealt with relativity, and which were repeated in 1920 in Leipzig, were expanded into a book (Haas 1920). He also lectured at University College in London in 1924, went on lecture tours in the United States in 1927 and 1931 and accepted a Visiting Professorship for the academic year 1935/1936 at Bowdoin College in Maine; his wife and two young sons soon joined him there and with Einstein's help (EA 12-008) he obtained a professorship at the University of Notre Dame in 1936.

General relativistic effects to order c^{-2} were considered by Leo Hufnagel, a student of the astronomer Oppenheim, in a dissertation devoted to the orbit of a particular comet, published in Hufnagel 1919, which concluded that the effects were too small to be observed. His later work, until his untimely death in 1933, was mainly concerned with observational astronomy. Another astonomical dissertation devoted to general relativity was written by Paul Lazarsfeld, who later became famous as a sociologist. The topic was suggested by Kottler, as acknowledged in Lazarsfeld 1926, the published version. It treated the advance of the perihelion of Mercury, calculated from the theory, and showed the invariance of the result in the first, but not in the next, approximation, a neglected result only rediscovered decades later. It appears that the only other Viennese dissertation devoted to general relativity in the period under consideration was Guido Beck's. It was published in Beck 1925 and contained among other results the first exact solution for energy-carrying gravitational waves. Although Beck was a student of Thirring, he appears to have worked entirely independently (Beck 1967). Thirring later asked him to contribute the article on general relativity to a volume of the *Handbuch der Physik* he was editing (Beck 1927). He only wrote one other article in this area (Beck 1926); his work is discussed in Eisenstaedt 1995 and Havas 1995.

3.3. Relevant mathematical work

The use of Riemannian geometry in the general theory had stimulated interest and work on various aspects of differential geometry among Viennese mathematicians. Walther Mayer, who had become Einstein's assistant in 1928, wrote a textbook of differential geometry with Adalbert Duschek of the TH (Duschek & Mayer 1930) as well as a paper on spaces of constant curvature (Duschek & Mayer 1931). With Einstein he published three papers on a new unified theory and two purely mathematical ones on semi-vectors (Einstein & Mayer 1930, 1931, 1932a, 1932b, 1934). These semi-vectors were used in a study of Dirac-type equations (Einstein & Mayer 1933a, 1933b).

Duschek also wrote the article on differential geometry (Duschek 1928) and Lense and Theodor Radakovic contributed the article on vector and tensor calculus and Riemannian geometry (Radakovic & Lense 1928) in volumes of the *Handbuch* edited by Thirring. The doyen of Austrian mathematicians Wilhelm Wirtinger weighed in with a lengthy proposal for a generalized infinitesimal geometry (Wirtinger 1922) which he apparently considered so important that he published it abroad in English, something quite unusual at that time. Its relation to experience is discussed, partly in a historical and philosophical context, in Wirtinger 1926. His student Hilda Geiringer discussed non-Euclidean geometry and the problem of space (Geiringer 1918). Roland Weitzenböck, whose numerous contributions to the theory of invariants were mostly written after he left Austria, wrote a study of the action principle in Weyl's theory as well as one of four-dimensional tensor analysis (Weitzenböck 1920–21, 1921).

3.4. *Volksbildung* and other popularizations

Anton Lampa had returned to Vienna in 1919. He had always been interested in adult education and devoted most of his energies to this area for the rest of his career, although he renewed his *venia legendi* in 1921 and was reappointed to a professorship the next year. From 1920 to 1922 he was *Referent* in the division for *Volksbildung* [adult education] of the Ministry of Education; in 1923 he became a member of the board of the Volksbildungshaus Urania, and from 1927 until 1934 he served as its president (Kleinert 1974). The Urania also had a planetarium and a small observatory in its centrally located building. Adult education was also provided within the Social Democratic Party as well as by the Volkshochschule ('people's university,' a name not allowed to be used officially under the monarchy) under the name of Verein Volksheim. It was founded in 1901 at the suggestion of a number of academics (including Mach) and others in Vienna's sixteenth district, a working class area, and soon expanded to five branches. (Its efforts were hailed in a letter addressed to it by Mach, 1911). By the school year 1920/1921 it offered 144 courses with 8079 participants (Fellinger 1969). A number of scientists lectured at these organizations both in general courses and on special occasions; Lampa gave a course on physics in the very first semester of the Volksheim in the spring of 1901.

Other organizations such as the Wiener Verein zur Förderung des physikalischen und chemischen Unterrichtes; the Verein zur Verbreitung naturwissenschaftlicher Kenntnisse in Wien, the Freie Vereinigung für technische Volksbildung, and academic clubs such as the Verein deutscher Mathematiker und Physiker and the Chemisch-Physikalische Gesellschaft similarly provided a forum for lectures and publications on relativity and related topics. In 1919, Fritz Beer gave six lectures on relativity at the Urania, designed to provide a popular, but scientifically accurate, introduction without the use of any formulas, which formed the basis of a book (Beer 1920) that went through at least seven editions and was also translated into Dutch. Another popular introduction was Quint 1922. Lectures by Hasenöhrl, Bauer, Flamm, Haas, and Sahulka were mentioned earlier. In 1921 Einstein gave a lecture under the auspices of the Urania at a larger auditorium on the general theory, which was reported at some length in the press (Anonymous 1921a). Other newspaper articles on the general theory were Flamm 1919, Frank 1920b and Thirring 1920 and 1929b.

4. Ernst Mach, the Vienna Circle, and other Viennese philosophers

Ernst Mach's ideas and investigations have been widely considered to be the spiritual precursors to the theory of relativity, including by Einstein himself. This relation has been put in doubt by Mach's apparent repudiation of the theory in his later years, in particular in the preface of his book on optics published posthumously (Mach 1921), allegedly written in July 1913. However, the authenticity of this preface has been questioned and the entire problem of Mach's attitude to relativity reexamined in great detail in Wolters 1987 and therefore this question will not be pursued here. Many of Mach's followers both in Austria and elsewhere, though not Jaumann, were friendly toward relativity. The attitude of philosophers of a number of schools everywhere, with incidental attention given to some Austrian ones, is surveyed in Hentschel 1990.

In 1922, Moritz Schlick was appointed to the chair in philosophy held earlier by Mach and by Boltzmann at the university. He had been a student of Max Planck in Berlin, had obtained his Dr. phil. in 1904, and then was Privatdozent at the University of Rostock and Professor at that of Kiel. He had written a lengthy article on the philosophical significance of the problem of relativity (Schlick 1915) and had sent the manuscript to Einstein for his opinion (EA 21-568), who responded that "it belongs to the best which has been written about relativity until now. From the philosophical side apparently nothing at all has been written on this subject approaching it in clarity" (EA 21-611). An article and a book on space and time in contemporary physics followed (Schlick 1917a,b), as well as an article on philosophy and the theory of relativity (Schlick 1922) the year he went to Vienna. The book went through several editions with additional material. Einstein and Schlick carried on an extended correspondence until Schlick's death in 1936, some of which has been discussed in Howard 1985 and Hentschel 1986.

Schlick's presence in Vienna attracted Herbert Feigl, then a student in Munich and discouraged by the prevailing anti-Semitism there. In 1922 he had written "a prizewinning (but unpublished) long essay on the philosophical significance of the theory of relativity," as he wrote later (Feigl 1974), which appears to have been lost. He became Schlick's student and also took courses with Thirring, the mathematician Hans Hahn, and others, and obtained his Dr. phil. in 1927. According to his recollections, he and his older friend Friedrich Waismann suggested to Schlick the formation of an evening discussion group. "Philosophically interested mathematicians and scientists, together with mathematically and scientifically trained philosophers, were to discuss issues in the logical and epistemological foundations of mathematics, physics and occasionally also of biology, psychology and the social sciences" (Feigl 1974). This was the beginning of the 'Vienna Circle.' It included Hahn, the sociologist and economist Otto Neurath, Philipp Frank, Rudolf Carnap (who had written two important studies on space and time, Carnap 1922, 1925), Kurt Gödel, Hans Thirring, Edgar Zilsel, and others. Frank 1923 gives a good picture of the prevailing attitude among this group in the early days. A much later contribution (Frank 1947) discussed the relation between Einstein, Mach and logical positivism in detail.

A discussion of Schlick's and the Vienna Circle's philosophy and impact is outside the area of this article. It is relevant, however, that "the Vienna Circle accepted the relational view of space and time, and the kinematic relativity of motion ... influential ... was the critique of the ether concept in Einstein's special theory." (Feigl 1969). The Vienna Circle expanded its area of influence beyond its professional kernel by founding the Verein Ernst Mach in 1928, with Schlick as its president, which initiated a series of public lectures as well as a series of publications and took over the journal *Annalen der Philosophie* under the new title *Erkenntnis*. Its origins and history are described in great detail in Stadler 1982. Among the lectures given only that of Wladimir Misar (Professor at the university) on "Problems of Astronomy" was concerned with general relativity; however, many members, Feigl and Zilsel in particular, also frequently lectured at the Volksheim. Zilsel (1927) provided an axiomatization of the time concept and criticized Reichenbach's relativistic axiomatization (Reichenbach 1924); a discussion of the time concept is also included in Schlick's article on causality (Schlick 1931).

Among the publications of the Verein's members, those of importance by containing discussions of relativity are Zilsel 1929, Frank's contribution on foundations of physics to the *Encyclopedia of Unified Science* (Frank 1939) and his discussions of causality in a book (Frank 1932) and in an address (Frank 1929) to the "1. Internationale Tagung für Erkenntislehre der exakten Wissenschaften" held in Prague, but organized by the Verein. Frank's book was discussed at length in the theoretical organ of the Social Democratic Party (Neurath 1932).

The Verein as well as its precursors and several adult education organizations were strongly influenced by the Social Democratic Party, although the Volksheim remained strictly neutral politically and Schlick himself was a conservative. This party was Marxist and generally sympathetic toward science. Its most prominent philosopher was Max Adler, who was a neo-Kantian, and one of the few Social Democrats and Jews who had been able to obtain a professorship at the university. He had lectured on and written a study of Mach and Marx (M. Adler 1913: chap. 9). Friedrich Adler (no relation), a fellow student of Einstein in Zurich, was a Machian and had written a number of articles on philosophical and scientific subjects, an obituary (F. Adler 1916) and later a book summarizing Mach's ideas (F. Adler 1918). He was the son of Victor Adler, one of the founders and the leader of the party until his death in 1918, and became the secretary of the party in 1911. His relations with Einstein (also described in Kann 1977) and his attitude toward relativity will be discussed later.

It is not surprising that, in the difficult years after the war, beliefs in the occult, in overcoming gravity by levitation (especially the exploits of a ten-year old, Willy Schneider, a native of Braunau, the birthplace of Hitler), and in other mystic phenomena, flourished in Vienna as elsewhere among all classes. This prompted Adler to write a two-part article in the party newspaper *Arbeiterzeitung* castigating the *Lasterhöhlen des Geistes* [dens of vice of the spirit], and quoting extensively from Mach's study of the "sense of the miraculous," contrasting the mindless belief in alleged miracles with the appreciation of the real wonders of nature (F. Adler 1924). While similarly critical of and actually exposing some fakes, Thirring, however, maintained the possibility of certain psychic phenomena he had been unable to explain, specifically including Willy Schneider's, some of whose "telekinetic" exploits were examined at the Physikalisches Institut of the university (Anonymous 1925). In 1927 he became the founding president of the Österreichische Gesellschaft für Psychische Forschung (later ... für Parapsychologie) and as late as 1964, in a five-page response to an inquiry by the Austrian Broadcasting Company, defended his belief in such a possibility (letter of 10 February 1964, Thirring *Nachlaß* B35-1461); he also mentioned that the board of the Gesellschaft had included, among other professors, Hans Hahn and Anton Lampa.

A number of well-known philosophers active in Vienna at the time, such as Friedrich Jodl, Karl Popper, Josef Popper-Lynkeus and Ludwig Wittgenstein, did not contribute to the discussion about relativity. A more obscure one, Franz Selety, had written a "critique of time" as a twenty-year old student (Selety 1913), which was critically reviewed in Willy 1914. After obtaining his Dr. phil. (under his original name of Jeiteles, which he changed officially in 1918) in 1915, he continued to take courses at the university and became interested in cosmology. He discussed his views on philosophy and cosmology in 1917 in two of the longest letters ever written to Einstein (EA 20-473 and 20-474). Unfortunately Einstein's answers are lost; they seem to have been fairly detailed and positive, judging from his comment (quoted in Selety's response, EA 20-475) that Selety's "report ... is of such clarity that I would regret if it would not be accessible to other friends of epistemology." Selety in turn asked for permission to publish Einstein's letters in a book he was working on, which was granted (EA 20-476). This book was apparently never completed. A long article on cosmology (Selety 1922) was commented on by Einstein (1922), to which Selety responded in Selety 1923a. A further paper on cosmology was submitted in June of the same year (Selety 1924), discussing Mach's principle among other topics. Shortly thereafter Selety went to Paris, where he published two further notes on cosmology in July (Selety 1923b, 1923c), which were presented to the French Academy by Émile Borel. A month later he disappeared, presumably victim of a (mountaineering?) accident.[9]

Another obscure philosopher (about whom I was not able to obtain any biographical information) was Heinrich Zlamal, who had planned an extensive study of "the relation of the theory of relativity to the aprioristic study of nature;" however, only the first of the planned four parts, on "the six possible interpretations of the theory of relativity before apriorism," critical, but not hostile, was published (Zlamal 1924), and then the project was abandoned, possibly for financial reasons. Another study was published a few years later (Zlamal 1932).

5. Anti-relativism in Vienna

In Germany there had been extensive opposition to the theory of relativity, both among scientists and non-scientists, in publications as well as some public protests, more and more tinged with anti-Semitism (Goenner 1992a, 1992b, 1993). In Austria the only (ex-)scientist opposing the theory in print and in technical terms was Friedrich Adler, who was a friend of Einstein and himself 'Jewish,' though baptized at birth. He had given up physics when he returned to Vienna from Switzerland to become secretary of the Social Democratic Party. The party had been opposed to all wars until the outbreak of World War I, but then, just like its German sister party, supported the war effort. Adler and a few other leaders remained opposed; Adler resigned as secretary in protest. All his protests against the war were fruitless, and in 1916 he decided that there was no legal avenue left to fight against it in parliament or by other public means. To give himself

[9] I am indebted to Dr. József Illy for biographical information on Selety.

a forum for public protest at his anticipated trial, he decided on and carried out the assassination of the then prime minister, Count Stürgkh, in November 1916.[10] This deed and his speech at his trial before the Ausnahmegericht had an enormous effect on the public and on the resistance to the war, as he had hoped. Einstein, however, showed little understanding of his motivation.

During his incarceration Adler wrote a book on Mach, mentioned earlier, and a criticism of the theory of relativity, which was published only after the war (F. Adler 1920). While writing the book, he corresponded at length with Einstein on various technical points, but remained unconvinced. He never returned to science after regaining his freedom at the collapse of the Empire, devoting himself entirely to politics, first in Austria and then as secretary of the Socialist International.[11]

While there was considerable anti-Semitism in Austria (protested against among others in a newspaper article by Mach, 1907),[12] and personal attacks on "the Jew Einstein" were frequent in Viennese right-wing newspapers in the twenties, these attacks never attained the intensity of the German ones. It should also be noted that, except for Frank who had left Vienna in 1912, none of the early workers in relativity was Jewish, contrary to a wide-spread impression even among scientists. When in 1922 right-wing students tried to stop Jewish students from attending classes at the university, Thirring posted a notice that he would not lecture if anybody was prevented from attending (Beck 1967). There are no published attacks against either Einstein or relativity by anybody in academia in Vienna. One can only speculate what would have happened if Philipp Lenard, who was born in Pressburg (Bratislava, the capital of Slovakia) a few dozen miles downstream from Vienna in 1862 and was briefly a student at the Vienna university, had stayed there.

The only professional philosophers who published criticisms of the theory of relativity in Vienna, both from outside Vienna, seem to have been Oskar Kraus, a professor in Prague who had published numerous articles there, and Josef Kremer of Graz who there had written a book critical of the theory (Kremer 1921). In Vienna, they published newspaper articles in the conservative press (O. Kraus 1920, 1923; Kremer 1922), and Kraus addressed "open letters" to Einstein and Laue in a booklet (Kraus 1925).

A most peculiar author, Th. Newest (a pseudonym for Hans Goldzier (1861–?), with both names listed in his books), who had published a series of booklets over the years on "world problems," attacking most established science, wrote a virulent

---

10 Ironically, and apparently unknown to Einstein, it was Stürgh, as the Minister of Education, who had approved his appointment in Prague.

11 Adler's first secretary at the International later married Pauli and Adler was a frequent guest at their house. In my interview with her it turned out that he had so completely withdrawn from science that in spite of the many conversations with Pauli there she did not realize that he had ever studied physics.

Although there have been several books and many studies of Friedrich Adler, none except an interview with Einstein (Joel 1917) are due to physicists, and most of them consciously avoid considering the scientific side of Adler; many also have a limited understanding of the political atmosphere of the early twentieth century in Austria. I hope to discuss Adler and his relation with Einstein in a future monograph.

12 In a report of the police to the Statthalterei in Prague of 8 August 1895 Mach had been listed as one of the open opponents of the anti-Semitic movement among the students there, according to Molisch 1939: 128, a book written from the Nazi point of view.

attack on Einstein and his theories (Newest 1921), and incidentally also Mach and Newton; it was prompted, as he stated in the almost hysterical preface of his book, by the fact that the Urania (most likely Lampa) as well as an unnamed scientific club had refused him their facilities for lectures "against Einstein," in response to pro-relativity lectures, possibly one by Einstein himself, or more likely those by Fritz Beer (see Section 3.4). He stated, however, that "mathematics had always remained my weakest side, therefore I am not able to verify the formulae and concede their absolute correctness to my adversaries" (p. 63), and simply argued that Einstein's ideas were absurd. Goldzier appears to have been an engineer and the owner of a printing shop in Vienna (Hamann 1996); his early antiscientific tracts left a deep impression on the young Hitler, then living in Vienna, who regurgitated some of Goldzier's ideas decades later in conversations with a confidant, Otto Wagener (Turner 1978).

Arguments as simpleminded as Goldzier's were presented in Weinmann 1922, 1923 and Mitis 1930, 1931a, 1931b. While these and other attacks on Einstein and his theories came from the right, another critical book, Eidlitz 1931, carried a dedication "To the Soviets." I was unable to unearth any biographical information on these three authors.

Attacks against Einstein and his theory as Jewish prompted an article by a Dr. Bruno Rosenberg in the first issue of a newsletter of the Verband nationaldeutscher Juden, an insignificant, clearly right-wing, Jewish organization, which was quoted at length in Anonymous 1921b, objecting against Jews who admired Einstein and defended the theory just because he was Jewish, without any understanding of it.

6. **Einstein and Vienna**

6.1. Visits and lectures

Einstein visited Vienna four times. In late September 1910 he went there apparently to discuss his appointment in Prague with the ministry of education and with Lampa. During that time he also visited Mach and Victor Adler, the father of his friend Friedrich. Both the timing and the purpose of his visit are discussed in some detail in Wolters 1987. He also wrote a brief statement about his visit to Adler on the occasion of a planned commemorative volume, which was only published many years later.

His next visit was on the occasion of his 1913 lecture, which was discussed in Section 3.2. After the war and the widely-heralded 1919 expedition led by Eddington to study the solar eclipse, which provided a test of the prediction of the general relativistic bending of light and made his name a household word, he was invited repeatedly to lecture in Vienna. In December 1919 Felix Ehrenhaft, at that time a.o. Professor, invited him in the name of the Chemisch-Physikalische Gesellschaft, which he headed at that time, to give a lecture as well as to be his guest, also mentioning that Lampa had returned to Vienna and would be at his disposal (EA 10-349). Einstein declined due to ill health and his mother being terminally ill, and suggested Pauli (then a first-year student in Munich), Schrödinger, and Thirring as possible substitutes (EA 10-352). A year later he accepted the invitation to lecture in January 1921 under the auspices of the Urania and of the Chemisch-Physikalische Gesellschaft (EA 10-368), and stayed with the Ehrenhafts (Frank 1947). He first gave a technical talk for scientists and invited guests at the Gesellschaft and two days later a popular one for the general public in the packed Grosses Konzerthaussaal. This talk was reported at length in the *Arbeiterzeitung* (Anonymous 1921a); a favorable report was also published a few weeks later in the conservative *Neue Freie Presse* by "F. B." (probably Fritz Beer), who also told of a longer meeting with Einstein "in the smallest circle," possibly at the Ehrenhafts (Anonymous 1921c). A report by a member of the audience is quoted in Broda 1979a.

Ten years later Einstein was invited by an obscure Komitee zur Veranstaltung von Gastvorträgen ausländischer Gelehrter der exakten Wissenschaften [Committee for arranging guest lectures of foreign scholars in the exact sciences], whose five members included Ehrenhaft, Lampa, and Hahn. He at first declined rather brusquely (EA 10-388), but, after receiving a pleading five-page letter by Ehrenhaft clarifying the situation in academia which Einstein had objected to (EA 10-389), he accepted and again stayed at the Ehrenhafts. His talk at the 1. Physikalisches Institut of the university on 14 October was reported at length the next day (Anonymous 1931); it appears to have concentrated on cosmology, in particular on the observations establishing the expansion of the universe, the Friedmann solution, and the five-dimensional theory of Einstein and Mayer. Conversations with Ehrenhaft were summarized in a letter (EA 10-394) at Ehrenhaft's request, which concisely described the main advance due to the theory of relativity as a "modification of the theoretical basis which decreases the number of logically independent premises," giving as examples the unification of electric and magnetic fields and of the terms involving spatial and temporal derivatives in Maxwell's equations in the special theory and his connecting inertia, gravitation, and metric in the general one.

After the spontaneous demonstration of 15 July 1927 by the Viennese workers protesting the acquittal of right-wing assassins, in which 89 persons were killed by the police (Gulick 1948) and the country was brought close to civil war, the government turned more and more to the right. This was reflected by the fact that Einstein's 1931 visit was officially ignored, as was stressed in a confidential report by the German embassy to the German foreign ministry (Broda 1979a). It stated that "it is characteristic for the way in which everything is treated in Vienna from a party political perspective that the official Austrian authorities showed special restraint toward Professor Einstein because he is a Jew and is considered to be on the left politically. Neither the Minister of Education nor the Rectors of the universities attended his talk and ... he was not received or invited by any official authority. This must have been even more noticeable since Prof. Picard, who was in Vienna at the same time, was received by the country's president. The Minister of Education, the President of the Academy of Science, and the Rector of the University contented themselves with accepting an invitation to a breakfast I gave today in Prof. Einstein's honor and to greet the great scholar at this opportunity ... " It also added that some of the reticence might have been due to the fact that Einstein's host (Ehrenhaft) was out of favor at the Ministry of Education, that the papers of the left had properly criticized the ignoring of Einstein's visit by official Austria, and that those of the right hardly took notice of his visit. It is remarkable that while there were many more public attacks on Einstein in Germany than in Austria, in this instance the official attitude of the authorities toward him was much more proper from the German than the Austrian side.

6.2. Other professional contacts

In a search for a successor to Hasenöhrl after his death in 1915 a commission had recommended Marian Smoluchowski in November, but he did not receive an absolute majority of faculty votes a month later. An expanded commission suggested the additional candidacies of Einstein, Max Laue, and Arnold Sommerfeld. Einstein was nominated *primo loco* in a faculty meeting of 11 March 1916. However, his name seems to have been withdrawn from consideration for unknown reasons soon thereafter, since in a letter of recommendation two months later, Planck only mentioned Laue, Sommerfeld, and Smoluchowski.[13] Smoluchowski died unexpectedly the next year, and the chair went to Gustav Jäger in 1918 and to Hans Thirring two years later.

The experimental chair had been occupied by Franz Exner, who retired in 1920. Both Jäger and Ehrenhaft wanted this chair. Ehrenhaft had done pioneering work on Brownian motion in gases and comparison with the theories of Einstein and Smoluchowski, and more recently on photophoresis. He had also tried to measure the charge of the electron by methods similar to those of Millikan, the famous oil drop experiments, but had believed that he had found various charges significantly smaller than the unique value found by Millikan. He became embroiled in a controversy about the supposed existence of "subelectrons," which lasted until his death.[14] In January, Anton Lampa cabled Einstein, asking him to send a report on Ehrenhaft's qualifications to Professor Wegscheider. Einstein responded immediately, asking for complete confidence (EA 10-354). He praised Ehrenhaft's originality and energy, but expressed strong disagreements with his interpretation of his measurements on electric charges. Nevertheless he considered him to be the most qualified Austrian experimentalist.

The chair went to Jäger. However, Einstein was approached again in June in a long letter by Thirring, which was countersigned by Flamm and Smekal and was also sent to Planck (EA 10-361). It expressed unhappiness about Ehrenhaft having pushed his candidacy in unorthodox fashion and not having been satisfied with suggested arrangements for giving him more space. It continued that "on his instigation a project was developed by the office for education, which in the opinion of the signatories, who by the way are speaking in the name of about 30 Austrian physicists, would have catastrophic consequences for Viennese physics.

---

13 I am indebted to Dr. Robert Schulmann for this information.

14 Details of the experiments of Millikan and Ehrenhaft and the subsequent controversy are discussed in Holton 1977. No mention is made, however, of the criticism of Ehrenhaft's work by several of his assistants, made after they left his institute.

It consists in abolishing the Institute for Theoretical Physics as such and to give it to Ehrenhaft together with part of the Second Phys. Institute as the Third Phys. Institute. The chair for Theoretical Physics would then be occupied by an *Extraordinarius*, who would have one or two rooms in the Second Phys. Institute at his disposal ... " Einstein responded immediately (EA 10-362), stating first that he had not spoken in the "delicate Ehrenhaft question" on his own initiative, but that he found "that the faculty had found a very fortunate solution ... without the danger of too great an influence of his personality and direction." He again distanced himself from Ehrenhaft's interpretation of his experiments, and wrote that it would be "particularly bad if the Viennese theoretical physicists would be restricted in their possibilities" and that their scientific weight surpassed that of the experimentalists.

At that time the Social Democrats were still in the government, and Ehrenhaft got his institute. The controversy also had an effect on Kottler, who wrote Einstein in August (of course ignorant of Einstein's role) that the Viennese physicists almost unanimously had fought Ehrenhaft's candidacy and that finally "it came to a sort of plebiscite ... where I was forced to take a position ... in writing that I considered Ehrenhaft to be equally qualified as the other Austrian physicists under consideration. All others voted not qualified." Kottler felt that in consequence of this he was passed over several times in favor of others of his age (EA 14-326).

In June 1917 Einstein and Wander de Haas (Lorentz's son-in-law) were awarded the A. Freiherr von Baumgartner Prize, a cash award, from the Academy of Sciences for their gyromagnetic experiments.[15] Einstein was never elected to membership, however.

Einstein, who had written an obituary of Mach (Einstein 1916b), was asked by Schlick to write a brief note on the occasion of the 1926 dedication of a monument to Mach in the Vienna Rathauspark (EA 21-592) adjoining the university. He responded with Einstein 1926, which was acknowledged with thanks by Schlick (EA 21-593, 21-594).

Einstein's published interchanges with Viennese physicists were discussed in Sections 2 and 3, as was some of the unpublished correspondence with physicists and philosophers in Sections 2–4. He also maintained contact with his former colleagues in Prague, Frank and Lampa, wherever they were, and had an extensive correspondence with Ehrenfest (one of the select few with whom he was on the familiar *Du* terms, although they remained on a last name basis to the end). Almost all other correspondence was initiated by the other side. It is remarkable that Einstein, in spite of all his duties and research efforts, almost always responded rapidly and sometimes at some length. Unfortunately, a number of his letters have not been found, and some of the letters to him also have been lost.

6.3. Political and other contacts

Although Einstein showed no understanding of the political motivation of Friedrich Adler's assassination of Count Stürgkh (as for example displayed in a letter to his

[15] The experiments are discussed in detail in Galison 1982.

friend Besso, EA 6-030), he immediately volunteered to be a character witness for him; he was not called to testify, however. He maintained a scientific correspondence with Adler during his imprisonment, as noted in Section 4.1. After Adler's release there was little correspondence because of Adler's complete immersion in the politics of the early postwar days; among other activities, he was instrumental in 1919 in stopping the Austrian workers' councils from joining the Communist regimes in neighboring Bavaria and Hungary (Gulick 1948). However, they remained in touch and Adler visited Einstein in Princeton during World War II.

After the defeat of the Hungarian Communists, the people's commissars were put on trial. Einstein was asked by Adler and others to join in an international appeal against the death sentence and for amnesty. He signed this appeal, together with Heinrich Mann, Sigmund Freud, Henri Barbusse and many others, which was published widely in December 1920, including the *Arbeiterzeitung* of the 15th. He was also approached by Jolán Kelen, then in Vienna, the wife of one of the commissars, who was a scientist. She asked him via Paul Hertz (EA 12-210, 12-211) to intervene. Einstein wrote a letter of support, which according to her letter of thanks (EA 12-215) was published in the *Neue Freie Presse* in Vienna with changes made by her without authorization. I have not yet been able to locate the published version, the version published in one of her books in Hungarian translation (Kelen 1976) is one she must have memorized since she wrote that the original letter was taken from her when she was incarcerated by the post-World War II Communist regime in Hungary. Her husband was condemned to life in prison, but with other commissars was exchanged to the Soviet Union, where he perished in the Stalin purges. In spite of these experiences she remained a Stalinist to the end.

According to K. Kraus 1929, quoting the *Rote Hilfe*, Einstein also joined an appeal by Karl Kraus (an influential Viennese writer and journalist) and Henri Barbusse to the Austrian Ministry of Justice against extradition of the political refugee Mavrak, who was subsequently only expelled to Germany.

In less political interventions, Einstein expressed support for the Freie Vereinigung für Technische Volksbildung (Einstein 1920) at the request of its president, Wilhelm Exner. He also agreed in 1921 to support a plan by Otto Neurath (EA 18-396) for a popular series of booklets, which does not appear to have come to fruition. Anonymous 1930 reports on a broadcast by the Vienna radio of talks by Bernard Shaw and Einstein; this appears to have been a rebroadcast of a rather general talk given by Einstein in London to a Zionist organization.

7. **Epilogue**

7.1. AUSTROFASCISM

In January 1933 Hitler was designated Chancellor by Hindenburg, the President of Germany. Einstein happened to be abroad; he was deprived of German citizenship and never returned to Germany. Accused of *Greuelpropaganda* [atrocity propaganda], he resigned from the Prussian Academy of Science in protest. The

Arbeiterzeitung, which had protested strongly against the *Kulturpolitik* of the Third Reich, including its treatment of Einstein (Anonymous 1933), printed his long letter of resignation in full on 13 April. Thirring immediately wrote a letter to his friend Schrödinger, then still a member of the Academy, and protested in the strongest terms against the behavior of the Academy (EA 23-061). A Viennese weekly dispatched a well-known journalist to Belgium for a long interview with Einstein (Lania 1933).

Making use of the reactionary situation in Europe after Hitler's rise to power, Engelbert Dollfuss, then the Chancellor of an Austria hemmed in between fascist regimes in Germany, Italy, and Hungary, in 1933 used patently illegal technicalities to stop the Austrian Parliament from meeting and to dissolve the Supreme Constitutional Court. The activities of the Social Democratic Party and its various cultural organizations were severely restricted, as were those of the Socialist city government of Vienna. On 12 February 1934 fighting broke out in Linz, where the Socialist workers resisted a search of the house of one of their organizations, and immediately spread to Vienna and other cities. Within a few days the government had triumphed, and immediately dissolved the Social Democratic Party and all its cultural organizations as well as the Vienna city government. One of the victims was the Verein Ernst Mach, which was ordered dissolved on the 23rd. Its president, Moritz Schlick, protested in a lengthy letter presenting his own conservative credentials, but to no avail (Stadler 1982). Most of the prominent members of the Verein had left Vienna earlier or did so within the next few years, mostly managing to relocate in the U.S. Hans Hahn, who had been the founder of the club of Socialist university professors, died in 1934. Schlick himself was murdered by a former student in June of 1936, allegedly for personal reasons. However, political motivations or instigations might have been behind it; in spite of martial law, the assassin was tried in a normal court in 1937 and, although sentenced to ten years in prison, was released the next year, a few months after the *Anschluss*, and then claimed political motives. Schlick had been widely considered as Jewish because of his name, although he was not; he had been christened Moritz in honor of one of his ancestors, Ernst Moritz Arndt, a nationalist icon. The court also excused a witness who might have shed light on possible instigators behind the murder (Siegert 1981a, 1981b; Cless-Bernert 1982a, 1982b).

While philosophy of science had flourished until then, research in relativity had stopped a decade earlier. Mayer had followed Einstein to the U.S., where Haas had settled before. Kottler had gone into optics around 1923, and most other early workers went into quantum mechanics or ceased any creative scientific activity, whether due to economic reasons or to burnout. Much of the responsibility for the lack of any research in relativity at the Institute for Theoretical Physics after 1925 falls on Hans Thirring, its head. He had made important pedagogical contributions, especially in the *Handbuch der Physik*. But as he himself wrote, after 1918 "apart from a few articles of a critical nature I have not accomplished any scientific achievement of any note in the area of my nominal subject, theoretical physics" because he thought that his real importance was in the area of applied psychology, he was trying to supplement his income abroad, and he was interested in technical inventions (as quoted in Zimmel & Kerber 1992). The situation was described by Schrödinger, writing to Max Born in 1958, sadly as

> Das Inst. f. theor. Phys. ist ja seit rund 30 Jahren unter der Führung meines lieben Freundes Hans Th. völlig vernachlässigt worden. Seitdem der hochbegabte Fritz Hasenöhrl sich 1914 freiwillig zur Front gemeldet hat und 1916 einer italienischen Granate zum Opfer fiel, treibt das Schifflein führerlos. ... Hans war für das Fach völlig unbegabt, hat meines Wissens seit seinen ersten zwei Arbeiten nichts ernsthaftes veröfftntlicht.[16] (Letter of 6 November 1958; Born *Nachlaß*, Staatsbibliothek Preussischer Kulturbesitz, Berlin)

Students at the university had been well aware of this for many years and most of those interested in modern physics had chosen other thesis advisors or decided to study abroad. Neither Thirring nor any other Viennese theoretical physicist who had once worked in relativity had established a school, any continuity was broken, and theory had to start afresh after the war.

Schrödinger had left Berlin in May of 1933 because of his dislike of the Nazi regime. After three years in Oxford he returned to Austria to accept a chair in Graz, a decision he regretted later (Schrödinger 1985). He also had the right to lecture in Vienna. He had been very interested in cosmology and had announced a course on "Das kosmische Problem der Quantenmechanik" at the Vienna University, to meet one hour a week starting in February 1938. I attended the first—and only—three classes; Schrödinger clearly had planned to elaborate on Eddington's ideas. But on 12 March, Hitler invaded Austria. Schrödinger was suspended and fled the country a few months later. This was the last gasp of relativity in Austria; it was only revived after the end of World War II.[17]

7.2. The *Anschluss* and its Aftermath

In addition to Schrödinger, Thirring and Duschek were also suspended and forced to retire because of their political views. Thirring sent a message to Einstein through a Swedish zoologist, asking whether there would be a possibility to find a position in the U.S. (EA 23-069). A further plea for help for him came via Arthur Haas, who felt that Thirring was in real danger (EA 23-070). Einstein, overwhelmed by requests for help from desperate Jews in imminent danger, clearly had no real understanding of the danger to which non-Jewish opponents of the Nazis might be exposed. He responded that he would like to help Thirring whom he valued as a human being, but did not see how; he also wrote that Thirring could not claim that he had done anything original as a theoretical physicist. Thirring sent another message after the outbreak of the war in Europe via the industrial physicist Mario Iona, stressing the danger in which he found himself (EA 23-072),

[16] "The Institute for Theoretical Physics has been totally neglected for about 30 years under the leadership of my dear friend Hans Th[irring]. Since the very gifted Fritz Hasenöhrl volunteered for the front in 1914 and became the victim of an Italian grenade in 1916 [actually 1915], the little ship is drifting leaderless. ... Hans was not at all gifted for the subject, has to my knowledge not published anything serious after his first two papers." A somewhat garbled translation of part of this passage is given in Moore 1989 and erroneously dated 6 January 1960.

[17] Moore 1989 only mentions a public lecture given by Schrödinger in Vienna on 18 February, but is not aware that he started teaching the course at the university.

but was forced to stay in Vienna, where he as well as Duschek worked in industry as consultants for the Elin- und Schorschwerke A. G. until the end of the war (EA 23-074; Pinl 1974). Ludwig Flamm of the TH, who was substituting for him at the university, let him keep a room there as well as the keys to the institute (Thirring 1965), however, a remarkable act of courage by a fundamentally apolitical man.

Ehrenhaft, who was Jewish, was fired immediately. He too turned to Einstein for help (EA 10-443), but managed to arrange to come to the U.S. on his own. Kottler, who was Lutheran and only Jewish on his mother's side (EA 14-329) and thus qualified as *Halbjude* under the Nuremberg laws, was forced into retirement with a reduced pension, as he wrote in a curriculum vitae he sent to Einstein as well as to Pauli with a plea for help. Einstein clearly did not realize the difference in treatment by the Nazis of Jews and half-Jews, many of whom survived in Vienna without being exposed to any persecution comparable to that of Jews. Although Einstein, contrary to his remarks in 1913, stated that Kottler had not "created anything of lasting value" (EA 14-331) he, considering Kottler to be a Jew, sounded the alarm and tried to obtain a position for him in several letters, in odd contrast to his reaction to Thirring's problems.[18] Whether through Einstein's help or otherwise, Kottler succeeded in reaching the U.S. before the outbreak of the war in 1939, where he worked at Eastman Kodak for many years.

Of course all the dismissed scientists who managed to reach the U.S. were already well known and also had many acquaintances abroad to whom they could turn for help. It was the younger 'Jewish' or politically exposed 'Aryan' generation without an established reputation which ran up against the restrictive immigration policies of the U.S. and other countries; many of them perished during the war years. But this is another story.

The professor of astronomy and director of the university observatory Kasimir Graff was also dismissed. He was replaced by the German astronomer Bruno Thüring, then only a Dozent in Munich. Within a year he organized a meeting on "Bekämpfung der jüdischen Relativitätstheorie" [Fight against the Jewish theory of relativity] at the observatory with the participation of several Nazi professors of experimental physics (Urban 1985). He also "had written idiotic articles with furious attacks against the theory of relativity," as Thirring wrote in 1946 in his first letter to Einstein after the war, a letter in which he also expressed the hope that Thüring would not be confused abroad with Thirring (EA 23-073). According to Paul Urban, then an Assistent at the Institute of Theoretical Physics, all library books on relativity or by Jewish authors were burned in the courtyard of the Institute; however, a later study (Tuscher 1989) argues that these were rather the books Ehrenhaft had to leave behind when he had to leave Vienna. Tuscher also lists a large number of articles by Einstein and other relativists which were cut out

[18] In a letter to Pauli (Einstein 1938) he wrote that "I have already received several letters concerning Herr Kottler. You can believe that with the unprecedented harshness of the present Jewish fate my readiness to help is unconditional ... ; no faculty will appoint a man over 50—and of course least of all a Jew. ... " The volume in which this letter is published also contains various letters by and to Pauli concerning the need to help Kottler and others. Einstein's judgement of the academic situation seems to have been based on a somewhat limited acquaintance with only a few of the more prestigious of the many hundreds of academic institutions of widely varying quality.

of the library journals with a razor, but anonymously returned to the library after the war.

Thirring, Duschek, Graff and Ehrenhaft regained their professorships in 1945 after the reestablishment of the Austrian Republic, as did the experimentalists Karl Przibram, whose brother, also a professor at the university, had perished in the Holocaust, and Stefan Meyer, who had survived in hiding. Schrödinger could have reclaimed his chair in Graz under the same law, but chose not to (Moore 1989). He returned to become professor in Vienna eleven years later; Kottler reclaimed his chair the same year. No attempt was made by the newly established authorities of the Second Republic to invite back any of the large number of younger intellectuals who had managed to escape and many of whom had established new careers, but who had not held official appointments in Austria before. This too is another story.

Acknowledgements

I am indebted to Drs. T. Hraba and Prenosil of Prague for locating documents concerning Einstein and Frank in Prague and providing me with copies, to Dr. Eckart Früh of the Dokumentation of the Arbeiterkammer für Wien for providing copies of numerous newspaper and magazine articles, to Dr. G. Wiemers of the Universitätsarchiv of the University of Leipzig for information and documents and to Dr. G. Haas for information concerning Arthur Haas, to Dr. W. Reiter of the Austrian Bundesmisterium für Wissenschaft, Forschung und Kunst, to Professor Wilhelm Frank of the TH for bringing the dissertation of Paul Lazarsfeld to my attention, and to Dr. W. Kerber and his staff at the Zentralbibliothek für Physik of the University of Vienna for help in tracing various aspects of the history of the physics institutes in Vienna, to Dr. Robert Schulmann of the *Collected Papers of Albert Einstein* at Boston University for help and advice in using its archives and to the Albert Einstein Archives, Hebrew University of Jerusalem, for permission to quote from many unpublished letters in these archives.

References

Abel, Emil. (1918). "Die spezielle Relativitätstheorie." *Österreichische Chemiker-Zeitung* **21**: 12–16, 32–36, 57–61.

Adler, Friedrich. (1916). "Ernst Mach." *Arbeiter-Zeitung*, 23 February.

——— (1918). *Ernst Machs Uberwindung des mechanischen Materialismus*. Vienna: Verlag der Wiener Volksbuchhandlung Ignaz Brand.

——— (1920). *Ortszeit; Systemzeit; Zonenzeit und das ausgezeichnete Bezugssystem der Elektrodynamik. Eine Untersuchung über die Lorentzsche und die Einsteinsche Kinematik*. Vienna: Verlag der Wiener Volksbuchhandlung.

——— (1924). "Lasterhöhlen des Geistes. 1. Die Aufhebung der Schwerkraft." *Arbeiterzeitung*, 9 January. "2. Die Zauberer und ihr Publikum." *Arbeiterzeitung*, 13 January.

Adler, Max. (1913). *Marxistische Probleme*. Stuttgart: Dietz.

Anonymous. (1921a). "Albert Einsteins Uraniavortrag." *Arbeiter-Zeitung*, 15 January.

——— (1921b). "Einstein = Rummel." *Arbeiter-Zeitung*, 2 October.

——— (1921c). "Einstein und die Zeitgenossen." By 'F. B.' *Neue Freie Presse*, 5 February.

——— (1925). "Ein Besuch in der Stätte der Medienentlarvung. Ein Gespräch mit Professor Thirring." *Neue Freie Presse*, Nr. 21755, 4 July.

——— (1930). "Shaw und Einstein im Rundfunk." *Arbeiter-Zeitung*, 3 November.

——— (1931). "Albert Einstein über seine neuen Forschungsergebnisse." *Arbeiter-Zeitung*, 15 October.

——— (1933). "Ist das deutsch? Die 'Kulturpolitik' des Dritten Reiches." By 'O. R.' *Arbeiter-Zeitung*, 16 March.

BAUER, Hans Adolf. (1918a). "Über die Energiekomponenten des Gravitationsfeldes." *Physikalische Zeitschrift* **19**: 163–165.

——— (1918b). "Kugelsymmetrische Lösungssysteme der Einstein'schen Feldgleichungen für eine ruhende, gravitierende Flüssigkeit mit linearer Zustandsgleichung." *Anzeiger der Akademie der Wissenschaften in Wien*: 135–137.

——— (1918c). "Kugelsymmetrische Lösungssysteme der Einstein'schen Feldgleichungen für eine ruhende, gravitierende Flüssigkeit mit linearer Zustandsgleichung." *Sitzungsberichte der Akademie der Wissenschaften in Wien, Abteilung IIa* **127**: 2141–2227.

——— (1920)). "Das Einsteinsche Prinzip der allgemeinen Relativität im Mittelschulunterricht." *Vierteljahrsberichte des Wiener Vereins zur Förderung des physikalischen und chemischen Unterrichtes* **23**: 22–39.

——— (1922). *Mathematische Einführung in die Gravitationstheorie Einsteins nebst einer exakten Darstellung ihrer wichtigsten Ergebnisse*. Leipzig & Vienna: Franz Deuticke.

——— (1928a). "Strenge Lösung der Einsteinschen Feldgleichungen für ein beliebiges kugelsymmetrisches, statisches Schwerefeld." *Physikalische Zeitschrift* **29**: 954–963.

——— (1928b). "Die Brechung der Materiewellen vom Standpunkt der speziellen Relativitätstheorie." *Zeitschrift für Physik* **52**: 221–224.

BECK, Guido. (1925). "Zur Theorie binärer Gravitationsfelder." *Zeitschrift für Physik* **33**: 713–728.

——— (1926). "La Propagation des ondes electromagnetiques dans la theorie de la relativité générale." *Archive des sciences physiques et naturelles* (5) **8**: 75–77.

——— (1927). "Allgemeine Relativitätstheorie." In *Handbuch der Physik*. H. Thirring, ed. Vol. 4: 299–407. Berlin: Springer.

——— (1967). "Interview." Center for History and Philosophy of Physics of the American Institute of Physics, 22 April.

BEER, Fritz. (1920). *Die Einsteinsche Relativitätstheorie und ihr historisches Fundament (6 Vorträge für Laien)*. Vienna: Moritz Perles.

BEHACKER, Max. (1913). "Der freie Fall und die Planetenbewegung in Nordstrøms Gravitationstheorie." *Physikalische Zeitschrift* **14**: 989–992.

BERNHEIMER, Walter E. (1926). "Astronomische Beobachtungsergebnisse und die ballistische Theorie der Lichtausbreitung." *Zeitschrift für Physik* **36**: 302–310.

——— (1929). "Der Bau des Kosmos." In *Handbuch der Physik*. H. Thirring, ed. Vol. 4: 577–657. Berlin: Springer.

BLACKMORE, John & HENTSCHEL, Klaus, eds. (1985). *Ernst Mach als Aussenseiter*. Vienna: W. Braumüller.

BRODA, Engelbert. (1979a). "Einstein und Österreich." *Österreichische Akademie der Wissenschaften, Veröffentlichungen der Kommission für die Geschichte der Mathematik, Naturwissenschaften und Medizin* Heft 33: 1–29.

——— (1979b). "Warum war es in Österreich um die Naturwissenschaft so schlecht bestellt?" *Wiener Geschichtsblätter* **34/3**: 89–107.

——— (1979c). "Der Einfluss von Ernst Mach und Ludwig Boltzmann auf Albert Einstein." In *Einstein Centenarium*. H.-J. Treder, ed. 227–237. Berlin: Akademie-Verlag.

——— (1981a). *The Intellectual Quadrangle: Mach–Boltzmann–Planck–Einstein*. Geneva: CERN.

——— (1981b). "Naturwissenschaftliche Leistungen im gesellschaftlichen Zusammenhang." In *Das geistige Leben Wiens in der Zwischenkriegszeit*. N. Leser, ed. 119–132. Vienna: Österreichischer Bundesverlag.

CARNAP, Rudolf. (1922). *Der Raum: ein Beitrag zur Wissenschaftslehre* (Kant-Studien, Ergänzungsheft Nr. 56). Berlin: Reuther & Reichard.

——— (1925). "Über die Abhängigkeit der Eigenschaften des Raumes von denen der Zeit." *Kant-Studien* **30**: 331–345.

CATTANI, Carlo & DE MARIA, Michelangelo. (1993). "Conservation Laws and Gravitational Waves in General Relativity (1915–1918)." In *The Attraction of Gravitation* (Einstein Studies, vol. 5). J. Earman, M. Janssen, and J. D. Norton, eds. 63–87. Boston: Birkhäuser.

CAVALLERI, G. (1968). "Solution of Ehrenfest's Paradox for a Relativistic Rotating Disk." *Nuovo Cimento* **53B**: 415–432.

CRESS-BERNERT, Trude. (1982a). "Der Philosoph und sein Mörder. Bericht einer Augenzeugin." *morgen* **22**: 83–85.

——— (1982b). "Der Mord an Moritz Schlick." *Zeitgeschichte* **9**: 229–234.

DUHEM, Pierre. (1892). "Commentaire aux principes de la thermodynamique." *Journal de mathematiques* **8**: 269–330.

DUSCHEK, Adalbert Ludwig. (1928). "Differentialgeometrie." In *Handbuch der Physik*. H. Thirring, ed. Vol. 3: 153–181. Berlin: Springer.

DUSCHEK, Adalbert Ludwig & MAYER, Walther. (1930). *Lehrbuch der Differentialgeometrie*. Vol. I. By A. Duschek. Vol. II: *Riemannsche Geometrie*. By W. Mayer. Leipzig & Berlin: Teubner.

——— (1931). "Räume konstanter Krümmung." *Rendiconti di circolo matematico di Palermo* **55**: 129–156.

EHRENFEST, Paul. (1907). "Die Translation deformierbarer Elektronen und der Flächensatz." *Annalen der Physik* **23**: 204–205.

——— (1909). "Gleichförmige Rotation starrer Körper und Relativitätstheorie." *Physikalische Zeitschrift* **10**: 918.

——— (1910). "Misst der Aberrationswinkel im Fall einer Dispersion des Äthers die Wellengeschwindigkeit?" *Annalen der Physik* **33**: 1571–1576.

——— (1912). "Zur Frage nach der Entbehrlichkeit des Lichtäthers." *Physikalische Zeitschrift* **13**: 317–319.

EIDLITZ, Otto. (1931). *Vom Syllogismus in der Relativitätstheorie, über Gravitation, und die Lösung des Welträtsels*. Vienna: Selbstverlag.

EINSTEIN, Albert. (1907). "Bemerkungen zu der Notiz von Hrn. Paul Ehrenfest: 'Die Translation deformierbarer Elektronen und der Flächensatz'." *Annalen der Physik* **23**: 206–208. [= EINSTEIN *CP2*: doc. 44].

——— (1911). "Zum Ehrenfestschen Paradoxon. Bemerkung zu V. Varičaks Aufsatz." *Physikalische Zeitschrift* **12**: 509–510. [= EINSTEIN *CP3*: doc. 22].

——— (1913). "Zum gegenwärtigen Stande des Gravitationsproblems." *Physikalische Zeitschrift* **14**: 1249–1266. [= EINSTEIN *CP4*: doc. 17].

——— (1914). "Nachträgliche Antwort auf eine Frage von Herrn Reissner." *Physikalische Zeitschrift* **15**: 108–110. [= EINSTEIN *CP4*: doc. 24].

——— (1916a). "Über Friedrich Kottlers Abhandlung 'Einsteins Äquivalenzhypothese und Gravitation'." *Annalen der Physik* **51**: 639–642. [= EINSTEIN *CP6*: doc. 40].

——— (1916b). "Ernst Mach." *Physikalische Zeitschrift* **17**: 101–104. [= EINSTEIN *CP6*: doc. 29].

——— (1918a). "Notiz zu E. Schrödingers Arbeit 'Die Energiekomponenten des Gravitationsfeldes'." *Physikalische Zeitschrift* **19**: 115–116.

——— (1918b). "Bemerkungen zu Herrn Schrödingers Notiz 'Über ein Lösungssystem der allgemein kovarianten Gravitationsgleichungen'." *Physikalische Zeitschrift* **19**: 165–166.

——— (1922). "Bemerkung zur Seletyschen Arbeit: Beiträge zum kosmologischen Problem." *Annalen der Physik* **69**: 436–438.

——— (1926). "Zur Enthüllung von Machs Denkmal." *Chronikbeilage der Neuen Freien Presse: Dem Gedächtnis Ernst Machs*, 12 June.

——— (1938). "[Einstein an Pauli]." Printed in Wolfgang Pauli: *Scientific Correspondence*. Vol. II: *1930–1939*. K. von Meyenn, ed. 600, [530]. Berlin: Springer (1985).

——— (*CP2*). *The Collected Papers of Albert Einstein*. Vol. 2: *The Swiss Years: Writings, 1900–1909*. J. Stachel, D. C. Cassidy, J. Renn, & R. Schulmann, eds. Princeton: Princeton University Press (1989).

——— (*CP3*). *The Collected Papers of Albert Einstein*. Vol. 3. *The Swiss Years: Writings, 1909–1911*. M. J. Klein, A. J. Kox, J. Renn & R. Schulmann, eds. Princeton: Princeton University Press (1993).

——— (*CP4*). *The Collected Papers of Albert Einstein*. Vol. 4. *The Swiss Years: Writings, 1912–1914*. M. J. Klein, A. J. Kox, J. Renn & R. Schulmann, eds. Princeton: Princeton University Press (1995).

——— (*CP5*). *The Collected Papers of Albert Einstein*. Vol. 5. *The Swiss Years: Correspondence, 1902–1914*. M. J. Klein, A. J. Kox, & R. Schulmann, eds. Princeton: Princeton University Press (1993).

——— (*CP6*). *The Collected Papers of Albert Einstein*. Vol. 6. *The Berlin Years: Writings, 1914–1917*. A. J. Kox, M. J. Klein, & R. Schulmann, eds. Princeton: Princeton University Press (1996).

Einstein, Albert, Lampa, Anton & Pick, Georg. (1912). "Report to the College of Professors of the Faculty of Philosophy of the German University of Prague." Printed in Einstein *CP5*: doc. 400.

Einstein & Mayer, Walther. (1930). "Zwei strenge statische Lösungen der Feldgleichungen der einheitlichen Feldtheorie." *Sitzungsberichte der Preussischen Akademie der Wissenschaften* (Berlin): 110–120.

——— (1931). "Einheitliche Theorie von Gravitation und Elektrizität." *Sitzungsberichte der Preussischen Akademie der Wissenschaften* (Berlin): 541–557.

——— (1932a). "Einheitliche Theorie von Gravitation und Elektrizität. (Zweite Abhandlung)." *Sitzungsberichte der Preussischen Akademie der Wissenschaften* (Berlin): 130–137.

——— (1932b). "Semi-Vektoren und Spinoren." *Sitzungsberichte der Preussischen Akademie der Wissenschaften* (Berlin): 522–550.

——— (1933a). "Dirac Gleichungen für Semi-Vektoren." *Proceedings of the Akademie van Wetenschappen* (Amsterdam) **36**: 497–516.

——— (1933b). "Spaltung der natürlichsten Feldgleichungen für Semi-Vektoren in Spinor-Gleichungen vom Diracschen Typus." *Proceedings of the Akademie van Wetenschappen* (Amsterdam) **36**: 615–619.

——— (1934). "Darstellung der Semi-Vektoren als gewöhnliche Vektoren von besonderem Differentiationscharakter." *Annals of Mathematics* **35**: 104–110.

Eisenstaedt, Jean. (1995). "Guido Beck in General Relativity." *Anais da Academia Brasileira de ciencias* (Supplemento 1). **67**: 49–66.

Eisenstaedt, Jean & Kox, Anne J., eds. (1992). *Studies in the History of General Relativity* (Einstein Studies, vol. 3). Boston: Birkhäuser.

Eötvös, Roland. (1953). *Gesammelte Arbeiten*. P. Selenyi, ed. Budapest: Akadémiai Kiadó.

Feigl, Herbert. (1969). "The Origin and Spirit of Logical Positivism." In *The Legacy of Logical Positivism*. P. Achinstein and S. F. Barker, eds. 3–24. Baltimore: Johns Hopkins University Press. [Reprinted: Herbert Feigl. *Inquiries and Provocations*. R. S. Cohen, ed. 1–20. Boston: D. Reidel.

——— (1974). "No Pot of Message." In *Mid-Twentieth Century Philosophy: Personal Statements*. P. A. Bertocci, ed. 120–139. Atlantic Highlands (NJ): Humanities Press. [Reprinted: Herbert Feigl. *Inquiries and Provocations: Selected Writings, 1929–1974*. R. S. Cohen, ed. 21–37. Boston: D. Reidel (1980)].

Fellinger, Hans. (1969). "Zur Entwicklungsgeschichte der Wiener Volksbildung." In *Zur Wiener Volksbildung*. 125–292. Vienna & Munich: Verlag Jugend & Volk.

Fermi, Enrico. (1922). "Über einen Widerspruch zwischen der elektrodynamischen und der relativistischen Theorie der elektromagnetischen Masse." *Physikalische Zeitschrift* **23**: 341–344.

Flamm, Ludwig. (1914a). "Die neuen Anschauungen über Raum und Zeit. Das Relativitätsprinzip." *Schriften des Vereins zur Verbreitung naturwissenschaftlicher Kenntnisse in Wien* **54**: 25–70.

——— (1914b). "Relativitätsprinzip in elementarer Darstellung." *Vierteljahresberichte des Wiener Vereines zur Förderung des physikalischen und chemichen Unterrichts* **19**: 4–21.

——— (1916). "Beiträge zur Einsteinschen Gravitationstheorie." *Physikalische Zeitschrift* **17**: 448–454.

——— (1919). "Albert Einstein und seine Lehre." *Neues Wiener Tagblatt*, 5 December.

——— (1958). "Hans Thirring zu seinem 70. Geburtstag am 23. März 1958." *Acta Physica Austriaca* **12**: 1–8.

——— (1960). "Adolf Gustav Smekal †." *Acta Physica Austriaca* **13**: 140–143.

Frank, Philipp. (1908a). "Das Relativitätsprinzip der Mechanik und die Gleichungen für die elektromagnetischen Vorgänge in bewegten Körpern." *Annalen der Physik* **27**: 897–902.

——— (1908b). "Relativitätstheorie und Elektronentheorie in ihrer Anwendung zur Ableitung der Grundgleichungen für die elektromagnetischen Vorgänge in bewegten ponderablen Körpern." *Annalen der Physik* **27**: 1059–1065.

——— (1909). "Die Stellung des Relativitäsprinzips im System der Mechanik und der Elektrodynamik." *Sitzungsberichte der kaiserlichen Akademie der Wissenschaften. Mathematisch-physikalische Klasse, Abteilung IIa* **118**: 373–446.

——— (1910). "Das Relativitätsprinzip und die Darstellung der physikalischen Erscheinungen im vierdimensionalen Raum." *Zeitschrift für physikalische Chemie* **74**: 466–495.

——— (1911a). "Das Verhalten der elektromagnetischen Feldgleichungen gegenüber linearen Transformationen der Raumzeitkoordinaten." *Annalen der Physik* **35**: 599–607.

——— (1911b). "Das Relativitätsprinzip und die Darstellung der physikalischen Erscheinungen im vierdimensionalen Raum." *Annalen der Naturphilosophie* **10**: 129–161.

——— (1911c). "Über den Zusammenhang von kinetischer Energie und transversaler Masse." *Physikalische Zeitschrift* **12**: 1112–1113.

——— (1911 d). "Eine neue Ableitung für die Dynamik der Relativtheorie." *Physikalische Zeitschrift* **12**: 1114–1115.

——— (1912). "Energetische Ableitung der Formeln für die longitudinale und transversale Masse des Massenpunktes." *Annalen der Physik* **39**: 693–703.

——— (1916). "Max Behacker." *Physikalische Zeitschrift* **17**: 41–43.

——— (1917). "Anwendung der Vektorrechnung auf die geometrische Optik in bewegten Körpern." *Annalen der Physik* **52**: 649–656.

——— (1920a). *Relativitätstheorie*. Leipzig: Teubner.

——— (1920b). "Absurditäten der Einsteinschen Theorie?" *Neues Wiener Tagblatt*, 22 October.

——— (1922). "Die Grundlagen der speziellen Relativitätstheorie." *Verhandlungen der Deutschen physikalischen Gesellschaft* **3**: 15–17.

——— (1923). "Was bedeuten die gegenwartigen physikalischen Theorien für die allgemeine Erkenntnislehre?" *Naturwissenschaften* **11**: 971–977 and 987–994.

——— (1924a). "Relativitätsprinzip nach Galilei und Newton." In *Physikalisches Handwörterbuch*. A. Berliner and K. Scheel, eds. Berlin: Springer.

——— (1924b). "Relativitätsprinzip nach Einstein." In *Physikalisches Handwörterbuch*. A. Berliner and K. Scheel, eds. Berlin: Springer.

——— (1928). "Über die 'Anschaulichkeit' physikalischer Theorien." *Naturwissenschaften* **16**: 121–128.

——— (1930a). "Relativitätsmechanik." In *Handbuch der physikalischen und technischen Mechanik*. Vol. 2. F. Auerbach and W. Hort, eds. 45–77. Leipzig: J. A. Barth.

——— (1930b). "Himmelsmechanik." In *Handbuch der physikalischen und technischen Mechanik*. Vol. 2. F. Auerbach and W. Hort, eds. 99–132. Leipzig: J. A. Barth.

——— (1932). *Das Kausalgesetz und seine Grenzen*. Vienna: Julius Springer.

——— (1939). "Foundations of Physics." *International Encyclopedia of Unified Science*. No. 6. [Reprinted: *International Encyclopedia of Unified Science*. Vol. 1, part 2: 423–504. Chicago: University of Chicago Press (1955)].

——— (1947). *Einstein. His Life and Times*. New York: Alfred A. Knopf.

——— (1951). "Einstein, Mach, and Logical Positivism." In *Albert Einstein: Philosopher-Scientist*. P. A. Schilpp, ed. 269–286. New York: Tudor Publishing Company.

Frank, Philipp & von Mises, Richard. (1930–35). *Die Differential- und Integralgleichungen der Mechanik und Physik*. 2nd edition. 2 vols. Braunschweig: Friedrich Vieweg.

Frank, Philipp & Rothe, Hermann. (1910a). "Über eine Verallgemeinerung des Relativitätsprinzips und die dazugehörige Mechanik." *Sitzungsberichte der kaiserlichen Akademie der Wissenschaften, Mathematisch-physikalische Klasse, Abteilung IIa* **119**: 615–630.

——— (1910b). "Zur Herleitung der Lorentztransformation." *Physikalische Zeitschrift* **13**: 750–753.

——— (1911). "Über die Transformation der Raumzeitkoordinaten von ruhenden auf bewegte Systeme." *Annalen der Physik* **34**: 825–855.

Galison, Peter. (1982). "Theoretical Predispositions in Experimental Physics: Einstein and the Gyromagnetic Experiments, 1915–1925." *Historical Studies in the Physical and Biological Sciences* **16**: 285–323.

Geiringer, Hilda. (1918). "Nicht-Euklidische Geometrien und das Raumproblem." *Naturwissenschaften* **6**: 635–641 and 653–638.

Goenner, Hubert. (1992a). "The Reaction to Relativity Theory, I: The Anti-Einstein Campaign in Germany in 1920." *Science in Context* **6**: 107–133.

——— (1992b). "The Reception of the Theory of Relativity in Germany as Reflected by Books Published Between 1908 and 1945." In Eisenstaedt & Kox 1992: 15–38.

——— (1993). "The Reaction to Relativity Theory in Germany, III: 'A Hundred Authors against Einstein'." In *The Attraction of Gravitation. New Studies in the History of General Relativity* (Einstein Studies, vol. 5). J. Earman, M. Janssen and J. D. Norton, eds. 248–273. Boston: Birkhäuser.

Gulick, Charles A. (1948). *Austria from Habsburg to Hitler.* 2 vols. Berkeley: University of California Press.

Haas, Arthur Erich. (1910). "Über die elektrodynamische Bedeutung des Planck'schen Strahlungsgesetzes und über eine neue Bestimmung des elektrischen Elementarquantums und der Dimensionen des Wasserstoffatoms." *Sitzungsberichte der kaiserlichen Akademie der Wissenschaften in Wien. Abteilung IIa* **119**: 119–144. [Reprinted: Haas *QA*].

——— (1913). "Ist die Welt in Raum und Zeit unendlich?" *Archiv für systematische Philosophie.*

——— (1918). "Über eine Beziehung der Gravitationskonstante zu den Grundgrössen der Elektrizitätstheorie." *Physikalische Zeitschrift* **19**: 338.

——— (1919). *Einführung in die theoretische Physik mit besonderer Berücksichtigung ihrer modernen Probleme.* 2 vols. Leipzig: Veit.

——— (1920). *Das Naturbild der neuen Physik.* Berlin & Leipzig: Walter de Gruyter.

——— (1927a). "Über die Ableitung der fundamentalen relativitäts-theoretischen Satze aus der Brogliesche Hypothese der Phasenwellen." *Physikalische Zeitschrift* **28**: 632–634.

——— (1927b). "Die Hypothese des elementaren Wirkungsquantums als Folge der Relativitätstheorie." *Physikalische Zeitschrift* **28**: 707–709.

——— (1927c). "Der Zusammenhang zwischen Relativitätstheorie und Quantentheorie." *Anzeiger der Akademie der Wissenschaften in Wien* **64**.

——— (1928). "Wellenmechanik und Relativitätstheorie." *Forschung und Fortschritt* **4**.

——— (1930). "Mittlere Massendichte des Universums." *Anzeiger der Akademie der Wissenschaften in Wien* **67**.

——— (1934). *Die kosmologischen Probleme der Physik.* Leipzig: Akademische Verlagsgesellschaft.

——— (*QA*). *Der erste Quantenansatz für das Atom* (Dokumente der Naturwissenschaft, Abteilung Physik, vol. 10). A. Hermann, ed. Stuttgart: Ernst Battenberg Verlag (1965).

Halpern, Otto. (1927). "Relativitätsmechanik." In *Handbuch der Physik.* R. Grammel, ed. Vol. 5: 578–616. Berlin: Julius Springer.

Hamann, Brigitte. (1996). *Hitlers Wien.* Munich & Zürich: Piper.

Hantsch, Hugo. (1953). "Die Nationalitätenfrage im alten Österreich." *Wiener historische Studien* **1**.

Hasenöhrl, Fritz. (1902). "Über die Grundgleichungen der elektromagnetischen Lichttheorie für bewegte Körper." *Sitzungsberichte der kaiserlichen Akademie der Wissenschaften. Mathematisch-naturwissenschaftliche Classe. Abtheilung IIa* **111**: 1525–1548.

——— (1904a). "Über die Veränderung der Dimensionen der Materie infolge ihrer Bewegung durch den Äther." *Sitzungsberichte der kaiserlichen Akademie der Wissenschaften. Mathematisch-naturwissenschaftliche Classe. Abtheilung IIa* **113**: 469–490.

——— (1904b). "Über die Reziprozität des Strahlenganges in bewegten Körpern. Thermodynamische Ableitung des Fresnel'schen Fortführungskoeffizienten." *Sitzungsberichte der kaiserlichen Akademie der Wissenschaften. Mathematisch-naturwissenschaftliche Classe. Abtheilung IIa* **113**: 493–500.

——— (1904c). "Zur Theorie der Strahlung bewegter Körper." *Sitzungsberichte der kaiserlichen Akademie der Wissenschaften. Mathematisch-naturwissenschaftliche Classe. Abtheilung IIa* **113**: 1039–1055.

——— (1904–05). "Zur Theorie der Strahlung in bewegten Körpern." *Annalen der Physik* **15**: 344–370 and **16**: 589–592.

——— (1907a). "Über den Lichtäther." *Schriften des Vereins zur Verbreitung naturwissenschaftlicher Kenntnisse in Wien* **47**: 297–318.

——— (1907b). "Zur Theorie der stationären Strahlung in einem gleichförmig bewegten Hohlraum." *Annalen der Physik* **22**: 791–792.

——— (1907–08). "Zur Thermodynamik bewegter Systeme." *Sitzungsberichte der kaiserlichen Akademie der Wissenschaften, Mathematisch-naturwissenschaftliche Klasse. Abteilung IIa* **116**: 1391–1405 and **117**: 207–215.

——— (1909). "Bericht über die Trägheit der Energie." *Jahrbuch der Radioaktivität und Elektronik* **6**: 297–318.

Havas, Peter. (1991). "Early Relativistic Theories of Gravitation Other Than Einstein's." Lecture at the Third International Conference in History and Philosophy of General Relativity, University of Pittsburgh, Johnstown (PA), 27–30 June.

——— (1995). "The Life and Work of Guido Beck: The European Years: 1903-1943." *Anais da Academia Brasileira de Ciencias* (Supplemento 1) **67**: 11–36.

Heller, Karl Daniel. (1964). *Ernst Mach: Wegbereiter der modernen Physik.* Vienna & New York: Springer.

Hentschel, Klaus. (1986). "Die Korrespondenz Einstein–Schlick zum Verhältnis der Physik zur Philosophie." *Annals of Science* **43**: 475–488.

——— (1990). *Interpretationen und Fehlinterpretationen der speziellen und der allgemeinen Relativitätstheorie durch Zeitgenossen Albert Einsteins* (Science Networks, vol. 6). Basel: Birkhäuser.

Hepperger, Josef. (1888). "Die Fortpflanzungsgeschwindigkeit der Gravitation." *Sitzungsberichte der kaiserlichen Akademie der Wissenschaften. Mathematisch-naturwissenschaftliche Classe. Abtheilung IIa* **97**: 337–362.

——— (1922). "Zur Ablenkung des Lichtes in Gravitationsfeldern." *Astronomische Nachrichten* **216**: 321–322.

Hermann, Armin. (1965). "Arthur Haas—Eine Biographie." In Haas *QA*: 7–25.

——— (1969). *Frühgeschichte der Quantentheorie (1899–1918).* Mosbach in Baden: Physik Verlag.

Holton, Gerald. (1977). "Electrons or Subelectrons? Millikan, Ehrenhaft and the Role of Preconceptions." In *History of Twentieth-Century Physics* (Proceedings of the International School of Physics "Enrico Fermi", course 57). C. Weiner, ed. 266–289. New York & London: Academic Press.

Howard, Don. (1984). "Realism and Conventionalism in Einstein's Philosophy of Science: The Einstein–Schlick Correspondence." *Philosophia Naturalis* **21**: 616–629.

Hufnagel, Leo. (1919). "Die Bahn des grossen Septemberkometen 1882 II unter Zugrundelegung der Einsteinschen Gravitationstheorie." *Sitzungsberichte der Wiener Akademie der Wissenschaften. Mathematisch-naturwissenschaftliche Klasse. Abteilung IIa* **128**: 1261–1270.

Illy, József. (1979). "Albert Einstein in Prag." *Isis* **70**: 76–84.

Jaumann, Gustav Andreas Johannes. (1906). Letter to Ernst Mach, 27 October. [Published: Thiele 1978: 227.

——— (1911). "Geschlossenes System physikalischer und chemischer Differentialgesetze." *Sitzungsberichte der kaiserlichen Akademie der Wissenschaften. Mathematisch-naturwissenschaftliche Klasse* **120**: 385–530.

——— (1912). "Theorie der Gravitation." *Sitzungsberichte der kaiserlichen Akademie der Wissenschaften. Mathematisch-naturwissenschaftliche Klasse* **121**: 95–182.

——— (1914). "Feststellung einer Priorität in der Gravitationstheorie." *Physikalische Zeitschrift* **15**: 159–160.

JOEL, Kurt. (1917). "Friedrich Adler als Physiker. Eine Unterredung mit Albert Einstein." *Vossische Zeitung* (Berlin), 23 May. [Reprinted in *Arbeiter-Zeitung*, 25 May.

KANN, Robert A. (1974). *A History of the Habsburg Empire 1526–1918*. Berkeley: University of California Press.

——— (1977). "Dokumente einer seltsamen Freundschaft." *Die Presse*, 2 and 3 July.

KELEN, Jolán. (1976). *Eliramlik az élet* .... Budapest: Kossuth könyvkiadó.

KIENLE, Hans. (1924). "Die astronomischen Prüfungen der allgemeinen Relativitätstheorie." *Ergebnisse der exakten Naturwissenschaften* **3**: 55–66.

KLEIN, Felix. (1927). *Vorlesungen über die Entwicklung der Mathematik im 19. Jahrhundert.* Part II: *Die Grundbegriffe der Invariantentheorie und ihr Eindringen in die mathematische Physik.* Berlin: Julius Springer. [Reprint: 1979].

KLEIN, Martin J. (1970). *Paul Ehrenfest.* Vol. 1. *The Making of a Theoretical Physicist.* Amsterdam & London: North Holland.

KLEINERT, Andreas. (1976). *Anton Lampa: 1868–1938* (Biobibliographien 4). Berlin: Deutscher Bibliotheksverband, Arbeitsstelle für das Bibliothekswesen.

KOHL, Emil Viktor. (1906). "Über eine Erweiterung der Stefanschen Entwickelung des elektromagnetischen Feldes für bewegte Medien." *Annalen der Physik* **20**: 1–34.

——— (1907). "Über die dielektrischen Verschiebungsgleichungen für schnelle Schwingungen in ruhenden Mitteln." *Annalen der Physik* **22**: 401–428.

——— (1909). "Über den Michelsonschen Versuch." *Annalen der Physik* **28**: 259–307 and 662.

KOTTLER, Friedrich. (1912). "Über die Raumzeitlinien der Minkowski'schen Welt." *Sitzungsberichte der kaiserlichen Akademie der Wissenschaften. Mathematisch-naturwissenschaftliche Klasse, Abteilung IIa* **121**: 1659–1750.

——— (1914a). "Relativitätsprinzip und beschleunigte Bewegung." *Annalen der Physik* **44**: 701–748.

——— (1914b and 1916a). "Fallende Bezugssysteme vom Standpunkte des Relativitätsprinzips." *Annalen der Physik* **45**: 481–516 and **50**: 600.

——— (1916b). "Beschleunigungsrelative Bewegungen und die konforme Gruppe der Minkowskischen Welt." *Sitzungsberichte der kaiserlichen Akademie der Wissenschaften. Mathematisch-naturwissenschaftliche Klasse* **125**: 899–919.

——— (1916c). "Über Einsteins Äquivalenzhypothese und die Gravitation." *Annalen der Physik* **50**: 955–972.

——— (1918). "Über die physikalischen Grundlagen der Einsteinschen Gravitationstheorie." *Annalen der Physik* **56**: 401–462.

——— (1921a). "Gravitation und Relativitätstheorie." In *Encyklopädie der mathematischen Wissenschaften* Vol. VI.2.2. *Astronomie.* K. Schwarzschild, S. Oppenheim, and W. v. Dyck, eds. 159–237. Leipzig: Teubner.

——— (1921b). "Rotierende Bezugssysteme in einer Minkowskischen Welt." *Physikalische Zeitschrift* **22**: 274–280, 392, 480–484, 519.

——— (1922a). "Newton'sches Gesetz und Metrik." *Sitzungsberichte der Wiener Akademie der Wissenschaften. Mathematisch-naturwissenschaftliche Klasse, Abteilung IIa* **131**: 1–14.

——— (1922b). "Maxwell'sche Gleichungen und Metrik." *Sitzungsberichte der Wiener Akademie der Wissenschaften. Mathematisch-naturwissenschaftliche Klasse, Abteilung IIa* **131**: 119–46.

——— (1924). "Considerations de critique historique sur la théorie de la relativité. 1. De Fresnel à Lorentz. 2. Henri Poincaré et Albert Einstein." *Scientia* **36**: 231–234, 301–316.

KRAUS, Karl. (1929). "Griff und Missgriff." *Die Fackel* Nr. 806-9: 21–22.

Kraus, Oskar. (1920). 'Philosophische Betrachtungen gegen die Relativitätstheorie." *Neue Freie Presse* Nr. 20130, 11 September.

——— (1923). "Über den gegenwärtigen Stand der Relativitätstheorie Einsteins." *Neue Freie Presse* Nrs. 21238 and 21240, 25 and 27 October.

——— (1925). *Offene Briefe an Albert Einstein und Max von Laue über die gedanklichen Grundlagen der speziellen und allgemeinen Relativitätstheorie*. Vienna & Leipzig: Wilhelm Braumüller.

Kremer, Josef. (1921). *Einstein und die Weltanschauungskrisis*. Graz & Vienna: Verlagsbuchhandlung "Styria".

——— (1922). "Die Relativität der Einsteinschen Relativitätstheorie." *Reichspost*, 10 December.

Lampa, Anton. (1934). "Wie erscheint nach der Relativitätstheorie ein bewegter Stab einem ruhenden Beobachter?" *Zeitschrift für Physik* **27**: 138–148.

von Lang, Viktor. (1911). "Zur Einführung in die Relativitätstheorie." *Vierteljahrsberichte des Wiener Vereines zur Förderung des physikalischen und chemischen Unterrichtes* **16**: 71–77.

Lania, Leo. (1933). "Der Staatenlose Nr. 14—Albert Einstein." *Bunte Woche* Nr. 40, 1 October.

Lazarsfeld, Paul. (1926). "Über die Berechnung der Perihelbewegung des Merkur aus der Einsteinschen Gravitationstheorie." *Zeitschrift für Physik* **35**: 119–128.

Lense, Josef. (1917a). "Das Newtonsche Gesetz in nichteuklidischen Räumen." *Sitzungsberichte der Akademie der Wissenschaften in Wien, Mathematisch-naturwissenschaftliche Klasse, Abteilung IIa* **126**: 1037–1063.

——— (1917b). "Das Newtonsche Gesetz in nichteuklidischen Räumen." *Astronomische Nachrichten* **205**: 241–248.

——— (1918a). "Über Relativitätseinflüsse in den Mondsystemen." *Astronomische Nachrichten* 206: 117–120.

——— (1918b). "Nichteuklidische Geometrie und Newtonsche Massenanziehung." *Weltall* **18**: 64–67.

Lense, Josef & Thirring, Hans. (1918). "Über den Einfluss der Eigenrotation der Zentralkörper auf die Bewegung der Planeten und Monde nach der Einsteinschen Gravitationstheorie." *Physikalische Zeitschrift* **19**: 156–163.

Lorentz, Hendrik Antoon. (1904). "Weiterbildung der Maxwellschen Theorie, Elektronentheorie." *Encyklopädie der mathematischen Wissenschaften*. Vol. V.2. *Physik*. A. Sommerfeld, ed. 151–280. Leipzig: Teubner.

Mach, Ernst. (1894). "Memorandum." Published in Thiele 1978: 229.

——— (1906). *Space and Geometry in the Light of Physiological, Psychological and Physical Inquiry*. LaSalle (IL): Open Court..

——— (1907). "Die Rassenfrage." *Neue Freie Presse*, 24 December. [Reprinted: Heller 1964: 99–102.

——— (1910). "Eine Betrachtung über Zeit und Raum." *Wissen für Alle* **10**/3. [Reprinted: E. Mach. *Populär-wissenschaftliche Vorlesungen*. 4th ed. Leipzig: J. A. Barth (1910): 492–508].

——— (1911). "Ernst Mach an das Volksheim." *Arbeiter-Zeitung*, 28 January.

——— (1921). *Die Prinzipien der physikalischen Optik Historisch und erkenntnispsychologisch entwickelt*. Leipzig: Johann Ambrosius Barth.

Mach, Ernst & Jaumann, Gustav. (1891). *Leitfaden der Physik für Studierende*. Prague & Vienna: Tempsky.

Minkowski, Hermann. (1908). "Raum und Zeit." (Lecture at the eightieth Versammlung Deutscher Naturforscher und Ärzte, 21 September 1908). Printed in *Physikalische Zeitschrift* **10** (1909): 104–111.

Mitis, Lothar. (1930). *Einsteins Grundirrtum.* Leipzig: Otto Hillmann.

——— (1931a). "Sachverhalt und Einstein." In *Hundert Autoren gegen Einstein.* H. Israel, E. Ruckhaber and R. Weinmann, eds. 34–35. Leipzig: R. Voigtländer.

——— (1931b). "Das Hauptargument gegen die Relativitätstheorie." *Die Quelle* **81**: 880–884.

Molisch, Paul. (1939). *Politische Geschichte der deutschen Hochschulen in Österreich von 1848 bis 1918.* Vienna & Leipzig: Wilhelm Braumüller.

Moore, Walter. (1989). *Schrödinger. Life and Thought.* Cambridge: Cambridge University Press.

Mosengeil, Kurt. (1906). "Theorie der stationären Strahlung in einem gleichförmig bewegten Hohlraum." Doctoral dissertation, Berlin.

——— (1907). "Theorie der stationären Strahlung in einem gleichförmig bewegten Hohlraum." *Annalen der Physik* **22**: 867–904. [A shortened and corrected version of Mosengeil 1906, edited by Max Planck].

Neurath, Otto. (1932). "Die 'Philosophie' im Kampf gegen den Fortschritt der Wissenschaft." *Der Kampf* **25**: 385–389.

Newest, Th. [Goldzier, Hans]. (1921). *Gegen Einstein: Die Erfahrung im Weltall.* Vienna, Leipzig & Bern: Frisch.

Oppenheim, Samuel. (1903). "Kritik des Newtonschen Gravitationsgesetzes." Programmabhandlung der K. K. Deutschen Staatsrealschule in Karolinenthal.

——— (1911). *Probleme der modernen Astronomie.* Leipzig: Teubner.

——— (1917). "Zur Frage nach der Fortpflanzungsgeschwindigkeit der Gravitation." *Annalen der Physik* **53**: 163–168.

——— (1919). "Theorie der Gleichgewichtsfiguren der Himmelskörper." In *Encyklopädie der mathematischen Wissenschaften.* Vol. VI.2.2. *Astronomie.* K. Schwarzschild, S. Oppenheim and W. v. Dyck, eds. 5–79. Leipzig: Teubner.

——— (1920a). "Kritik des Newtonschen Gravitationsgesetzes." In *Encyklopädie der mathematischen Wissenschaften.* Vol. VI.2.2. *Astronomie.* K. Schwarzschild, S. Oppenheim and W. v. Dyck, eds. 80–158. Leipzig: Teubner.

——— (1920b). "Gravitation." In *Die Kultur der Gegenwart.* Vol. III.3.3. *Astronomie*: 598–630.

——— (1923). "Kometen." In *Encyklopädie der mathematischen Wissenschaften.* Vol. VI.1.1. *Astronomie.* K. Schwarzschild and S. Oppenheim, eds. 899–939. Leipzig: Teubner.

Oppolzer, Egon R. (1902). "Erdbewegung und Aether." *Sitzungsberichte der mathematisch-naturwissenschaftlichen Classe der kaiserlichen Akademie der Wissenschaften. Abtheilung IIa* **111**: 244–254. [Reprinted: *Annalen der Physik* **8** (1902): 898–907].

Pauli, Wolfgang. (1919a). "Über die Energiekomponenten des Gravitationsfeldes." *Physikalische Zeitschrift* **20**: 1–3.

——— (1919b). "Zur Theorie der Gravitation und der Elektrizität von Hermann Weyl." *Physikalische Zeitschrift* **20**: 457–467.

——— (1919c). "Mercurperihelbewegung und Strahlenablenkung in Weyls Gravitationstheorie." *Verhandlungen der Deutschen Physikalischen Gesellschaft* **21**: 741–750.

——— (1921). "Relativitätstheorie." In *Encyklopädie der mathematischen Wissenschaften.* Vol. V.1. *Physik.* A. Sommerfeld, ed. 539–775. Leipzig & Berlin: Teubner. [Also printed separately].

Pinl, Maximilian. (1974). "Kollegen in einer dunklen Zeit. Schluss." *Jahresbericht der deutschen Mathematikervereinigung* **75**: 166–208.

Quint, Heinz. (1922). *Die Relativitätstheorie.* Vienna: Anzengruber-Verlag.

RADAKOVIC, Theodor & LENSE, Josef. (1928). "Vektor- und Tensorrechnung, Riemannsche Geometrie." In *Handbuch der Physik*. Vol. 3: 182–214. H. Thirring, ed. Berlin: Julius Springer.

REICHENBACH, Hans. (1924). *Axiomatik der relativistischen Raum-Zeit-Lehre*. Braunschweig: Friedrich Vieweg.

ROTHE, Hermann. (1924). *Einführung in die Tensorrechnung*. Vienna: L. W. Seidel.

RÜGER, Alexander. (1988). "Atomism from Cosmology: Erwin Schrödinger's Work on Wave Mechanics and Space-Time-Structure." *Studies in History and Philosophy of Science* **18**: 377–401.

SAHULKA, Johann. (1907). *Erklärung der Gravitation, der Molekularkräfte, der Wärme, des Lichtes, der magnetischen und elektrischen Erscheinungen aus gemeinsamer Ursache auf rein mechanischem, atomistischen Wege*. Vienna & Leipzig: Carl Fromme.

——— (1909). "Über die bisherigen Versuche, die Gravitation aus dem Stosse der Ätherteilchen zu erklären." *Schriften des Vereins zur Verbreitung naturwissenschaftlicher Kenntnisse in Wien* **49**: 331–363.

SCHECHTMAN, Josef B. (1962). *Postwar Population Transfers in Europe 1945–1955*. Philadelphia: University of Pennsylvania Press.

SCHLICK, Moritz. (1915). "Die philosophische Bedeutung des Relativitätsproblems." *Zeitschrift für Philosophie und philosophische Kritik* **159**: 129–175.

——— (1917). *Raum und Zeit in der gegenwartigen Physik. Zur Einführung in das Verständnis der Relativitäts- und Gravitationstheorie*. Berlin: Julius Springer.

——— (1922). "Die Relativitätstheorie in der Philosophie." *Verhandlungen der Gesellschaft der Naturforscher und Ärzte* **87**: 58–69.

——— (1931). "Die Kausalitat in der gegenwartigen Physik." *Naturwissenschaften* **19**: 143–162.

SCHRÖDINGER, Erwin. (1918a). "Die Energiekomponenten des Gravitationsfeldes." *Physikalische Zeitschrift* **19**: 4–7.

——— (1918b). "Über ein Lösungssystem der allgemein kovarianten Gravitationsgleichungen." *Physikalische Zeitschrift* **19**: 20–22.

——— (1934). "Über die Unanwendbarkeit der Geometrie im Kleinen." *Naturwissenschaften* **22**: 518–520. (1985). *Mein Leben, Meine Weltansicht*. Vienna: Paul Zsolnay.

SELETY, Franz. (1913). "Die wirklichen Tatsachen der reinen Erfahrung, eine Kritik der Zeit." *Zeitschrift für Philosophie und philosophische Kritik* **152**: 78–93.

——— (1922). "Beiträge zum kosmologischen Problem." *Annalen der Physik* **68**: 281–334.

——— (1923a). "Erwiderung auf die Bemerkungen Einsteins über meine Arbeit 'Beiträge zum kosmologischen Problem'." *Annalen der Physik* **72**: 59–66.

——— (1923b). "Une distribution des masses avec une densité moyenne nulle, sans centre de gravité." *Comptes Rendus de l'Académie des Sciences* (Paris) **177**: 104–106.

——— (1923c). "Possibilité d'un potentiel infini, et d'une vitesse moyenne de toutes les étoiles égale à celle de la lumière." *Comptes Rendus de l'Académie des Sciences* (Paris) **177**: 250–252.

——— (1924). "Unendlichkeit des Raumes und allgemeine Relativitätstheorie." *Annalen der Physik* **73**: 291–325.

SCHÜTZ, J. R. (1897). "Prinzip der absoluten Erhaltung der Energie." *Nachrichten der Akademie der Wissenschaften zu Göttingen*: 110–123.

SIEGERT, Michael. (1981a). "Mit dem Browning philosophiert." *Forum* July/August: 18–25.

——— (1981b). "Der Mord an Professor Moritz Schlick." In *Attentate die Österreich erschütterten*. L. Spira, ed. 123–131. Vienna: Locker.

SMEKAL, Adolf Gustav. (1920a). "Spezielle Relativitätstheorie und Probleme des Atomkerns." *Naturwissenschaften* **8**: 206–207.

——— (1920b). "Über die Abweichungen vom Coulombschen Gesetze in grosser Nähe der elementaren Ladungen." *Verhandlungen der Deutschen Physikalischen Gesellschaft* **1**: 55–58.

——— (1920c). "Über die Dimensionen der Partikel und die Abweichungen vom Coulombschen Gesetze in grosser Nähe elektrischer Ladungen." *Sitzungsberichte der Akademie der Wissenschaften, Mathematisch-naturwissenschaftliche Klasse. Abteilung IIa* **129**: 455–481.

——— (1921). "Atomgewichte und Relativitätstheorie." *Verhandlungen der Deutschen Physikalischen Gesellschaft* **2**: 19.

ŚREDNIAWA, Bronisław. (1985). *History of Theoretical Physics at Jagellonian University in the XIXth Century and in the First Half of the XXth Century* (Universitatis Jagellonicae Folia Physica, fascicule 24). Cracow: Jagellonian University.

——— (1987). "The Reception of the Theory of Relativity in Poland." In *The Comparative Reception of Relativity*. T. F. Glick, ed. 327–350. Dordrecht: Reidel.

——— (1991a). "The Evolution of the Concept of Ether and the Early Development of Relativity on the Example of the Research in Physics at Cracow University." Preprint PTJU-9/91. Cracow: Jagellonian University.

——— (1991b). "Early Investigations in the Foundations of General Relativity in Poland." Preprint PTJU-25/91. Cracow: Jagellonian University.

——— (1994). *Three Essays on the History of Relativity in Cracow* (Universitatis Jagellonicae Folia Physica, fascicule 37). Cracow: Jagellonian University.

STADLER, Friedrich. (1982). *Vom Positivismus zur "Wissenschaftlichen Weltauffassung."* Vienna & Munich: Locker Verlag.

STADLER, Karl R. (1971). *Austria*. New York: Praeger Publishers.

SYNGE, John Lighton. (1935). "Angular Momentum, Mass-Center and the Inverse Square Law of Special Relativity." *Physical Review* **47**: 760–767.

THIELE, Joachim. (1978). *Wissenschaftliche Kommunikation. Die Korrespondenz Ernst Machs*. Kastellaun: A. Henn Verlag.

THIRRING, Hans. (1918a and 1921a). "Über die Wirkung rotierender ferner Massen in der Einsteinschen Gravitationstheorie." *Physikalische Zeitschrift* **19**: 33–39 and **22**: 29–30.

——— (1918b). "Über die formale Analogie zwischen den elektromagnetischen Grundgleichungen und den Einsteinschen Gravitationsgleichungen erster Näherung." *Physikalische Zeitschrift* **19**: 204–205.

——— (1918c). "Über die Relativität der Rotationsbewegung in der Einsteinschen Gravitationstheorie." *Vierteljahrsberichte des Wiener Vereines zur Föderung des physikalischen und chemischen Unterrichts* **21**: 17–29.

——— (1920). "Die Gravitationstheorie Einsteins." *Neue Freie Presse* Nr. 19897, 18 January: 3-4; Nr. 19911, 1 February: 4; Nr. 19918, 8 February: 4.

——— (1921b). *Die Idee der Relativitätstheorie*. Berlin: Julius Springer.

——— (1921c). "Über das Uhrenparadoxon in der Relativitätstheorie." *Naturwissenschaften* **9**: 209–212.

——— (1921d). "Ziele und Methoden der theoretischen Physik." *Naturwissenschaften* **9**: 1023–1028.

——— (1921e). "Erwiderung." *Naturwissenschaften* **9**: 481–483 and 551.

——— (1921f). "Die Relativitätstheorie Einsteins." *Zeitschrift des österreichischen Ingenieur- und Architekten-Vereines* **73**: 189–192.

——— (1922). "Relativitätstheorie." *Ergebnisse der exakten Naturwissenschaften* **1**: 26–59.

——— (1923). "Lenard, P. Über Äther und Uräther." *Naturwissenschaften* **11**: 228–230.

——— (1924). "Bemerkung zu einem Einwand gegen die spezielle Relativitätstheorie." *Zeitschrift für Physik* **30**: 63–65.

——— (1925a). "Über die empirische Grundlage des Prinzips der Konstanz der Lichtgeschwindigkeit." *Zeitschrift für Physik* **31**: 133–138.

——— (1925b). "Bemerkung zur Arbeit Herrn Tomascheks über die Aberration." *Zeitschrift für Physik* **33**: 153–154.

——— (1925c). "Aberration und Relativitätstheorie." *Zeitschrift für technische Physik* **6**: 561–563.

——— (1925d). "Relativität und Aberration. Ein Dialog." *Naturwissenschaften* **13**: 445–447.

——— (1926a). "Kritische Bemerkungen zur Wiederholung des Michelsonversuches auf dem Mount Wilson." *Zeitschrift für Physik* **35**: 723–731.

——— (1926b). "Prof. Miller's Ether Drift Experiments." *Nature* **118**: 81–82.

——— (1926c). "Neuere experimentelle Ergebnisse zur Relativitätstheorie." *Naturwissenschaften* **14**: 111–116.

——— (1927). "Elektrodynamik bewegter Körper und spezielle Relativitätstheorie." In *Handbuch der Physik.* Vol. 12: 245–348. W. Westphal, ed. Berlin: Julius Springer.

——— (1929a). "Begriffssystem und Grundgesetze der Feldphysik." In *Handbuch der Physik.* Vol. 4: 81–177. H. Thirring, ed. Berlin: Julius Springer.

——— (1929b). "Zum fünfzigsten Geburtstag Albert Einsteins." *Neue Freie Presse,* Nr. 23167, 13 March: 17.

——— (1931). "Halblaien und Ignoranten gegen Einstein." *Neue Freie Presse*, Nr. 23852, 8 February: 12.

——— (1965). "Ludwig Flamm." *Acta Physica Austriaca* **24**: 1–5.

TURNER, Henry A., Jr. (ed.) (1978). *Hitler aus nächster Nähe. Aufzeichnungen eines Vertrauten 1929–1932.* Berlin: Ullstein.

TUSCHER, Engelbert. (1989). "Von den Handbibliotheken der physikalischen Institute der Wiener Universität zur Zentralbibliothek für Physik in Wien." Unpublished memoir. Vienna: Zentralbibliothek für Physik der Universität.

URBAN, Paul. (1985). "Erinnerungen." Unpublished autobiography. Vienna: Zentralbibliothek für Physik der Universität.

URBANTKE, Helmuth. (1992). "Schrödinger and Cosmology." In EISENSTAEDT & KOX 1992: 453–459. Boston: Birkhäuser.

WEINMANN, Rudolf. (1922). *Gegen Einsteins Relativierung von Raum und Zeit.* Munich: Oldenburg.

——— (1923). *Anti-Einstein.* Leipzig: Otto Hillmann.

WEITZENBÖCK, Roland. (1920–21). "Über die Wirkungsfunktion in der Weyl'schen Physik." *Sitzungsberichte der Akademie der Wissenschaften, Mathematisch-naturwissenschaftliche Klasse. Abteilung IIa* **129**: 683–696, 697–708; **130**: 15–23.

——— (1921). "Zur vierdimensionalen Tensoranalysis." *Sitzungsberichte der Akademie der Wissenschaften, Mathematisch-naturwissenschaftliche Klasse. Abteilung IIa* **130**: 31–45.

——— (1929). *Der vierdimensionale Raum.* Braunschweig: Friedrich Vieweg.

WESSELY, Karl. (1920–21). "Bemerkung zu den Grundlagen der Relativitätstheorie." *Physikalische Zeitschrift* **21**: 349–350 and **22**: 310–312.

WILLY, Rudolf. (1914). "FRANZ SELETY (Wien). Die wirklichen Tatsachen der reinen Erfahrung, eine Kritik der Zeit." *Zeitschrift für positivistische Philosophie* **2**: 149–151.

WIRTINGER, Wilhelm. (1922). "On a General Infinitesimal Geometry, in Reference to the Theory of Relativity." *Transactions of the Cambridge Philosophical Society* **22**: 439–449.

——— (1926). "Allgemeine Infinitesimalgeometrie und Erfahrung." *Abhandlungen aus dem mathematischen Seminar der Hamburgischen Universität* **4**: 178–200. [Also published separately in *Hamburger mathematische Einzelschriften* **3** (1926)].

WOLTERS, Gereon. (1987). *Mach I, Mach II, Einstein und die Relativitätstheorie*. Berlin: Walter de Gruyter.

ZILSEL, Eduard. (1927). "Über die Asymmetrie der Kausalität und die Einsinnigkeit der Zeit." *Naturwissenschaften* **15**: 280–286.

——— (1929). "Philosophische Bemerkungen." *Der Kampf* **22**: 178–188.

ZIMMEL, Brigitte & KERBER, Gabriele, eds. (1992). *Hans Thirring. Ein Leben für Physik und Frieden*. Vienna, Cologne & Weimar: Böhlau Verlag.

ZLAMAL, Heinrich. (1924). *Das Verhältnis der Einsteinschen Relativitätstheorie zur exakten Naturforschung*. 1. Heft: *Die phänomenalistische und die sophistische Auffassung und Bedeutung der Relativitätstheorie*. Vienna & Leipzig: Wilhelm Braumüller.

——— (1932). *Die Lehre von der Zustandsbeeinflussung des Weltäthers durch die Anwesenheit und Bewegung ponderabler Körper als klassisches Correlatum der Einsteinschen Relativitätstheorie*. Vienna: Selbstverlag.

Controversies in the History of the Radiation Reaction Problem in General Relativity

Daniel Kennefick

BEGINNING IN THE EARLY NINETEEN-FIFTIES, experts in the theory of general relativity debated vigorously whether the theory predicted the emission of gravitational radiation from binary star systems. For a time, doubts also arose on whether gravitational waves could carry any energy. Since radiation phenomena have played a key role in the development of twentieth-century field theories, it is the main purpose of this paper to examine the reasons for the growth of scepticism regarding radiation in the case of the gravitational field. Although the focus is on the period from the mid-1930s to about 1960, when the modern study of gravitational waves was developing, some attention is also paid to the more recent and unexpected emergence of experimental data on gravitational waves which considerably sharpened the debate on certain controversial aspects of the theory of gravity waves. I analyze the use of the earlier history as a rhetorical device in review papers written by protagonists of the "quadrupole formula controversy" in the late 1970s and early 1980s. I argue that relativists displayed a lively interest in the historical background to the problem and exploited their knowledge of the literature to justify their own work and their assessment of the contemporary state of the subject. This illuminates the role of a scientific field's sense of its own history as a mediator in scientific controversy.[1]

This paper makes considerable use of interviews with participants in the events discussed. As yet, no arrangements have been made for a permanent disposal of the materials from these interviews, and the tapes have not yet been made into complete transcripts. In the meantime, anyone interested in their contents should contact the author directly, who will be happy to oblige any requests, subject to the agreement of the interviewee. In one or two cases, such as the interview with S. Chandrasekhar, no tape is available, only notes taken at the interview.

[1] The construction of history as part of the self-definition of a field of science is an important topic in the history of science. For an excellent discussion in a different context, see Barkan 1992.

H. Goenner, J. Renn, J. Ritter, T. Sauer (eds.)
The Expanding Worlds of General Relativity
(Einstein Studies, volume 7) pp. 207–234

1. The Einstein-Rosen paper

In a letter to to his friend Max Born, probably written sometime during 1936, Albert Einstein reported:

> Together with a young collaborator, I arrived at the interesting result that gravitational waves do not exist, though they had been assumed a certainty to the first approximation. This shows that the non-linear general relativistic field equations can tell us more or, rather, limit us more than we have believed up to now.[2] (Born 1971: 125)

The young collaborator was Nathan Rosen, with whom Einstein had been working for some time, producing papers on several topics. They had submitted a paper to the *Physical Review* based on the work referred to in Einstein's letter to Born under the title "Do Gravitational Waves Exist?"[3] and the answer they proposed to give, as the letter states, was no. It is remarkable that at this stage in his career, Einstein was prepared to believe that gravitational waves did not exist, all the more so because he had made them one of the first predictions of his theory of general relativity. In his autobiography Leopold Infeld, who arrived in Princeton in 1936 to begin an important collaboration with Einstein, described his surprise on hearing of the result (Infeld 1941: 239). Despite his initial scepticism, Infeld soon allowed himself to be convinced by Einstein's arguments, and even came up with his own version of the proof, which reinforced his belief in the result (Infeld 1941: 243). However, not everyone was so easily convinced. When Einstein sent the paper to the *Physical Review* for publication, it was returned to him with a critical referee's report (EA 19-090), accompanied by the editor's mild request that he "would be glad to have your reaction to the various comments and criticisms the referee has made." (John T. Tate to Einstein, 23 July 1936, EA 19-088). Instead, Einstein wrote back in high dudgeon, withdrawing the paper, and dismissing out of hand the referee's comments:

> Dear Sir,
>
> We (Mr. Rosen and I) had sent you our manuscript for *publication* and had not authorized you to show it to specialists before it is printed. I see no reason to address the—in any case erroneous—comments of your anonymous expert. On the basis of this incident I prefer to publish the paper elsewhere.
>
> Respectfully,
>
> P.S. Mr. Rosen, who has left for the Soviet Union, has authorized me to represent him in this matter.[4] (Einstein to Tate, 27 July 1936, EA 19-086; emphasis in the original)

---

[2] "Ich habe zusammen mit einem jungen Mitarbeiter das interessante Ergebnis gefunden, daß es keine Gravitationswellen gibt, trotzdem man dies gemäß der ersten Approximation für sicher hielt. Dies zeigt, daß die nichtlinearen allgemeinen relativistischen Feldgleichungen mehr aussagen, bezw. einschränken, als man bisher glaubte." (Born 1969: 173) The translation is by Irene Born, from the English language edition.

[3] Although the original version of Einstein and Rosen's paper probably no longer exists, its original title is referred to in the report by the *Review*'s referee (EA 19-090).

[4] "Sehr geehrter Herr:
Wir (Herr Rosen und ich) hatten Ihnen unser Manuskript zur Publikation gesandt und Sie nicht autorisiert, dasselbe Fachleuten zu zeigen, bevor es gedruckt ist. Auf die—übrigens irrtümlichen—

To this Tate replied that he regretted Einstein's decision to withdraw the paper, but stated that he would not set aside the journal's review procedure. In particular, he "could not accept for publication in THE PHYSICAL REVIEW a paper which the author was unwilling I should show to our Editorial Board before publication." (Tate to Einstein, 30 July 1936, EA 19-089). Einstein must have continued in his dislike of the REVIEW's editorial policy (which in fairness may have been unfamiliar to him, the practice of German journals being less fastidious),[5] for he never published there again.[6] The paper with Rosen was, however, subsequently accepted for publication by the *Journal of the Franklin Institute* in Philadelphia.[7]

What had led Einstein to the conclusion which so surprised Infeld? He and Rosen had set out to find an exact solution to the field equations of general relativity that described plane gravitational waves, and had found themselves unable to do so without introducing singularities into the components of the metric describing the wave. As a result, they felt they could show that no regular periodic wavelike solutions to the equations were possible (Rosen 1937 and 1955). However, in July of 1936, the relativist Howard Percy Robertson returned to Princeton from a sabbatical year in Pasadena and subsequently struck up a friendship with the newly arrived Infeld. He told Infeld that he did not believe Einstein's result, and his scepticism was much less shakeable. Certain that the result was incorrect, he went over Infeld's version of the argument with him, and they discovered an error (Infeld 1941: 241). When this was communicated to Einstein, he quickly concurred and made changes in proof to the paper which was then with the *Franklin Journal*'s publisher (Infeld 1941: 244 and letter, Einstein to editor of the *Franklin Journal*, 13 November 1936, EA 20-217).[8]

Ausführungen Ihres anonymen Gewährsmannes einzugehen sehe ich keine Veranlassung. Auf Grund des Vorkommnisses ziehe ich es vor, die Arbeit anderweitig zu publizieren.

Mit vorzüglicher Hochachtung

P.S. Herr Rosen, der nach Sowjet-Russland abgereist ist, hat mich autorisiert, ihn in dieser Sache zu vertreten." The translation from the original German is by Diana Barkan.

[5] In a letter to Einstein in March 1936, Cornelius Lanczos remarks on "the rigorous criticism common for American journals," such as the *Physical Review* (translated and quoted in Havas 1993: 112). Infeld claims that the German attitude, by contrast, was "better a wrong paper than no paper at all." (Infeld 1941: 190). Jungnickel and McCormmach (1986) describe the editorial workings of the *Annalen der Physik* in the first decade of this century in some detail. They note that "the rejection rate of the journal was remarkably low, no higher than five or ten percent," and describe the editors' reluctance to reject papers from established physicists (p. 310). As this was the time and place in which Einstein began his published career, the "rigorous criticism" he was to experience very shortly after receiving Lanczos' letter must have come as something of a shock.

[6] Einstein's bibliography to 1949, given in Schilpp 1949, lists no papers by him appearing in *Physical Review* after 1936, and the index of the *Physical Review* from then until his death refers only to one short note of rebuttal, mentioned by Pais (1982: 494–495) in his brief account of the rejection of the Einstein-Rosen paper.

[7] The paper appeared in the *Franklin Journal* under a different title and with radically altered conclusions in early 1937. That it had previously been accepted in its original form is indicated by a letter from Einstein to its editor on 13/11/36 (EA 20-217), explaining why "fundamental"changes in the paper were required because the "consequences" of the equations derived in the paper had previously been incorrectly inferred.

[8] Curiously, Infeld states that when he communicated to Einstein his discovery with Robertson of an error in his (Infeld's) version of the proof, Einstein replied that he had coincidentally and independently uncovered a (more subtle) error in his own proof the night before (Infeld 1941: 245). He does tell

Although a footnote attached to the published version acknowledges Robertson's help, it does not indicate its nature (Einstein & Rosen 1937). However, it appears that his chief contribution was to observe that the singularity could be avoided by constructing a cylindrical wave solution. In this way the offending singularity would be relegated to the infinitely long central symmetry axis of the wave, where it was less objectionable, being identifiable with a material source (Rosen 1955). In view of this, Einstein might have been better advised not to dismiss the referee's report so hastily, as the anonymous reviewer also observed that, by casting the Einstein-Rosen metric in cylindrical coordinates the apparent difficulty with the metric was removed, and it was easily seen to be describing cylindrical waves (Referee's report, EA 19-090: 2, 3, 5).[9] That Robertson was familiar with the referee's criticisms is shown by his letter to Tate of 18 February 1937 (Caltech archives, Robertson papers, folder 14.6) in which he says

> You neglected to keep me informed on the paper submitted last summer by your most distinguished contributor. But I shall nevertheless let you in on the subsequent history. It was sent (without even the correction of one or two numerical slips pointed out by your referee) to another journal, and when it came back in galley proofs was completely revised because I had been able to convince him in the meantime that it proved the opposite of what he thought.
>
> You might be interested in looking up an article in the *Journal of the Franklin Institute*, January 1937, p. 43, and comparing the conclusions reached with your referee's criticisms.

This suggests that, in spite of himself, Einstein did benefit from the referee's advice in the end, by a very circuitous route.

In fact the cylindrical wave solution presented in the revised paper had been previously published by the Austrian physicist Guido Beck in 1925, but his paper has been largely overlooked since. In a 1926 paper by Baldwin and Jeffrey, and in the referee's report on Einstein's paper, there was discussion of the fact that singularities in the metric coefficients are unavoidable when describing plane waves with infinite wave fronts, but although there is some distortion in the wave, "the field itself is flat" at infinity, as the referee noted (EA 19-090: 9). In any case, the Einstein-Rosen paper, as published, contains no direct reference to any other paper

---

us that Einstein's position still had to evolve from that of demolishing his proof, to that of reversing it (by showing an exact solution for cylindrical waves), and this was Robertson's key contribution according to Rosen's paper of 195 Unfortunately, Infeld gives us no details of the false proofs and their correction in his account, which was intended for a popular audience. He does relate the amusing detail that Einstein was due to give a lecture in Princeton on his new "result," just one day after completely reversing his conclusions on its validity. He was forced to lecture on the invalidity of his proof, concluding by stating that he did not know whether gravitational waves existed or not (Infeld 1941: 246).

[9] The identity of the Review's referee is unfortunately not known. Few records of the journal exist for this period, and the report has only survived amongst Einstein's own papers. It is 10 pages long and shows an excellent, if not perfect, familiarity with the literature on gravitational waves (the referee knew of Baldwin and Jeffrey's 1926 paper, but not Beck's of 1925). The copy forwarded to Einstein is typewritten and the spelling follows American practice ("behavior" rather than "behaviour", "neighborhood" rather than "neighbourhood"). It is likely, therefore, that the author was an American with a strong interest in general relativity, not a very inclusive category at this time. It is tempting to suspect Robertson himself, especially in view of his subsequent remarks to Tate, quoted below.

whatever. Rosen published a paper in 1937 in a Soviet journal, carrying through what is presumably the chief argument of the original version of the Einstein-Rosen paper, in order to show that plane gravitational waves were an impossibility due to the ineradicability of singularities in the metric. In the immediate post-war period, other papers suggested that plane waves were not permitted in general relativity (for example, McVittie 1955). Felix Pirani and Hermann Bondi were both partly motivated by these papers to work on the problem of gravitational waves.[10] In the mid-fifties, Ivor Robinson independently rediscovered the plane wave metric and, together with Bondi and Pirani, published the seminal work on the subject. They were familiar with Rosen's paper, and noted that his regularity conditions for the metric were unnecessarily severe by post-war standards. "In effect, Rosen did not distinguish sufficiently between coordinate singularities and physical singularities, which could, in principle, be detected experimentally" (Bondi, Pirani & Robinson 1959).[11]

2. Gravitational radiation since Einstein

In 1916, in a paper exploring the physical implications of the final version of his general theory of relativity, Einstein proposed the existence of gravitational radiation as one of its important consequences (Einstein 1916). Although both Maxwell and Poincaré have been cited as anticipating the idea of gravitational waves (Havas 1979 and Damour 1987a), Einstein produced the first concrete description in a relativistic field theory. In a subsequent paper of 1918, Einstein corrected some errors in his previous description of the waves, and went on to calculate the flux of energy carried by the waves far from their source (Einstein 1918). Appealing to the principle of conservation of energy, he assigned an equivalent loss of energy to the source system, an effect already familiar from electromagnetic theory, nowadays known variously as "radiation reaction,""back reaction" or, in cases involving the decay of periodic motion such as orbital motion, "radiation damping." Because Einstein's formula for the energy emission depended on changes in the mass quadrupole moment of the source, it became known as the quadrupole formula. In deriving the formula, Einstein made use of a linearized version of his field equations both for ease of manipulation and because of its strong analogy to the field equations of electromagnetism. Not surprisingly, therefore, his quadrupole formula was itself similar in form to the multipole radiation formulas of electromagnetism, in which field, however, the lowest order of emission is the dipole.

Einstein was not the first to discuss gravitational radiation reaction. In 1908, Poincaré had suggested that planetary orbits must slowly lose energy to wave emission in the gravitational field and indicated that any such effect would be

---

[10] Interviews by the author with Hermann Bondi (7 November 1994) and Felix Pirani (25 October 1994). Pirani reviewed the McVittie 1955 paper for *Mathematical Reviews* and was dissatisfied with its conclusions (Pirani 1955).

[11] In their work, Bondi, Pirani and Robinson followed the new approach of Lichnerowicz in imposing regularity conditions on the metric (Lichnerowicz 1955). For a thorough review of the tangled history of plane gravitational waves, see Schwimming 1980.

too small to explain the perihelion shift of Mercury (Poincaré 1908). As early as 1776, Pierre Laplace had considered the problem of an orbital damping force arising from a finite speed of propagation of gravity. His aim was to discover an explanation for the observed decrease of the Moon's orbital period with respect to ancient eclipse observations (Laplace 1776).

In general there are two distinguishable approaches to the back reaction problem. The first, and generally the simpler is the energy balance argument used by Einstein in his 1918 paper. This approach has been criticized in principle on several counts in the context of general relativity, but was an obvious choice for a first approximation. The second approach, more direct but much more complex, is to iteratively calculate the effect of the source's own field (changing because of the source's motion), upon the source's motion, corrections to which can then be reapplied to calculate the field more accurately. This iteration is carried through one or more steps until it is judged that the reaction effects have been calculated to the desired level of accuracy. This problem is part of a more general one known as the problem of motion. Laplace's method, which took into account the deflection of the Newtonian central force on an orbiting body as a result of the time lag in propagation, was a "one-step" calculation of this type. A key issue in this approach is the fact that the field, in the case of finite propagation, is "retarded", which is to say that the field experienced at a given point in space, at a given time is not that produced by the source at that time, but that of the source at an earlier time, where the difference between the two times is the time of propagation of the field changes from the source's retarded position to the field point in question. As Laplace showed, an orbital decay would be one consequence of introducing retarded propagation instead of dealing with instantaneous propagation. His ultimate conclusion, however, was that the lunar orbital decay could be explained by other, conservative gravitational effects. Therefore finite propagation times had no observable effect in real systems, and the instantaneous action-at-a-distance hypothesis of the day was justified (Laplace 1805: 325–326).[12]

3. Later work on radiation

Arthur Stanley Eddington is associated with the remark that gravitational waves propagate with "the speed of thought" (Eddington 1922). Despite the scepticism this implies, Eddington was arguing only that certain classes of gravity waves, the "transverse-longitudinal" and "longitudinal-longitudinal" waves analyzed by Weyl (1921: 228; 1922: 252) and Einstein (1918) were unphysical. As mere coordinate effects they could be propagated with any velocity desired by the human mind. In the linearized theory at least, Eddington could show that transverse-transverse

[12] See also Damour 1982 for a brief but interesting discussion of Laplace's"radiation reaction" calculation. It is now known, from laser rangefinding, that the moon is receding from the earth, not approaching it. But the increased lunar orbital angular momentum is gained at the expense of earth's rotational velocity, by tidal friction. The resultant lengthening of the earth's day gives the appearance of quickening to all celestial motions, including the lunar orbital period (i.e., although the month has lengthened, it is shorter in terms of days, since the day has also grown longer).

waves could carry energy, and he reproduced Einstein's quadrupole formula while correcting an erroneous factor of two in Einstein's early version (Eddington 1922: 279). He noted, at the same time, that the linearized theory was invalid for sources such as binary stars, in which the system was held together by gravitational forces (Eddington 1922: 280). In 1941, the Russian physicists Lev Landau and Yevgeniy Lifschitz published a back reaction calculation that did treat a binary star system, including its gravitational binding, in the slow-motion weak-field case (Landau & Lifschitz 1951). Their analysis has been influential, although some have felt that it took too much for granted, a problem worsened by the book's terse style.

Although the main topic of the Einstein-Rosen paper had nothing explicitly to do with the back reaction problem, it is very noteworthy as the first serious (if abortive) attempt to disprove the existence of gravitational waves. In an interesting passage addressing radiation reaction, the published paper suggests that one is not compelled to the conclusion that waves emitted by a source must damp the source's motion, if one supposes that any outbound radiant energy is matched by a second system of incoming waves, impinging on the source. In short, they observed that the use of half-advanced plus half-retarded potentials will avoid motion damping in the source system even if the waves exist. "This leads to an undamped mechanical process which is embedded in a system of standing waves," in the author's words (Einstein & Rosen 1937: 48). The paper refers cryptically to the work of Ritz and Tetrode "in former years" relating to the question of advanced versus retarded potentials (in which the field at time t is that produced by the source from a *future* or a *past* position respectively), and it appears that Einstein often quoted Ritz approvingly in this context (Infeld & Plebanski 1960: 201).

Walter Ritz, a Swiss contemporary and friend of Einstein's had complained in his criticism of Lorentz's electrodynamics that advanced potentials were admitted as solutions of the equations of electrodynamics just as well as the retarded potentials (Ritz 1908). To Ritz, this defied the principle of causality, since effect preceded cause. Just as abhorrent to Ritz were combinations of the two potentials, such as the average of advanced and retarded fields (half-advanced plus half-retarded) which allowed "perpetual" motion because, like the instantaneous interaction, it produced no motion damping due to back reaction. Ironically, what Ritz regarded as so damning, Einstein appears to imply might have a positive virtue, in the context of gravitation.[13]

The Dutch physicist Hugo Tetrode, also an acquaintance of Einstein, discussed the half-advanced plus half-retarded potential in a paper of 1922. At the time this solution to the classical wave equations seemed a possible explanation for the failure of orbiting atomic electrons to radiate continuously. Furthermore, as Tetrode pointed out, in the quantum regime, the emission and absorption of radiation seemed to each depend on the other, rather than emission being required for absorption, but not the reverse. This suggested to him that the classical aversion

[13] Since general relativity is a non-linear theory,the fact that two potentials (the advanced and retarded) satisfy the field equations does not imply that their linear combination (half-advanced plus half-retarded) would, as it does in electromagnetism. In linearized gravity, however, this obviously does follow.

to making absorption a requirement for emission should be discarded. As he put it, "the Sun would not shine if it were alone in the universe"[14] (Tetrode 1922: 325; translation by J. Dörling). In their paper, Einstein and Rosen appear to share Tetrode's preference for this potential, if not for his full action-at-a-distance program.

The story, in any case, is of particular concern to us, because of the project upon which Einstein and Infeld now embarked together with Banesh Hoffman. They wished to develop the post-Newtonian theory of the problem of motion, an ambitious project involving intensive calculations (Einstein, Infeld and Hoffman 1938). Since the non-linear field equations of relativity are too complex to be solved exactly for dynamical systems of masses, approximation schemes are required. In general relativity, two different schemes have been commonly employed. The post-Newtonian expansion makes corrections to the Newtonian motion of the system. Since the Newtonian limit is only valid for weak fields and slow motion, the expansion is in powers of the field strength (expansion parameter $(G/c^2)(m/r)$, where G is the gravitational constant, c the speed of light, and m and r represent internal masses and distances of the source) and the source velocities (expansion parameter v/c, where v represents small velocities of the source). An alternative approach is to make corrections to the linearized equations of motion, in an expansion based on powers of the field strength alone. Because it was not limited to small velocities, the second approach became known as the fast motion approximation (and since the 1970s as "post-linear" or "post-Minkowski"). For a modern review of approximation methods in the problem of motion, see Damour 1987b.

The problem of motion had been previously tackled by Einstein and others,[15] but the post-Newtonian Einstein-Infeld-Hoffman (EIH) method, was to be one of the more influential, in a very general way. Einstein particularly wished to vindicate his conjecture that in general relativity the allowed motions of the particles were completely determined by the field equations (Einstein & Grommer 1927), in contrast to other field theories where a separate force law is invoked.

Not long after the work was successfully completed, Infeld, who had with Robertson's help secured a position at the University of Toronto, put his graduate student Phillip Wallace to work applying the EIH formalism to the problem of motion in electrodynamics. In their paper, as also in the EIH paper itself (where radiation effects were not considered), we see a preference for the averaged potential, "half-advanced plus half-retarded". Infeld and Wallace state that this solution "does not specify a privileged direction for the flow of time" and is besides the simplest for their method (Infeld & Wallace 1940). They note that this solution does not damp orbital motion, and further state that "the addition of radiation seems from this point of view arbitrary," since one must choose the retarded potential to obtain it. This viewpoint partly reflects Einstein's own. The solutions which admit radiation damping are objectionable because they involve an arbitrary imposition

[14] "Es würde die Sonne nicht strahlen, falls sie allein im Weltall vorhanden wäre."

[15] See Havas 1989 for an excellent review.

of the arrow of time into field theories which are otherwise time-symmetric. Although Ritz had pointed out how this arbitrariness was an unsatisfactory feature of electrodynamics, his conclusion had been that one must choose the retarded potential to make any sense of it, until a theory which imposed it could be found. Einstein however, felt that time asymmetry had no business in field theories and that its origins lay solely in probability theory (Einstein & Ritz 1909). His views may have influenced Infeld, who preferred the half-advanced plus half-retarded potential, with its standing wave solution, as the most natural choice in the EIH approximation. In the case of the gravitational field, where the existence of radiation could not be experimentally proven, Infeld may have felt there was no compulsion to impose the arrow of time, as one would in electromagnetism, knowing from experiment that radiation existed in that field.

In the 1970s, Rosen returned to the problem of the arrow of time in gravitational radiation theory, in a paper whose title notably echoed that of his rejected 1936 submission to *Physical Review* with Einstein (Rosen 1979). In "Does Gravitational Radiation Exist?" he adapted the Wheeler-Feynman absorber theory to gravitation, and concluded that as the gravitational force interacted much less strongly with matter than the electromagnetic field, a source system would not undergo radiation reaction for lack of a sufficiently strong absorber field. In the Wheeler-Feynman theory it is the field of the absorbers, back-reacting on the source, that breaks the time symmetry of the source field. (Wheeler & Feynman 1945 and 1949). However, Rosen's arguments do not appear completely convincing even to himself, since towards the end of the paper he retreats to a more Tetrode-like position, conceding that an absorber (such as a gravity wave detector) could presumably act so as to draw energy from the source at a distance. In any case, his paper did not excite much debate on the subject.

4. **Post-war work**

The first post-Newtonian attempts to deal with gravitational radiation reaction via the problem of motion had to wait until after the war. In 1946 Ning Hu, a Chinese graduate of Caltech, presented results based on a scheme inspired by the EIH method to the Royal Irish Academy in Dublin, reporting an energy loss disagreeing with the quadrupole formula in the case of an equal mass binary system in a circular orbit (Hu 1947). Shortly before publication, however, he added a note in proof after finding a calculational error which changed the sign of his result, giving anti-damping instead of damping. In other words, the system would gain, rather than lose energy as the result of emitting radiation. The binary would therefore slowly increase, not decrease in radius. In Canada, Infeld and his student, Adrian Scheidegger, worked on the problem of gravitational radiation reaction in the EIH formalism (Infeld & Scheidegger 1951). They concluded that the most natural treatment of the scheme, employing the standing wave boundary condition, led to a no-radiation-reaction result. It was possible, they conceded, to find terms at certain large odd powers of v/c which appeared to correspond to back-reaction terms, but they contended that these could always be transformed away by a suitable choice of coordinates. The result, when announced at an American Physical Society meeting in 1950, "gave rise to a considerable flow of discussion," as Scheidegger put it (Scheidegger 1951). That same year Infeld left Canada, after a McCarthyite campaign against him organized in the press and in parliament, absurdly alleging that he was in possession of atomic secrets. He returned to his native Poland, while Scheidegger continued to argue the no-damping position in North America in his absence, before leaving the field of general relativity for that of geophysics in the mid-fifties.

In 1955 came two futher contributions. Joshua Goldberg, a student of Peter Bergmann (who had criticized the Infeld and Scheidegger results), examined the reaction problem in the EIH formalism (Goldberg 1955). His conclusions were twofold. On the one hand, he denied that the slow motion approach tended to exclude the possibility of damping (arguing that coordinate transformations that removed some back-reaction terms, would reintroduce other reaction terms of odd order in v/c), but on the other hand, he determined that it was poorly suited to the back reaction problem, principally because of the restriction to slow motions of the source. In fact, it was generally agreed that radiation reaction terms did not enter into the post-Newtonian equations of motion until terms of order at least $(v/c)^5$ beyond Newtonian order. Since leading order post-Newtonian effects ($(v/c)^2$ order), such as those obtained by EIH, were both small and difficult to calculate, the expansion method seemed unpromising for studying radiation in that it had to be pushed to high order to succeed. A couple of years later Goldberg was introduced to Peter Havas, a physicist with experience in the problem of radiation in special relativity, who shared his interest in developing a fast motion expansion in general relativity. Having each worked on the problem independently, they began a collaboration based on this approach.[16]

Also in 1955, the Russian physicist Vladimir Fock treated the orbital damping problem in his book *Spacetime and Gravitation* (Fock 1959: sect. 90). He made use of a slow-motion expansion which he had developed independently of EIH, coupled with "no-ingoing wave" boundary conditions in the past of the system. His results were in agreement with those of Landau and Lifschitz. His work was not translated into English for four years, and even then wielded little influence in the west, perhaps because of Fock's unorthodox views on general covariance. He employed so-called "harmonic coordinates" in his calculations, and claimed a special status for them in physical theory. His views in this regard were vigorously opposed by Infeld and most other relativists then and since. Furthermore, Fock himself regarded his back-reaction result as merely demonstrating that wave phenomena played an inconsequential role in the problem of motion in gravity, due to the small size of the effect for known astronomical systems.

---

16 Interviews with Joshua Goldberg (10 April 1995) and Peter Havas (5 April 1995).

5. The Bern and Chapel Hill conferences

Between the War and the Bern conference of 1955 marking the 50th anniversary of special relativity, general relativity was at a low ebb (Eisenstaedt 1986a, b). Work on the radiation problem seemed confused and controversial, leading only to some consensus that the problem required closer attention. At the Bern conference Rosen, returning to the cylindrical wave solution of his 1937 paper with Einstein, adduced evidence backing up Scheidegger's position by proposing the possibility that gravitational waves did not transport energy (Rosen 1955). It is a peculiar characteristic of general relativity that the energy contained in the gravitational field, and thus the energy in gravitational radiation, is not described in a coordinate invariant way. This energy is considered to be real enough, and can be converted into other forms of energy which can be expressed invariantly, but the principle of equivalence prevents one from doing this for field energy in gravity. The reason is that any observer in a gravitational field is always entitled to imagine himself in a locally Lorentz (that is zero gravity) freely falling frame of reference which, locally, contains no field energy. Of course, one is not free to transform away the entire field energy of a planet but one can always choose coordinates on an infinitesimally small portion of its surface so as to eliminate the field energy in that region. Thus it is said that gravitational field energy is non-localizable.[17] This problem of defining field-energy had led Einstein, Landau and Lifschitz and others to employ a non-invariant quantity known as a pseudo-tensor to describe energy in the wave flux in their back reaction calculations. Rosen now observed that each of these (slightly different) definitions of the pseudo-tensor showed no energy at all when applied to the cylindrical waves of his 1937 paper with Einstein in cylindrical coordinates. Although drawing conclusions on the tentative basis of the pseudo-tensor was regarded as dangerous, Rosen observed that the result seemed to support the view of Infeld and Scheidegger. This cast further doubt on the uncertain status of wave phenomena in gravitation theory.[18]

---

[17] In 1968, Richard Isaacson discovered an invariant tensorial quantity that described wave energy in a local sense, by averaging over a wavelength of the wave. Thus, using this approach, gravitational wave energy can be localized within a wavelength, but no further (Isaacson 1968).

[18] As we shall see, Rosen's paper was soon answered in a manner convincing to most relativists. He himself revised his opinion on this matter in a letter to the *Physical Review* (Rosen 1958), after realising that, when using Cartesian coordinates, the pseudo-tensor did show energy in the cylindrical waves. His new calculations on the energy content of cylindrical waves did not appear until after some delay (Rosen & Virbhadra 1993). The issue was addressed in some depth in the fifties, however (Stachel 1959). The problem of the pseudo-tensor in the study of gravitational waves was not new then, nor has it entirely ceased to be the subject of debate since. In recent years, Fred Cooperstock has suggested that, based on the hypothesis that the preferred frames of reference when describing the field energy should be those that eliminate the pseudo-tensor, the gravitational field energy should be described only by an invariant tensor quantity. The result of this would be that the conservation relation in relativity would require that no field energy be present where there was no matter, preventing gravitational waves from propagating energy through empty space (Cooperstock 1992). In the very early days of general relativity Levi-Civita had made a proposal with somewhat similar (but more drastic) consequences, in response to the confusing and incorrect results derived by Einstein in his 1916 paper on gravitational waves (Levi-Civita 1917). For a very interesting discussion of this episode, which includes some revealing comments reflecting the initial unease about gravitational radiation brought on by Einstein's early errors (including the mistaken conclusion of his 1916 paper that spherically symmetric motions of matter could generate gravitational waves), see Cattani & De Maria 1993.

The Bern conference is remembered as an important stimulus to the field of relativity. The discussions there, and the interest taken by Felix Pirani, prompted Hermann Bondi to take up the problem of gravitational radiation.[19] Bondi brought an open mind to the issue, in the sense that he was sceptical enough of the existence of gravitational waves. He was influenced in this by Eddington, from whose writings he had learned relativity. Eddington's emphasis on a coordinate invariant approach, making use of tensorial quatities such as the Riemann curvature tensor, had enabled him to show that certain classes of gravitational waves "in existence" before 1922 were spurious (Eddington 1922). Bondi, like some relativists of the day, was not impressed by the existing radiation reaction work, finding Landau's and Lifschitz' treatment "a little glib."[20] At the same time, gravitational waves seemed like an attractive topic within gravitational theory, since in this area the predictions of general relativity diverged radically from those of Newtonian gravitational theory. Up to this time, most work in relativity, outside of cosmology, had been devoted to deriving small corrections to Newtonian theory, such as the famous perihelion shift of Mercury, a more precise estimation of which was one of the goals of the EIH paper (Robertson 1938). The study of gravitational waves, if they existed, seemed likely to generate more 'new physics' than simply adding terms to Newton's theory. Now, as Infeld himself observed when writing of his surprise at Einstein's "proof" that waves did not exist, no respectable modern field theorist would, under normal circumstances, deny the existence of radiation in a field theory. The mere fact that the force was propagated in the field rather than by action-at-a-distance, a basic tenet of all relativistic field theories, seemed to imply the existence of radiation. Einstein also remarked, in his letter to Born, of the "certainty" which the analogy between the linearized Einstein equations and electromagnetism had inspired concerning the existence of a gravitational analogue to the Maxwellian wave equation. Bondi nevertheless seized on a key argument made by Infeld and Scheidegger, which seemed to him important.

As Scheidegger observed, relativity occupied a "peculiar place" amongst classical field theories (Scheidegger 1953). One important peculiarity is that the equations of motion are constrained by the field equations, as Einstein had noted. In electrodynamics, where this was not the case, one was perfectly free to demonstrate damping effects by moving the particles around in whatever fashion, and showing that this gave rise, when the field equations were invoked, to radiation and loss of energy from the local system. In relativity, it was necessary to show that the motions in question were allowed by the same field equations. This was all the more important when one considered the question of what *type* of motion gave rise to radiation. One obvious example was an accelerating charge in electrodynamics. What of the apparently equivalent case of a falling mass? It was clearly accelerating with respect to the person who dropped it, but in a relativistic

[19] Interviews by the author with Felix Pirani (25 October 1994) and Hermann Bondi (7 November 1994).

[20] Interview with Hermann Bondi (7 November 1994).

sense, it was merely following a geodesic, doing what came naturally, as it were.[21] In terms of the local space-time, the particle that was really being *accelerated* was the one still being held in the observer's other hand, which was prevented from falling freely. Which one of these particles *ought* to radiate? This question had no immediately obvious answer that the relativists of the day could agree upon.[22]

At the Chapel Hill conference of 1957 and elsewhere at that time, Bondi pointed out the distinction between two masses being waved about at the end of someone's arms,[23] clearly not following geodesics, and clearly emitting gravitational waves (but vanishingly weak ones!), and two masses in a binary star system, following geodesics and, if Infeld and Scheidegger were right, not radiating anything (DeWitt 1957: 33). Since gravitational forces were likely to be the only forces capable of moving large masses very quickly, the issue of whether purely gravitational systems could give rise to radiation was an issue of whether such radiation would ever be detectable. That issue, to the surprise of most theorists, was soon to become one of some practical interest.

The Chapel Hill conference on "The Role of Gravitation in Physics" brought together relativists and theoretical physicists interested in then new topics such as quantum gravity. The session on gravitational radiation was lively and varied. Felix Pirani presented important new work on wave theory (DeWitt 1957: 37). Influenced by the Irish relativist John Synge during a year spent in Dublin (and also by the work of Petrov (1955) and Lichnerowicz (1955)),[24] Pirani drew attention to the Riemann curvature tensor, whose importance had previously been stressed by Eddington in his 1922 paper, as an invariant geometrical quantity which was well suited to the description of the behavior of gravitational waves. Using the geodesic deviation description of gravitational effects advocated by Synge, he showed how particles in the path of a wave were moved about relative to each other by the space-time curvature of the passing wave. In this view, gravitational waves were depicted as ripples in the fabric of space-time itself, whose physical effects were observable by monitoring the relative motion of two adjacent particles during the passage of a wave.

Later in the conference an interesting exchange took place during the section on quantization of gravity. During Richard Feynman's presentation on the need for a quantum theory of gravity, Rosenfeld made the following remark:

> It seems to me that the question of the existence and absorption of waves is crucial for the question whether there is any meaning in quantizing gravitation. In electrodynamics the whole idea of quantization comes from the

---

[21] The question of whether particles following geodesics should radiate, given that they are behaving "naturally" in a gravitational field, seems intriguingly Aristotelian.

[22] Interview with E. T. Newman (11 April 1995). He relates how J. A. Wheeler once asked a roomful of relativists to vote on the answer to the two particle question and recalls the room being fairly equally divided. This seems to be a rare example of the 'democratic' approach to science.

[23] A number of those interviewed by the author recalled Bondi vigorously demonstrating this method of generating gravitational waves.

[24] Interview with Felix Pirani (25 October 1994).

> radiation field, and the only thing we know for sure how to quantize is the pure radiation field. (DeWitt 1957: 141)

Feynman demurred somewhat from the premise, arguing that there existed a quantum theory of electrostatics, but agreed that some of his arguments in favor of quantization depended on the existence of waves. Bondi was moved to note that "this vexed question of the existence of gravitational waves does become more important for this reason." Feynman then presented an argument based on Pirani's earlier talk. Appealing to the equation of geodesic deviation, he argued that a particle lying beside a stick would be rubbed back and forth against the stick by a passing wave, and the friction would generate heat, so that energy would have been extracted from the wave. Furthermore, he felt that any system which could be an absorber of waves, could also be an emitter. For these reasons, he expected gravitational waves to exist (DeWitt 1957: Supplement).[25]

This line of argument, suggested by Pirani's new work, was also elaborated in two papers published that same year. In a letter to *Nature*, Bondi used a slightly different version of it to refute Rosen's argument of 1955 on energy transport (Bondi 1957), as did Joseph Weber and John Wheeler in a more detailed paper (Weber & Wheeler 1957). Weber demonstrated real confidence in the physicality of gravitational waves by embarking within a few years on an experimental program to detect them, using large resonant metal bars as antennae (Weber 1960). 'Quixotic' is probably not quite the word contemporary theorists would have used to describe Weber's aim.[26] The wave theory, in so far as it existed at all, with no particular notion as to potential astrophysical sources or signals, would be better described as a "disabling" rather than an enabling theory for experiment. The quadrupole formula, the only guide to source strength and signal amplitude, suggested that any waves reaching the detector would be very weak. With no theory of sources, the question of what frequency to search at was theoretically undetermined.[27] It is remarkable that the field of gravity wave detection began at a time when the theoretical state of the subject was in such disarray.

6. The rebirth of relativity

An important requirement for the development of any scientific field is funding. The field of gravitational wave theory was fortunate in this regard in that, from 1956

[25] Cooperstock's 1992 paper (see note 10) contains an argument based on a counter-example to the Feynman-Bondi thought experiments, which claims that no energy is deposited in the "absorber" despite the motion locally induced by the wave. His hypothesis would imply that waves exist in general relativity, are detectable by certain types of instruments, but carry no energy. However, this paper and its conclusions have provoked little debate. This perhaps reflects the difficulty in physics of reopening an argument considered closed by most in the field. At some point, the premise of the paper becomes sufficient grounds for dismissal. However, the problem may be simply due to the fact that papers outside the current thrust of research interests are unlikely to receive much attention, whatever their conclusions.

[26] Several interviews (especially one with Joseph Weber, 20 June 1995) and anecdotal recollections, as well as the impression given by conference proceedings, agree that the reaction to Weber's initial efforts to detect gravitational waves in the 1960s ranged from polite scepticism to derision.

[27] Interview with Joseph Weber (20 June 1995).

to 1963, Joshua Goldberg was responsible for United States Air Force support of research in general relativity, based at the Aeronautical Research Lab at Wright-Patterson Air Force Base in Ohio. At this time, and up until the passage by Congress in 1969 of the Mansfield Amendment prohibiting the Department of Defense from sponsoring basic scientific research, the US armed forces provided considerable financial support for even very esoteric subjects in theoretical physics. Goldberg was active himself in the study of gravitational radiation, as we have seen, and did much to encourage groups such as that of Bondi and Pirani at King's College, London. Although support was available for groups outside the US, it was not permitted to support scientists based in Communist countries, inhibiting the use of these funds to facilitate travel between the London group and Infeld's group in Warsaw, who interacted extensively.[28] The Air Force laboratory itself was home to an active group until the 1970s. With one of his earliest grants, Goldberg was able to support the Chapel Hill conference organized by Bryce DeWitt with Air Force money, and this important meeting became the forerunner of the successful General Relativity and Gravitation (GRG) series of conferences, which continues today. For a valuable account of this unlikely episode in the history of general relativity,see Goldberg (1988).

Following the Mansfield Amendment, research in relativity theory in the US depended primarily for its support on the National Science Foundation (NSF). From 1973 to the present, the chief advisor on funding for gravitation physics at the NSF has been Richard Isaacson, like Goldberg a relativist who has made important contributions to the theory of gravitational waves. Isaacson had also previously worked at the Air Force laboratory on the Wright-Patterson base. By good fortune then, despite the overall decrease in funding for theoretical physics precipitated by the Mansfield Amendment, the principal source of funds for research on gravitational wave theory remained in sympathetic and knowledgeable hands.[29]

As interest in relativity grew in the period after Chapel Hill, the reaction problem was pursued with renewed vigor. The EIH approximation was adopted by Andrzej Trautman, a student in Infeld's group in Warsaw, who departed from Infeld's approach in adopting "outgoing wave only" boundary conditions. He also confirmed

[28] Interview with Felix Pirani (25 October 1995).

[29] The advantage of having an insider at the primary funding agency did not ensure that everyone in the field was sponsored to the extent that they desired or felt necessary. Complaints about the funding choices made and its effect on research directions were very noticeable on the experimental side, where groups and research programs depended very heavily on the munificence of different (usually governmental) funding agencies. But even on the theoretical side, work on the problem of motion or radiation reaction was computationally so intensive that funding for postdocs and assistants could make a big difference to a group or research program. It may be that less popular research programs suffered in this regard (such as fast motion approximations versus slow motion ones), but it is difficult to assess the extent of this factor. This partial assessment is based on interviews by the author with Richard Isaacson (7 April 1995), Joshua Goldberg (10 April 1995), Peter Havas (5 April 1995) and Joseph Weber (20 June 1995). Given the importance of debates during conference sessions, it is also worth noting the complaint that, because of the influence of slow motion advocates such as Infeld on the organizing committee, the fast motion approximation was not discussed at any of the GRG conferences, such as Warsaw 1962 (Peter Havas, private communication). Thus, the growth of this research program may have been retarded by a lack of exposure.

Goldberg's earlier claim that the net back-reaction effect could not be transformed away, merely moved from one point in the expansion to another. He found positive damping, although differing somewhat from the quadrupole formula result (Trautman 1958a, b). Infeld himself stuck to his earlier opinion, despite the contrary views of his students. In his 1960 book, *Motion and Relativity*, he included a detailed argument against the existence of back reaction in freely falling systems (Infeld & Plebanski 1960), without the knowledge or agreement of his co-author and former student, Jerzy Plebanski.[30] Another effort at this time, by Peres, initially found anti-damping, as had Hu, but this was corrected shortly after, and his new result agreed with that of Landau and Lifschitz for circular binary orbits (Peres 1959, 1960). Peres' second paper has been referred to as containing the first correct back reaction calculation (Thorne 1989). Nevertheless, the perceived arbitrariness of the slow-motion approach in imposing the wave zone boundary conditions from one step in the expansion to the next, which seemed reflected in the wildly differing results produced by the method, gave rise to arguments that the approach was hopeless (Bonnor 1963).

While conceptually more appealing in some ways, the alternative fast-motion approach, as developed by Havas and Goldberg (Havas & Goldberg 1962) and others (for example, Bertotti & Plebanski 1960, Kerr 1959, Westpfahl 1985) was also proving frustrating. It was a difficult task to go beyond the leading order corrections to the linearized theory and the results of applying that step to the reaction problem, published by Smith and Havas, again showed an energy gain in the source (Smith & Havas 1965). Therefore many theorists at the time concluded that the question of whether freely falling sources experienced damping remained unsettled.

Bondi, who with his collaborators had done much to improve the understanding of wave propagation far from the source (see especially Bondi, van der Burg & Metzner 1962 and Sachs 1962) made this point at the Warsaw conference of 1962 (Bondi 1962).[31] However, there were those, like Feynman, who viewed the relativists' caution with impatience. Feynman was "surprised to find a whole day at the conference devoted to this question" (of whether gravity waves could carry energy), as far back as Chapel Hill (letter from R. P. Feynman to Victor

---

30 Interview with Jerzy Plebanski (30 June 1995). In general, however, Infeld proved reasonably tolerant of the opposing viewpoints within his group. Indeed, in the late sixties, shortly before his death, he was finally won over by his students' arguments (interview with Andrzej Trautman, 17 October 1994).

31 Bondi, van der Burg and Metzner (1962) and Sachs (1962) showed that when a certain function (known as the Bondi news function) was present, an isolated system would lose mass to the emission of gravitational waves. At Warsaw in 1962, in the discussion with Bergmann and Feynman (Bondi 1962), Bondi stressed the importance of dealing with specific equations of state in the components of the binary system, because it was his opinion that binaries composed entirely of pressure free dust would not radiate, as all particles would follow geodesics and there would be no possibility of"news", in the form of a departure from geodesic motion. In the case of a real physical system, even if no deviation from geodesic motion occurs, this is news, since no news, if not good news, is still news, if news was expected. Bondi eventually decided against his position that idealized dust-filled binaries might not radiate (interview with Bondi, 7 November 1995). In the same discussion in the Warsaw proceedings (Bondi 1962) Feynman gave a brief account of his own unpublished calculations which convinced him that gravity waves exist.

Weisskopf, 11 February 1961),[32] and was caustic in his appraisal of the discussions at the Warsaw conference, noting they were "not good for my blood pressure" in a letter to his wife (Feynman & Leighton 1988). Bondi's lecture, however, inspired the astrophysicist Subrahmanyan Chandrasekhar to take up the problem.[33] Throughout the 1960s, Chandrasekhar developed his own slow-motion formalism, dealing with extended fluid bodies (as opposed to point masses) at one post-Newtonian order after another (Chandrasekhar 1965). By the end of the decade he had advanced far enough in the expansion (to order $(v/c)^5$ beyond Newtonian order) to describe reaction effects. His conclusion agreed with the quadrupole formula result (Chandrasekhar & Esposito 1970). At about this time William Burke, a student of Kip Thorne's at Caltech, introduced improvements to the slow-motion approach that removed much of the arbitrariness in imposing the boundary conditions. Burke made use of the applied mathematics technique of matched asymptotic expansions, which allowed one to determine the solution to the problem of motion in the zone near the source, by matching it through an intermediate zone, to the "outgoing wave only," or other potential of choice, in the far zone of the waves. In this way the chosen boundary condition could be unambiguously applied to the solution of the near zone problem, thus addressing the arbitrariness that bedeviled the slow-motion approach up to this time (Burke 1969). Employing Burke's novel approach, Burke and Thorne also derived the quadrupole formula for emission from binary systems (Burke & Thorne 1970).

During the sixties, great progress had been made on many fronts in the description of wave propagation and interaction with matter. Possible astrophysical sources, such as supernovæ and binary neutron stars, began to be suggested, inspired at first by Weber's work (Dyson 1963).[34] Some experts were of the opinion that the subject was maturing and furthermore the prospect of some real astrophysical application for gravity waves, seemed to emerge with the discovery of the quasi-stellar ("quasar") radio sources (Fowler 1964; Robinson, Schild & Schucking 1965; and Cooperstock 1967). Then, to the great surprise of the theoreticians, Weber announced in 1969 that he was detecting gravitational waves (Weber 1969). Although his results, which confounded all theoretical predictions of source strengths then and since, were eventually discounted amidst much controversy, they focused much attention on the subject, and sparked a great increase in the number of experimentalists working on gravitational waves. (See Collins 1975 and 1981 for a detailed account and Franklin 1994 for an alternative viewpoint). On the theoretical front, research in the 1960s on black holes, cosmology and other topics had made the field of relativity very relevant to astrophysics. Gravitational waves shared somewhat in this popularity, and seemed likely to continue to grow

---

32 A copy of this letter was kindly supplied to the author by Kip Thorne. Copies are also kept amongst the Feynman papers at Caltech.

33 Interview with Subrahmanyan Chandrasekhar (12 July 1995).

34 Interview with Joseph Weber (20 June 1995). Weber recalls that Freeman Dyson suggested asymmetric collapse of stars during supernova events as one possible source for his detectors in the early 1960s.

in practical importance as experimental interest waxed. The discovery of the first binary pulsar (PSR 1913+16) by Hulse and Taylor in 1975 (Hulse & Taylor 1975) crystalized the excitement in the field, providing the first testbed for strong field effects of general relativity, although there were doubts at first that the system would exhibit measurable orbital damping effects (Damour & Ruffini 1974).[35]

7. The quadrupole formula controversy

The successes in improving the slow motion approximation, and the increasing likelihood of practical applications of gravitational radiation theory encouraged some experts, such as Kip Thorne, to suggest that the reaction problem was now well understood,[36] and the multipole formalism could be used with confidence in astrophysical applications to give approximate estimates of source strength, much as one would in electromagnetic wave theory (Thorne 1980). This viewpoint however, was sharply opposed by some others who were still seriously dissatisfied with the state of the field.[37] They were particularly concerned that the quadrupole formula would be used as a reliable formula in contexts in which its results might be wholly misleading. One of these was Havas who was still very unhappy with the various slow motion results (Havas 1973).[38] One of his students, Arnold Rosenblum, brought the unsatisfactory state of affairs to the attention of the mathematical physicist Jürgen Ehlers, who also took up the cause of alerting the relativity community to the dangers of complacency on the matter.[39]

The alarm sounded by Havas, Rosenblum and Ehlers had the effect of again focusing attention on the reaction question, and this interest was redoubled by the announcement, in 1980 of observations of orbital decay in the binary pulsar. Taylor and coworkers, after years of careful observation of the system, were able to announce an orbital period decrease in line with the predictions of the quadrupole formula with an accuracy of measurement of about 20% (Taylor & McCulloch 1980). The warnings of the unverifiability of the quadrupole formula within the theory now had their effect. A chief use of the binary pulsar data since its discovery had been as a test of general relativity against rival theories of gravity. Agreement between observation and the quadrupole formula could only constitute a test of

[35] Interview with Thibault Damour (11 October 1994).

[36] Interview with Kip Thorne (17 July 1995). Thorne recalls first putting this view forward at a meeting in Paris, June 1967.

[37] Interview with Kip Thorne (17 July 1995). He recalls Havas taking issue with his comments at the Paris meeting, June 1967.

[38] The non-linearities still continued to bedevil the problem in some people's minds. At Caltech, despite Thorne's complacency, Burke noted in early versions of his work that his approach was not guaranteed to work outside of linearizable systems, and therefore could not settle the issue for freely gravitating systems. There is still on display at Caltech the record of a wager between Burke and Thorne on whether non-linear effects would "significantly affect the radiation in the lowest order" from sources in free-fall motion. Thorne gave odds of 25–1 for this bet, which Burke conceded in 1970.

[39] Interview with Jürgen Ehlers (14 October 1995).

general relativity theory if the quadrupole formula was established as a prediction of the theory.[40] Not all relativists were of the opinion that it was so established.[41] There was a surge in interest in the problem of motion and in back-reaction in particular, including by some who had not previously worked on the problem. The great majority of the new results vindicated the use of the formula. A further round of sharp debate ensued as these results, via a wider variety of approaches then ever before, and in greater detail than ever before, convinced many that the issue was at last settled, pushing the remaining sceptics into an embattled minority. As the eighties advanced, and the increasingly convincing experimental data continued to agree solidly with the theoretical work carried out in close parallel by Thibault Damour and his collaborators (Damour 1983), the debate slowly died away. At present the focus in the field is on calculating higher order contributions to the waveforms produced by binary systems for use in conjunction with data extraction techniques in the next generation of wave detectors, indicating a high degree of confidence in most quarters in the basic theoretical position. A direct detection of gravitational waves is still being sought, and this is one of the principle goals of the planned new detectors.

An important feature of the radiation reaction debate in the seventies and eighties was the series of review papers by different authors, each employing the history of the subject to illustrate a particular view of the contemporary state of the field. These papers show that relativists were keenly aware of the history of their field and they were able to draw lessons from their reading of history which reinforced the points they wished to make. The earliest of these papers was that of Ehlers, Rosenblum, Goldberg and Havas whose argument was that previous attempts to deal with the back-reaction problem were all inadequate in one way or another. In consequence, they advanced an outline of a program which would overcome these past failings (Ehlers, Rosenblum, Goldberg & Havas 1976). Essentially an attempt to formulate a research program for the subject, their paper was followed by an Enrico Fermi summer school in Varenna organized by Ehlers, whose aim was also to foster new work in the field along more rigorous lines than before (Ehlers 1979).

Martin Walker and Clifford Will in 1980 took a very different tack, addressing the problem of non-reproducibility which had plagued the subject (Walker & Will 1980). They argued that a basic iterative algorithm, applicable for both fast motion and slow motion methods, could be followed to recover the quadrupole formula

---

[40] The rate of energy emission predicted by the quadrupole formula can be writen as

$$\frac{dE}{dt} = \frac{1}{5}\left\langle \left(\frac{d^3\mathcal{I}_{jk}}{dt^3} \right)^2 \right\rangle,$$

where $\mathcal{I}_{jk}$ is the Newtonian mass quadrupole moment of the source (evaluated at retarded time). The square of the third time derivative of this quantity, averaged over several wavelengths (the meaning of the angular brackets), determines the total energy flux in the waves from the source.

[41] Interview with James Anderson (3 April 1995).

from reaction calculations. They presented an analysis of a cross section of well-known calculations, dating back to the paper of Hu in 1947, and argued that those that had advanced through sufficient steps in the iteration recovered the quadrupole formula, and that others, with fewer steps did not (except for a couple which found the result with the aid of compensating errors). In this view of the history of the field, there existed a definitive method by which the standard results could be recovered in a reliable way. This was in stark contrast to the views expressed by Ehlers et al., which were to advocate a more general prescription, whose outcome was not yet known. Yet another view was put forward by Cooperstock and Hobill in 1982. They refused to set forward a general scheme or advocate a particular result, instead arguing against preconceived notions (Cooperstock & Hobill 1982). Their history, as befitted their standpoint, was more descriptive than prescriptive, celebrating the diversity in the development of the field. Another protagonist with an interest in and excellent knowledge of the field's history was Damour. His papers were often prefaced with a discussion setting his work in a historical context (for example Damour 1982). In this role, the object of history was to motivate the new work being presented, and the focus was on the previous failings which were being addressed by the new contributions (see, for instance, Damour 1983). A more active role for the historical literature was found in the account of James Anderson, who returned to the Einstein-Infeld-Hoffmann scheme complete with its surface integral method, and married it to the matched asymptotic expansions of Burke, with further additions of his own, to produce another influential derivation of the quadrupole formula (Anderson 1987).

A very significant aspect of the debate in the seventies and eighties was the problem of when theory ends.[42] As we have seen, different authors could look at the same history and give very different answers to this question. One answer might be, it already has ended, we really know the answer ("Conservative"). Another is, it has just ended now, with this paper, for the issues addressed ("Technocratic"). A third is, it will end, as soon as the general program we advance is carried through ("Marxist"). A fourth is that it can never end, and it is best that it should not ("Anarchist"). Finally there is the view that the answer is hidden in the past, waiting to be extracted and pieced together from the literature ("Archaeological"). It is interesting that just as there was agreement on the details of the history (and the debate was largely a historical debate), opinions diverged on the matter of *interpretation*. The lesson of history was different for everyone. This is still the case, but the debate having lost its impetus, the individual perception of history has lost its public relevance once more. The dynamic of the debate is that some level of consensus must be found for the resolution of an existing problem, and yet progress seems to be measured by many scientists by the extent to which an issue can be settled, allowing the next problem to be addressed. A field like general relativity has historical memories of the isolation that may be the fate of a discipline which does not progress in this way. The remarks of Feynman at Chapel Hill (DeWitt 1957: 150), express the view of the progressives, when he says "the second choice

---

42 The analogy to the problem of *How Experiments End* (Galison 1987) should be obvious.

of action is to ... drive on," to "make up your mind [whether gravitational radiation exists] and calculate without rigor in an exploratory way." He concludes with the advice, "don't be so rigorous or you will not succeed."[43] The contrast in attitude suggested here may explain why the debate tended to become more vitriolic in its last stages, as a consensus developed for many, with some still arguing that the matter was unsettled.[44]

In studying the controversy following Weber's announcement of gravitational wave detections, Harry Collins (1985) has introduced the concept of the 'Experimenter's Regress'. This describes the difficulty faced by experimenters when confronted with a dispute over non-confirmation of claimed results. Since none of the experiments will exactly duplicate the others' behavior, achieving consensus is hampered by the problem that the device that is working properly should get the correct result, but the correct result can only be known from the output of a properly operating device. Although Collins' view has been criticized (Franklin 1994), it seems to provide a useful model for understanding the Weber controversy. In the theoretical controversy surrounding gravitational waves, one seems to observe a similar phenomenon, the 'Theoretician's Regress'. The complex, tedious calculations designed to approximate to the full general relativity theory can be thought of as experiments, with the theory itself in the role of a notional "reality." These experiments constituted a delicate technical apparatus, designed to probe this "reality," aided by the craft and mathematical skill of the theorists. "Experimental error" was impossible to account for fully, whether as systematic error in the form of an inappropriate expansion scheme or failure to properly control errors from neglected terms (a difficult problem which was rarely addressed programmatically), or as accidental error in the form of simple calculational mistakes amidst the welter of terms which had to be collected.

As with the experimentalists, direct replication of another method was rarely even attempted. Even the best known schemes, such as EIH, were employed with improvements designed to simplify the calculations or overcome objections in principle, such as the use of point mass sources (Anderson 1995). Therefore, the array of review papers, conference workshops and other social efforts to achieve consensus had to overcome the cycle of regression constructed by the fact that the right scheme would be the one that gave the right result, but the right result was the answer given by the right scheme. The difference in emphasis between those who gave weight to having the right answer, and those who preferred to rely on method alone gave rise to further disagreement. One event that helped to partially break the cycle was the advent of the binary pulsar data. Initially this gave rise to more activity and more disagreement, but it also lent external support to the

[43] The alert reader will have guessed that I have just described as "progressives" the same class of people whose historical outlook I earlier labelled "conservative". In this case, conserving and defending the orthodox historical account plays a crucial role in the progressive agenda, discouraging debate on topics that are regarded as settled and directing energy towards problem solving work within the established paradigm.

[44] Interview with Fred Cooperstock (26 June 1995). A number of other interviewees recalled rather heated exchanges taking place at conferences in the early 1980s during the quadrupole formula controversy.

preferred 'right result' given by the quadrupole formula. It did not however, put an end to disagreements about the correctness of various methods, except in so far as it tended to rule out methods that disagreed with the canonical result. This was enough to gradually bring an end to the public side of the quadrupole formula controversy.

Throughout all this, one notes the tensions within the field over technical matters, especially regarding the level of rigor required to inspire confidence in a particular result. Relativity has a tradition that places it towards the mathematical end of the spectrum in this regard amongst branches of theoretical physics. Yet from the sixties on, astrophysics and relativity became relevant to each other, even spawning the new field of relativistic astrophysics. Theoretical astrophysics stands at the opposite extreme from relativity, preferring a more 'physical' approach, eschewing not only mathematical rigor, but also dependence on exact results. Order of magnitude calculations and heuristic arguments are common. Such arguments, for instance, might be used to identify the 'correct' result, as a guide when undertaking longer calculations.[45] Within relativity there were those whose practice tended towards each approach, and it was naturally difficult for them to agree on the question of standards of proof.[46] For practical purposes results such as the binary pulsar measurements were obviously welcome, but at issue was on whose terms a given result was to be accounted a prediction of general relativity: the"astrophysicists" or the "mathematicians."

A second, less significant, mixing of fields concerned attempts to quantize gravity. Especially in the fifties, it was argued by some that the existence of radiation was a crucial matter for this project (Rosenfeld in DeWitt 1957: 141; Rosen 1979). In fact, Felix Pirani's view was that "the primary motivation for the study of [gravitational radiation] theory is to prepare for quantization of the gravitational field." (Pirani 1965: 368). The uncertain position of general relativity as an independent, yet thriving field, seems to have played into fears and attitudes concerning the radiation problem. Relativists' own practices and their own opinions of what the problems were in the field may have seemed endangered by the twin possibilities of classical relativity becoming a mere adjunct to astrophysics, and the theory as a whole being submerged by a unified quantum field theory of gravity.[47] The emergence of relativity into the mainstream of physics had a highly ambivalent aspect for relativists, in that it brought with it the danger that the character of the small stream would be lost in the larger current. The fears and hopes which this dual prospect raised for scientists who had consciously chosen the field for its own beauty and intimacy no doubt helped shape attitudes in the debate.

---

45 This question of 'style' in physics seems an important one. Chandrasekhar suggests that the greatest physicists (such as Newton) employed both of these styles equally well. He relates that Fermi would say that he would not believe a physical argument without a mathematical derivation, nor would he believe the mathematics without a physical explanation. Interview with S. Chandrasekhar (12 July 1995).

46 Interviews with Kip Thorne (14 June 1995) and Jürgen Ehlers (14 October 1995).

47 See remarks by Roger Penrose in Lightman & Brawer 1990: 429.

8. A final note

I have concentrated, in this paper, on the emergence of certain important issues that contributed to the uncertainty and controversy which at times surrounded the theoretical development of the subject of gravitational radiation. In doing so, not only have I focused here on the immediate post-war period up to about 1960, but I have deliberately not attempted to cover the entire breadth of the literature for any period. I have tried to illustrate how the debate on the existence of gravitational waves came to arise as a serious discussion, and how the subject's own history was used as a rhetorical and motivational tool in subsequent debates after its emergence as the subject of experimental and not just theoretical research. I have left many interesting aspects of the history of this problem for another time. For now, I have tried to give a sense of how the thought that gravitational waves might *not* exist first arose, and then became a serious issue in the field of relativity, and how the issues raised fed into later debate as the subject matured. It is worth noting that *scepticism* about their existence encouraged important scientists, such as Bondi, to focus attention on gravitational waves, at a time when those who were certain of their existence dismissed their effects as insignificant (Landau and Lifschitz and Fock, for instance). If the attempt to detect gravitational radiation is now a multi-million dollar field, thanks to the pioneering work of Weber on the experimental side, some credit must go to those who thought the theory of the subject worth advancing for reasons of principle many decades ago.

Acknowledgements

My most grateful thanks go to Diana Barkan, for her help and support with this research, and to Kip Thorne, at whose suggestion I undertook it. Thanks also Peter Havas, John Stachel, Jean Eisenstaedt and Martin Krieger for valuable discussions. I am also grateful for permission granted by the Albert Einstein Archives, The Hebrew University of Jerusalem, as well as by the Einstein papers project and John Tate, Jr., to quote from the Einstein-Tate correspondence, and to Robert Schulman for his kind help at the Einstein papers project in Boston. I am indebted to the Archives at the California Institute of Technology for permission to quote from Robertson's letter to Tate. To the organizers of the Berlin conference, for their hospitality, kindness and generosity, especially to Tilman Sauer and Jürgen Renn, my thanks. I am also the grateful recipient of a Doctoral Dissertation Improvement grant (No. SBR-9412026) from the National Science Foundation, which enabled me to travel to consult archival material and conduct interviews. Finally, I thank all of those who agreed to be interviewed for their time and help.

References

Anderson, James L. (1987). "Gravitational Radiation Damping in Systems with Compact Components." *Physical Review D* **36**: 2301–2313.

——— (1995). "Conditions of Motion for Radiating Charged Particles." Preprint. Stevens Institute of Technology, Hoboken (NJ).

Baldwin, O. R. & Jeffery, G. B. (1926). "The Relativity Theory of Plane Waves." *Proceedings of the Royal Society of London A* **111**: 95–104.

Barkan, Diana Kormos. (1992). "A Usable Past: Creating Disciplinary Space for Physical Chemistry" In *The Invention of Physical Science.* M. J. Nye, J. L. Richards and R. H. Stuewer, eds. 175–202. Boston: Kluwer Academic.

BECK, Guido. (1925). "Zur Theorie binärer Gravitationsfelder." *Zeitschrift für Physik* **33**: 713–728.

BERTOTTI, Bruno & PLEBANSKI, Jerzy. (1960). "Theory of Gravitational Perturbations in the Fast Motion Approximation." *Annals of Physics* **11**: 169–200.

BONDI, Hermann. (1957). "Plane Gravitational Waves in General Relativity." *Nature* **179**: 1072–1073.

——— (1964). "Radiation from an Isolated System." In *Relativistic Theories of Gravity* (Proceedings of the Warsaw Conference, 25–31 July 1962). Leopold Infeld, ed. 120–121. Paris: Gauthier-Villiers.

BONDI, Hermann, PIRANI, Felix A. E. & ROBINSON, Ivor. (1959). "Gravitational Waves in General Relativity III. Exact Plane Waves." *Proceedings of the Royal Society of London A* **251**: 519–533.

BONDI, Hermann, VAN DER BURG, M. G. J. & METZNER, A. W. K. (1962). "Gravitational Waves in General Relativity VII: Waves from Axi-symmetric Isolated Systems." *Proceedings of the Royal Society of London A* **269**: 21–52.

BONNOR, William B. (1963). "Gravitational Waves." *British Journal of Applied Physics* **14**: 555–562.

BORN, Max, ed. (1969). *Albert Einstein–Max Born. Briefwechsel.* Munich: Nymphenburger.

——— (1971). *The Einstein–Born Letters.* Irene Born, trans. London: MacMillan. [English translation of BORN 1969].

BURKE, William. (1969). "The Couling of Gravitational Radiation to Nonrelativistic Sources." Ph.D. thesis, California Institute of Technology, Pasadena (CA).

BURKE, William & THORNE, Kip S. (1970). "Gravitational Radiation Damping." In *Relativity.* M. Carmeli, S. I. Fickler and L. Witten, eds. 209–228. New York: Plenum Press.

CATTANI, Carlo & DE MARIA, Michelangelo. (1993). "Conservation Laws and Gravitational Waves in General Relativity (1915–1918)." In *The Attraction of Gravitation: New Studies in the History of General Relativity* (Einstein Studies, vol. 5). J. Earman, M. Janssen and J. D. Norton, eds. 63–87. Boston: Birkhäuser.

CHANDRASEKHAR, Subrahmanyan. (1965). "The Post-Newtonian Equations of Hydrodynamics in General Relativity." *The Astrophysical Journal* **142**: 1488–1512.

CHANDRASEKHAR, Subrahmanyan & ESPOSITO, F. P. (1970). "The $2\frac{1}{2}$-Post-Newtonian Equations of Hydrodynamics and Radiation Reaction in General Relativity." *The Astrophysical Journal* **160**: 153–179.

COLLINS, Harry M. (1975). "The Seven Sexes: A Study in the Sociology of a Phenomenon, or the Replication of Experiments in Physics." *Sociology* **9**: 205–24.

——— (1981). "Son of Seven Sexes: The Social Destruction of a Physical Phenomenon." *Social Studies of Science* **11**: 33–63.

——— (1985). *Changing Order.* London: Sage Publications.

COOPERSTOCK, Fred I. (1967). "Energy Transfer via Gravitational Radiation in the Quasi-stellar Sources." *Physical Review* **163**: 1368–1373.

——— (1992). "Energy Localization in General Relativity: A New Hypothesis." *Foundations of Physics* **22**: 1011–1024.

COOPERSTOCK, Fred I. & HOBILL, D. W. (1982). "Gravitational Radiation and the Motion of Bodies in General Relativity." *General Relativity and Gravitation* **14**: 361–378.

DAMOUR, Thibault. (1982). "Gravitational Radiation and the Motion of Compact Bodies." In *Rayonnement gravitationelle.* N. Deruelle & T. Piran, eds. 59–144. Amsterdam: North Holland.

——— (1983). "Gravitational Radiation Reaction in the Binary Pulsar and the Quadrupole-Formula Controversy." *Physical Review Letters* **51**: 1019–1021.

——— (1987a). "An Introduction to the Theory of Gravitational Radiation." In *Gravitation in Astrophysics* B. Carter and J. B. Hartle, eds. 3–62. New York: Plenum Press.

——— (1987b). "The Problem of Motion in Newtonian and Einsteinian Gravity." In *300 Years of Gravitation*. S. Hawking and W. Israel, eds. 128–198. Cambridge: Cambridge University Press.

DAMOUR, Thibault & RUFFINI, R. (1974). "Sur certaines vérifications nouvelles de la relativité générale rendues possibles par la découverte d'un pulsar membre d'un systéme binaire" *Comptes Rendu de l'Academie des Sciences A* (Paris) **279**: 971–973.

DEWITT, Cecile M. (1957). *Conference on the Role of Gravitation in Physics*. (Proceedings of the Chapel Hill Conference, 18–23 January 1957). Wright Air Development Center (WADC) technical report 57-216, United States Air Force, Wright-Patterson Air Force Base, Ohio. [A supplement with an expanded synopsis of Feynman's remarks was also distributed to participants; a copy can be found, for example, in the Feynman papers at Caltech].

DYSON, Freeman. (1963). "Gravitational Machines." In *Interstellar Communications*. A. G. W. Cameron, ed. 115–120. New York: Benjamin.

EDDINGTON, Arthur Stanley. (1922). "The Propagation of Gravitational Waves." *Proceedings of the Royal Society of London A* **102**: 268–282.

EHLERS, Jürgen, ed. (1979). *Isolated Gravitating Systems in General Relativity* (Proceedings of the 1976 International Summer School of the Italian Physical Society "Enrico Fermi," course LXVII). Amsterdam: North Holland.

EHLERS, Jürgen, ROSENBLUM, Arnold, GOLDBERG, Joshua N. & HAVAS, Peter. (1976). "Comments on Gravitational Radiation Damping and Energy Loss in Binary Systems." *The Astrophysical Journal* **208**: L77–L81.

EINSTEIN, Albert. (1916). "Näherungsweise Integration der Feldgleichungen der Gravitation." *Sitzungsberichte der Königlichen Preussischen Akademie der Wissenschaften* (Berlin): 688–696. [= EINSTEIN *CP6*: doc. 32].

——— (1918). "Über Gravitationswellen." *Sitzungsberichte der Königlichen Preussischen Akademie der Wissenschaften* (Berlin): 154–167.

——— (*CP2*). *The Collected Papers of Albert Einstein*. Vol. 2: *The Swiss Years: Writings, 1900–1909*. J. Stachel, D. C. Cassidy, J. Renn, & R. Schulmann, eds. Princeton: Princeton University Press (1989).

——— (*CP6*). *The Collected Papers of Albert Einstein*. Vol. 6. *The Berlin Years: Writings, 1914–1917*. A. J. Kox, M. J. Klein, & R. Schulmann, eds. Princeton: Princeton University Press (1996).

EINSTEIN, Albert & GROMMER, Jakob. (1927). "Allgemeine Relativitätstheorie und Bewegungsgesetz." *Sitzungsberichte der Preussischen Akademie der Wissenschaften* (Berlin): 2–13.

EINSTEIN, Albert, INFELD, Leopold & HOFFMANN, Banesh. (1938). "The Gravitational Equations and the Problem of Motion." *Annals of Mathematics* **39**: 65–100.

EINSTEIN, Albert & RITZ, Walter. (1909). "Zum gegenwärtigen Stand des Strahlungsproblems." *Physikalische Zeitschrift* **10**: 323–324. [= EINSTEIN *CP2*: doc. 57. English translation: *Collected Papers of Albert Einstein*. Vol. 2: *The Swiss Years: Writings 1900–1909*. A. Beck, trans. 376. Princeton: Princeton Univ. Press].

EINSTEIN, Albert & ROSEN, Nathan. (1937). "On Gravitational Waves." *Journal of the Franklin Institute* **223**: 43–54.

EISENSTAEDT, Jean. (1986a). "La Relativité générale à l'étiage: 1925–1955." *Archive for the History of Exact Sciences* **35**: 115–185. (1986b). "The Low Water Mark of General Relativity, 1925–1955." In *Einstein and the History of General Relativity* (Einstein Studies, vol. 1). D. Howard and J. Stachel, eds. 277–292. Boston: Birkhäuser.

Feynman, Richard P. & Leighton, Ralph. (1988). *What Do You Care What Other People Think? Further Adventures of a Curious Character*. New York: Norton. [Remark quoted appears on p. 91 of the paperback edition: New York: Bantam (1989)].

Fock, Vladimir A. (1959). *The Theory of Space, Time and Gravitation*. 1st English ed. Oxford: Pergamon.

Fowler, William A. (1964). "Massive Stars, Relativistic Polytropes and Gravitational Radiation: Gravitational Waves as Trigger for Radio Galaxy Emissions." *Reviews of Modern Physics* **36**: 545–555.

Franklin, Alan. (1994). "How to Avoid the Experimenters' Regress." *Studies in the History and Philosophy of Science* **25**: 463–491.

Galison, Peter. (1987). *How Experiments End*. Chicago: University of Chicago Press.

Goldberg, Joshua N. (1955). "Gravitational Radiation" *Physical Review* **99**: 1873–1883.

——— (1988). "US Air Force Support of General Relativity: 1956–1972." In *Studies in the History of General Relativity* (Einstein Studies, vol. 3). J. Eisenstaedt and A. J. Kox, eds. 89–102. Boston: Birkhäuser.

Havas, Peter. (1973). "Equations of Motion, Radiation Reaction, and Gravitational Radiation." In *Ondes et radiation gravitationelles* (Colloques internationaux du CNRS 220). Proceedings of meeting, Paris, 18–22 June 1973. Y. Choquet-Bruhat, ed. 383–392. Paris: Éditions du Centre national de la recherche scientifique.

——— (1979). "Equations of Motion and Radiation Reaction in the Special and General Theory of Relativity." In Ehlers 1979: 74–155.

——— (1989). "The Early History of the Problem of Motion in General Relativity." In *Einstein and the History of General Relativity* (Einstein Studies, vol. 1). D. Howard and J. Stachel, eds. 234–276. Boston: Birkhäuser.

——— (1993). "The Two-Body Problem and the Einstein-Silberstein Controversy." In *The Attraction of Gravitation: New Studies in the History of General Relativity* (Einstein Studies, vol. 3). J. Earman, M. Janssen and J. D. Norton, eds. 88–125. Boston: Birkhäuser.

Havas, Peter & Goldberg, Joshua N. (1962). "Lorentz-Invariant Equations of Motion of Point Masses in the General Theory of Relativity." *Physical Review* **128**: 398–414.

Hu, Ning. (1947). "Radiation Damping in the General Theory of Relativity." *Proceedings of the Royal Irish Academy* **51A**: 87–111.

Hulse, R. A. & Taylor, J. H. (1975). "Discovery of a Pulsar in a Binary System." *The Astrophysical Journal* **195**: L51–L53.

Infeld, Leopold. (1941). *Quest—The Evolution of a Physicist*. London: Gollancz.

Infeld, Leopold & Plebanski, Jerzy. (1960). *Motion and Relativity*. Oxford: Pergamon.

Infeld, Leopold & Scheidegger, Adrian E. (1951). "Radiation and Gravitational Equations of Motion." *Canadian Journal of Mathematics* **3**: 195–207.

Infeld, Leopold & Wallace, Philip. (1940). "The Equations of Motion in Electrodynamics." *Physical Review* **57**: 797–806.

Isaacson, Richard. (1968). "Gravitational Radiation in the Limit of High Frequency II: Nonlinear Terms and the Effective Stress Tensor." *Physical Review* **166**: 1272–1280.

Jungnickel, Christa & McCormmach, Russell. (1986). *Intellectual Mastery of Nature II: The Now Mighty Theoretical Physics 1870–1925*. Chicago: University of Chicago Press.

Kerr, Roy P. (1959). "On the Lorentz-Invariant Approximation Method in General Relativity III: The Einstein-Maxwell Field." *Nuovo Cimento* **13**: 673–689.

Landau, Lev D. & Lifschitz, Yevgeniy M. (1951). *Classical Theory of Fields*. M. Hammermesh, trans. Cambridge (MA): Addison-Wesley.

LAPLACE, Pierre. (1776). "Recherches sur le principe de la gravitation universelle, et sur les inégalités séculaires des planètes qui en dépendent." *Mémoires de mathématique et de physiaue, présentés ... par divers sçavans* 1773/6: 37–232. [Reprinted: *Œuvres complètes de Laplace*. Vol. 8, 198–275. Paris: Gauthier-Villars (1891)].

——— (1805). *Traité de mécanique céleste*. Paris: Courcier. [Reprinted: *Œuvres complètes de Laplace*. Vol. 4. Paris: Gauthier-Villars (1891)]].

LEVI-CIVITA, Tullio. (1917). "Sulla espressione analitica spettante al tensore gravitazionale nella teoria di Einstein." *Rendiconti Accademia dei Lincei* (ser. 5) **26**: 381–391 [Reprinted: *Opere mathematiche*. Vol. IV: 47–58. Bologna: Zanichelli (1960)].

LICHNEROWICZ, André. (1955). *Théories relativistes de la gravitation et de l'électromagnétisme*. Paris: Masson.

LIGHTMAN, Alan & BRAWER, Roberta. (1990). *Origins*. Cambridge (MA): Harvard University Press.

MCVITTIE, George C. (1955). "Gravitational Waves and One-Dimensional Einsteinian Gas Dynamics." *Journal of Rational Mechanics and Analysis* **4**: 201–220.

PAIS, Abraham. (1982). *Subtle is the Lord ... : The Science and Life of Albert Einstein*. Oxford: Clarendon.

PERES, Asher. (1959). "Gravitational Motion and Radiation II." *Nuovo Cimento* **11**: 644–655.

——— (1960). "Gravitational Radiation." *Nuovo Cimento* **15**: 351–369.

PETROV, A. Z. (1955). "O prostranstvakh maksimal'noy podvizhnosti opredelyayushchikh polya tyagoteniya." *Doklady Akademiy Nauk SSSR* **105**: 905–908.

PIRANI, Felix A. E. (1955). [Review of MCVITTIE 1955]. *Mathematical Reviews* **16**: 1165.

——— (1965). "Introduction to Gravitational Radiation Theory." In A. Trautman, F. A. E. Pirani and H. Bondi. *Lectures on General Relativity* (Brandeis Summer School in Theoretical Physics, vol. 1). 249–373. Englewood Cliffs (NJ): Prentice-Hall.

POINCARÉ, Henri. (1908). "La Dynamique de l'électron" *Revue générale des Sciences pures et appliqués* **19**: 386–402 [Reprinted: *Œuvres de Henri Poincaré*. Vol. 9: 551–586. G. Petiau, ed. Paris: Gauthier-Villars (1954); English translation: "The New Mechanics." In H. Poincaré. *Science and Method*. F. Maitland, trans. 199–250. New York: Dover (1952)].

RITZ, Walter. (1908). "Recherches critiques sur l'électrodynamique générale." *Annales de chemie et de physique* **13**: 145–275. [English translation: *Critical Researches on General Electrodynamics*. R. S. Fitzius, trans. n. p. (1980)].

ROBERTSON, Howard P. (1938). "Note on the Preceding Paper: The Two-Body Problem in General Relativity." *Annals of Mathematics* **39**: 101–104.

ROBINSON, Ivor, SCHILD, Alfred & SCHUCKING, E. L., eds. (1965). *Quasi-Stellar Sources and Gravitational Collapse, Including the Proceedings of the First Texas Symposium on Relativistic Astrophysics*. Chicago: University of Chicago Press.

ROSEN, Nathan. (1937). "Plane Polarized Waves in the General Theory of Relativity." *Physikalische Zeitschrift der Sowjetunion* **12**: 366–372.

——— (1955). "On Cylindrical Gravitational Waves." In *Jubilee of Relativity Theory* (*Helvetica Physica Acta*, supplement 4). Proceedings of the Fiftieth Anniversary Conference, Bern, 11–16 July 1955. André Mercier and Michel Kervaire, eds. 171–175. Basel: Birkhäuser.

——— (1958). "Energy and Momentum of Cylindrical Gravitational Waves." *Physical Review* **110**: 291–292.

——— (1979). "Does Gravitational Radiation Exist?" *General Relativity and Gravitation* **10**: 351–364.

ROSEN, Nathan & VIRBHADRA, K. S. (1993). "Energy and Momentum of Cylindrical Gravitational Waves." *General Relativity and Gravitation* **25**: 429–433.

SACHS, Rainer K. (1962). "Gravitational Waves in General Relativity VIII: Waves in Asymptotically Flat Space-Time." *Proceedings of the Royal Society of London A* **270**: 103–126.

SCHEIDEGGER, Adrian E. (1951). "Gravitational Transverse-Transverse Waves." *Physical Review* **82**: 883–885.

——— (1953). "Gravitational Motion." *Reviews of Modern Physics* **25**: 451–468.

SCHILPP, Paul A., ed. (1949). *Albert Einstein, Philosopher-Scientist*. 2–95. LaSalle (IL): Open Court.

SCHWIMMING, Rainer. (1980). "On the History of the Theoretical Discovery of the Plane Gravitational Waves." Preprint. Karl Marx University, Leipzig.

SMITH, Stanley F. & HAVAS, Peter. (1965). "Effects of Gravitational Radiation Reaction in the General Relativistic Two-body Problem by a Lorentz-invariant Method." *Physical Review* **138**: 495–508.

STACHEL, John. (1959). "Energy Flow in Cylindrical Gravitational Waves." Master's thesis, Stevens Institute of Technology, Hoboken (NJ).

TAYLOR, Joseph H. & MCCULLOCH, P. M. (1980). "Evidence for the Existence of Gravitational Radiation from Measurements of the Binary Pulsar PSR 1913+16." In *Proceedings of the Ninth Texas Symposium on Relativistic Astrophysics*. J. Ehlers, J. Perry and M. Walker, eds. 442–446. New York: New York Academy of Sciences.

TETRODE, Hugo M. (1922). "Über den Wirkungszusammenhang der Welt. Eine Erweiterung der klassischen Dynamik." *Zeitschrift für Physik* **10**: 317–328.

THORNE, Kip S. (1980). "Multipole Expansions of Gravitational Radiation." *Reviews of Modern Physics* **52**: 299–339.

——— (1989). "Gravitational Radiation: A New Window onto the Universe." Unpublished manuscript. [Includes an historical review of the radiation reaction controversy].

TRAUTMAN, Andrzej. (1958a). "Radiation and Boundary Conditions in the Theory of Gravitation." *Bulletin de l'Académie polonaise des sciences, séries des Sciences mathematiques* **6**: 407–412.

——— (1958b). "On Gravitational Radiation Damping." *Bulletin de l'Académie polonaise des sciences, séries des Sciences mathématiques* **6**: 627–633.

WALKER, Martin & WILL, Clifford M. (1980). "The Approximation of Radiative Effects in Relativistic Gravity: Gravitational Radiation Reaction and Energy Loss in Nearly Newtonian Systems." *The Astrophysical Journal* **242**: L129–L133.

WEBER, Joseph. (1960). "Detection and Generation of Gravitational Waves." *Physical Review* **117**: 306–313.

——— (1969). "Evidence for Discovery of Gravitational Radiation." *Physical Review Letters* **22**: 1320–1324.

WEBER, Joseph & WHEELER, John A. (1957). "Reality of the Cylindrical Gravitational Waves of Einstein and Rosen." *Reviews of Modern Physics* **29**: 509–515.

WESTPFAHL, Konradin. (1985). "High Speed Scattering of Charged and Uncharged Particles in General Relativity." *Fortschritte der Physik* **33**: 417–493.

WEYL, Hermann. (1921). *Raum, Zeit, Materie*. 4th ed. Berlin: Springer.

——— (1922). *Space, Time, Matter*. H. L. Brose, trans. London: Methuen. [English translation of WEYL 1921; Reprinted: New York: Dover (1950)].

WHEELER, John A. & FEYNMAN, Richard P. (1945). "Interaction with the Absorber as the Mechanism of Radiation." *Reviews of Modern Physics* **17**: 157–181.

——— (1949). "Classical Electrodynamics in Terms of Direct Interparticle Action." *Reviews of Modern Physics* **21**: 425–433.

The Penrose-Hawking Singularity Theorems: History and Implications

John Earman

THE HISTORY OF SINGULARITY THEOREMS in Einstein's general theory of relativity (GTR) is very far from an idealized textbook presentation where an analysis of space-time singularities is followed by theorems about the existence of singularities in solutions to the Einstein field equations (EFE);[1] indeed, crucial advances in the understanding of the concept of space-time singularity were driven by a need to understand what various singularity theorems did, and did not, demonstrate. The seminal singularity theorems of Roger Penrose and Stephen Hawking relied on a new and mathematically precise definition of singularities, although this was not clear from the first publications of the results and it may not have then been entirely clear to the authors themselves. These theorems did succeed in convincing the general relativity community that singularities, in one sense of that term, are a generic feature of solutions to EFE. However, these theorems and their subsequent generalizations did not settle the debate about the correct definition of singularities; in fact, it has become increasingly clear that there is no one 'correct'

[1] Einstein's field equations (EFE), with cosmological constant term, read:

$$R_{\mu\nu} - \tfrac{1}{2} R\, g_{\mu\nu} + \Lambda\, g_{\mu\nu} = 8\pi\, T_{\mu\nu}$$

where $R_{\mu\nu}$ is the Ricci tensor, R is the Riemann curvature scalar, Λ is the cosmological constant, and $T_{\mu\nu}$ is the stress-energy tensor. An equivalent form of the field equations is

$$R_{\mu\nu} - \Lambda\, g_{\mu\nu} = 8\pi\, (T_{\mu\nu} - \tfrac{1}{2}\, g_{\mu\nu}\, T)$$

where T is the trace of $T_{\mu\nu}$. In what follows I have changed notation in the original papers to conform to the $(+\,+\,+\,-)$ signature for the space-time metric. In keeping with the style of the times, I have used the component notation for tensors instead of the abstract index notation now in vogue.

H. Goenner, J. Renn, J. Ritter, T. Sauer (eds.)
The Expanding Worlds of General Relativity
(Einstein Studies, volume 7) pp. 235–267

definition and that the term 'space-time singularity' points to a sizable family of distinct though interrelated pathologies that can infect relativistic space-times.

Many general relativists view singularities as being intolerable, and those who do tend to see theorems proving the prevalence of singularities among solutions to EFE as showing that GTR contains the seeds of its own destruction. As a result there have been attempts either to modify classical GTR so as to avoid singularities or else to combine GTR with quantum mechanics in the hope that the quantum version of gravity will smooth away the singularities. A discussion of these foundations issues is beyond the scope of the present paper.[2]

The aim here is the modest one of tracing the route to the Penrose-Hawking theorems. Even so the entirety of all of the relevant literature is so large that a definitive treatment would have to be book length. The selection principle used in the present survey is to concentrate on results which contributed directly to the debate about whether or not singularities in GTR are only artifacts of the idealizations of models of cosmology and gravitational collapse. Although the present survey is far from definitive, the brevity of the treatment allows the main themes to stand out from a very cluttered background.

The singularity theorems of interest are naturally divided into four sets: a group of results by Richard Tolman, Georges Lemaître, and others from the 1930s; results by Amalkumar Raychaudhuri and Arthur Komar from the 1950s; a transitional result by Lawrence Shepley from the early 1960s; and finally the Penrose-Hawking theorems from the mid to late 1960s. The theorems from the 1930s were recognizably singularity theorems independently of the then extant controversies about how to define singularities. Nor were any of the key results from the 1930s or 1950s motivated by a desire to understand the status of singularities in the Schwarzschild solution, the De Sitter solution, or other cosmological models. It would be a mistake, however, to neglect these matters. For Raychaudhuri's first engagement with singularities in GTR stemmed from a desire to clarify the status of the $r = 2M$ Schwarzschild singularity; and, more generally, the critical reaction to the main results of Raychaudhuri and Komar reflected the unsettled state of opinion about how best to analyze singularities. Thus in what follows overviews of attempts to define and understand the nature of singularities in general-relativistic space-times will be interwoven with discussions of singularity theorems.

1. The struggle to understand space-time singularities: 1916–1939

Singularities began to demand attention soon after Einstein's general theory was codified in its final form. The Schwarzschild (1916) exterior solution was not only the first exact solution of EFE, it was also the basis of the three classical tests of the theory. In "Die Grundlagen der Physik" David Hilbert (1917) not only confronted the singularity structure of the Schwarzschild solution, but used the opportunity to give what amounts to the first general definition of singularities in GTR. In

---

[2] See Earman 1996 for a discussion of these issues.

coordinates introduced by Johannes Droste (1916), the line element is

$$ds^2 = \left(1 - \frac{2M}{r}\right)^{-1} dr^2 + r^2(d\theta^2 + \sin^2\theta\, d\phi^2) - \left(1 - \frac{2M}{r}\right) dt^2. \qquad (1)$$

Hilbert pronounced that this metric is singular (or "not regular") at both $r = 0$ and $r = 2M$:

> For α [$2M$] $\neq 0$, it turns out that $r = 0$ and (for positive α) also $r = \alpha$ are points at which the line element is not regular. By that I mean that a line element or a gravitational field $g_{\mu\nu}$ is *regular* at a point if it is possible to introduce by a reversible one-one transformation a coordinate system, such that in this coordinate system the corresponding functions $g'_{\mu\nu}$ are regular at that point, i.e., they are continuous and arbitrarily differentiable at the point and in a neighborhood of the point, and the determinant g' is different from 0.[3] (Hilbert 1917: 70–71)

Hilbert was, of course, correct that the apparent singularity at $r = 2M$ (the 'Schwarzschild radius') cannot be removed by a coordinate transformation that is required to be smooth and invertible not only for $r > 2M$ but also at $r = 2M$. Later critics charged that there is no need to restrict attention to coordinate transformations that are smooth at the Schwarzschild radius; on the contrary, it was said, one would expect the transformation to be singular at this radius since the Droste coordinates 'go bad' there, as evidenced by the fact that it takes an infinite amount of t-coordinate time to reach $r = 2M$ from $r > 2M$ even though only a finite amount of proper time elapses. But by itself such a remark would hardly have been found decisive; indeed, two decades later Einstein was to appeal to a similar feature of isotropic coordinates as evidence that there *is* a real singularity at the Schwarzschild radius (see below).[4]

Einstein's initial concern with the exterior Schwarzschild solution focused not on the problem of singularities but rather on the perceived anti-Machian character of the solution. As John Stachel has written, it was for Einstein a "scandal that a solution to his field equations should exist which corresponds to the presence of a single body in an otherwise 'empty universe'" (Stachel 1979: 440). The scandal seemed to worsen with the introduction of the cosmological constant term into the field equations and Willem De Sitter's (1917a, b) discovery of a solution with the line element

$$ds^2 = dr^2 + R^2 \sin^2\left(\frac{r}{R}\right)(d\chi^2 + \sin^2\psi\, d\theta^2) - \cos^2\left(\frac{r}{R}\right) dt^2, \qquad (2)$$

---

[3] "Für $\alpha \neq 0$ erweisen sich $r = 0$ and bei positivem α auch $r = \alpha$ als solche Stellen, an denen die Maßbestimmung nicht regulär ist. Dabei nenne ich eine Maßbestimmung oder ein Gravitationsfeld $g_{\mu\nu}$ an einer Stelle *regulär*, wenn es möglich ist, durch umkehrbar eindeutige Transformation ein solches Koordinatensystem einzuführen, daß für dieses die entspreschenden Funktionen $g'_{\mu\nu}$ an jener Stelle regulär, d.h. in ihr und in ihrer Umgebung stetig und beliebig oft differezierbar sind und eine von Null verschiedene Determinante g' haben." In the second volume of *Die Relativitätstheorie* Max von Laue concurred with Hilbert (von Laue 1921: 215).

[4] An illuminating account of early attempts to understand the mysteries of the Schwarzschild solution is to be found in Eisenstaedt 1982.

where R is a positive constant (*not* the Riemann curvature scalar). The metric of this line element can be considered to be a solution to Einstein's vacuum field equations with cosmological constant $\Lambda = 3/R^2$. Here Einstein found the issues of Mach's principle and singularities joined, for he wanted to treat the violation of the former as *ersatz* on the grounds that mass concentrations were hiding in the $r = (\pi R)/2$ singularity of (2). This required an analysis of what it meant for a space-time to be singular, which was duly supplied. For Einstein, a space-time was to be counted as *non*singular if in the finite realm the covaraiant and contravariant components of the metric are continuous and differentiable and, thus, the determinant g never vanishes. A space-time point p is said to be in the finite realm if it can be joined to an arbitrary origin point p_0 by a curve of finite length. Einstein apparently had in mind using proper length along space-like or time-like curves. But if no bound is put on the acceleration of time-like curves, points 'at infinity' will be counted as lying at a finite distance. Later writers would use affine distance along a geodesic to judge what is at a finite distance. It seemed to Einstein in 1918 that the discontinuity in (2) at $r = (\pi R)/2$, which lies at a finite distance, could not be removed by any choice of coordinates and, thus, represented a real singularity.[5] In this he was mistaken, as emerged from the work of Felix Klein (1918), Kornel Lanczos (1922, 1923), and Arthur S. Eddington (1923). Referring to Einstein's idea that the De Sitter singularity represents a "'mass horizon' or ring of peripheral matter," Eddington responded that "A singular ds^2 does not necessarily indicate material particles, for we can introduce or remove such singularities by making transformations of coordinates" (Eddington 1923: 165). But then he proceeded to express his uncertainty about how to handle the general situation: "It is impossible to know whether to blame the world-structure [space-time metric] or the appropriateness of the coordinate system" (ibid.).

The following year Eddington (1924) produced a coordinate transformation that showed that the $r = 2M$ Schwarzschild singularity is a coordinate artifact, although he apparently was not aware of the significance of his result.[6] The first self-conscious demonstration of the 'fictive' nature of the $r = 2M$ singularity was due to Georges Lemaître (1932), who compared this singularity to the $r = (\pi R)/2$ singularity of the De Sitter line element (2). Although a definite advance in understanding had been achieved—not only had the nature of $r = 2M$ been clarified but the Lemaître's extension of the Schwarzschild solution also revealed the nonstationary character that extensions must possess—neither the Eddington nor the Lemaître treatment revealed the full singularity structure of the Schwarzschild solution. For, as will become clear, singularities implicate the global structure of space-time, and neither of these treatments indicate all of the relevant properties

---

[5] It is not clear whether Einstein, like Hilbert, meant to require that an allowed coordinate transformation must be smooth and invertible not only up to but also at the (apparent) singularity. Some evidence that he did not comes from the fact that he explicitly recognized that the $\psi = 0$ singularity in (2) is "*nur scheinbar*" because it can be removed by transforming from spherical to Cartesian coordinates; but the transformation is singular at the north pole.

[6] The transformation is $t' = t - 2M \ln(r - 2M)$. (Actually Eddington's formula was missing a factor of 2.) Note that the transformation is singular at $r = 2M$. David Finkelstein (1958) rediscovered Eddington's transformation.

of the maximal analytic extension of the exterior Schwarzschild solution.

Although Lemaître's paper appeared in two Belgian journals that were not widely read, it was known to a few key actors, including Howard P. Robertson, who had produced his own transformation for removing the $r = 2M$ singularity.[7] In 1939 Robertson was at Princeton and in contact with Einstein. In view of the fact that discussions with Robertson are explicitly acknowledged in Einstein's "On a Stationary System with Spherical Symmetry Consisting of Many Gravitating Masses" (1939), it is more than a little surprising to see how Einstein treats the $r = 2M$ singularity. Einstein chooses to write the Schwarzschild line element in isotropic coordinates:

$$ds^2 = \left(1 + \frac{\mu}{2r'}\right)(dx^2 + dy^2 + dz^2) - \left(\frac{1 - \frac{\mu}{2r'}}{1 + \frac{\mu}{2r'}}\right)^2 dt^2, \tag{3}$$

where $r'^2 = x^2 + y^2 + z^2$, r' is related to the Droste radial coordinate r by $r = \mu + r' + \mu^2/4r'$, and μ is the mass. In this form of the line element the blow up behavior of g_{11} in (1) is removed, but as noted by Einstein, g_{44} in (3) vanishes at $r' = \mu/2$.[8]

> [This] means that a clock kept at this place would go at a zero rate. Further it is easy to show that both light rays and material particles take an infinitely long time (measured in "coordinate time") in order to reach the point $r' = \mu/2$ when originating from a point $r' > \mu/2$. In this sense the sphere $r' = \mu/2$ constitutes a place where the field is singular. (Einstein 1939: 922)

To modern eyes what this behavior indicates is not the presence of a singularity but that, as with the 'mass horizon' in the De Sitter solution (2), $r' = \mu/2$ is an event horizon.[9]

The purpose of Einstein's paper was to argue that the $r' = \mu/2$ (or $r = 2M$) Schwarzschild singularity does "not exist in physical reality." The argument given is that for the special case of a spherically symmetric cluster of particles moving in circular orbits under the influence of their mutual gravitational field, the radius of the cluster cannot be smaller than its Schwarzschild radius. It is surprising that

---

[7] Synge (1950: 84) reports a transformation, attributed to a 1939 lecture of Robertson's, for removing the $r = 2M$ singularity.

[8] I have taken the liberty of changing Einstein's r to r'.

[9] Although the nature of event horizons was not clarified until much later, Lanczos (1923) was clear that $r = (\pi R)/2$ in the De Sitter solution corresponds to an event horizon and not to a singularity. J. Robert Oppenheimer and Snyder (1939) recognized that in the context of gravitational collapse the Schwarzschild radius acts as an event horizon, although they did not use this terminology; see Section 2 below. An event horizon for a system of observers is the boundary between the region of space-time from which those observers can receive causal signals and the region from which they cannot receive signals. In the maximal extension of the exterior Schwarzschild solution $r = 2M$ has an absolute status as an event horizon: it is the boundary between that portion of space-time that can be seen from future null infinity and that portion that cannot be so viewed. This is the basis of the modern definition of 'black hole.'

Einstein produced an elaborate fifteen page calculation to reach this conclusion, for it was known that in the Schwarzschild exterior field there are no circular geodesics for $r < 3M$. It is not just surprising but nearly inexplicable that Einstein thought that a static analysis, which did not allow for collapse of matter, could yield the desired impossibility result. Einstein's model, which has the particles in orbit rather than headed for a common origin, seems unconsciously chosen to yield the wanted result. In the very year that Einstein's paper appeared, Oppenheimer and Snyder (1939) showed that the gravitational collapse of matter could lead to the uncovering of the Schwarzschild radius (see Section 2).

In sum, at the end of the 1930s not only was there no agreement on how to define singularities, there was not even a consensus about the status of singularities in the key test case, the Schwarzschild solution. There were examples, such as the De Sitter solution, which showed the need to distinguish between genuine and apparent singularities; but again, there was no consensus on how this distinction was to be drawn. There was, however, at least tacit agreement on this much: if some relevant physical variable, such as energy density or a curvature scalar, becomes unbounded, and this behavior occurs at a finite distance (to use Einstein's (1918) phrase), then a genuine singularity is implicated.[10]

2. Singularity theorems of the 1930s

This Section reviews singularity theorems in the cosmological setting by Tolman and Morgan Ward (1932), Lemaître (1932), Tolman (1934a), and John Lighton Synge (1934), and results by Oppenheimer and Volkoff (1939) and Oppenheimer and Snyder (1939) for gravitational collapse.

Tolman (1930a, 1930b, 1930c) studied non-static, homogeneous, and isotropic models whose line element turned out the be equivalent to that of what are now called the Friedmann-Lemaître-Robertson-Walker (FLRW) models (see Tolman 1930d). We now know that the symmetries of the space-time metrics in question force the stress-energy tensor to have the form of a perfect fluid: $T^{\mu\nu} = (\rho + p)\, U^{\mu} U^{\nu} + p\, g^{\mu\nu}$, where ρ is the density of matter, p is the pressure, and U^{μ} is the normed four-velocity of the fluid. Tolman and Ward (1932) examined the case where the cosmological constant is set to zero and where $\rho > 0$ and $p \geq 0$. They showed that, as a consequence of EFE, if the volume of space is initially finite, there is a finite upper bound beyond which the model cannot expand; further, this bound is reached in a finite time, after which the model contracts to zero volume, also in a finite time.

This is arguably the first significant singularity theorem for GTR. Once having achieved this significant result, Tolman and Ward lost no time in trying to negate it in two ways. First, they argued that despite what the mathematical analysis says, it is plausible on physical grounds to expect that the contraction to zero volume

---

[10] Although I cannot cite specific passages from the literature of the 1930s to substantiate this claim, the fact that the theorems discussed in Section 2 were accepted as singularity theorems is strong indirect support for the claim.

would "be followed by renewed expansion, thus leading to a continued succession of somewhat similar expansions and contractions" (Tolman & Ward 1932: 842). This conclusion was supported by a citation to a previous paper of Tolman's which contains the stronger assertion that "it is evident physically that contraction to a zero volume could only be followed by another expansion" (Tolman 1931: 1765). The idea of an irremovable singularity—one which cannot be removed by any suitable extension of the space-time—was evidently one which Tolman did not want to contemplate.[11] Second, Tolman and Ward, citing the authority of Einstein (1931), opined that "it is possible the idealization upon which our considerations have been based should be regarded as failing in the neighborhood of zero volume" (p. 842). Specifically, they thought that the perfect fluid idealization might break down at very small volumes. Einstein, however, laid the finger of blame on symmetry assumptions. Speaking of the initial singularity in the Friedmann model, Einstein wrote: "Here one can try to get out of the difficulty by pointing out that the inhomogeneity of stellar matter makes illusory our approximate treatment" (Einstein 1931: 237).[12] Other cosmologists, such as Robertson (1932) and De Sitter (1933), joined the chorus that sang that the initial singularity in the FLRW models is an artifact of the unrealistic symmetry assumptions.

Although Tolman initially sang in harmony, he soon produced some discordant evidence. Tolman (1934a) investigated inhomogeneous dust filled models ($T^{ab} = \rho\, U^a U^b$) which exhibit spherical symmetry. Now called Tolman-Bondi models, they are more properly called Lemaître-Tolman-Bondi models since they were introduced by Lemaître (1932). Tolman (1934a) cites Lemaître, and Hermann Bondi (1947) in turn cites Tolman. The line element of this model can be written in the form

$$ds^2 = e^{\lambda} dr^2 + e^{\omega}(d\theta^2 + \sin^2\theta d\phi^2) - dt^2,$$

where λ and ω are functions of t and the radial coordinate. Applying EFE with cosmological constant yields a second order equation for ρ:

$$\frac{\partial^2 \ln \rho}{\partial t^2} = 4\pi\rho - \Lambda + \frac{1}{3}\left(\frac{\partial \ln \rho}{\partial t}\right)^2 + \frac{2}{3}\left(\frac{\dot{\omega}'}{\omega'}\right)^2, \tag{4}$$

where the prime and dot denote respectively differentiation with respect to r and t. Consider regions where $4\pi\rho - \Lambda > 0$. If ρ is initially increasing, it follows from (4) that "reversal in the process of condensation would not occur short of arrival at a singular state involving infinite density or a breakdown of our simplified equations" (Tolman 1934a: 173). What Tolman did not say, but can be easily proved to follow from (4), is that ρ becomes infinite in a finite amount of time. Again the tendency to blame the singularity on the idealizations of the model is

---

[11] The trick is to specify what a 'suitable' extension is. If no continuity requirements are put on an extension, then any singularity is removable. For reflections on what continuity conditions are appropriate to GTR, see Earman 1995a: chap. 2.

[12] "Hier kann man der Schwierigkeit durch den Hinweis darauf zu entgehen suchen, daß die Inhomogeneität der Verteilung der Sternmaterie unsere approximative Behandlung illusorisch macht."

noteworthy. Results similar to Tolman's were obtained by Synge (1934) by means of another technique.[13]

Ancther interesting singularity result was sketched by Lemaître (1932) for a class of homogeneous but possibly non-isotropic models, which in the modern classification scheme belong to the Bianchi Type I class. The line element can be written as

$$ds^2 = \sum_{\alpha=1}^{3} g_{\alpha\alpha}(t)\, dx^\alpha dx^\alpha - dt^2. \tag{5}$$

Defining $R^3 \equiv \sqrt{-g} \equiv \sqrt{-\det(g_{ij})}$, Lemaître argued that, as a consequence of EFE without cosmological constant term, "If, at a certain moment, $\dot{R}$ is negative, it follows that R attains a value of zero and thus that the volume is annulled" (Lemaître 1932: 84).[14]

The connection between the vanishing of g and a space-time singularity is not evident at first glance;[15] but it is known from hindsight that a genuine singularity is involved. For example, in the vacuum case EFE imply that either the metric (5) is flat or else

$$g_{\alpha\alpha} = (t - t_0)^{2\sigma_\alpha} C_\alpha \qquad \text{(no summation,)}$$

where the σ_α and C_α are constants, and the σ_α must satisfy

$$\sigma_1 + \sigma_2 + \sigma_3 = 1$$

$$\sigma_1{}^2 + \sigma_2{}^2 + \sigma_3{}^2 = 1$$

(see Ryan & Shepley 1975: sect. 9.5). It follows that either the metric (5) is flat or else that there is a curvature singularity at $t = t_0$ since $R^{\mu\nu}{}_{\delta\eta} R^{\delta\eta}{}_{\mu\nu} \longrightarrow \infty$ as $t \longrightarrow t_0{}^+$ where $R_{\mu\nu\delta\eta}$ is the Riemann curvature tensor.

Although incomplete as a singularity theorem, Lemaître's result embodied two remarkably prescient features. First, he did not assume, as was common at the time, that matter was in the form of a dust or a perfect fluid but only that the stress-energy tensor satisfied a reasonable energy condition. Second, his argument that g goes to zero is an early form of the Raychaudhuri effect. Raychaudhuri's seminal 1955 paper will be discussed below in Section 4. It makes no reference to Lemaître. But Raychaudhuri was familiar with Lemaître's work since there is a reference to it in Raychaudhuri 1953. Thus, it is plausible that Lemaître's construction was an unconscious inspiration for Raychaudhuri's work.

Singularities reared their heads not only in cosmology but also in stellar dynamics. The most general static and spherically symmetric metric has a line element of the form

$$ds^2 = e^\lambda\, dr^2 + r^2\, d\theta^2 + r^2 \sin^2\theta\, d\phi^2 - e^\nu\, dt^2,$$

---

[13] For more details on Synge's technique, see Eisenstaedt 1993.

[14] "Si donc à un certain moment R' [$\dot{R}$] est négatif, il faut que R atteigne la valeur zéro et donc que le volume s'annule."

[15] As we will see in Sections 4 and 5, this problem became important in the 1950s and 1960s.

where λ and ν are functions of r alone. If matter is assumed to act as a perfect fluid, EFE without the cosmological constant term imply three ordinary differential equations:

$$\begin{cases} 8\pi p = e^{-\lambda}\left(\frac{\nu'}{r}+\frac{1}{r^2}\right)-\frac{1}{r^2} \\ 8\pi \rho = e^{-\lambda}\left(\frac{\lambda'}{r}-\frac{1}{r^2}\right)+\frac{1}{r^2} \\ \rho' = \frac{(p+\rho)\nu'}{2}, \end{cases} \tag{6}$$

as had been shown by Tolman (1934b). It was assumed that outside of the matter, $\rho = p = 0$, and that in this exterior region the line element takes on its Schwarzschild form

$$e^{-\lambda(r)} = 1 - \frac{2M}{r}, \qquad e^{\nu(r)} = 1 - \frac{2M}{r}.$$

Adjoining an equation of state $\rho(p)$ to (6) gives four equations for four unknowns. Oppenheimer and Volkoff (1939) used an equation of state designed to model a cold Fermi gas, which they took as a reasonable first approximation for a neutron star. They found that for masses greater than $(3/4)M_\odot$, static solutions for the mass distribution do not exist.

> There would then seem to be only two answers possible to the "final" behavior of very massive stars: either the equation of state we have used so far fails to describe the behavior of highly condensed matter that the conclusions reached above are qualitatively misleading, or the star will continue to contract indefinitely, never reaching equilibrium. Both alternatives deserve serious consideration. (Oppenheimer & Volkoff 1939: 380–381)

A heuristic discussion of the possible deviations from the Fermi equation of state left them confident that the first possibility was not plausible.

The need to examine non-static solutions was recognized, and in a subsequent paper Oppenheimer and Snyder (1939) provided such an analysis which bypassed the question of equation of state by studying the case of the free gravitational collapse of a dust ball ($p = 0$). In effect, the Lemaître-Tolman-Bondi solution for a spherically symmetric dust ball is being used as the interior solution joined to an exterior Schwarzschild solution.[16] It was found that after a finite proper time as measured by an observer comoving with the matter, no light signals could be sent from the star to external observers: "the cone within which a signal can escape has closed entirely" (Oppenheimer & Snyder 1939: 459). This is the first explicit and unequivocal prediction from GTR of the formation of what was to become known as a black hole. That the black hole would contain an infinite density singularity was not explicitly mentioned, but it was a clear consequence of their analysis.

[16] Oppenheimer and Synder (1939: 457) cite Tolman 1934a. But as Eisenstaedt (1993) demonstrates, it is really Lemaître who deserves the credit.

The paper ended with an acknowledgment that "actual stars would collapse more slowly than the example studied analytically because of the effect of the pressure of matter, of radiation, and of rotation" (p. 459). But they expected that the same kind of behavior would be found for "all collapsing stars which cannot end in a stable stationary state" (ibid.) A considerable effort would be needed before general agreement on this confident pronouncement could be secured. Oppenheimer and Snyder did not draw any implications from the study of gravitational collapse for the status of the $r = 2M$ Schwarzschild singularity. It was left to Raychaudhuri (1953) to make this connection (see Section 3).

At the close of the 1930s the cumulative evidence from various singularity theorems was sufficient to suggest that singularities play more than an incidental role in GTR. Yet there were few, if any, research workers who seem to have been aware of all the evidence. In the case of Lemaître's (1932) paper this is understandable since it appeared in two obscure journals. It is less easy to understand how the Oppenheimer-Volkoff and Oppenheimer-Snyder papers could be ignored since they appeared in the *Physical Review*, but ignored they were (see Section 3). And in any case the available evidence was compatible with the attitude that simplifying assumptions of symmetry and idealized forms of matter were responsible for the singularity results.

3. Further attempts to understand singularities: 1940s and early 1950s

Neither the journal literature nor textbooks from the 1940s provided much of an advance in understanding of singularities. The only reference to singularities in Peter Bergmann's *Introduction to the Theory of Relativity* (1942) is to the Schwarzschild solution, and there Bergmann endorses Einstein's (1939) claim that the $r = 2M$ singularity cannot occur in nature. However, his uncertainty about the status of the $r = 2M$ singularity is revealed by his comment on Robertson:

> Robertson has shown that, if the Schwarzschild field could be realized, a test body which falls freely towards the center would take only a finite proper time to cross the "Schwarzschild singularity," even though the coordinate time is infinite; and he has concluded that at least part of the singular character of the surface $r = 2M$ must be attributed to the coordinate system. (Bergmann 1942: 203)

A reader might well have been puzzled by this remark. How can part of the singular character of $r = 2M$ be attributed to the choice of coordinates and part not? If, as Robertson showed (see Section 2), there is a coordinate transformation which removes the $r = 2M$ singularity, is not the singularity thereby shown to be due wholly to the choice of coordinate system?[17]

Many of the features of the Schwarzschild solution were clarified in a remarkable and remarkably undercelebrated paper by John Lighton Synge (1950). From the

[17] There are two possibilities. Either Robertson did not inform Einstein and Bergmann about his transformation. Or else he did, but Bergmann thought that it did not show that the singularity could be entirely removed because the transformation is singular at $r = 2M$.

work of Lemaître and Robertson, Synge was aware that the $r = 2M$ singularity was only a coordinate artifact. He modestly described his own contribution as follows: "I have removed the Schwarzschild [$r = 2M$] singularity in a different way" (Synge 1950: 84). In fact, what Synge produced was the maximal analytic extension of the exterior Schwarzschild solution. This extension was rediscovered a decade later by C. Fronsdal (1959) and Martin Kruskal (1960), the latter of whom made an advance on Synge's analysis by presenting the metric of the maximal extension in a single global coordinate system.

Synge felt the need for a general analysis of singularities and was apologetic for not being able to supply one.

> It was hoped that at this point there might be given a brief but thorough discussion of singularities of space-time in general, and that the ideas there developed might be applied in particular to the line element [of Synge's extension]. However, the further one looks into the question of singularities, the more difficult the situation appears. ... Obviously, before we talk of singularities at all we should define them, but there are difficulties here which may not appear on the surface. ... Thus we must content ourselves for the present with definitions dependent on the coordinate system employed. (Synge 1950: 100)

Synge proceeded to define the notions of "component singularity" and "determinant singularity" in a fashion that is dependent on the choice of coordinate system and is, thus, useless from the modern point of view.

The lack of a satisfactory definition of singularity was also decried by Abraham Taub (1951), who wanted to test the validity of Mach's principle in GTR. The version of Mach's Principle at issue stated that "the nature of space-time is determined by the matter present. The latter is described either by the singularities in $g_{\mu\nu}$ ... or by the stress-energy tensor $T_{\mu\nu}$" (Taub 1951: 472). On this reading of Mach's Principle, a space-time that is empty ($T_{\mu\nu} = 0$) and is singularity-free should be flat. To conform to this principle, Einstein's field equations (without cosmological constant term) should have the property that any singularity-free solution of $R_{\mu\nu} = 0$ should be flat: $R_{\mu\nu\sigma\eta} = 0$. Taub tested this constraint for the case of solutions of $R_{\mu\nu} = 0$ admitting a three-parameter group of motions and was able to establish a restricted version of the Principle:

> **Theorem** (Taub). A spatially homogeneous space-time whose three-parameter group of motions is the group of Euclidean translations, for which $R_{\mu\nu} = 0$, and for which the $R_{\mu\nu}$ are bounded for all points with finite coordinates, is a flat space.

(This is a version of the Lemaître result mentioned in Section 2 above for Bianchi Type I models.) However, Taub also produced potential counterexamples to Mach's Principle by showing that if the components $R_{\mu\nu\sigma\eta}$ are not required to be bounded for all points with finite coordinates, then $R_{\mu\nu} = 0$ does not imply $R_{\mu\nu\sigma\eta} = 0$. But to decide whether or not the counterexamples are effective requires a decision as to whether the unboundedness of the curvature components in Taub's coordinate system corresponds to an essential singularity. But no generally accepted criterion for making the decision existed.

Amalkumar Raychaudhuri's interest in singularities was awakened by reading Bergmann's *Introduction to the Theory of Relativity* and more specifically by Bergmann's report of Einstein's (1939) attempt to show that the Schwarzschild $r = 2M$ singularity is unattainable by matter. In 1953 he showed how to join an exterior Schwarzschild field to a non-static solution of EFE representing a spherically symmetric cluster of particles moving radially towards the center of symmetry. The interior solution chosen was the Tolman form of the FLRW models. Since this form requires that the density of matter is a function of t alone, the analysis is less general than that of Oppenheimer and Snyder (1939), who allowed the density to depend on r.[18] But Raychaudhuri opined that "A spatially non-uniform distribution of particles (retaining spherical symmetry) would, however, lead to the same results so far as our investigation is concerned" (Raychaudhuri 1953: 418, n. 7). He concluded from his analysis that:

> No singularity corresponding to the [$r = 2M$] Schwarzschild singularity appears at any phase in the exterior for any arbitrary finite concentration in the cluster. The Schwarzschild singularity thus appears to be only a property of particular coordinate systems, and there seems to be no theoretical limit to the degree of concentration [of matter]. (Raychaudhuri 1953: 418).

The difference between Raychaudhuri's cluster, which can go on contracting indefinitely, and Einstein's (1939) cluster, which cannot, was explained by the fact that a "null sphere" lies beyond $r = 2M$; since Einstein's particles are supposed to move in circles, their orbits must lie outside $r = 2M$. (In fact, as noted above, they must lie outside $r = 3M$.) Lemaître (1932) had already realized that a Friedmann solution could be joined to an exterior Schwarzschild solution and, thus, that a mass could have a radius smaller than its Schwarzschild radius. In fact, this realization was what motivated him to search for a coordinate transformation that removed the $r = 2M$ singularity in the Schwarzschild-Droste coordinates.[19] An irony of the construction that led Raychaudhuri to conclude that there is a merely fictitious singularity at $r = 2M$ is that this construction also led to the prediction of a real singularity: in the collapse model "there is a singularity at a finite time, the whole region [occupied by matter] collapsing to a zero volume" (1953: 421).[20]

To summarize, a completely knowledgeable observer in the 1950s would have been able to give an account of the status of the $r = 0$ and $r = 2M$ Schwarzschild singularities. But owing to the obscurity of the journals in which the papers of

---

[18] The Oppenheimer-Snyder paper is not referenced in Bergmann's book, and consequently Raychaudhuri was unaware of it in 1953 (private communication from A. Raychaudhuri).

[19] For details, see Eisenstaedt 1993.

[20] Raychaudhuri, like most of his contemporaries, was reluctant to accept the possibility of a singularity in nature. But rather than blaming the singularity in his model on the idealizations of the analysis, he was prepared to fault GTR. "What happens after that [the infinite density singularity], our equations cannot say. It appears, indeed, that while we can trace the history of the birth of a particle, we cannot tell what happens when the particle is actually born. This perhaps can be attributed to the fact, as remarked by Einstein, that the general theory of relativity would break down under such stringent conditions" (Raychaudhuri 1953: 421). The reference here to Einstein is to *The Meaning of Relativity* (1950), where Einstein contemplated a break down of the field equations of GTR in order to avoid the big bang singularity of the FLRW models.

Lemaître (1932) and Synge (1950) appeared, few such observers existed. No progress had been made during the 1940s and early 1950s towards a general definition of singularities. Synge and Taub found this situation embarrassing, but the need for a definition was not urgent as long as it was felt that singularities in GTR arise only as artifacts of unrealistic assumptions. That impression was to be challenged by the theorems proved in the mid 1950s.

4. The singularity theorems of Raychaudhuri and Komar

Further evidence was produced in 1955 that, in the cosmological context at least, singularities in general relativistic space-times are not artifacts of symmetry assumptions. The evidence came not from one of the havens of general relativity in America or Europe, but from Calcutta. In 1953 Raychaudhuri produced an analysis of cosmological singularities that was to have a profound effect on later developments, but because it ran into difficulties with referees, the paper did not appear until 1955.[21] "Relativistic Cosmology I" contained a brief and modest abstract:

> The paper presents some general relations obtaining in relativistic cosmology. It appears from these that a simple change over to anisotropy without the introduction of spin does not solve any of the outstanding difficulties of isotropic cosmological models. (Raychaudhuri 1955: 1123)

The "outstanding difficulties" were twofold. First, the age of the universe estimated from the isotropic models and the observation of nebular distances was too short—it was not even consistent with the estimated age of the earth. Second, the models led to an "original singularity" or "creation in the finite past," which many found repugnant. The purpose of Raychaudhuri's paper was to show that these difficulties would not automatically disappear if the assumptions of homogeneity and isotropy were dropped.

Raychaudhuri's analysis assumed that cosmic matter may be treated as dust. Einstein's field equations then imply that the world lines of the dust particles are geodesics. In a coordinate system which is adapted to the dust ($x^\alpha =$ constant, $\alpha = 1, 2, 3$, for the dust particles) and in which coordinate time measures proper time along the geodesics, the line element is of the form

$$ds^2 = g_{\alpha\beta}\, dx^\alpha\, dx^\beta + 2g_{\alpha 4}\, dt\, dx^\alpha - dt^2 \qquad (\alpha, \beta = 1, 2, 3), \tag{7}$$

where $\partial g_{\alpha 4}/\partial t = 0$. If the $g_{\alpha 4} = 0$, the coordinate system is called 'synchronous'. In the present case where the coordinate system is adapted to the flow lines of the

[21] The results were first presented in April 1953 in a letter to *Physical Review*. This letter was rejected for publication because the referee could not understand how the results were derived. A full paper was submitted to the *Astrophysical Journal*. This too was rejected, now on the grounds that it was too speculative. A modified version was sent to *Physical Review*. After hearing nothing for several months, Raychaudhuri sent queries to the editor but received no reply. After waiting for a year with no response, he sent the manuscript to *Zeitschrift für Physik*. He received a prompt rejection. Finally in February 1955 the acceptance from *Physical Review* came. The Editor, R. A. Goudsmidt, wrote: "We endeavor to choose as referees those colleagues who accept this task conscientiously. We regret in this case, there was an extensive delay." Raychaudhuri 1955 was entitled "Relativistic Cosmology I." The contemplated second part never appeared. I am most grateful to Prof. Raychaudhuri for sharing these details with me.

dust, (7) will have synchronous form if and only if the matter is nonrotating. Raychaudhuri did not assume nonrotation at this point in his analysis, but he noted that without loss of generality the $g_{\alpha 4}$ can be made to vanish along one of the flow lines of the dust matter, in which case the only non-vanishing component of the stress-energy tensor is $T_4^4 = \rho$, where ρ is the density of the dust. Einstein's field equations (with cosmological constant term) then imply that

$$R_4^4 = \Lambda - 4\pi\rho.$$

Combining this with a direct calculation of the Ricci tensor in the coordinate system of (7) yields

$$\frac{1}{G}\frac{\partial^2 G}{\partial t^2} = \frac{\Lambda - 4\pi\rho - \phi^2 + 2\omega^2}{3},$$

where $G^6 = -g$, ω is the magnitude of rotation of matter, and ϕ is a function of the metric potentials that vanishes if and only if the expansion or contraction of matter is isotropic.

In the case of non-rotating matter ($\omega = 0$), two consequences can be drawn from the previous equation, or so Raychaudhuri claimed. First, if $\Lambda = 0$, G cannot have a minimum so that "one has to start from a singularity at a finite time in the past as in isotropic models" (Raychaudhuri 1955: 1125). Second, the time scale from this singular state to the present is a maximum for isotropic models. The upshot is that dropping the assumption of isotropy does not by itself solve either of the outstanding difficulties of cosmology. However, it was also noted that the changeover to anisotropy would allow Λ to escape the lower bound on its value set by observational data in the case of isotropy, and that a higher value for Λ can give a longer time scale and even lead to an avoidance of the original singularity. But the use of this "arbitrary parameter," Raychaudhuri opined, "robs the theory of much of its appeal" (Raychaudhuri 1955: 1125).

It needs to be emphasized that Raychaudhuri's result is largely geometrical and that EFE enter at only one juncture. Consider a congruence of time-like geodesics, and let V^μ be the normed ($V^\mu V_\mu = -1$) tangent vector field of this congruence. The relevant geometrical quantities are defined as follows: the expansion $\theta \equiv \nabla_\mu V^\mu$, where ∇_μ is the derivative operator associated with the space-time metric $g_{\mu\nu}$; the shear $\sigma_{\mu\nu} \equiv \nabla_{(\mu} V_{\nu)} - (1/3)\,\theta\, h_{\mu\nu}$, where $h_{\mu\nu} \equiv g_{\mu\nu} + V_\mu V_\nu$ is the space metric of the hyperplane orthogonal to V^μ; and the rotation $\omega_{\mu\nu} \equiv \nabla_{[\mu} V_{\nu]}$. What is now called 'Raychaudhuri's equation' is a purely geometrical identity which states that

$$\dot{\theta} = -\frac{1}{3}\theta^2 - \sigma_{\mu\nu}\sigma^{\mu\nu} + \omega_{\mu\nu}\omega^{\mu\nu} - R_{\mu\nu} V^\mu V^\nu,$$

where the dot denotes differentiation with respect to proper time. Specializing to the case where there is no rotation ($\omega_{\mu\nu} = 0$), or equivalently, where the geodesic congruence is hypersurface orthogonal,[22] it follows that *if* $R_{\mu\nu}\, V^\mu V^\nu \geq 0$, then

$$\dot{\theta} + \frac{1}{3}\theta^2 \leq 0.$$

Integrating this yields the result that if the initial expansion θ_0 is negative (i.e., the geodesic congruence is initially converging), then $\theta \longrightarrow -\infty$ within a proper time $\leq 3/|\theta_0|$. Thus far Einstein's field equations have played no role. They now come on stage because they imply that $R_{\mu\nu}\, V^\mu V^\nu = 8\pi(T_{\mu\nu}\, V^\mu V^\nu + (1/2)T) + \Lambda$. If $\Lambda = 0$, then to assure that $R_{\mu\nu}\, V^\mu V^\nu \geq 0$ all that is needed in addition is that $T_{\mu\nu}\, V^\mu V^\nu + (1/2)\, T \geq 0$, which is an instance of the strong energy condition (which requires that this last inequality holds for every unit time-like V^μ). In the case of a dust model with $T_{\mu\nu} = \rho\, V_\mu V_\nu$, the strong energy condition is automatically fulfilled if ρ is non-negative, which was implicitly assumed all along. The connection with Raychaudhuri's theorem is made by noting that in a synchronous coordinate system adapted to the dust flow,

$$\theta = \frac{\partial}{\partial t} \ln(\sqrt{-g}),$$

so that $\theta \longrightarrow -\infty$ corresponds to $g \longrightarrow 0$.

But in what sense does $\theta \longrightarrow -\infty$ or, equivalently, $g \longrightarrow 0$, implicate a space-time singularity? For the moment, take $\theta \longrightarrow -\infty$ to imply the crossing of geodesics which are orthogonal to some initial space-like hypersurface. Such a crossing, however, need not indicate a genuine space-time singularity since, for example, it can happen even in Minkowski space-time with the appropriate choice of initial hypersurface.[23] In the case where matter consists of pure dust, the world lines of the dust particles are (as noted above) time-like geodesics so that crossing of these geodesics does imply an infinite density singularity. This is also seen from the conservation law $\nabla_\mu T^{\mu\nu} = 0$, which in a synchronous coordinate system adapted to the matter world lines gives $\rho\sqrt{-g} = \text{constant}$, so that $g \longrightarrow 0$ implies that $\rho \longrightarrow \infty$. However, the dust idealization is unrealistic, and if, for example, pressure effects are included, matter will not follow geodesics. Furthermore, one can wonder whether even a small amount of rotation for matter will lead to an avoidance of infinite densities. Thus, in the general setting the implications of a Raychaudhuri $g \longrightarrow 0$ singularity for a space-time singularity is left unsettled.

Actually, there is a more subtle gap in the above argument that was not fully appreciated until nine years later when Shepley (1964) was able to fill the gap. One way of exposing the gap is to note that $\theta \longrightarrow -\infty$ does not necessarily mean that the geodesics cross (a specific example will be given below). The argument that, for dust matter, $\rho\sqrt{-g} = \text{constant}$ and, therefore, that $\rho \longrightarrow \infty$ as $g \longrightarrow 0$

---

[22] This equivalence follows from Frobenius' theorem. If the geodesic congruence is initially non-rotating, it will remain so.

[23] This is a point emphasized by the Russian reaction to the results of Raychaudhuri and Komar; see Section 5.

assumes the validity of the synchronous coordinate system adapted to the dust flow; but this coordinate system may break down. Of course, a new synchronous coordinate system can be resurrected in its stead if the world lines of the dust remain orthogonal to a family of space-like hypersurfaces; but this second system may break down even more quickly than the first. A third system can be erected in place of the second, but it may break down even more quickly than the second, etc. Geometrically what may go wrong is that the hypersurfaces orthogonal to the geodesic flow may change from space-like to null (again a specific example will be considered below).

Arthur Komar (1956), apparently unaware of Raychaudhuri's paper, published similar but seemingly more general results. Komar did not assume that matter is nonrotating nor that matter is in the form of dust. But he did employ a synchronous coordinate system. Geometrically such a system is derived by adapting coordinates to a congruence of time-like geodesics orthogonal to some initial space-like hypersurface; initially the rotation is zero and, one can show, will remain zero. This congruence may or may not represent the flow of matter. Using a synchronous coordinate system, Komar defined a symmetric tensor field (now called the extrinsic curvature) $\chi_{\alpha\beta} \equiv \partial g_{\alpha\beta}/\partial t$ and showed that if $\Lambda = 0$ and $T_{44} + (1/2)\,T \geq 0$, χ^{α}_{α} will diverge at a finite time unless it is initially zero. (The relation to Raychaudhuri's result is seen from the fact that in a synchronous coordinate system, $\chi^{\alpha}_{\alpha} = \partial \ln \sqrt{-g}/\partial t$.) What did Komar take the significance of this formal result to be? The title of the paper, "Necessity of Singularities in the Solutions of the Field Equations of General Relativity," was explained in the introductory section.

> The question naturally arises whether such singular points [as the initial singularity in the Friedmann model] are a consequence of the particular symmetry proposed in Friedmann's model, or whether perhaps for more general distributions of matter one need not expect instants of creation or annihilation of the universe. The purpose of this paper is to show that singularities in the solution of the field equations of general relativity are to be expected under very general hypotheses ... (Komar 1956: 544)

In the penultimate section of the paper this confident pronouncement was taken back with the acknowledgment that the connection between the divergence of χ^{α}_{α} and "instants of creation or annihilation" is not pellucid.

> We should note that one cannot easily determine whether the singularity is in the coordinate system or whether the space itself is singular. Taub[24] has pointed out that there is as yet no well-defined way of determining what constitutes an essential singularity within the general theory of relativity. (Komar 1956: 546)

And yet in the conclusion of the paper the initial pronouncement is partially restated.[25] The condition $\chi^{\alpha}_{\alpha} = 0$, together with the other assumptions of the analysis, imply that space-time is flat. Thus, if space-time is not flat,

---

[24] The reference here is to Taub 1951; see Section 3 above.

[25] It seems plausible to conjecture that the reference to Taub was added in the revised version of the paper prepared in reaction to the referee's report.

> we must be prepared either: (A) to allow for singularities (as in the Schwarzschild or Friedmann solutions); (B) to permit the possibility of a cosmological term or a negative pressure term ... ; or (C) to consider spaces which do not have the simplifying property of containing a set of geodesically parallel space-like hypersurfaces for all times. (Komar 1956: 546)

Komar was correct in dropping the worry about whether the singularity is only in the coordinate system. Although his argument is framed in coordinate terms, the point is purely geometrical. If non-flat solutions to Einstein's field equations are considered, Λ is set to 0, and $T_{\mu\nu} V^{\mu} V^{\nu} + (1/2) T \geq 0$ for all time-like V^{μ}, then the space-time cannot be covered by a 'geodesically parallel' family of space-like hypersurfaces. For if there were such a family, that would mean that there is an everywhere defined congruence of non-rotating time-like geodesics. One could then apply the Raychaudhuri-Komar effect to the unit four-velocity field V^{μ} of the geodesic congruence to conclude that the geodesics cross at some finite time, yielding a contradiction, at least if it is assumed that the geodesics of the congruence can be prolonged indefinitely. To repeat, this result does not assume that matter is in the form of dust nor that matter is non-rotating—the V^{μ} need not be the four-velocity of matter.[26] But for this very reason the generality of Komar's result is achieved at the expense of severing the connection with singularities in the sense of an infinite matter density since now $\theta = \nabla_{\mu} V^{\mu} \longrightarrow -\infty$ need not imply a matter density singularity.

To probe further the significance of Komar's result, suppose that in order to escape the contradiction Komar derived, one drops either the assumption that space-time is covered by a one-parameter family of geodesically parallel space-like hypersurfaces or that the geodesics of normal congruence can be extended indefinitely far (as measured in affine distance). What does this portend for singularities in the sense then in play? Taub-NUT space-time[27] is a vacuum solution to Einstein's field equations with $\Lambda = 0$. It trivially satisfies the strong energy condition, and it is non-flat. It is covered by a one-parameter family of hypersurfaces on which the metric is homogeneous. The family members covering the initial or Taub portion of the space-time are space-like. But the member at the boundary of the Taub and NUT regions is null. And yet there is no singularity in the sense with which Raychaudhuri, Komar, and their interlocutors were concerned—there is no infinite density of matter (the Taub-NUT universe being a vacuum solution), nor is there any curvature blow up. Furthermore, this example shows why taking $\theta \longrightarrow -\infty$ to mean that the geodesics cross is only a loose way of speaking. For the time-like geodesics orthogonal to an initial space-like surface of homogeneity in the Taub portion of the space-time, θ does become infinitely negative as the Taub-NUT boundary is approached; but these geodesics do not cross but rather asymptote to a null hypersurface. As a result the geodesics cannot be extended to

---

26 In a note responding to Komar's paper, Raychaudhuri (1957) claimed that Komar's result was "explicitly given by the present author [i.e., Raychaudhuri 1955]". This is true if Raychaudhuri's construction is interpreted in the modern way as applying to any time-like geodesic congruence and not just to the world lines of dust matter.

27 See Misner 1963 and Misner & Taub 1968 for a description of this space-time.

indefinitely large values of affine parameter. In this sense Taub-NUT space-time *is* singular. But the adoption of this conception of singularities required a new way of thinking about singularities, a way that was and still is controversial (see Sections 6 and 9).

These points were certainly not clear to the researchers at the time. Nevertheless, the Raychaudhuri and Komar results were attacked, but from another direction.

5. The Russian reaction to the theorems of Raychaudhuri and Komar

The specter of singularities rampant among general relativistic cosmological models had been raised by the results of Raychaudhuri and Komar. The Russian school of Lev Landau, Yevgeniy Lifshitz, and coworkers sought to slay it. Before soft pedaling the implications of the Raychaudhuri-Komar results, the Russian school laid a priority claim. Lifshitz and Khalatnikov (1960b) claimed that Landau had "long ago" proved that in a synchronous coordinate system the determinant g of the metric vanishes within a finite time.[28] They went on to note that the vanishing of g need not indicate a singularity in the space-time itself but only a breakdown in the coordinate system due to the crossing of the geodesics orthogonal to some initial space-like hypersurface. For the Russian school, a "true physical singularity in the metric" is "one which belongs to the space-time itself and is not connected with the character of the reference system." By this they meant that "Such a singularity is characterized by scalar quantities, such as density of matter and the invariants of the curvature tensor, becoming infinite" (Lifshitz & Khalatnikov 1963: 190–191).[29] To settle the question of whether singularities are a general feature of solutions to Einstein's field equations, the Russian school proposed to implement the following program. First, write down the form of the most general solution of the field equations in the neighborhood of a singularity. Second, count the number of arbitrary functions of coordinates in such a solution. Third, count the number of arbitrary functions needed to fix an initial distribution of matter and the state of the free gravitational field. Fourth and finally, compare the two counts and conclude that singularities are not a general feature of general relativistic space-times just in case the latter count is larger, meaning that the subset of singular solutions is of 'measure zero' in the full set of solutions. By the Russian reckoning, the number of arbitrary functions of coordinates in a singular solution is always one less than for a general solution. They thus felt justified in stating:

---

[28] They refer to Komar 1956 but not to Raychaudhuri 1955, perhaps because the proof given by the Russian school is closer in style to Komar's. Lifshitz & Khalatnikov 1963 does refer to "an analogous result of Raychaudhuri."

[29] The Russians were not alone in this characterization of true or physical singularities. Thus, John Graves and Dieter Brill spoke of the need to distinguish between "true geometric singularities at which invariants of the Riemann curvature tensor become singular, and 'pseudo-singularities,' which are due to an unfortunate choice of coordinate system" (1960: 1507). As noted above, more careful writers might have added that the blow up behavior must occur 'at a finite distance,' the point being that if the blow up only happens as spatial or temporal infinity is approached, no singularity in the space-time itself is indicated. However, Penrose's concept of naked singularities covers behavior which happens, so to speak, only at infinity; see Section 9 below.

> All the foregoing leads to the fundamental conclusion that the presence of a time singularity is not an essential property of the cosmological model of the general theory of relativity, and the general case of an arbitrary distribution of matter and gravitational field does not lead to the appearance of a singularity. (Lifshitz, Sudakov, & Khalatnikov 1961: 1301)

This claim was repeated in a communication to *Physical Review Letters* (Khalatnikov, Lifshitz, & Sudakov 1961), in a review article by Lifshitz and Khalatnikov (1963), and in the revised second edition of *Classical Theory of Fields* by Landau and Lifshitz (1962).

The Russians eventually recanted, but not until after the singularity theorems of Penrose (1965), Hawking (1965, 1966a–d) and Robert Geroch (1966). A silent recantation took place with the 1967 Russian edition of *Classical Theory of Fields* which omitted § 110 ("The absence of singularities in the general cosmological solution"). A more explicit recantation come in 1970 in the form of a communication to *Physical Review Letters*. Khalatnikov and Lifshitz frankly admitted the limitations and pitfalls of their method:

> Since there exists no systematic method for examining the singularities of the solutions of Einstein's equations, our search for increasingly more general solutions of this kind proceeded essentially by trial and error. A negative result from such a procedure could of course never be completely conclusive by itself ... (Khalatnikov & Lifshitz 1970: 78)

As will become clear later in our story, the Russian recantation has a curious flavor because the Russian school was concerned with one sense of singularity while the theorems of Penrose et al. were concerned with a different sense.

6. Analysis of singularities in the early to mid-1960s

The difficulties in interpreting the results of Raychaudhuri and Komar pointed to the need for a better understanding of the elusive concept of space-time singularity. A plea for more clarity on these matters was made by Charles Misner in 1963. Misner's analysis was to have a crucial influence on the development of singularity theorems, though the influence lay as much in what was ignored in Misner's analysis as what was taken from it.

> The first step is to find some clearly stated problems, and the clue to clarity is to refuse even to speak of a singularity but instead to phrase everything in terms of the properties of differentiable metric fields on manifolds. If one is given a manifold and on it a metric which does not at all points satisfy the necessary differentiability requirements, one simply throws away all the points of singularity. The starting point for any further discussion is then the largest submanifold on which the metric is differentiable. ... The first problem then is to select a criteria which will identify in an intuitively acceptable way a "non-singular space." ... The problem ... is to recognize the holes left in the space where singular (or even regular) points have been omitted. (Misner 1963: 924)

For a Riemannian space $(M, k_{\mu\nu})$ with a positive definite metric, the recognition of "holes" can be achieved by investigating the metric topology. If $d(p, q)$, $p, q \in$

M, is defined as the greatest lower bound on the $k_{\mu\nu}$ length of paths joining p and q, it is easily seen that $(M, d(,))$ obeys the axioms for a metric space. The deletion of points leaving behind holes can then be detected by the incompleteness of the distance function $d(,)$ in the sense that not every Cauchy sequence of points converges.[30] For a relativistic space-time $(M, g_{\mu\nu})$ this criterion of incompleteness is unworkable since $g_{\mu\nu}$ does not define a distance function. However, in the Riemann case, Cauchy completeness is equivalent to geodesic completeness, the latter meaning that any geodesic can be extended to an arbitrarily great value of an affine parameter. This suggests using geodesic completeness as the defining characteristic of a non-singular space-time. Misner partially endorsed this suggestion by taking geodesic completeness as a *sufficient* condition for non-singularity of a space-time. His grounds were that a geodesically complete space-time is maximal, i.e., cannot be isometrically imbedded as a proper subset of a larger space-time. But this only shows that the space-time in question cannot have been obtained from a larger space-time by deleting *regular* points and leaves open the possibility that *singular* points were deleted. In fact, some later analyses counted some geodesically complete space-times as singular.[31] On the other hand, Misner rejected geodesic completeness as a *necessary* condition for non-singularity of a space-time. No compact manifold (without boundary) can be imbedded as a proper subset of another (Hausdorff) manifold of the same dimension. Thus, for either a compact Riemann space or a compact relativistic space-time, there are no holes that arise from deleting regular or singular points. For the Riemann case all is well since if M is compact, $(M, k_{\mu\nu})$ is geodesically complete. But Misner showed that all is not well for the relativistic space-time case by producing an example of a space-time $(M, g_{\mu\nu})$ that is geodesically incomplete despite the fact that M is compact.[32] More generally, if the incomplete geodesics are contained in a compact set, then the space-time is not counted as singular according to Misner's point of view; in particular, Taub-NUT space-time is seen as non-singular (see Misner & Taub 1968). Misner also acknowledged the generally accepted sufficient condition for a singular space-time; namely, a curvature scalar becomes unbounded along an incomplete geodesic. But he also warned that "It is not to be presumed that all spaces which should be called 'essentially singular' are identified by this criteria" (Misner 1963: 926).

The upshot of Misner's discussion was partly encouraging and partly discouraging. The encouraging part was that there is a clear project: for $(M, g_{\mu\nu})$ with $g_{\mu\nu}$ defined differentiable (to whatever degree you like) at all points of M, define what it means for $(M, g_{\mu\nu})$ to be non-singular. The discouraging part was that

---

[30] p_i, $i = 1, 2, 3, \ldots$ is a Cauchy sequence just in case, for any $\epsilon > 0$, there is an N such that for any $m, n > N$, $d(p_m, p_n) < \epsilon$.

[31] In the b-boundary approach of Bernd Schmidt, a space-time is counted as singular if it contains inextendible curves of finite generalized affine length (see Hawking & Ellis 1973: 283–284). This can happen even if the space-time is geodesically complete.

[32] Examples of this sort had been discussed by Lawrence Marcus (1962), from whom Misner said that he had "borrowed heavily" (Misner 1963: 924, n. 4). Another example was produced by Robert Hermann (1964).

although there are a number of ideas waiting to be used, there was no obvious way to combine these ideas into a precise, relatively simple, and intuitively appealing definition.

György Szekeres (1960), like Synge (1950) and Taub (1951) before him, also decried the lack of an adequate definition of a singularity of a Lorentzian manifold. He implicitly embraced part of the sentiments Misner was to express three years later; namely, the starting point for analysis is a differentiable manifold M and a Lorentz metric $g_{\mu\nu}$ defined and C^∞ at every point of M. But Szekeres not only wanted to say what it is for $(M, g_{\mu\nu})$ to be singular (or non-singular); for singular space-times he also wanted to be able to speak of singularities, and took the first steps towards a local characterization of these objects. Using an equivalence relation on geodesics, he defined a boundary point of the space-time as an equivalence class of incomplete geodesics. Such a boundary point was called a singularity just in case it remains a boundary point in every extension of the space-time. Singularities were then classified as 'ordinary' (or 'non-ordinary') according as some derivative (respectively, no derivative) of the metric in a normal coordinate system along a geodesic fails to approach a limit as the singularity is approached. Szekeres' paper seem to have passed virtually unnoticed by those engaged in debate of the early to mid-1960s over the existence and nature of singularities in GTR.[33]

In sum, in the mid-1960s, as in the 1950s and the beginning of the 1960s, there existed no adequate and generally accepted analysis either of the notion of a space-time singularity or a singular space-time. One might think that this lacuna would make it difficult to prove general theorems about the existence of singularities in solutions to Einstein's field equations—for how can a mathematical theorem be established for a mathematically ill defined object? Such a question presupposes a naive view of how science actually works. In the event, uncertainties about the definition of singularities allowed room for maneuvering. After the dust had settled, one could work backwards from the theorems to a definition of singularities. This is not necessarily a self-aggrandizing procedure, at least not if the theorems are beautiful enough and powerful enough. They were.

7. The singularity theorem of Shepley

While the Russians were trying to exorcise the specter of singularities raised by the Raychaudhuri-Komar results, Lawrence Shepley, a student of John Wheeler and Charles Misner, was working to make the specter more threatening. Like Raychaudhuri, Shepley (1964) focused on dust models but of a sort that allowed the dust to be rotating. The models were assumed to be spatially homogeneous in the sense that the space-times admit the three-dimensional Lie group SO(3, **R**) as a symmetry.[34]

[33] Geroch 1968a is the only relevant reference from the period that I have found to cite Szekeres 1960.

[34] SO(3, **R**) is the group of unit determinant, 3×3 orthogonal, real matrices. The $k = +1$ (spatially closed) FLRW models belong to this class. But the class is much broader in that it includes anisotropic models.

Any such space-time is covered by a one-parameter family $H(t)$ of surfaces (which are topologically S^3) of homogeneity. It was further assumed that initially the $H(t)$ are space-like. If they remained forever space-like then a contradiction would result by the Raychaudhuri-Komar construction since the $H(t)$ are geodesically parallel. It would seem that the only way out is to conclude that for some value of t, $H(t)$ turns from being space-like to null, as in the Taub-NUT example. But Shepley established that such a change in character of the $H(t)$ is not possible for SO(3, $\mathbf{R}$) dust models. Thus, for a special class of models, Shepley showed that one of the gaps in the Raychaudhuri-Komar analysis could be filled.

At first glance, however, Shepley's result is puzzling. With the gap filled, why doesn't a genuine contradiction result, showing that such models are impossible? Shepley took the escape hatch to be the existence of singularities. He stated his result as

> **Theorem** (Shepley). All dust filled cosmological models of general relativity which have the symmetry SO(3, $\mathbf{R}$) and which have at least one space-like invariant three-dimensional hypersurface have a point singularity.

But what sense of singularity is indicated, and exactly how does the existence of such a singularity resolve the contradiction?

The answer lies in the fact that the contradiction results from assuming that the geodesics normal to the initial space-like hypersurface of homogeneity can be prolonged to an arbitrarily large value of an affine parameter. The escape hatch lies in the conclusion that the space-time is singular in the sense of having incomplete time-like geodesics. Here we have the first air-tight geodesic incompleteness singularity theorem, although some hindsight is needed to produce this reading of Shepley's paper. Shepley defines a "point of singularity" as a point "which can be reached by a geodesic of finite total length from other points of the space-time manifold, where the metric is degenerate or otherwise irregular (for example, a point where a curvature scalar is infinite)" (1964: 1403). The definition suggests, but does not say, that geodesic incompleteness is *the* identifying characteristic feature of a singular space-time. Having helped to prepare Misner's 1963 paper, Shepley was aware of both the attractions and the pitfalls of such a suggestion.[35] And possibly as a result, Shepley's definition is a superposition of the old ideas of curvature blow up and geodesic incompleteness.

As noted by Shepley himself, the connection of his singularity result to a singularity in the infinite density sense is far from evident. The conservation law $\nabla_\mu(\rho V^\mu V^\nu) = 0$ yields $\rho V^4\sqrt{-g}$ = constant in a synchronous coordinate system. In the Raychaudhuri case of non-rotating matter, the synchronous coordinate system can be chosen to be comoving with the matter and $V^4 = -1$, so that, if the synchronous coordinate system remains valid, $g \longrightarrow 0$ implies that ρ diverges. But in Shepley's more general case of rotating dust, it could conceivably happen that V^4 becomes infinite while ρ stays finite.

---

[35] Misner 1963: 924, n. 4 thanks Shepley for preparing the review of the literature on singularities.

8. The singularity theorems of Penrose, Hawking, and Geroch

The major turning point in the study of singularities was undoubtedly Penrose's (1965) article "Gravitational Collapse and Space-Time Singularities." The importance of this article is belied by its brevity—it occupied less than three pages of *Physical Review Letters*. Although Penrose's argumentation was somewhat obscure (as will be discussed shortly), it quickly became clear to the experts that singularities of gravitational collapse could no longer be dismissed as artifacts of the symmetries or the special conditions on matter assumed in the Oppenheimer-Volkoff-Snyder analysis (see Section 2). The article also set off a flurry of activity that, within a few short years, resulted in the generally accepted opinion that, if Einstein's field equations are correct, singularities are to be expected in generic circumstances in both gravitational collapse and cosmology.

Because of the importance of Penrose's theorem, it is worth going through the assumptions in some detail. It was assumed, first, that the space-time $(M, g_{\mu\nu})$ is temporally orientable so that the null half-cones can be continuously divided into 'past' and 'future.' This is not a restrictive assumption since if it fails for $(M, g_{\mu\nu})$ it can be secured by passing to a covering space-time. A second, and very strong, assumption is that $(M, g_{\mu\nu})$ possesses a Cauchy surface (a space-like hypersurface which is intersected exactly once by every time-like curve without endpoint) that is non-compact. Spatially closed universes are thus excluded from the purview of the theorem. And the Cauchy property precludes any hint of acausal structure for the space-time; it is, for example, even stronger than the property of stable causality which says (intuitively) that the null cones can be widened out without closed causal loops resulting.[36] Third, it is assumed that $R_{\mu\nu} K^{\mu} K^{\nu} \geq 0$ for any null vector K^{μ}. If Einstein's field equations hold, with or without cosmological constant, this condition will be fulfilled as long as matter obeys the condition $T_{\mu\nu} K^{\mu} K^{\nu} \geq 0$ for any null vector K^{ν}, which by continuity is a consequence of the weak energy condition: $T_{\mu\nu} V^{\mu} V^{\nu} \geq 0$ for all time-like V^{μ}. (Penrose's paper stipulated that $(R_{\mu\nu} - (1/2) R g_{\mu\nu} + \Lambda g_{\mu\nu}) V^{\mu} V^{\nu} \geq 0$ for all time-like V^{μ}, which is stronger than the theorem required.) Fourth, there must be a trapped surface $\mathcal{T}$, i.e., a space-like two-surface such that the outgoing and ingoing null geodesics that intersect it orthogonally are both converging. In the Oppenheimer-Volkoff-Snyder model of gravitational collapse, such a surface will form when the body contracts within its Schwarzschild radius—the two-sphere r = constant, t = constant for $r < 2M$ being an example. But—and this is the crucial point—the concept of a trapped surface does not presuppose the spherical symmetry of the Oppenheimer-Volkoff-Snyder analysis. Finally, according to Penrose, it needs to be assumed that the space-time is future null geodesically complete. "The existence of a singularity can never be inferred ... without an assumption such as completeness of the manifold under consideration" (Penrose 1965: 58).

---

[36] For a precise definition, see Hawking & Ellis 1973: 198. The existence of a Cauchy surface is a necessary condition for the global version of determinism, according to which initial data on a time slice uniquely fix the entire future (and past).

When Penrose summarized his results as showing that "*deviations from spherical symmetry cannot prevent space-time singularities from arising*" (Penrose 1965: 58) the reader in 1965 would naturally have taken this to mean that given the above assumptions, it follows that gravitational collapse eventuates in a physical singularity in the sense understood in the debate over the prevalence of physical vs. coordinate singularities—i.e., infinite density or an infinite curvature scalar. The caption of Penrose's figure 1, showing what is presumably a density or curvature singularity in spherical gravitational collapse, also encouraged this interpretation: "The diagram essentially serves for the discussion of the asymmetrical case" (Penrose 1965: 58, fig. 1). In fact, however, nothing about a density or curvature singularity is part of or is proved by the theorem. Null geodesic completeness is an assumption of the proof only in the reductio sense. In the now accepted reconstruction of the theorem (see Hawking & Ellis 1973: 263–264 and Wald 1984: 239–240) the *reductio* assumption is used to prove that the boundary of $I^+(\mathcal{T})$, the chronological future of the trapped surface,[37] is compact. Then the compactness of $I^+(\mathcal{T})$ is shown to contradict the existence of a non-compact Cauchy surface. Thus, given the other assumptions, the space-time is null geodesically incomplete. We have a singularity theorem if and only if null geodesic incompleteness is taken as a sufficient condition of a singular space-time.

Six months after Penrose's seminal paper appeared in print, Hawking and George Ellis (1965) submitted a note to *Physics Letters* containing a new singularity theorem that generalized Shepley's (1964) result for homogeneous cosmologies. Shepley had assumed dust matter; Hawking and Ellis treated the more general case of a perfect fluid. Shepley had assumed spatial homogeneity in the form of partition of space-time by a one-parameter family of hypersurfaces of homogeneity; Hawking and Ellis assumed only that there is at least one space-like hypersurface on which a three-parameter group of motions acts transitively. Following Penrose's lead, Hawking and Ellis also assumed that the models in question were null and time-like geodesically complete: "Any models in which this were not the case would not seem to be reasonable models of the universe" (Hawking & Ellis 1965: 246). They then proceeded to demonstrate the existence of a "physical singularity" in the form of an infinite matter density. Of course, if the matter density becomes unbounded "at a finite distance," the space-time structure breaks down and geodesic completeness results. Unlike Penrose's (1965) result, however, the Hawking and Ellis theorem is not a pure *reductio* proof of geodesic incompleteness since it supplies the reason for the incompleteness.

The unclarity over what was being demonstrated in the singularity theorems persisted for at least another few months. On 16 August 1965 *Physical Review Letters* received a communication from Hawking entitled "Occurrence of Singularities in Open Universes." Hawking noted that in open ($k = 0$ or $k = -1$) FLRW models there are trapped surfaces. He argued that universes that are similar on a large scale to FLRW universes but are not homogeneous or isotropic locally would

---

[37] For a space-time $(M, g_{\mu\nu})$ and a set $S \subset M$, $I^+(S)$ is defined as the set of all $p \in M$ such that there is a future directed time-like curve from S to p.

still possess a trapped surface. Penrose's (1965) theorem could then be applied to deduce the existence of a singularity. This theorem was characterized by Hawking as follows: "Penrose has shown that either a physical singularity must occur or space-time is incomplete if there is a closed trapped surface" (Hawking 1965: 689).

The 22 August 1966 issue of *Physical Review Letters* published communications from Hawking (1966a) and Geroch (1966) containing new singularity theorems. Both authors were now quite specific that geodesic incompleteness is to be taken as a defining characteristic of a singular space-time—time-like incompleteness for Hawking, time-like or null incompleteness for Geroch. When giving their definitions of a singular space-time both authors referred to Misner 1963, which is somewhat ironic since one of the purposes of Misner's paper was to argue that geodesic incompleteness is *not* a sufficient condition for labeling a space-time as singular (see Section 6). And while taking geodesic incompleteness as the definition of a singular space-time served to clarify the otherwise murky logic of the previous singularity theorems—Shepley 1964, Penrose 1965, Hawking & Ellis 1965, and Hawking 1965—and to make the new theorems mathematically precise, it also undercut the claim that what was being demonstrated was the existence of "physical singularities" in the sense which was then in play in the literature and which had been the core of the debate about the prevalence of singularities in solutions to Einstein's field equations. Nevertheless, the shift of focus to geodesic incompleteness was entirely justified as a piece of opportunism. The techniques of Penrose, Hawking, and Geroch could be used to prove rigorous and powerful results about geodesic incompleteness. *Pace* Misner, even if geodesic incompleteness does not always signal singularities in the originally intended sense, it surely is a pathology worth noting; and one could suppose that in typical circumstances this pathology would be a symptom of density or curvature singularities. But these were matters that could be sorted out later. Opportunism demanded that the theorems be proved; their exact physical significance would become apparent in the fullness of time.

The Penrose-Hawking singularity results were extended and codified in a series of three articles by Hawking, published in the *Proceedings of the Royal Society* (Hawking 1966b, 1966c, 1967); in Hawking's (1966d) Adams Prize Essay "Singularities and the Geometry of Space-Time"; in Penrose's (1966) Adams Prize Essay "An Analysis of the Structure of Space-Time";[38] in a joint article by Hawking and Penrose (1970); in Penrose's (1972) monograph *Techniques of Differential Topology in General Relativity*; and in Hawking and Ellis' (1973) *The Large Scale Structure of Space-time.* The Raychaudhuri-Komar effect plays an important role in some of the theorems, but now this effect is explicitly recognized to function as part of a *reductio* demonstration of geodesic incompleteness rather than as part of a demonstration of a 'physical' singularity.

---

[38] Portions of this essay were published in Penrose (1968).

9. What is a space-time singularity?

The Penrose-Hawking theorems focused attention on geodesic incompleteness as the mark of a singular space-time. In part, there was a good reason for this choice: it provides a mathematically precise criterion that corresponds to intuitive judgments in a number of paradigm cases. The choice was also partly a matter of expediency: the proof techniques developed by Penrose and Hawking lent themselves to this definition. But, powerful and elegant as they are, the Penrose-Hawking theorems did not settle the debate about the correct definition of space-time singularity. A strong indication of the situation shortly after several of the key theorems had appeared in print was the publication of Geroch's (1968b) "What Is a Singularity in General Relativity?" The body of the paper is in the form of a Galilean dialogue. Although such a format is unusual for *Annals of Physics*, it was nicely tailored to revealing the uncertainties and ambiguities which existed at the time. What Geroch's article and subsequent analysis revealed was a situation of daunting complexity.[39]

Begin, as Misner (1963) recommended, with a relativistic space-time $(M, g_{\mu\nu})$, where M is a differentiable manifold and $g_{\mu\nu}$ is a Lorentz signature metric which is defined and differentiable on all of M. There are then two tasks. The first is to find a criterion that will detect when $(M, g_{\mu\nu})$ is singular (despite the fact that $g_{\mu\nu}$ is everywhere well-defined and smooth and, in that sense, nonsingular). If one wants to speak not only of a singular space-time but also of singularities, then the second task is to provide a means that would justify talking about these things as localizable objects. This would involve constructing a set of idealized points—to represent the singularities—and (at least) a topology for the manifold M + idealized points. Unfortunately, extant procedures yield counterintuitive results, e.g. the singular points may not be Hausdorff separated from the interior points of M; and there is reason to believe that such results will be common to all procedures which conform to some seemingly natural requirements (see Geroch, Liang, & Wald 1982). It remains to be seen whether or not 'object talk' about singularities can be given an expression that is at once mathematically precise, intuitively appealing, and useful in classifying singular space-times.

Returning to the prior task of demarcating singular space-times, four or more families of ideas compete for attention. The first starts from the intuitive idea that motivated most of the pre-Penrose-Hawking singularity theorems; namely, a singular space-time is one in which a relevant physical quantity blows up. But the technical elaborations of this idea have gone far beyond anything considered by the pioneers of singularity theorems. For instance, a space-time may be considered to be singular even if all curvature scalar polynomials remain bounded if some component of the Riemann curvature tensor, as measured in an orthonormal frame parallelly propagated along a geodesic, becomes unbounded at a finite affine distance.

---

[39] Technical elaborations of the concepts mentioned in this Section can be found in a number of sources: Hawking & Ellis 1973; Ellis & Schmidt 1977; Wald 1984; Scott & Szekeres 1994. For an overview of the literature, see Earman 1995a: chap. 2.

A second family of ideas identifies a singular space-time as one which exhibits some form of incompleteness. Geodesic incompleteness—the criterion used in the Penrose-Hawking theorems—belongs to this family, but there are many other members. Geroch (1968b) produced an example of a space-time which is time-like, space-like, and null geodesically complete but which contains inextendible time-like curves of bounded acceleration and finite proper length. Such a curve might, for example, correspond to the world line of a rocket ship whose motor uses only a finite amount of fuel. The pilot of such a ship might, not unnaturally, complain that he inhabits a singular space-time. An even more demanding notion of completeness can be formulated using the concept of generalized affine length which is available for all differentiable curves, not just geodesics or curves of bounded acceleration.

There is obviously a solid connection in one direction between these first two families of ideas: the unboundedness of, say, a curvature scalar at a finite distance, as measured by affine length along a geodesic, entails geodesic incompleteness. But the implication in the other direction can fail. An interesting example of the failure is to be found in the dispute between Ludwik Silberstein and Einstein,[40] although the principals were apparently unaware of the fact. Silberstein (1936) claimed to have produced a stationary and axi-symmetric solution to the source-free EFE, with two singularities lying on the otherwise singularity-free axis of symmetry. If correct, this claim would have been an embarrassment for Einstein's program of deriving the geodesic postulate from the field equations by treating a test particle as a singularity of the field, for due to the stationary character of the solution the two "mass centers" do not move towards one another, in contradiction to "man's most ancient primitive experience" (Silberstein 1936: 270). Einstein's response was that Silberstein's solution (which was in fact the Curzon (1924a, 1924b) bipolar solution) is singular along the axis joining the "mass centers" (Einstein & Rosen 1936). He was correct in that the metric cannot be smoothly extended to cover the axis, and consequently, the solution is geodesically complete. However, all the components of the Riemann curvature tensor remain well-behaved as the axis is approached.[41]

This gap between the first two families also serves to undercut the sincerity of the Russian school's recantation of the claim that singularities are absent from a general solution to EFE. In reaction to the singularity theorems of Penrose et al., Khalatnikov and Lifshitz wrote:

> The situation has changed since the discovery of Penrose (and later by Hawking and Geroch), of new theorems which reveal a connection between the existence of a singularity (of unknown type) and some very general properties of the equations, which bear no relation to the choice of reference system. (Khalatnikov & Lifshitz 1970: 78)

---

[40] See Havas 1993 for a detailed account of this controversy.

[41] Neither Einstein nor Silberstein nor Nathan Rosen realized that the singularities of the Curzon-Silberstein solution are not of the simple pole type but have a complicated topological structure. The bizarre global structure of the Curzon monopole solution was untangled only recently by Susan Scott and Peter Szekeres (1986).

The paper ended with a charming flourish: "The new developments finally clarify the problem of singularities in general solutions and remove all previous contradictions" (Khalatnikov & Lifshitz 1970: 78). Marx could have told them otherwise. The dialectic of singularities was only getting in full swing. Khalatnikov and Lifshitz were explicit that they were concerned with singularities in the sense of "infinite density of matter or (in empty space) infinite curvature invariants" (Khalatnikov & Lifshitz 1970: 76). The theorems of Penrose et al. did not deal with singularities in this sense but rather in the sense of geodesic incompleteness. In the year before the recantation in *Physical Review Letters*, Belinskiy and Khalatnikov once again advanced considerations that argued "in favor of the absence of a physical singularity in the general cosmological solution of Einstein's equations" (Belinskiy & Khalatnikov 1969: 911). They mentioned Penrose's theorem that "there exists (under very natural assumptions) a singularity." But, they added, this singularity is one "whose character, however, no one has succeeded in establishing and which, apparently, is so weak that it does not appear in the invariants of the curvature tensor" (Belinskiy & Khalatnikov 1969: 911). Under special conditions geodesic incompleteness can be shown to entail the blow up of a component of the Riemann curvature tensor in a parallelly propagated frame (see Clarke 1975). But the general conditions under which one can move from singularities in the sense of incompleteness to curvature singularities remains obscure.

A third family of ideas is rooted in Misner's (1963) notion that a nonsingular space-time is one without any holes or missing points. Because of the lack of the appropriate technical elaboration of this notion,[42] it is not possible at the present time to make any general statements about the relations to the first two families, except to say that they will surely turn out to be complex. Misner's (1963) examples of cases of incomplete geodesics contained in compact sets show that a space-time can be singular in the incompleteness sense even though there are no missing points.

A fourth family of ideas is centered on Penrose's notion of a naked singularity, which in turn can be defined as a violation of 'cosmic censorship.' Cosmic censorship comes in many varieties, the strongest of which requires that a space-time contain a Cauchy surface. On this version of cosmic censorship anti-De Sitter space-time is counted as nakedly singular even though it is geodesically complete and displays no curvature blow up.[43] Contrary to Einstein's (1918) idea of a singularity, what happens at infinity as well as what happens at a finite distance determines whether a naked singularity is present. In the other direction, the big bang singularity of the FLRW models is not counted as nakedly singular even though it is highly visible and involves curvature blow up and geodesic incompleteness. The hypothesis that naked singularities do not develop from regular initial data in generic solutions of EFE was put forward over a quarter of a century ago by Penrose (1969). Despite all of the effort devoted to it, the cosmic censorship

[42] The work of Scott and Szekeres (1994) can be seen as lending itself to this task.

[43] See Hawking & Ellis 1973: 131–134 for a description of this space-time.

hypothesis remains controversial.[44]

The failure of strong cosmic censorship is a failure of causality: unless a space-time possesses a Cauchy surface, Laplacian determinism in its global form cannot hold. There are many other forms of causal pathologies which can be exhibited by solutions to EFE, up to and including the existence of closed time-like curves. These causal pathologies may, or may not, be associated with curvature blow up or geodesic incompleteness. Attempts have been made to prove that, consistent with EFE and energy conditions, closed time loops cannot be produced by a finite device. Such theorems (e.g., Hawking 1992), while formally correct, are less than convincing as proof of the impossibly of operating a 'time machine' (see Earman 1995b) in GTR.

In the light of the developments sketched above, it seems pointless to try to produce a simple formula that will count as *the* correct definition of a singular space-time. When we try to explore our naive conception of singularities in the setting of relativistic space-times, we encounter a wide array of phenomena. We are still far from an understanding of the interrelations of these phenomena and the roles they play in GTR. In that sense, the Penrose-Hawking theorems are a starting point rather than the endpoint for the study of singularities in GTR. The conclusion of Geroch's Ph.D. dissertation is as appropriate today as it was thirty years ago:

> What a strange little object is the singularity with its strange properties and nonexistent definition. Yet the singularity promises to remain one of the most intriguing and disturbing aspects of gravitation theory for a long time to come. Here is a problem with which we must some day come to grips— at least if we are ever to understand this phenomenon called gravitation. (Geroch 1967: 145)

Acknowledgements

I am grateful to Jean Eisenstaedt, Al Janis, and Andrzej Krasiński for helpful comments on an earlier draft of this paper.

References

Belinskiy, V. A. & Khalatnikov, I. M. (1969). "On the Nature of the Singularities in the General Solution of the Gravitational Equations." *Journal of Experimental and Theoretical Physics (USSR)* **56**: 1701–1712; *Soviet Physics JETP* **29**: 911–917.

Bergmann, Peter G. (1942). *Introduction to the Theory of Relativity*. New York: Prentice-Hall.

Bondi, Hermann. (1947). "Spherically Symmetric Models in General Relativity." *Monthly Notices of the Royal Astronomical Society* **107**: 410–425.

Clarke, C. J. S. (1975). "Singularities in Globally Hyperbolic Space-Time." *Communications in Mathematical Physics* **41**: 65–78.

[44] See Earman 1995a: chap. 3 for an overview on the pros and cons of cosmic censorship.

Curzon, H. E. J. (1924a). "Bipolar Solutions of Einstein's Gravitation Equations." *Proceedings of the London Mathematical Society* **23**: xxix.

——— (1924b). "Cylindrical Solutions of Einstein's Gravitation Equations." *Proceedings of the London Mathematical Society* **23**: 477–480.

De Sitter, Willem. (1917a). "On the Curvature of Space." *Koninklijke Akademie van Wetenschappen te Amsterdam. Proceedings of the Section of Sciences* **20**: 229–242.

——— (1917b). "On Einstein's Theory of Gravitation and its Astronomical Consequences. Third Paper." *Monthly Notices of the Royal Astronomical Society* **78**: 3–28.

——— (1933). "On the Expanding Universe and the Time Scale." *Monthly Notices of the Royal Astronomical Society* **93**: 628–634.

Droste, Johannes. (1916). "The Field of a Single Center in Einstein's Theory of Gravitation, and the Motion of a Particle in that Field." *Koninklijke Akademie van Wetenschappen te Amsterdam. Proceedings of the Section of Sciences* **19**: 197–215.

Earman, John. (1995a). *Bangs, Crunches, Whimpers, and Shrieks: Singularities and Acausalities in Relativistic Spacetimes*. New York: Oxford University Press.

——— (1995b). "Outlawing Time Machines: Chronology Protection Theorems." *Erkenntnis* **42**: 125–139.

——— (1996). "Tolerance of Spacetime Singularities." *Foundations of Physics* **26**: 263–640.

Eddington, Arthur Stanley. (1923). *The Mathematical Theory of Relativity*. Cambridge: Cambridge University Press.

——— (1924). "A Comparison of Whitehead's and Einstein's Formulæ." *Nature* **113**: 192.

Einstein, Albert. (1918). "Kritisches zu einer von Hrn. De Sitter gegebenen Lösung der Gravitationsgleichungen." *Sitzungsberichte der Königlichen Preussischen Akademie der Wissenschaften* (Berlin): 270–272.

——— (1931). "Zum kosmologischen Problem der allgemeinen Relativitätstheorie." *Sitzungsberichte der Preussischen Akademie der Wissenschaften* (Berlin): 235–237.

——— (1939). "On a Stationary System with Spherical Symmetry Consisting of Many Gravitating Masses." *Annals of Mathematics* **40**: 922–936.

——— (1950). *The Meaning of Relativity*. 4th ed. London: Meuthen.

Einstein, Albert & Rosen, Nathan. (1936). "Two-Body Problem in General Relativity Theory." *Physical Review* **49**: 404–405.

Eisenstaedt, Jean. (1982). "Histoire et singularités de la solution de Schwarzschild (1915–1923)." *Archive for History of Exact Sciences* **27**: 157–198.

——— (1993). "Lemaître and the Schwarzschild Solution." In *The Attraction of Gravitation: New Studies in the History of General Relativity* (Einstein Studies, vol. 5). J. Earman, M. Janssen and J. Norton, eds. 353–389. Boston: Birkhäuser.

Ellis, George F. R. & Schmidt, Bernd G. (1977). "Singular Space-Times." *General Relativity and Gravitation* **8**: 915–953.

Finkelstein, David. (1958). "Past-Future Asymmetry of the Gravitational Field of a Point Particle." *Physical Review* **110**: 965–967.

Fronsdal, C. (1959). "Completion and Embedding of the Schwarzschild Metric." *Physical Review* **116**: 778–781.

Geroch, Robert P. (1966). "Singularities in Closed Universes." *Physical Review Letters* **17**:445–447.

——— (1967). "Singularities in the Space-Time of General Relativity: Their Definition, Existence, and Local Characterization." Ph.D. Thesis, Princeton University.

——— (1968a). "Local Characterization of Singularities in General Relativity." *Journal of Mathematical Physics* **9**: 450–465.

——— (1968b). "What is a Singularity in General Relativity?" *Annals of Physics* **48**: 526–540.

Geroch, Robert P., Liang, C. & Wald, Robert M. (1982). "Singular Boundaries of Space-Times." *Journal of Mathematical Physics* **23**: 432–435.

Graves, John C. & Brill, Dieter R. (1960). "Oscillatory Character of Reissner-Nordström Metric for an Ideal Charged Wormhole." *Physical Review* **120**: 1507–1513.

Grishchuk, L. P. (1966). "Some Remarks on the Singularities in the Cosmological Solutions of the Gravitational Equations." *Journal of Experimental and Theoretical Physics (USSR)* **51**: 475–481; *Soviet Physics JETP* **24** (1967): 320–324.

Havas, Peter. (1993). "The General Relativistic Two-Body Problem and the Einstein-Silberstein Controversy." In *The Attraction of Gravitation: New Studies in the History of General Relativity*. (Einstein Studies, vol. 5). J. Earman, M. Janssen and J. D. Norton, eds. 88–125. Boston: Birkhäuser.

Hawking, Stephen W. (1965). "Occurrence of Singularities in Open Universes." *Physical Review Letters* **15**: 689–690.

——— (1966a). "Singularities in the Universe." *Physical Review Letters* **17**: 444–445.

——— (1966b). "The Occurrence of Singularities in Cosmology." *Proceedings of the Royal Society of London A* **294**: 511–521.

——— (1966c). "The Occurrence of Singularities in Cosmology II." *Proceedings of the Royal Society of London A* **295**: 490–493.

——— (1966d). "Singularities and the Geometry of Space-Time." *Adams Prize Essay*, mimeo..

——— (1967). "The Occurrence of Singularities in Cosmology III." *Proceedings of the Royal Society of London A* **300**: 187–201.

——— (1992). "Chronology Protection Conjecture." *Physical Review D* **46**: 603–611.

Hawking, Stephen W. & Ellis, George F. R. (1965). "Singularities in Homogeneous World Models." *Physics Letters* **17**: 246–247.

——— (1973). *The Large Scale Structure of Space-Time*. Cambridge: Cambridge University Press.

Hawking, Stephen W. & Penrose, Roger. (1970). "The Singularities of Gravitational Collapse and Cosmology." *Proceedings of the Royal Society of London A* **314**: 529–548.

Hermann, Robert. (1964). "An Incomplete Compact Homogeneous Lorentz Metric." *Journal of Mathematics and Mechanics* **13**: 497–501.

Hilbert, David. (1917). "Die Grundlagen der Physik: zweite Mitteilung." *Nachrichten von der Königlichen Gesellschaft der Wissenschaften zu Göttingen*: 53–76.

Khalatnikov, I. M. & Lifshitz, Yevgeniy M. (1970). "General Cosmological Solution of the Gravitational Equations with a Singularity in Time." *Physical Review Letters* **24**: 76–79.

Khalatnikov, I. M., Lifshitz, Yevgeniy M. & Sudakov, V. V. (1961). "Singularities of the Cosmological Solutions of Gravitational Equations." *Physical Review Letters* **6**: 311–313.

Klein, Felix. (1918). "Über die Integralform der Erhaltungssätze und die Theorie der räumlich-geschlossenen Welt." *Nachrichten von der Königlichen Gesellschaft der Wissenschaften zu Göttingen*: 394–423.

Komar, Arthur. (1956). "Necessity of Singularities in the Solution of the Field Equations of General Relativity." *Physical Review* **104**: 544–546.

Kruskal, Martin D. (1960). "Maximal Extension of Schwarzschild Metric." *Physical Review* **119**: 1743–1745.

Lanczos, Kornel. (1922). "Bemerkung zur De Sitterschen Welt." *Physikalische Zeitschrift* **23**: 539–543.

——— (1923). "Über die Rotverschiebung in der Sitterschen Welt." *Zeitschrift für Physik* **17**: 168–189.

LANDAU, Lev D. & LIFSHITZ, Yevgeniy M. (1965). *Classical Theory of Fields*. Revised 2nd ed. Reading, MA: Addison-Wesley.

VON LAUE, Max. (1921). *Die Relativitätstheorie*. Vol. 2. Braunschweig: F. Vieweg.

LEMAÎTRE, Georges. (1932). "L'Univers en expansion." *Publication du Laboratoire d'Astronomie et de Géodésie de l'Université de Louvain* **9**: 171–205. Also in *Société Scientifique de Bruxelles. Annales A* **53** (1933): 51–85.

LIFSHITZ, Yevgeniy M. & KHALATNIKOV, I. M. (1960a). "On the Singularities of Cosmological Solutions of the Gravitational Equations I." *Journal of Experimental and Theoretical Physics (USSR)* **39**: 149–157; *Soviet Physics JETP* **12** (1961): 108–113.

——— (1960b). "On the Singularities of Cosmological Solutions of the Gravitational Equations." *Journal of Experimental and Theoretical Physics (USSR)* **39**: 800–808; *Soviet Physics JETP* **12** (1961): 558–563.

——— (1963). "Investigations in Relativistic Cosmology." *Advances in Physics* **12**: 185–249.

LIFSHITZ, Yevgeniy M., SUDAKOV, V. V. & KHALATNIKOV, I. M. (1961). "Singularities of Cosmological Solutions of Gravitational Equations. III." *Journal of Experimental and Theoretical Physics* (USSR) **40**: 1847–1855; *Soviet Physics JETP* **13** (1961): 1298–1303.

MARCUS, Lawrence. (1962). Mimeographed lecture notes, American Mathematical Society Summer Institute, Santa Barbara (CA).

MISNER, Charles W. (1963). "The Flatter regions of Newman, Unti, and Tamburino's Generalized Schwarzschild Space." *Journal of Mathematical Physics* **4**: 924–937.

MISNER, Charles W. & TAUB, Abraham H. (1968). "A Singularity-Free Empty Universe." *Journal of Experimental and Theoretical Physics (USSR)* **55**: 233–255; *Soviet Physics JETP* **28** (1969): 122–133.

OPPENHEIMER, J. Robert & SNYDER, H. (1939). "On Continued Gravitational Contraction." *Physical Review* **56**: 455–459.

OPPENHEIMER, J. Robert & VOLKOFF, G. M. (1939). "On Massive Neutron Cores." *Physical Review* **54**: 374–381.

PENROSE, Roger. (1965). "Gravitational Collapse and Space-Time Singularities." *Physical Review Letters* **14**: 57–59.

——— (1966). "An Analysis of the Structure of Space-Time." *Adams Prize Essay*.

——— (1968). "Structure of Space-Time." In *Battelle Rencontres. 1967 Lectures in Mathematics and Physics*. C. M. DeWitt and J. A. Wheeler, eds. 121–235. New York: W. A. Benjamin.

——— (1969). "Gravitational Collapse: The Role of General Relativity." *Revisita del Nuovo Cimento,* Serie I, 1, Numero Speciale: 252–276.

——— (1972). *Techniques of Differential Topology in General Relativity*. Philadelphia: SIAM.

RAYCHAUDHURI, Amalkumar. (1953). "Arbitrary Concentrations of Matter and the Schwarzschild Singularity." *Physical Review* **89**: 417–421.

——— (1955). "Relativistic Cosmology I." *Physical Review* **98**: 1123–1126.

——— (1957). "Singular State in Relativistic Cosmology." *Physical Review* **106**: 172–173.

ROBERTSON, Howard P. (1932). "The Expanding Universe." *Science* **76**: 221–226.

RYAN, Michael P. & SHEPLEY, Lawrence C. (1975). *Homogeneous Relativistic Cosmologies*. Princeton (NJ): Princeton University Press.

SCHWARZSCHILD, Karl. (1916). "Über das Gravitationfeld eines Massenpunktes nach der Einsteinschen Theorie." *Sitzungsberichte der Königlichen Preussischen Akademie der Wissenschaften* (Berlin): 189–196.

SCOTT, Susan & SZEKERES, Peter. (1986). "The Curzon Singularity II. Global Picture." *General Relativity and Gravitation* **18**: 571–583.

——— (1994). "The Abstract Boundary—a New Approach to Singularities on Manifolds." *Journal of Geometry and Physics* **13**: 223–253.

SHEPLEY, Lawrence C. (1964). "Singularities in Spatially Homogeneous, Dust-Filled Universes." *Proceedings of the National Academy of Sciences* **52**: 1403–1409.

SILBERSTEIN, Ludwik. (1936). "Two-Centers Solution of the Gravitational Equations, and the Need for a Reformed Theory of Matter." *Physical Review* **49**: 268–270.

STACHEL, John. (1979). "The Genesis of General Relativity." In *Einstein Symposion, Berlin* (Lecture Notes in Physics, vol. 100). H. Nelkowski, A. Hermann, H. Posner, R. Schrader and R. Seiler, eds. 428–442. Berlin: Springer-Verlag.

SYNGE, John Lighton (1934). "On the Expansion or Contraction of a Symmetrical Cloud Under the Influence of Gravity." *Proceedings of the National Academy of Sciences* **20**: 635–640.

——— (1950). "The Gravitational Field of a Particle." *Proceedings of the Royal Irish Academy* **53**: 83–114.

SZEKERES, György. (1960). "On the Singularities of a Riemannian Manifold." *Publicationes Mathematicæ* (Debrecen) **7**: 285–301.

TAUB, Abraham H. (1951). "Empty Space-Times Admitting a Three-Parameter Group of Motions." *Annals of Mathematics* **53**: 472–490.

TOLMAN, Richard C. (1930a). "The Effect of the Annihilation of Matter on the Wave-Length of Light from the Nebulæ." *Proceedings of the National Academy of Sciences* **16**: 320–337.

——— (1930b). "More Complete Discussion of the Time-Dependence of the Non-Static Line Element for the Universe." *Proceedings of the National Academy of Sciences* **16**: 409–420.

——— (1930c). "On the Estimation of Distances in a Curved Universe with Non-Static Line Element." *Proceedings of the National Academy of Sciences* **16**: 511–520.

——— (1930d). "Discussion of Various Treatments Which Have Been Given to the Non-Static Line Element for the Universe." *Proceedings of the National Academy of Sciences* **16**: 582–594.

——— (1931). "On the Theoretical Requirements for a Periodic Behavior of the Universe." *Physical Review* **38**: 1758–1771.

——— (1934a). "The Effect of Inhomogeneity on Cosmological Models." *Proceedings of the National Academy of Sciences* **20**: 169–176.

——— (1934b). *Relativity, Thermodynamics and Cosmology*. Oxford: Oxford University Press.

TOLMAN, Richard C. & WARD, Morgan. (1932). "On the Behavior of Non-Static Models of the Universe When the Cosmological Constant Term Is Omitted." *Physical Review* **39**: 835–843.

WALD, Robert M. (1984). *General Relativity*. Chicago: University of Chicago Press.

Part III

Relativity at Large

The Cosmological Woes of Newtonian Gravitation Theory

John D. Norton

In November 1894, the astronomer Hugo Seeliger sent the journal *Astronomische Nachrichten* a short article pointing out that Newton's law of gravitation could not be applied without modification to an infinite universe with a roughly uniform matter distribution. The problem Seeliger described was exceedingly simple. As we shall see in Section 1 below, it could be developed with just a few moments thought. Thus it is no surprise that Seeliger was not the first to notice this problem. But he was the first to lay it out with sufficient vigor that the need for a solution of some sort could not be avoided. Over the decades to follow, the problem lurked quietly in the corners of gravitational and cosmological theory, with proposals for its resolution reflecting whatever were the then current trends in physical theory.

The simplest solution—possibly that of Newton himself—was just to deny that Seeliger's argument was valid. This untenable 'no-solution-needed' solution was in the minority of those views expressed in the historical record. The majority felt that the problem revealed that one or other of the commitments of Newtonian cosmology required modification. These commitments can be collected into three groups:

> **Cosmological.** Space is infinite, Euclidean and filled with a (near) uniform mass distribution.
> **Gravitational.** All bodies attract one another according to Newton's inverse square law of gravitation.
> **Kinematical.** Newton's three laws of motion.

The earliest escapes were sought in minute adjustments of Newton's inverse square law of gravitational attraction. Seeliger and Neumann proposed augmenting the

H. Goenner, J. Renn, J. Ritter, T. Sauer (eds.)
The Expanding Worlds of General Relativity
(Einstein Studies, volume 7) pp. 271–322

law with a tiny correction term whose effect would only become apparent at cosmic distances. More diverse escapes were also sought. Kelvin proposed that ethereal matter may not gravitate, allowing at least this form of matter to be spread uniformly through space. August Föppl suggested that there may be negative gravitational masses. Charlier, Selety and others proposed that the distribution of matter in the universe may have a hierarchical structure that allowed a vanishing mean matter density yet without concentrating all matter in a central island. Soon virtually every supposition buried in the 'cosmological' or 'gravitational' groups had been weighed and its modification proposed. Lense even explored the possibility of an escape through an alternative geometry for space.

The problem achieved its moment of greatest glory when Einstein (1917a) thrust it into his first attempts at a relativistic cosmology. Agreeing with Seeliger, Einstein saw the problem as revealing a need for adjustment of Newton's inviolable law of gravitation. He used the adjustment as a foil to motivate the introduction of a cosmological term in the gravitational field equations of general relativity. However he also used the paradox to pose a dilemma for Newtonian cosmology: either the universe was homogeneous and gravitationally paradoxical or its matter was concentrated in a physically untenable island universe. Selety soon showed, however, that this was a false dilemma. The work on hierarchical cosmologies had already shown an escape between the horns of the dilemma.

This work in the 1920s marked the end of the first phase of the problem posed by Seeliger; my purpose in this paper is to review the course of this first phase.[1] In a sequel I will review the later phase initiated by the discovery of the expansion of the universe and the advent of dynamical cosmologies. There it is found that the most satisfying escape from the problem lies not in modification of either 'cosmological' or 'gravitational' assumptions of Newtonian cosmology. Rather it lay in a modification of its kinematical core. The resolution depends on a hitherto obscured sense of relativity of acceleration in Newtonian cosmology and finds its fullest expression in the connection between gravitation and space-time curvature in Newtonian space-time theory.

1. The non-convergence of gravitational force in Newtonian cosmology

In order to fix our topic, it will be helpful to give a brief and simple derivation of the problem that exercised Seeliger. In a Newtonian universe, the gravitational force exerted on a test body of unit mass is the resultant of the forces exerted by all the masses of the universe, which we shall assume to be distributed uniformly in space with mass density ρ. This force is computed by an integration over all these masses. This integral fails to converge. It can take on any value according to how we approach the limit of integration over all space.

To see this lack of convergence, picture the uniform matter density ρ as distributed in concentric spherical shells of very small thickness Δr all centered on

---

[1] I gratefully acknowledge the assistance of the many before me who have mapped out various parts of the story in greater and lesser detail, especially Jaki (1979), Jammer (1961: 127–29), North (1965: 16–23, 180–185) Oppenheim (1920: 86–87), Sklar (1976) and Zenneck (1901: 51–53).

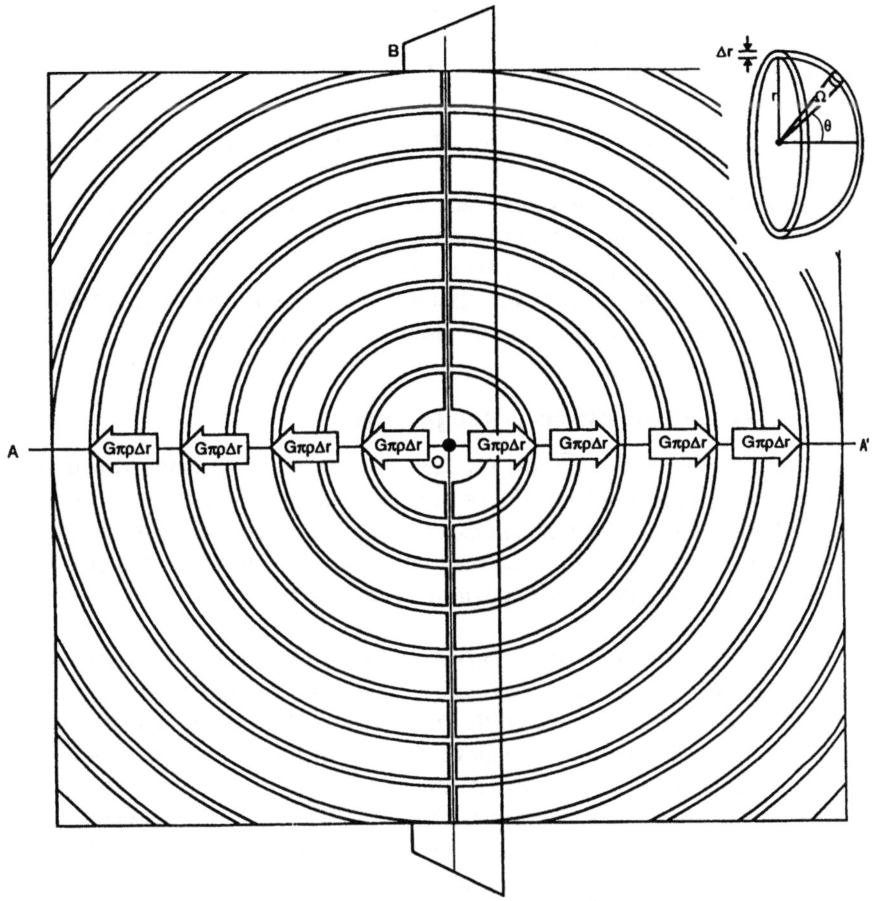

Figure 1. Non-convergence of force on a test mass in Newtonian cosmology.

the unit test mass at O, as shown in Figure 1. Choose some arbitrary axis AA' and divide all the spherical shells into hemispherical shells by passing a plane B through O and perpendicular to AA'. Each hemispherical shell exerts a force on the test mass in direction AA' and its magnitude is independent of the radius r of the shell. To see the independence from r, consider how much matter in a shell at radius r is subtended by some small solid angle Ω at O. That amount of matter increases with r^2, but the gravitational force it exerts on the test mass decreases with $1/r^2$. So, overall, this force will be independent of r. This holds for each element in the shell, so we conclude that the net force exerted by the entire shell is a constant, independent of its radius r. The direct calculation reveals that the value of the constant is $G\pi\rho\,\Delta r$.[2] Thus the net force F on the test mass along an arbitrarily chosen axis AA' is given by an infinite series, each term representing the force due to one hemispherical shell

$$F = G\pi\rho\,\Delta r - G\pi\rho\,\Delta r + G\pi\rho\,\Delta r - G\pi\rho\,\Delta r + G\pi\rho\,\Delta r - \ldots$$

The series has alternating signs since shells on alternating sides exert a force in alternating direction. This alternating series is well known not to converge. According to how one groups and reduces the terms in the series, the sum can have many different values.[3] Each corresponds to a different way of approaching the limit of infinitely many masses in the associated integration.

2. Seeliger's formulations of the problem

Seeliger's papers, especially his (1895a), contain the most detailed and general development of the problem. The price is that his exposition is the most cumbersome of all expositions; he alone resorts to infinite series expansions in Legendre polynomials and includes tidal forces in the analysis. Virtually all later commentators managed to reduce the exposition of the core difficulty to one or two lines of formulae.

Seeliger initiated his discussion in his Seeliger 1895a (p. 129) by asking whether Newton's law of gravitation holds exactly for masses separated by "immeasurably great distances" [*unermesslich große Entfernung*]. While observational astronomy gives the strongest reasons to believe the law within our planetary system, we have no similar foundation in experience for the law on the larger, cosmic scale. Nonetheless, he urged, the matter can be decided by applying the law to "simple and obvious examples" [*einfachen und naheliegenden Beispielen*] on the cosmic scale. It turns out that "thoroughly possible and conceivable assumptions lead to quite impossible or unthinkable conclusions" so that

---

[2] The total mass in a ring with an angular position θ and an angular thickness of $d\theta$ in the shell is $\rho r\, d\theta\, 2\pi r \sin\theta\, \Delta r$. Therefore the total force exerted by the hemisphere at r is

$$2\pi G\rho\,\Delta r \int_0^{\pi/2} \sin\theta\, \cos\theta\, d\theta = G\pi\rho\,\Delta r.$$

[3] For example

$$\begin{aligned}(G\pi\rho\,\Delta r - G\pi\rho\,\Delta r) + (G\pi\rho\,\Delta r - G\pi\rho\,\Delta r) + (G\pi\rho\,\Delta r - G\pi\rho\,\Delta r) + \ldots \\ = 0 + 0 + 0 + \ldots \\ = 0\end{aligned}$$

and

$$\begin{aligned}G\pi\rho\,\Delta r - (G\pi\rho\,\Delta r - G\pi\rho\,\Delta r) - (G\pi\rho\,\Delta r - G\pi\rho\,\Delta r) - \ldots \\ = G\pi\rho\,\Delta r - 0 - 0 - \ldots \\ = G\pi\rho\,\Delta r.\end{aligned}$$

> Newton's law, applied to the immeasurably extended universe, leads to insuperable difficulties and irresolvable contradictions if one regards the matter distributed through the universe as infinitely great.[4] (Seeliger 1895a: 132)

To arrive at these difficulties, Seeliger asked after the gravitational effect of the masses of the universe at a point O in space. To compute these effects, he laid out a spherical coordinate system (r, θ, γ) centered on O. He represented the discontinuously distributed masses of the universe by an equivalent continuous distribution with density ρ. He found the gravitational effect of the masses between radial coordinate values $r = R_0$ and $r = R_1$ to be

$$\text{(1a)} \qquad \varphi = -G \iiint_{\text{space}} \frac{\rho}{r}\, dV = -G \int_0^{2\pi} \int_0^{\pi} \int_{R_0}^{R_1} \rho r \sin\gamma\, dr\, d\gamma\, d\theta,$$

$$\text{(1b)} \qquad F_x = -\frac{d\varphi}{dx} = G \int_0^{2\pi} \int_0^{\pi} \int_{R_0}^{R_1} \rho \sin\gamma \cos\gamma\, dr\, d\gamma\, d\theta,$$

$$\text{(1c)} \qquad Z_x = -\frac{d^2\varphi}{dx^2} = 2G \int_0^{2\pi} \int_0^{\pi} \int_{R_0}^{R_1} \frac{\rho}{r}\, dr \sin\gamma\, \frac{3\cos^2\gamma - 1}{2}\, d\gamma\, d\theta,$$

where φ is the gravitational potential at O and dV a volume element of space. F_x is the gravitational force in the direction of the x coordinate axis, which aligns with angular coordinate $\gamma = 0$. Z_x, called "strain" [*Zerrung*] by Seeliger, is the tidal gravitational force acting in the x direction on neighboring masses located on the x axis. It is defined as the difference in gravitational force acting between two such bodies per unit distance of separation. For further details, including a synopsis of Seeliger's derivation, see Appendix A.

The integrals (1b) and (1c), remain well defined as long as R_1 is finite. However, Seeliger observed, if they are applied to an infinite matter filled universe, they cease to be well defined:

> If ρ is a finite magnitude for infinitely large regions, then, in general, F_x and Z_x are completely undetermined, as long as one makes no definite assumption on the way in which the finite values of R_1 become infinitely great. Therefore both quantities can equally well become infinite or remain finite.[5] (Seeliger 1895a: 131)

Seeliger's claim is that integrals of (1b) and (1c) give no definite results in the limit as R_1 goes to infinity; they vary according to the path taken to the limit.

To make good on this claim, Seeliger applied the formulae to the case of a universe filled with a homogeneous matter distribution of everywhere constant

---

[4] "durchaus mögliche und vorstellbare Annahmen zu ganz unmöglichen oden undenkbaren Consequenzen führen ... das Newton'sche Gesetz auf das unermesslich ausgedehte Universum angewandt auf unüberwindtliche Schwierigkeiten und unlösbare Widerspruche führt, wenn man die im Weltall verstreute Materie als unendlich groß ansieht."

[5] "Wenn dann [ρ] eine endliche Größe für unendlich große Bezirke ist, dann werden im Allgemein [F_x] und [Z_x] völlig unbestimmt, so lange man über die Art, wie man von endlichen Werthen R_1 zu den unendlich großen gelangt, keine bestimmte Voraussetzung macht. Beide Größen können ebensogut unendlich werden, wie endlich beliben."

density ρ. To approach the limit, he took the region of integration bounded by R_1 to be a sphere, centered on an arbitrary point *other than* O, and the sphere was then allowed to grow infinitely large, as shown in Figure 2. If r is the distance of the point O from the center of the sphere, Seeliger reported that the force[6]

$$F_x \text{ is proportional to } r\,\rho, \tag{1b'}$$

and is directed towards the center of the sphere. The tidal force

$$Z_x \text{ is proportional to } \rho. \tag{1c'}$$

Since the location of the center of the sphere is arbitrary, the gravitational force is also arbitrary, taking any desired value, according to where one locates the center of the sphere.

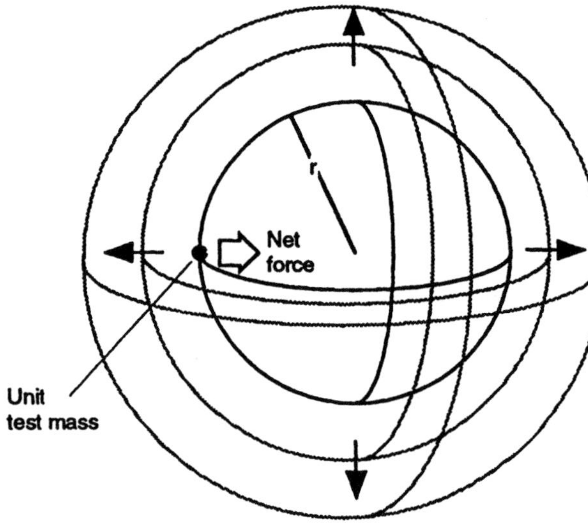

Figure 2. Seeliger's spheres.

[6] If $\mathbf{r}$ is the radius vector from the center of the sphere, one quickly sees that Seeliger's result is given more fully by the force $\mathbf{F} = -(4/3)G\pi\rho\,\mathbf{r}$ (as Seeliger pointed out in effect in his later Seeliger 1896: 379). To see this, note that once Seeliger's sphere has grown so that it touches the test mass, any extra matter added will exert no net gravitational force on the test mass. All extra matter will be added in spherical shells and a well-known theorem of Newtonian gravitation theory tells us that such shells exert no net force. Another familiar theorem helps us find the force exerted by the matter in the sphere that has just grown to touch the test mass. That force is the same force that would be exerted if all the matter in the sphere were concentrated at its center. That would be a mass $(4/3)\pi r^3\rho$ at a distance r from the test mass. It exerts a force on the unit test mass of magnitude $(4/3)G\pi\rho\,r$ towards the center of the sphere which gives the vector result $\mathbf{F} = -(4/3)G\pi\rho\,\mathbf{r}$. From this result, we can also derive an expression for the tidal force. If we imagine that the result defines a force field through space, then the differential force on two unit test masses separated by small distance $\Delta\mathbf{r}$ is given as $\Delta\mathbf{F} = -(4/3)G\pi\rho\Delta\mathbf{r}$. From this, we read off the tidal force as $Z_x = -(4/3)G\pi\rho$ and see that the result is the same along all axes. For a simple 'lines of force' development of this argument, see Norton 1993a.

Seeliger's analysis could well have ended there were it not for the nuisance that the tidal force (1c′) does not turn out to be indeterminate, in apparent contradiction with Seeliger's claim; its value is independent of the disposition of the spheres used in the integration. This is probably the reason Seeliger proceeded to his second example, a different infinite matter distribution. He considered the gravitational and tidal forces at the apex of a double cone of very small solid angle ω and constant mass density ρ, as shown in Figure 3.

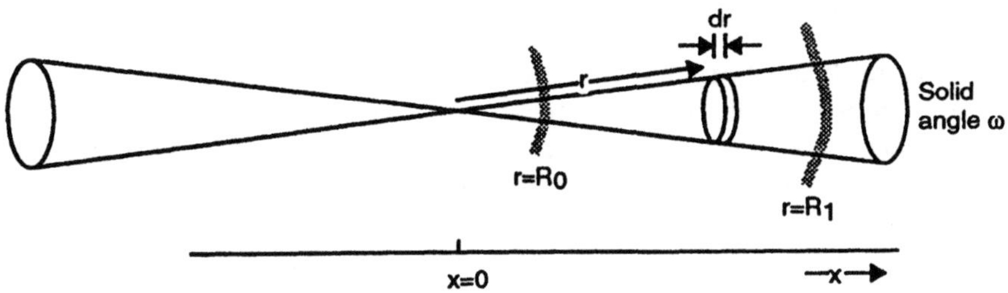

Figure 3. Seeliger's double cones.

The gravitational force due to each individual cone is infinite. Thus, he noted, the gravitational force exerted by both cones at the apex is of the form "$\infty - \infty$" and is indeterminate. The tidal force due to one cone acting in the direction of the cone's axis follows directly from substitution into (1c) with $\gamma = 0$. It is given by[7]

$$Z_x = 2\omega G \int_{R_0}^{R_1} \frac{\rho}{r}\, dr \qquad (1c'')$$

and becomes infinite in the limit of infinite R_1. Since both cones produce the same tidal force—an expansion along the axis of the cone—the effect of both cones is double that of a single cone, so that the tidal force at the apex of the double cone is infinite.

This example of the double cone completed Seeliger's first analysis. It was unsatisfactory in so far as Seeliger had only shown non-convergence of tidal force in a rather contrived example of an infinite matter distribution, the infinite double cone. He had still not shown that there was a problem with tidal forces in a universe homogeneously filled with matter. This deficiency was remedied when Seeliger (1896) returned to review his results. There he showed that there was a way of approaching the limits in the integrals (1b) and (1c) for such a universe so that the gravitational force F_x vanished, but the tidal force became infinite. In concept the method was simple. Seeliger's double cones had yielded an infinite tidal force,

[7] More directly, the volume element at distance r to $r + dr$ from the apex contains mass $\rho\,\omega\,r^2\,dr$ and it exerts a gravitational force $F_x = G\rho\,\omega\,r^2\,dr/(r-x)^2$ on masses at coordinate position x on its axis. The tidal force is $(d^2F_x/dx^2)|_{x=0} = 2G\rho\,\omega\,dr/r$. Integration over one cone yields (1c″).

but they did not correspond to a matter distribution filling all space. All Seeliger needed was a shape similar to the double cone that would fill all space as it grew to infinite size. If this shape were well chosen, the tidal force integral (1c) would diverge along this path to the infinite limit, demonstrating that tidal forces were also ill-behaved, converging or diverging according to the way the limit in the integration is taken. Seeliger had no trouble in finding such a shape. It is given by

$$\log \frac{R_1}{R_0} = am + mP_2(\cos\gamma), \tag{2}$$

where R_0 and R_1 are the limits of integration of the radial coordinate r, $P_2(\cos\gamma) = (1/2)(3\cos^2\gamma - 1)$ is the second Legendre polynomial in $\cos\gamma$, a is some constant greater[8] than $1/2$ and m is a parameter which becomes infinite as the shape grows to infinite size. This expression corresponds to the shape shown in Figure 4. For convenience I will call the shape 'Seeliger's peanut.' As the shape grows, it becomes more elongated so that the tidal force at the center of the peanut grows without limit. However the peanut also grows in width so that in the limit it fills all of space.

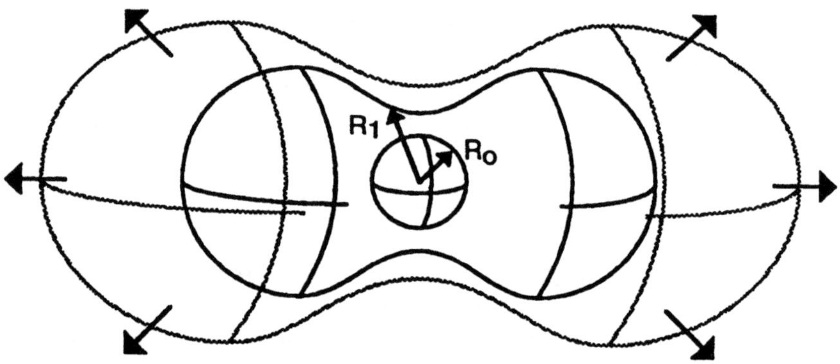

Figure 4. Seeliger's peanut.

The family of shapes specified in (2) is not the only one that has the property Seeliger sought. It does have the fortunate property, however, that the integrations of (1b) and (1c) become especially easy. The Legendre polynomial introduced in (2) combines with that already in (1c) to enable easy evaluation of the integral from known integrals. The force F_x is obviously zero from the symmetry of the shapes. For some fixed m, the tidal force is given by substitution of (2) into (1c)

[8] The need for this restriction on the value of a becomes apparent if the expression in (2) is rewritten as $am + mP_2(\cos\gamma) = m(a - 1/2 + (3/2)\cos^2\gamma)$.

yielding

$$Z_x = 2G\rho \int_0^{2\pi} \int_0^{\pi} \log \frac{R_1}{R_0} P_2(\cos\gamma) \sin\gamma \, d\gamma \, d\theta = \frac{8}{5} G\pi\rho \, m,$$

where the $\gamma = 0$ axis is aligned with the axis of rotational symmetry of the peanut. It diverges in the limit of infinite m.

Seeliger (1896: 380–382) completed his discussion with a qualitatively different type of demonstration of the problems faced by gravitation theory in a universe containing an infinite matter distribution. The argument will reappear often enough for it to be worth us labeling it the 'flux argument.' Seeliger considered some closed surface F in space which enclosed n masses $m_1, m_2, \ldots, m_n$. Through an application of Green's theorem, he arrived at the result for the gravitational potential φ

$$\int \frac{\partial\varphi}{\partial n} ds = -4\pi G(m_1 + \ldots + m_n),$$

where the integration extends over the entire surface of F and n is an inward pointing, unit, normal vector.[9] Seeliger then imagined the masses $m_1, m_2, \ldots, m_n$ distributed uniformly within the volume enclosed by F with constant density ρ. He rewrote his result in the form

$$M\left(\frac{\partial\varphi}{\partial n}\right) = -4\pi G\rho \frac{R}{S}, \tag{3}$$

where S is the surface area of F, R is its volume and the operator M returns the value of $\partial\varphi/\partial n$ averaged over the surface area of F. By selection of a large surface F enclosing sufficiently many masses, the right hand side of (3) can be made arbitrarily large. Therefore the average value of $\partial\varphi/\partial n$ can also be made arbitrarily large. Thus individual values of $\partial\varphi/\partial n$ can be made arbitrarily large as well. This already yields an undesirable result since it corresponds to arbitrarily large field strengths. Seeliger elaborated on its undesirability:

> It follows from potential theory that there must be in the *universe unlimited (infinitely) great accelerations* and indeed with every conceivable mass distribution. Therefore there are motions that begin with finite speed and lead to infinitely great speeds in finite time. This in itself already contains something absolutely inadmissible, if one does not want to call all of mechanics into question.[10] (Seeliger 1896: 382; emphasis in original)

---

[9] As in Appendix A1, I have replaced Seeliger's potential V by the now standard $-\varphi$ (note sign change) and restored G which Seeliger had set to one. Similarly, below, I replace Neumann's φ by $-\varphi$.

[10] "Nach der Potentialtheorie müssen demzufolge im *Universum unbegrenzt (unendlich) große Beschleunigungen* vorkommen und zwar bei jeder denkbaren Massenverteilung. Das sind also Bewegungen, die mit endlicher Geschwindigkeit beginnend in endlicher Zeit zu undendlich großen Geschwindigkeiten führen, was an sich schon eine absolute Unzulässigkeit enthält, wenn man nicht die ganze Mechanik in Frage stellen will." Here as elsewhere Seeliger slides rather too hastily from the conclusion that the field strength can be set without upper bound to the conclusion that it is actually infinite.

3. The puzzle of Neumann's elusive priority claim

Shortly after Seeliger had published his first note, he was answered with a cry from Carl Neumann that he had already seen the problem some twenty years before. Neumann's 1896 book is a lengthy treatise in electrical field theory, published after Seeliger's first communication. Its focus was the notion of electrostatic equilibrium, such as arises for electric charges in a conductor, if the forces between the charges are given by a Coulomb potential, which Neumann wrote as $\varphi(r) = -1/r$. The existence of such equilibrium was elevated (Neumann 1896: viii) to the "principle" (*Princip*) or "axiom" (*Axiom*) at the basis of the treatise and its project became the investigation of alternative forms for the potential function $\varphi(r)$ compatible with his axiom of equilibrium. To motivate his occupation with alternative forms of the potential, Neumann sought to cast doubt on what he called the "Newtonian potential function (*Newton'sche Potentialfunction*) $\varphi(r) = -1/r$" or sometimes "Newton's Law" (*Newton'sche Gesetz*). In the first chapter, he briefly reviewed objections to it, including one based in gravitation theory whose potential also conformed to "Newton's law." After mentioning the possibility that the Newton's law may require modification in the domain of very small distances, he proceeded:

> However a modification of Newton's law might also be called for not only in the very *small* but also in the very *large*, at least in case one entertains the usual representation that all of universal space is filled with stars to infinity in roughly uniform distribution. For then the universe would be looked upon in the relevant aspect as an infinitely great sphere of roughly constant density. And this infinitely great homogeneous sphere, representing the universe in all its entirety, would obviously, on the foundation of Newton's law, strive to draw in the individual heavenly bodies, such as f[or] ex[ample] the Sun, Mercury, Venus, Earth, Mars, etc., towards its center. Further, the intensity of the corresponding forces would be proportional to the displacements of the individual heavenly bodies from that center.
>
> Now, however, the surface of the universal sphere under discussion lies fully in the infinite. Therefore its center has a completely *undetermined position*. And so those forces, exerted by this universal sphere on the individual heavenly bodies, would be likewise completely *undetermined* in their direction and strength—which obviously would be absurd.[11] (Neumann 1896: 1–2; emphasis in original)

Neumann here has recapitulated one of the arguments that Seeliger gave for the in-

[11] "Aber nicht nur für sehr kleine, sondern auch für sehr größe Entfernungen dürfte eine Modification des Newton'schen Gesetzes geboten sein, wenigstens, falls man der gewöhnlichen Vorstellung sich hingiebt, daß der ganze Weltraum ins unendliche hin, in einigermassen gleichförmiger Vertheilung, von Sternen erfüllt sei. Denn als dann würde das Universum der Hauptsache nach anzusehen sein als eine unendlich große Kugel von einigermassen constanter Dichtigkeit. Und diese das Universums im Großen und Ganzen repräsentierende unendlich große homogene Kugel würde offenbar, bei Zugrundelegung des Newton'schen Gesetzes, die einzelnen Himmelskörper, wie z.B. Sonne, Mercur, Venus, Erde, Mars u.s.w., nach ihrem Centrum hinzuziehen bestrebt sein. Auch würden die Intensität der betreffenden Kräfte proportional sein mit den Abständen der einzelnen Himmelskörper von jenem Centrum.

Nun liegt aber die Oberfläche der in Rede stehenden Universalkugel überall im Unendlichen. Folglich hat ihr Centrum eine völlig unbestimmte Lage. Und es würden daher jene von dieser Universalkugel auf die einzelnen Himmelskörper ausgeübten Kräfte, ihrer Richtung und Stärke nach, ebenfalls völlig unbestimmt sein;—was offenbar absurd wäre."

determinacy of gravitational force in a universe filled homogeneously with matter. In Seeliger's hands, the result arose from using a sphere that grows to infinity to evaluate the integral (1b), recovering (1b′). Neumann's treatment was far briefer. But it really did not need to be any more elaborate. Neumann was merely calling to mind quite standard results: A test body within a homogeneous sphere is drawn to its center by a gravitational force of magnitude proportional to the distance from the center. Allowing the sphere to grow by adding layers to it does not alter the result, presumably even if the sphere is allowed to grow infinitely large.

Aside from this brief application, the cosmological problems of Newton's theory of gravitation played no role in Neumann's 292 page tome. A footnote to the last sentence of the passage quoted above contained mention of Seeliger's work and Neumann's claim of priority:

> Already a long time ago these matters were remarked on by me in the *Abhandl[ungen] der K[öniglichen] Sächs[ischen] Ges[ellschaft] d[er] Wiss[enschaften]*, 1874, page 97, 98. Otherwise something similar has also been remarked on by the astronomer Seeliger (in Munich) in *Astron[omische] Nachrichten*, 1894, Vol. 137, page 3272.[12] (Neumann 1896: 2n)

This priority claim produced an effusive and apologetic response from Seeliger:

> Since [publication of Seeliger 1895a] it has become known to me that Carl Neumann[13] had remarked already on difficulties of a similar kind that may be represented as special cases of the arguments advanced by me. With agreement with such an outstanding researcher and also the circumstance that the considerations advanced by me were indeed also expressed in another form, but in doing so without changing their essential content, it might appear superfluous to return to this subject. On the other hand ...[14] (Seeliger 1896: 373)

Neumann's priority and the citation Neumann himself gives have been routinely repeated in historical surveys.[15] But Seeliger need not have been so effusive. The work Neumann cites proves rather hard to find. It is usually taken to be Neumann 1873, whose title page is dated 1873 but which appeared in a volume assembled in 1874. The first problem is that the pages Neumann cites—pages 97–98—are not within the pagination of Neumann 1873, which occupies pages 417–524 of

[12] "Schon vor langer Zeit is von mir auf diese Dinge aufmerksam gemacht worden, in den Abhandl. der K. Sächs. Ges. d. Wiss., 1874, Seite 97, 98. Uebrigens ist Aehnliches neuerdings auch von dem Astronomen Seeliger (in München) bemerkt worden, in den Astron. Nachrichten, 1894, Bd. 137, Seite 3272."

[13] Seeliger's footnote at this point reads "Comp[are] (*Vergl.*) Carl Neumann, Allgemeine Untersuchungen über das Newton'sche Princip etc. Leipzig 1896, S. 1" which is Neumann (1896: 1).

[14] "Seitdem ist mir bekannt geworden, daß Carl Neumann schon früher auf Schwierigkeiten ählicher Art aufmerksam gemacht hat, die sich als specielle Fälle der von mir vorgebrachten Argumente darstellen dürften. Die Zustimmung eines so herrvorragenden Forschers und auch der Umstand, daß sich die von mir angestellten Ueberlegungen zwar auch in anderer Form aussprechen lassen, daß hierdurch aber ihr wesentlicher Inhalt nicht sich ändert, könnte es überflüssig erscheinen lassen, auf diesen Gegenstand zurückzukommen. Andererseits ... "

[15] For example Zenneck's (1901: 51) authoritative Teubner encyclopedia article allows that Neumann had "first hinted" (*hat wohl zuerst darauf hingewiesen*) at an undetermination of the gravitational force in his work of 1874. Mention of independent discovery is given also by Oppenheim (1920: 86) in his Teubner encyclopedia article. Jammer (1961: 127) gives Neumann credit as "the first to call attention to such difficulties [of Newton's law of gravitation applied to the universe as a whole]."

the relevant volume. (Pages 97–98 of the same volume are filled with the work of another author.) The second is that it is by no means obvious where in the paper Neumann addresses gravitational problems of cosmology. I cannot say that the discussion is not there, buried somewhere in the paper's lengthy and technical discussion of electrodynamic forces. But I can say that I could not find it and that, if it is there, it is not given any prominence whatever.

Thus Neumann's claim presents something of a puzzle. Did Neumann really lay out the problem decisively in the 1870s? Perhaps the incorrect pagination 97–98 stems from Neumann citing his own work from a separatum. The pagination of separata used not to match that of the journal printing. Each article in separatum would start at page 1. Since Neumann 1873 exceeded 100 pages in length, a separatum may well include pages 97–98. This is a kind construal, however, especially since nothing pertaining to the gravitation problem seems to appear on the pages of the article that would correspond to pages 97–98 of a separatum. More plausible is that Neumann just gave an incorrect citation. In the same footnote in which he cited his own work of 1874, he also cited Seeliger 1895a. As the reader can see from the above quote, the citation is incorrect. Neumann gives the year of publication as 1894. The article was signed "München 1894 November" but appeared in Volume 137 of 1895. Similarly his citation is to *page* (*Seite*) 3272. The correct pagination is pages 129–136. The article appears in *issue* number 3273—even the issue number is off by one! This is strong evidence that Neumann's citations are inaccurate.

If Neumann's citation is incorrect, should we look elsewhere, perhaps more widely in the 1874 volume of *Abhandlungen der mathematisch-physischen Classe der Königl. Sächsichen Gesellschaft der Wissenschaften* cited? That volume contains nine articles submitted and printed in the period 1871–1874. Only one is by Neumann and is Neumann 1873. We cannot rule out the possibility of another of Neumann's publications containing the work claimed. If that other work exists, however, later reviewers and historians have given no hint of it. Zenneck (1901: 51) cites the relevant work as "Leipz. Abh. 1874", Oppenheim (1920: 86) gives it as "Leipzig. Ber. 26 (1874), p. 97" and Tolman (1934: 322) as "*Abh. d. Kgl. Sächs. Ges. d. Wiss. zu Leipzig, math.-nat Kl.* **26**, 97 (1874)" suggesting none looked further than Neumann's own (1896: 1) citation (although Oppenheim and Tolman's "26" is a mystery). Jammer (1961: 127) cites Neumann 1873 directly. And Seeliger (1896: 1) himself does not cite any work of Neumann from the 1870s; he merely cites Neumann (1896: 1) and we may wonder what subtlety is masked by Seeliger's " ... also expressed in another form. ... "

While I cannot resolve the issue of Neumann's priority, we can at least be fairly confident that he deserved far less than Seeliger conceded. At best in his work of the 1870s he may have noticed that the relevant integral for gravitational force fails to converge. But this is not to say that he recognized that failure to amount to a very significant theoretical problem, one that deserved detailed and prominent investigation in its own right and one that ought to be brought with vigor to much wider notice. It was Seeliger who did that. Indeed the inclusion of the problem in the introductory pages of Neumann's later work is something of an oddity since that work has nothing to do with gravitation and cosmology. It reads like an overreaching attempt to secure priority for a result that Neumann may never quite have published—but certainly wished that he had. As it turns out, the bulk of citation to the problem in the decades following Seeliger's work attribute the result to Seeliger, either explicitly or implicitly and without mention of priority in Neumann's work of the 1870s.[16]

One final irony remains. While Seeliger's development of the problem involved some detailed calculation, its existence, as Neumann showed, can be made apparent through quite simple reflection. Thus it would be surprising if Seeliger and Neumann were the only ones to see it. We may expect that anyone who reflects seriously on the gravitational properties of an infinite matter distribution would run into one or other form of the problem. And we know that this happened in at least two cases. As we shall see below, forms of the problem were hit upon by both Richard Bentley (who put them to Newton in their famous correspondence) and Kelvin in the context of his ether theory. For different reasons, neither saw in the problem a serious challenge for Newton's theory of gravitation. Presumably many more researchers found and dismissed the result in similar ways—but they remain unknown to us since they did not publish or, if they did, in such an obscure way that it remains hidden to us now. The Neumann of the 1870s would surely fall into this class were it not for his own efforts of 1896 to draw attention to his work. What Neumann seems not have had is a sense of the importance of his result in the 1870s—or surely he would then have drawn more attention to it. It was Seeliger who had the courage to insist on the importance of the problem. Here Seeliger seems to be alone.[17]

4. **Kelvin finds the flux argument**

Lord Kelvin (William Thomson) also hit upon the cosmological problems of Newtonian gravitation theory independently, it seems, of Seeliger and Neumann—or at least he acknowledges no debt to them. Moreover we can see in the development of his work a line of thought that would take him directly to the problem. In the fall of 1884, Kelvin delivered twenty lectures at Johns Hopkins University on wave theory and molecular dynamics. In the original lectures (recorded stenographically now reproduced in Kargon and Achinstein 1987), in Lecture XVI (p. 162), Kelvin addressed the question of whether the luminiferous ether was imponderable, that is, has no weight.

[16] Charlier (1908: 3) wrote "It is Seeliger (Astr. Nachr. No. 3273, 1895) who first drew attention [to the cosmological difficulties of Newton's law]." ("Es ist Seeliger, der zuerst hierauf aufmerksam gemacht hat (Astr. Nachr. No. 3273, 1895) ... ") Einstein (1954: chap. XXX) wrote of a "fundamental difficulty ... , which, to the best of my knowledge, was first discussed in detail by the astronomer Seeliger." ("eine ... prinzipielle Schwierigkeit ... , welche meines Wissens zuerst von dem Astronomen Seeliger ausführlich diskutiert wurde.") See also Föppl 1897: 93, Arrhenius 1909: 223, Charlier 1922: 3, Selety 1922: 282, Findlay-Freundlich 1951: 5.

[17] This is one of the outcomes of Jaki (1979) who surveyed the awareness of the problem from the time of Newton to Seeliger. Many stood on the problem's threshold in one form or another. None gave it the uncompromising formulation of Seeliger.

> We are accustomed to call [the ether] imponderable. How do we know it is imponderable? If we had never deal with air except by our senses, air would be imponderable to us. But we can show that the weight of a column of air is sufficient to cause a difference of pressure on the two sides of a glass receiver. We have not the slightest reason to believe the luminiferous aether to be imponderable; it is just as likely to be attracted to the sun as air is. I do not like to make too many statements of that kind. At all events, the onus of proof rests with those who assert that it is imponderable. I think we shall have to modify our ideas of what gravitation is if we have a mass spreading through space with mutual gravitations between its parts without being attracted by other bodies.

This image of the ether as a mass with gravitational attraction between its parts and spreading through infinite space takes Kelvin directly to the cosmological problem. For he need only ask how strong might these gravitational forces be to arrive directly at the problem. At that was just the sort of question Kelvin would ask. His ether was no mysterious medium beyond normal physical considerations. The paragraph immediately following his pronouncement that the ether has weight contains a computation of the pressure exerted by a column of ether of infinite height on the surface of the sun.

Kelvin edited his Baltimore Lectures for subsequent publication. The edited Lecture XVI appeared in *Philosophical Magazine* as Kelvin 1901 and then in the final edition, Kelvin 1904. From parenthetical inserts we know that Kelvin had completely changed his mind by 1899 on the question of whether the ether is ponderable. One insert after the sentence "We have not the slightest reason ... " read (1901: 166; 1904: 266)

> Nov. 17, 1899. I now see that we have the strongest possible reason to believe that the aether is imponderable.

An insert with the same date at the end of the paragraph gave Kelvin's reasoning. Gravitational attraction between parts of the infinite ether would lead to infinitely great pressures in the ether, so that it would collapse unless capable of infinite resistance. His argument for these infinite pressures ran as follows:

> Suppose that aether is given uniformly spread through space to infinite distances in all directions. Any large enough spherical portion of it, if held with its surface absolutely fixed, would by the mutual gravitation of its parts become heterogeneous; and this tendency could certainly not be counteracted by doing away with the supposed rigidity of its boundary and by the attraction of aether extending to infinity outside it. The pressure at the center of a spherical portion of homogeneous gravitational matter is proportional to the square of the radius, and therefore, by taking the globe large enough, may be made as large as we please, whatever be the density. In fact, if there were mutual gravitation between its parts, homogeneous aether extending through all space would be essentially unstable, unless infinitely resistant against compressing or dilating forces.

Kelvin's argument depended on a result—that the gravitationally induced pressure at the center of a sphere of matter grew in proportion to the radius. This result in turn depended on Kelvin's version of Seeliger's flux argument, which Kelvin laid out several paragraphs later (1901: 168–169; 1904: 267–268).[18]

> Let V be any volume of space bounded by a closed surface, S, outside of which and within which there are ponderable bodies; M the sum of the masses of all these bodies within S; and ρ the mean density of the whole matter in the volume V. We have
>
> $$M = \rho V.$$
>
> Let Q denote the mean value of the normal component of the gravitational force at all points of S. We have
>
> $$QS = 4\pi M = 4\rho V \qquad [(4)]$$
>
> by a general theorem discovered by Green seventy-three years ago regarding force at a surface of any shape, due to matter (gravitational, or ideal electric, or ideal magnetic) acting according to the Newtonian law of the inverse square law of the distance. ... If the surface is wholely convex, the normal component force must be everywhere inwards.
>
> Let now the surface be spherical of radius r. We have
>
> $$S = 4\pi r^2; \quad V = (4/3)\pi r^3; \quad V = (1/3)rS$$
>
> Hence, for a spherical surface, [(4)] gives
>
> $$Q = (4/3)\pi r\rho = M/r^2 \qquad [(5)]$$
>
> This shows that the average normal component force over the surface S is infinitely great, if ρ is finite and r infinitely great, which suffices to prove [the earlier claim of infinite force on bodies in a universe filled with a non-zero density of ponderable matter].

5. Einstein's assault on Newtonian cosmology

The flaws of the old regime are never clearer than after it has fallen and a new power has taken its place. These flaws become all the more incontestable when that new power undertakes to explain just how debased was its predecessor. And the explanation is often so oversimplified that it fails closer scrutiny. Such was the fate of Newtonian cosmology at the hands of Einstein in 1917 when he reported his efforts to apply general relativity, his new theory of gravitation, to cosmology. And we shall see below how Einstein dangerously oversimplified in his explanation. Where Seeliger had seen an arcane technical complication that required a modest technical solution, Einstein saw a symptom of a deeper and fatal ailment. In introducing Newtonian cosmology in his celebrated 1917 "Cosmological Considerations on the General Theory of Relativity," he observed that a uniform matter distribution extending to spatial infinity is incompatible with Newtonian gravitation theory as it is usually applied:

---

[18] The result Kelvin needs is that the gravitational force per unit mass at radial position r in a sphere of matter density ρ is $(4/3)G\pi\rho r$, Kelvin's result (5). If we now assume that this force is counterbalanced by an isotropic pressure $P(r)$, we derive its dependence on r as follows. Consider a small volume of radial thickness Δr and unit area normal to the radius. It exerts a gravitational force of $(4/3)G\pi\rho r \cdot \rho\Delta r$ on the matter below it. Therefore the pressure $P(r)$ diminishes according to $dP(r)/dr = -(4/3)G\rho^2 r$, with increasing r. Integration, assuming that the pressure P drops to zero at $r = R$, gives the pressure $P(r) = (2/3)G\pi\rho^2(R^2 - r^2)$. At the center $r = 0$, the pressure is $(2/3)G\pi\rho^2 R^2$ which exhibits the dependence on the square of radius R invoked by Kelvin.

> It is well known that Newton's limiting condition of the constant limit for [gravitational potential] φ at spatial infinity leads to the view that the density of matter becomes zero at infinity. For we imagine that there may be a place in universal space round about which the gravitational field of matter, viewed on a large scale, posesses spherical symmetry. It then follows from Poisson's equation [$\nabla^2\varphi = 4\pi G\rho$] that, in order that φ may tend to a limit at infinity, the mean density ρ must decrease towards zero more rapidly that $1/r^2$ as the distance r from the center increases.[19] In this sense, therefore, the universe according to Newton is finite, although it may possess an infinitely great total mass.[20] (Einstein 1917a: 177–178)

This was the same problem Seeliger had identified, as Einstein made clear in his parallel treatment of the same issue in his popularization of relativity theory. He wrote of a "fundamental difficulty attending classical celestial mechanics, which, to the best of my knowledge, was first discussed in detail by the astronomer Seeliger." He considered the possibility of a roughly uniform matter density throughout infinite space. However:

> This view is not in harmony with the theory of Newton. The latter theory rather requires that the universe should have a kind of center in which the density of the stars is a maximum, and that as we proceed outwards from this center the group-density of the stars should diminish, until finally, at great distances, it is succeeded by an infinite region of emptiness. The stellar universe ought to be a finite island in the infinite ocean of space.[21] (Einstein 1917b: 105)

A footnote to the last sentence described the result Einstein had in mind:

> *Proof.* According to the theory of Newton, the number of 'lines of force' which come from infinity and terminate in a mass m is proportional to the mass m. If, on the average, the mass density ρ_0 is constant throughout the universe, then a sphere of volume V will enclose the average mass $\rho_0 V$. Thus the number of lines of force passing through the surface F of the sphere into its interior is proportional to $\rho_0 V$. For unit area of the surface of the sphere the number of lines of force which enters the sphere is thus proportional to $\rho_0 V/F$ or to $\rho_0 R$. Hence the intensity of the field at the surface would ultimately become infinite with increasing radius R of the

---

19 Einstein's footnote here reads: "ρ is the density of matter, calculated for a region which is large as compared with the distance between neighboring fixed stars, but small in comparison with the dimensions of the whole stellar system." ("ρ ist die mittlere Dichte der Materie, gebildet für einen Raum, der groß ist gegenüber der Distanz benachbarter Fixterne, aber klein gegenüber den Abmessungen des ganzen Sternsystems.")

20 "Es ist wohlbekannt, daß die Newtonsche Grenzbedingung des konstanten Limes für φ im räumlich Unendlichen zu der Auffassung hinführt, daß die Dichte der Materie im Unendlichen zu null wird. Wir denken uns nämlich, es lasse sich ein Ort im Weltraum finden, um den herum das Gravitationsfeld der Materie, im großen betrachtet, Kugelsymmetrie besitzt (Mittelpunkt). Dann folgt aus der Poissonschen Gleichung, daß die mittlere Dichte ρ rascher als $1/r^2$ mit wachsender Entfernung r vom Mittelpunkt zu null herabsinken muß, damit φ im Unendlichen einem Limes zustrebe*. In diesem Sinne ist also die Welt nach Newton endlich, wenn sie auch unendlich große Gesamtmaße besitzen kann."

21 "haftet der klassischen Himmelsmechanik ... eine ... prinzipielle Schwierigkeit an, welche meines Wissens zuerst von dem Astronomen Seeliger ausführlich diskutiert wurde. ... Diese Auffassung ist mit der Newtonschen Theorie unvereinbar. Letztere verlangt vielmehr, daß die Welt eine Art Mitte habe, in welcher die Dichte der Sterne eine maximale ist, und daß die Sterndichte von dieser Mitte mach außen abnehme, um weit außen einer unendlichen Leere Platz zu machen. Die Sternwelt müsste eine endliche Insel im unendlichen Ozean des Raumes bilden."

sphere, which is impossible.[22]

Einstein here gives his version of Seeliger's and Kelvin's flux argument. Seeliger's equation (3) and Kelvin's (4) captures the same result as Einstein. But, where Seeliger and Kelvin used the technical machinery of Green's theorem, Einstein used the equivalent but more vivid image of lines of force.

Einstein now administered the coup de grâce. Newtonian theory allows an island of stars only for our cosmology. But not even this is satisfactory. Such an island loses radiation to infinite space. Likewise, as energy of motion is distributed statistically among the stars of the island, some, we may suspect, will acquire enough velocity to escape the island's gravitational pull. Over time, would not all the stars eventually use this mechanism of escape, so that an island universe provides no stable system of stars? To answer, Einstein, master of statistical physics, could simply call to mind Boltzmann's analysis of the statistical physics of gas molecules in a gravitational field. Its results, Einstein could see, would hold equally for a gas of molecules or a cluster of stars. The equilibrium distribution required a finite ratio of densities at the gravitational center and at infinity, so that a vanishing density at infinity entailed a vanishing density at the center. In short, Einstein could conclude that an island universe of stars would evaporate, in apparent contradiction with the static character he presumed (notoriously in error) for the stars on the largest scale.

How was Einstein's analysis oversimplified? We shall soon see that the choice he sought to force between an evaporating island universe and an ill-behaved infinite matter distribution was a false dilemma.

Seeliger, Neumann and Einstein posed a problem that had to be solved. In the remaining Sections of the paper, I will review the various escapes entertained in the decades following, prior to 1930. We shall see that, at one time or another, virtually every facet of every assumption in the cosmological and gravitational commitments listed above were held up for scrutiny and it was urged that the rejection of each provided the escape from the problem.

6. The no-solution solution

The simplest of all responses to the problem was just to deny that there was a problem. Somehow the bad behavior of gravitational force in Newtonian cosmology was an illusion that would be dispelled by closer thought. Symmetry considerations, it would seem, must override all else: the only force distribution that is compatible with the homogeneity and isotropy of the Newtonian cosmology is a vanishing force. The problem Seeliger identified is at best a mathematical oddity that deserves little attention in the world of physics. This symmetry based

[22] "Begründung. Nach der Newtonschen Theorie enden in einer Masse m eine Anzahl 'Kraftlinien,' welche aus den Unendlichen kommen, und deren Zahl der Masse m proportional sind. Ist die Dichte ρ_0 der Masse in der Welt im Mittel konstant, so umschließt eine Kugel vom Volumen V im Durchschnitt die Masse $\rho_0 V$. Die Zahl der durch die Oberfläche F ins Innere der Kugel eintretenden Kraftlinien ist also proportional $\rho_0 V$. Durch die Oberflächeneinheit der Kugel treten also Kraftlinien ein, deren Zahl $\rho_0 V / F$ oder $\rho_0 R$ proportional ist. Die Feldstärke an der Oberfläche würde also mit wachsendem Kugelradius R ins Unendliche wachsen, was unmöglich ist."

'no-solution solution' appears very infrequently in published discussions. This infrequency of publication, however, cannot be used to establish that the view is unpopular. For if one holds to this escape, one is less likely to seek publication; infrequency of publication is compatible with both a popularity and lack of popularity of the no-solution solution.

At worst, this no-solution solution is simply a blunder. As long as one holds that the gravitational force on a test body is the sum of forces exerted by all other masses, then the indeterminacy of the sum is a serious problem. It does not evaporate just because one happens to like one of the possible values over all others; one cannot wish the others away. At best, the no-solution solution is an unfulfilled promise. Since symmetry considerations do dictate one particular value for the sum, it would seem that there must be some fallacy that allows the sum to take other values. The no-solution solution, in effect, directs one to find the fallacy. However the solution can hardly be satisfactory without clearer indication of where the fallacy lies.

With the infallible wisdom of hindsight, this symmetry driven solution looks even less satisfactory. For in 1934, in the hands of Milne (1934) and Milne and McCrea (1934), Newtonian cosmology was reborn as a cosmology that mimics the expanding universe cosmologies of general relativity. This neo-Newtonian cosmology is predicated on the assumption that the gravitational force field in a homogeneous, infinite matter distribution is *not* homogeneous after all. Its inhomogeneity turns out to be what gives the cosmology is interesting properties.

6.1. Newton stumbles

Isaac Newton himself is the best known proponent of the no-solution solution. In 1692, Newton entered into a correspondence with the theologian, Richard Bentley. The latter had undertaken to inaugurate the Boyle Lectures. These would be a series of eight lecture-sermons, defending Christian religion and refuting atheism. Bentley wrote to Newton for assistance in determining how much comfort he might find in Newton's work.[23] Bentley's queries turned to the infinitude of the universe. In response, Newton gave us a portrait of how he envisaged the accommodation of an infinite distribution of masses in his system. His second letter began by considering the gravitational collapse of matter initially scattered through a finite portion of space. In such collapse, he agreed with Bentley, that it is enormously unlikely to suppose that there would be one central mass so perfectly placed that it maintained equal forces of attraction on all sides and remain at rest "a supposition fully as hard as to make the sharpest needle stand upright on its point upon a looking-glass."[24] Newton then turned to describe a cosmology with an infinite, homogeneous matter distribution. Such a distribution is possible, he asserted, but it would be (in modern language) pseudostable:

> And much harder it is to suppose all the particles in an infinite space should be so accurately poised one among another, as to stand still in perfect

[23] See Koyré 1965: chap. 4; 1957: chap. 7.

[24] These letters are reprinted as Bentley 1756.

> equilibrium. For I reckon this as hard as to make, not one needle only, but an infinite number of them (so many as there are particles in an infinite space) stand accurately poised upon their points. Yet I grant it possible, at least by a divine power; and if they were once to be placed, I agree with you that they would continue in that posture without motion for ever, unless put into new motion by the same power. When, therefore, I said that matter evenly spread through all space would convene by its gravity into one or more great masses, I understood it of matter not resting in an accurate poise. (Bentley 1756: 208)

While denial of such a static, pseudostable cosmology would become the basis of the neo-Newtonian cosmology of Milne and McCrea, Newton here described the cosmology expected by everyone to arise from Newtonian theory through to the 1920s—including Seeliger and Einstein.[25] Newton then turned directly to Bentley's formulation of the problem Seeliger later identified.

> But you argue, in the next paragraph of your letter, that every particle of matter in an infinite space has an infinite quantity of matter on all sides, and, by consequence, an infinite attraction every way, and therefore must rest in equilibrio, because all infinites are equal. Yet you suspect a paralogism in this argument; and I conceive the paralogism lies in the position, that all infinites are equal. The generality of mankind consider infinites no other ways than indefinitely; and in this sense they say all infinites are equal; though they would speak more truly if they should say, they are neither equal nor unequal, nor have any certain difference or proportion one to another. In this sense, therefore, no conclusions can be drawn from them about the equality, proportions, or differences of things; and they that attempt to do it usually fall into paralogisms. So, when men argue against the infinite divisibility of magnitude, by saying, that if an inch may be divided into an infinite number of parts, the sum of those parts will be an inch; and if a foot may be divided into an infinite number of parts, the sum of those parts must be a foot; and therefore, since all infinites are equal, those sums must be equal, that is, an inch equal to a foot.
>
> The falseness of the conclusion shews an error in the premises; and the error lies in the position, that all infinities are equal. (Bentley 1756: 208–209)

The difficulty Bentley imagined is instantiated by the argument for the non-convergence of gravitational force given in Section 1 above. The shells to the left and right of the mass at *O* exert an infinite force to the left and again to the

---

[25] However we have Newton's admission later, in the fourth letter, that he had not thought much about this cosmology:

> The hypothesis of deriving the frame of the world by mechanical principles from matter evenly spread through the heavens, being inconsistent with my system, I had considered it very little before your letters put me upon it ... (Bentley 1756: 215)

Here he informs Bentley that he had not devoted much attention to the possibility of our planetary system arising through gravitational collapse from a homogeneous matter distribution. In his *Principia*, he had already given a slightly more robust explanation of the stability of the fixed stars than pseudostability:

> And lest the system of the fixed stars should, by their gravity, fall on each other, he [God] hath placed those systems at immense distances from one another. (Newton 1687: 544)

right. Presumably Bentley wished to conclude that these two forces are equal so that the mass at O remains in equilibrium—but he was loath to do so, fearing that the comparison of competing infinite forces cannot be made without fallacy ("paralogism"). Newton affirmed Bentley's worry. The "generality of mankind" are unable to compare infinites without disastrous consequences. To make his point, he recalled a classic paradox of measure, which, Newton urges, depends on the incorrect assumption that all infinites are equal.

While Newton could not agree with the argument that the two infinite forces balance, he did wish to retain the conclusion that the mass they act on remains in equilibrium. To assure Bentley of this conclusion, he indicated that there are consistent ways of comparing infinites:

> There is, therefore, another way of considering infinites used by mathematicians, and that is, under certain definite restrictions and limitations, whereby infinites are determined to have certain differences or proportions to one another. Thus Dr. Wallis considers them in his *Arithmetica Infinitorum*, where, by the various proportions of infinite sums, he gathers the various proportions of infinite magnitudes: which way of arguing is generally allowed by mathematicians, and yet would not be good were all infinities equal. According to the same way of considering infinites, a mathematician would tell you, that though there be an infinite number of infinite little parts in an inch, yet there is twelve times that number of such parts in a foot; that is, the infinite number of those parts in a foot is not equal to, but twelve times bigger than the infinite number of them in an inch. (Bentley 1756: 209)

Newton recalled the work of Wallis' *Arithmetica Infinitorum* of 1655. In order to solve problems of quadrature, Wallis needed to sum infinite series. For example, to find the area under a cubic, Wallis needed to employ the infinite sum $0^3 + 1^3 + 2^3 + 3^3 + \ldots$. While this sum is infinite, Wallis noticed that the ratio of this sum to other infinite sums was well behaved and finite.[26] Thus he found

$$\frac{0^3+1^3}{1^3+1^3}=\frac{1}{4}+\frac{1}{4} \qquad \frac{0^3+1^3+2^3+3^3}{3^3+3^3+3^3+3^3}=\frac{1}{4}+\frac{1}{12}$$

$$\frac{0^3+1^3+2^3}{2^3+2^3+2^3}=\frac{1}{4}+\frac{1}{8} \qquad \frac{0^3+1^3+2^3+3^3+4^3}{4^3+4^3+4^3+4^3+4^3}=\frac{1}{4}+\frac{1}{16}$$

etc. As the number of terms grew without limit, the ratio of the two series approaches $1/4$. This example illustrates Newton's claim that infinites can be compared (by mathematicians!) and were compared by Wallis and that they can come out to be unequal. The infinite sums of the numerator and denominator prove to have the ratio of $1:4$—the infinite numerator is only one fourth the size of the infinite denominator. And this is the whole result, for, in Wallis' hands, it becomes (in later notation) $\int_0^1 x^3\,dx = 1/4$.

So far Newton's analysis of Bentley's paralogism is impeccable. He then fell into error. Having recalled for us that there are perfectly good methods of comparing

---

[26] This example is given in the fragment from Wallis' *Arithmetica Infinitorum* in Struik 1986: 244–245.

infinites by means of limits, Newton seemed not to have applied them himself to the problem at hand. For, had he done so, he would surely have noticed that there was no determinate way of balancing the infinites. In the example of Section 1, he would need to find a value for the non-convergent series. Instead, apparently, he *presumed* the result: the net force on a body in an infinite, homogeneous matter distribution is zero. He then proceeded to consider how the equilibrium of the body might be disturbed by the addition of more masses:

> And so a mathematician will tell you, that if a body stood in equilibrio between any two equal and contrary attracting infinite forces, and if to either of these forces you add any new finite attractive force, that new force, how little soever, will destroy their equilibrium, and put the body into the same motion into which it would put it were those two contrary equal forces but finite, or even none at all: so that in this case the two equal infinities, by the addition of a finite to either of them, become unequal in our ways of reckoning; and after these ways we must reckon, if from the considerations of infinites we would always draw true conclusions. (Bentley 1756: 209)

It is hard to understand how Newton could make such a mistake. His mathematical and geometric powers are legendary. Perhaps Newton was so sure of his incorrect result from the symmetry considerations that he did not deem it worthwhile the few moment's reflection needed to see through to a final result.[27]

6.2 ... Arrhenius too

Arrhenius 1909 is a gentle survey of the problem of the infinity of the universe with some effort to shield the reader from technicalities. In addition to extensive discussion of Charlier's hierarchic universe (see Section 9 below), Arrhenius directly addressed the problem for an infinite universe which had been pointed out by Seeliger and which played a role in Charlier's postulation of the hierarchical universe. Arrhenius was unable to see that Seeliger had identified a real problem. He could not see any difficulty in the infinite gravitational potential Seeliger foresaw:

> Why may the potential [φ] not become infinite? The answer is: since then the speed of a star coming in 'from the outside' would be infinite at the point in question, and we observe no excessively high speeds among the stars. ... But if one assumes with the ancient philosophers a roughly infinite distribution of stars, then there is no 'outside' in relation to the world of stars and there exists no danger of infinite speeds.[28] (Arrhenius 1909: 224–225)

He then gave a brief synopsis of Seeliger's discussion of the indeterminacy of gravitational force in Newtonian cosmology, including the derivation of (1b′) by

[27] Newton could expect no saving correction from Bentley, who, presumably, falls into the mathematically ignorant "generality of mankind." Certainly, in his Boyle Lectures, Bentley (1756: 171) was quite happy to affirm for an infinite matter distribution that "An infinite attraction on all sides of all matter is just equal to no attraction at all. ... "

[28] "Warum darf das Potential [φ] nicht unendlich werden? Die Antwort ist: weil dann die Geschwindigkeit eines 'von Außen' gekommenen Sterns an dem betreffenden Punkt unendlich wäre, und wir beobachten keine übermäßig hohe geschindigkeiten bei den Sternen. ... Wenn man aber mit den alten Philosophen eine ungefähr gleichmäßige Verteilung der Sterne im unendlichen Raum annimt, so giebt es kein 'Außen' in Bezug auf die Sternenwelt und die Gefahr der unendlichen Geschwindigkeiten existiert nicht."

means of a sphere allowed to grow infinitely large. Arrhenius then laid out a clear statement of the 'no-solution solution,' with its foundation in a symmetry argument:

> Accordingly, it is very much understandable that Seeliger's argumentation is frequently construed as conflicting with the infinity of the world. This, however, is not true. The difficulty lies in that the attraction of a body surrounded by infinitely many bodies is undetermined according to Seeliger's way of calculation and can take on all possible values. *This, however, only proves that one cannot carry out the calculation by this method.* Also, how can one think of an infinitely great sphere, containing the stars, as surrounded by an infinitely great empty space? If a body is in an infinite space, in which other bodies are roughly uniformly distributed, then, if one ignores nearby bodies, its attraction is equally great in all directions, as is easy to see for reasons of symmetry. Consequently, these attractions cancel one another; and the body behaves just as if it were under the influence of nearby bodies or collections of bodies and the distant bodies were not present at all; therefore [it would be] exactly as if an absorption of gravitational force took place.
>
> So there really is no valid reason for why the world would not be sown roughly uniformly with stars.[29] (Arrhenius 1909: 226, my emphasis)

6.3 Bach's no-solution non-solution

Bach's (1918) response to the problem can be conveniently mentioned here although it is not properly a no-solution solution. Bach believed that the problem could be escaped by allowing for random non-uniformities in the cosmic matter distribution that was, in some sense, uniform overall. He computed the gravitational force on a test body due to stars surrounding it, all stars presumed to be of the same mass. The nearest star, the second nearest star, ... the nth nearest star, ... were assumed to lie at unit distance, distance $\sqrt[3]{2}$, ... distance $\sqrt[3]{n}$, ... respectively from the test body. This prescription ensured that there would be exactly n stars enclosed in a sphere of radius n centered on the test body, for all n. Now each star lies on the surface of a sphere of the radius specified, centered on the test mass. Bach introduced a random element by assuming that each star might be found with equal probability on any location of its sphere's surface. He then proceeded to compute the sequence of scalar forces on the test body due to the nearest star; the nearest and the second nearest; the nearest, second and third nearest; etc. This calculation was carried to the limit of infinitely many stars. Because of the random

29 "Es is demnach sehr wohl verständlich, daß die Seeliger'sche Beweisfuhrung häufig als gegen die Unendlichkeit der Welt streitend angeführt wird. Dies trifft aber nicht zu. Die Schwierigkeit sollte darin bestehen, daß die Anziehung eines Körpers in einer Umgebung von unendlich vielen Körpern nach der Seeliger'schen Berechnungweise unbestimmt wird und alle mögliche Werte annehmen kann. Dies beweist aber nur, daß man nach dieser Methode nicht die Berechnung ausführen kann. Wie kann man sich auch eine unendliche große Kugel, die Sterne enthält, von einem unendlich großen leeren Raum umgeben denken? Wenn ein Körper in einem unendlichen Raum sich befindet, wo andere Körper ungefahr gleichmäßig verteilt sind, so ist seine Anziehung, wenn man von den nächstliegenden Körpern absieht, nach allen Richtungen gleich groß, wie aus Symmetriegründen leicht ersichtlich. Diese Anziehungen heben einander infolgedessen auf, und der Körper verhält sich ganz so, als ob er nur unter dem Einfluß der nächstliegenden Körper oder Körpersammlungen stände und die entfernteren Körper gar nicht vorhanden wären, also genau wie wenn eine Absorption der Schwerkraft stattfände.
Es giebt also tatsächlich kein triftiger Grund, weshalb das Weltall nicht ungefähr gleichförmig mit Sternen besäet wäre."

element introduced, Bach could compute only a probability distribution $p(x)$ for the scalar force x and, because of the complication of the calculation, his result was arrived at by numerical methods. Bach found the limit probability distribution p massed near $x = 2$, where unit force is set as the force due to the nearest star, with the probability of much larger forces very small. This dominance of finite forces, Bach believed, resolved the problem.

Bach's analysis, however, fails to resolve the problem. It is based on the simple misunderstanding that the problem resides in the supposedly infinite values of gravitational forces in a uniform Newtonian cosmology. However the difficulty is not the infinity of force but its lack of convergence. Any force—finite or infinite—can be recovered by choosing a suitable approach to the limit in the integral that represents the total force. All that Bach has shown is that some approaches in the cases he considers can yield finite results. This does not preclude others yielding quite different results for the same mass distribution, in which case the problem remains.[30]

7. Escape by modifying Newton's inverse square law of gravitation

While Newton and Arrhenius thought Newtonian cosmology in no need of repair, Seeliger was convinced otherwise. Arrhenius' (1909) discussion came to his notice during the printing of his Seeliger 1909, to which he added a dismissive appraisal of Arrhenius' paper in a footnote (p. 7). "I content myself with the affirmation of the fact that Herr Arrhenius has totally misunderstood my exposition."[31] Seeliger, like many others, thought the cosmological problem he had laid out called for a modification of Newton's inverse square law of gravitation.

7.1. Seeliger's modification

Seeliger (1895a: 132–133) saw the problem posed by his work as a dichotomy: either Newton's inverse square law is not the exact expression for gravitational force; or "the total matter of the universe must be finite or, more exactly expressed, infinitely great parts of space may not be filled with masses of finite density."[32] Seeliger was inclined to accept the first horn of the dilemma.[33] He saw nothing sacred in Newton's law. It was merely "a purely empirical formula and assuming its exactness would be a new hypothesis supported by nothing."[34] Thus he pro-

[30] I am grateful to Michel Janssen for drawing this paper to my attention and for the information that "Rudolf Bach" is actually a pseudonym of Rudolf Förster, according to Förster's claim in a letter of February 16, 1918, to Einstein, in which the paper Bach 1918 is foreshadowed.

[31] "Ich benüge mich mit der Konstatierung der Thatsache, daß Herr Arrhenius meine Darlegungen total mißverstanden hat."

[32] "muß die Gesamtmaterie des Weltalls endlich sein oder genauer ausgedrückt, es dürfen nicht endlich große Theile des Raumes mit Masse von endlicher Dichtigkeit erfüllt sein."

[33] Seeliger (1896: 384) explained his preference for the first since "one is saved from awkward metaphysical considerations on the finitude or infinitude of matter." ("man hierdurch mißlichen metaphysischen Betrachtungen über die Endlichkeit oder Unendlichkeit der Materie entrückt ist.")

[34] "Nun ist weiter das Newton'sche Gesetz eine rein empirische formel, deren Genauigkeit als eine absolute anzunehmen eine neue und durch nichts gestützte Hypothese wäre."

posed the addition of extra terms to Newton's law that would solve the difficulty while remaining compatible with observations of planetary motions. The difficulty was that one could find infinitely many such admissible modifications. They were merely constrained by the requirement that they lead to converging integrals corresponding to (1). The indeterminacy of these three integrals depended on the fact that the three integrals

$$\int_{R_0}^{R_1} \rho\, r\, dr, \qquad \int_{R_0}^{R_1} \rho\, dr, \qquad \int_{R_0}^{R_1} \rho \frac{1}{r}\, dr \tag{6}$$

all diverged to infinity in the limit of infinitely large R_1 (with ρ a constant). A satisfactory modification of Newton's law would be one that led to finite values for these three integrals. Seeliger proposed a particular modification—but, he admitted, more as an example of what such modification may be than as an attempt for a deeper physical viewpoint. Since the notion that gravitation is an unmediated action at a distance had fallen from favor, he raised the possibility of a very slight absorption of gravitational force in space. On the analogy with agents such as light, this led to an attenuation factor of the form $\exp(-\lambda r)$ for gravitation passing through a distance r, for λ some positive constant.[35]

Seeliger chose to apply this attenuation factor to the expression for gravitational force F between two bodies of mass m and m', which became

$$F = -Gmm' \frac{e^{-\lambda r}}{r^2}. \tag{7}$$

This modification immediately solved the problem since the three integrals corresponding to (6) became

$$\int_{R_0}^{R_1} \rho\, e^{-\lambda r} r\, dr, \qquad \int_{R_0}^{R_1} \rho\, e^{-\lambda r}\, dr, \qquad \int_{R_0}^{R_1} \rho \frac{e^{-\lambda r}}{r}\, dr,$$

and they were finite in the limit of infinite R_1.

Seeliger was also interested in estimating the value of λ. The simplest choice was to make λ so small that no observation of gravitational phenomena in astronomy could distinguish it from zero—any non-zero value at all, no matter how small, resolved the cosmological problem. But Seeliger also raised the possibility that one could place a value on λ by assuming that it was responsible for known anomalies in the motion of the planets. The anomalous (and soon to be famous) advance of the perihelion of Mercury of about 40″ per century could be accounted for, it turned out, by setting λ to the modest value 0.000 000 38. However he was concerned that this value of λ led to slight but definite perihelion motion in the remaining planets. In the first of his 1895 papers, he could proceed no further in comparing these motions with known motions. He had to await the results of Newcomb's

[35] Seeliger clearly intended that the agent lose intensity λ per unit distance by absorption, so that intensity $I(r)$ at distance r satisfies $dI/dr = -\lambda I$, which solves to yield $I(r) = I(0)\exp(-\lambda r)$.

work on planetary motion then in progress. In any case, Seeliger did not take very seriously the possibility that he had found an explanation for the anomalous motion of Mercury. The astronomical anomalies were only superficially connected with his cosmological considerations. If it works, he remarked, it does so by chance, only formally and without deeper foundation.

Seeliger must have been relieved that he took this modest view. Shortly, in Seeliger 1896: 387–389, with Newcomb's (1895) work in hand, he reported the failure of his attempt to read off a value of λ from astronomical anomalies. If its value were set from the motion of Mercury, then it predicted perihelion motions for Venus, Earth and Mars well outside the actual range Newcomb found compatible with observation. A similar fate befell the attempt to relate Neumann's proposed modification of Newton's law (see Section 7.2 below). However a proposal by Hall fared much better. He had proposed that Newton's inverse square law be replaced by a law with a distance dependence of

$$\frac{1}{r^{2+\alpha}} \tag{8}$$

with $\alpha = .00000016$, and this modification turned out to give perihelion motions for these same planets within Newcomb's error bounds. Seeliger was unmoved by Hall's success. He felt it resulted from chance. In any case, in his (1895a: 136), he had already pointed out that Hall's modification fails to resolve the cosmological problem. The three integrals corresponding to (6) are

$$\int_{R_0}^{R_1} \rho\, r^{1-\alpha}\, dr, \qquad \int_{R_0}^{R_1} \rho\, \frac{dr}{r^{\alpha}}, \qquad \int_{R_0}^{R_1} \rho\, \frac{dr}{r^{1+\alpha}},$$

and the first two still diverge for infinite R_1. This result, incidentally, shows that the cosmological problems are somewhat robust. They do not depend entirely on taking exactly Newton's law for gravitational attraction. Some obvious modifications still yield the problem. The second integral still diverges for any $\alpha \leq 1$ and the first for any $\alpha \leq 2$.

7.2. Neumann's modification

Seeliger's modification (7) was offered as one of many. It was intended to show that there exists at least *some* modification of Newton's law that allows an escape from the cosmological problem, without necessarily finding the one correct such modification. Finding this one correct modification was not possible solely on the basis of the requirement that it escape the cosmological problem, for very many modifications succeed. Some further condition would be needed to pick among them.

Neumann's 1896 modification of Newton's law was introduced in quite another spirit. The purpose of the work was to find alternatives to the Newtonian potential function $\varphi(r) = -1/r$ of electrostatics, with gravitation only an incidental interest. Neumann had found a quite precise condition to restrict his search for alternative forms. The Newtonian form was distinctive in so far as it allowed a system of charges to come to equilibrium when confined to electrical conductors. It turned out that most other forms for $\varphi(r)$ were incompatible with such equilibrium. Neumann showed (p. 16), for example, that charges mediated by the potential $\varphi(r) = -(1 - \exp(-\alpha r))/r$, for α some constant, if placed in a metal sphere "can never come to rest, but must remain in ceaseless motion."[36] Chapters 3 and 4 contained his principal theorem: if a system of electric charges can come to equilibrium in a conductor, then the charges must interact via a potential of the form

$$\varphi(r) = -\frac{Ae^{-\alpha r}}{r} - \frac{Be^{-\beta r}}{r} - \frac{Ce^{-\gamma r}}{r} - \ldots, \tag{9}$$

with the Newtonian $\varphi(r) = -1/r$ a special case, and (conversely), if the constants $\alpha, \beta, \gamma, \ldots$ are all positive and the constants $A, B, C, \ldots$ all have the same sign, then any system of charges in a conductor mediated by a potential (9) will admit equilibrium.

All these considerations and Neumann's extensive calculations proceeded essentially independently of the cosmological problem of gravitation. It made a brief reappearance on p. 122, where Neumann showed that the problem was resolved by a single term special case of his general potential

$$\varphi(r) = -\frac{Ae^{-\alpha r}}{r}, \tag{9$'$}$$

for any positive α, no matter how small, so that the escape still succeeds if the difference of the law from Newton's original is, as he put it, "vanishingly small" [*verschwindend kleine*]. This followed from a result on the previous page (p. 121). There he showed that the potential φ due to form (9′) inside a homogeneous sphere of charge density ϵ and radius R at position $\rho_1 < R$ was given by

$$\begin{aligned} \varphi &= -4\pi R^2 \epsilon \cdot A \left(\frac{1}{\alpha^2 R^2} - \frac{(1+\alpha R)e^{-\alpha R}}{\alpha^2 R^2} \frac{e^{\alpha \rho_1} - e^{-\alpha \rho_1}}{2\alpha \rho_1} \right) \\ &\longrightarrow -\frac{4\pi\epsilon \cdot A}{\alpha^2} \quad \text{as} \quad R \longrightarrow \infty. \end{aligned} \tag{10}$$

Thus, if the matter filling a homogenous, infinite universe were conceived as an infinite sphere, then the exponential potential law (9′) predicted a constant gravitational potential and no gravitational force on a test body.

Our impression that this gravitational problem was of the least concern to Neumann is reinforced by the footnote he embedded in its discussion:

> Incidentally, that objection is only a *hypothetical* one. For it is based on the hypothetical representation that all of universal space, into the infinite, is filled with stars in roughly *uniform* distribution.[37] (Neumann 1896: 122; emphasis in original)

---

36 "niemals zur Ruhe kommen können, sondern in unaufhörlicher Bewegung sich befinden"

37 "Uebrigens ist jener Einwand nur eine *hypothetischer*. Denn er basirt auf der hypothetischen Vorstellung, daß der ganze Weltraum ins Unendliche hin, in einigermaßen *gleichförmiger* Verteilung, von Sternen erfüllte sei."

He had noted earlier (p. 114) that his exponential potential (9′) led to a gravitational force between masses m_1 and m_2 at distance r of

$$F = -Gm_1m_2 \frac{1+\alpha r}{r^2} e^{-\alpha r}.$$

He *conjectured*, but apparently did not think it worth the effort of calculating, that this force formula would also be able to explain the anomalous advance of Mercury's perihelion if Seeliger's force (7) could. He also introduced Seeliger's force expression (7) as to due to Laplace, mentioning Seeliger only briefly at the end of his discussion.

That Laplace had used the formula was allowed by Seeliger (1896: 386) (without mention of Neumann). In spite of his effusive welcome of Neumann's attention, Seeliger must have grown weary of Neumann's attempts to minimize Seeliger's priority.[38] Such weariness may explain Seeliger's (1896: 385–386) strained observation that Neumann's general exponential potential (9) need not resolve the gravitational problem. For in the limit of infinitely many terms, the sum becomes an integral, which can yield a potential $\varphi(r) = -A/r^{1+\alpha}$. This potential corresponds to Green's formula (8) and does not solve the gravitational problem.[39]

7.3. Einstein and the cosmological constant

Neumann had his own purpose in calling attention to the cosmological problems of Newtonian theory. He wanted to motivate consideration of quite specific, alternative forms for the potential for his work on electrostatics. Einstein (1917a), too, had his own very similar purpose in recalling the cosmological difficulties of Newtonian theory. He hinted at it when he indicated how a modification of Newton's law of gravitation (written in the field form of Poisson's equation $\nabla^2\varphi = 4\pi G\rho$) could resolve the problems.[40]

> We may ask ourselves the question whether [these difficulties] can be removed by a modification of the Newtonian theory. First of all we will indicate a method which does not in itself claim to be taken seriously; it merely serves as a foil for what is to follow. In place of Poisson's equation we write
>
> $$\nabla^2\varphi - \lambda\varphi = 4\pi G\rho \qquad [(11)]$$

[38] Neumann's obsession with priority includes his claim (1896: 225) that he introduced the much used word *ponderomotorisch*.

[39] The observation is, in fact, strained since, in replacing the sum (9) by an integral,

$$-\int_{\alpha_1}^{\alpha_2} g(\alpha)\exp(-\alpha r)/r\, d\alpha,$$

one has to choose a quite specific density function $g(\alpha)$ and limits α_1, α_2 to recover Green's formula—for example, $g(\alpha) = -A \ln r\, r^{-\alpha} e^{\alpha r}$ with $\alpha_2 = \alpha$ and $\alpha_1 = \infty$.

[40] For uniformity of notation I have represented Einstein's gravitational constant κ by G and his Δ by ∇^2.

> where λ denotes a universal constant. If ρ_0 be the uniform density of a distribution of mass, then
>
> $$\varphi = -\frac{4\pi G}{\lambda}\rho_0 \qquad [(12)]$$
>
> is a solution of equation [(11)]. This solution would correspond to the case in which the matter of the fixed stars was distributed uniformly through space, if the density ρ_0 is equal to the actual mean density of the matter in the universe.[41] (Einstein 1917a: 179)

Einstein's escape is delivered with his customary flair for simplicity. Seeliger and Neumann offered modifications of Newton's law, but one could only see after tedious calculation that they resolved the cosmological difficulty. In Einstein's case, one sees merely by effortless inspection that the modified law (11) admits a cosmological solution (12) with uniform matter distribution. Einstein does not relate his proposal (11) to Seeliger's or Neumann's proposals. In his popular treatment (Einstein 1954: 106), he does mention Seeliger's slight adjustment of Newton's law. However a reader would be mistaken in equating Seeliger's adjustment (7) with Einstein's proposal (11). Rather Einstein's proposal coincided with Neumann's single term proposal (9′), as comparison of (12) with Neumann's limiting case (10) suggests.[42] Neumann (1896: 81–85) had calculated the differential field equations that corresponded with his potential (9) for the cases of forms with one, two, three and higher terms. They were of successively greater complication as the number of terms increased. The equation for the one term case (12′) was

$$\nabla^2\varphi - \alpha^2\varphi = 4\pi A\epsilon,$$

which corresponds to Einstein's (11) if we set $\lambda = \alpha^2$, $\rho = \epsilon$ and $G = A$.

Einstein's deeper purpose in introducing (11) became clear as the paper unfolded. In this paper, Einstein sought to develop a relativistic model of space-time that is spatially closed and with a static matter distribution that we now know as the Einstein universe. The problem is that this model is not a solution of his gravitational field equations of general relativity, unless they are augmented by the notorious 'cosmological' term $\lambda g_{\mu\nu}$; that is, unless his original field equations

$$G_{\mu\nu} = \kappa T_{\mu\nu}$$

become

$$G_{\mu\nu} - \lambda g_{\mu\nu} = \kappa T_{\mu\nu},$$

[41] "Man kann sich die Frage vorlegen, ob sich dieselben durch ein Modifikation der Newtonschen Theorie beseitigen lassen. Wir geben hierfür zunächst einen Weg an, der an sich nicht beansprucht, ernst genommen zu werden; er dient nur dazu, das Folgende besser hervorzutreten zu lassen. An die Stelle des Poissonschen Gleichungen setzen wir ... wobei λ eine universelle Konstante bedeutet. Ist ρ_0 die (gleichmäßige) Dichte einer Massenverteilung, so ist ... eine Lösung der Gleichung [(11)]. Diese Lösung entspräche dem Falle, daß die Materie der Fixsterne gleichmäßig über den Raum verteilt wäre, wobei die Dichte ρ_0 gleich der tätsachlichen mittleren Dichte der Materie des Weltraumes sein möge."

[42] To see the equivalence, set Einstein's and Neumann's quantities into correspondence according to $G = A$, $\rho_0 = \epsilon$ and $\lambda = \alpha^2$.

where $G_{\mu\nu}$ is the Einstein tensor, $T_{\mu\nu}$ is the stress-energy tensor, $g_{\mu\nu}$ the metric tensor, κ and λ constants. But how is Einstein to give independent motivation for the addition of the cosmological term? The way had been prepared by his analysis of the problems of Newtonian cosmology. At the moment of its introduction (p. 186) Einstein could announce that the modified field equation of general relativity "is perfectly analogous to the extension of Poisson's equation given by equation [(11)]."[43] (The analogy is less than perfect—something Einstein may have found it expedient to overlook in order not to compromise his introduction of the cosmological term. As Trautman (1965: 230) pointed out, Einstein's augmented gravitational field equation reduces in Newtonian limit to a field equation other than (11): $\nabla^2\varphi + \lambda c^2 = 4\pi G\rho$.)

A small puzzle remains. Why did Einstein characterize his modification of Poisson's equation in Newtonian theory as something that " ... does not in itself claim to be taken seriously; it merely serves as a foil for what is to follow. ... " Is it not a perfectly proper resolution of a problem in Newtonian cosmology? We may entertain several explanations. Perhaps Einstein, like Seeliger, was worried that the modification is by no means uniquely determined by the condition that it resolve the cosmological problem; it is just a haphazard selection among many viable choices. Or perhaps he was not able to take such modification of Newtonian theory seriously since it still left the theory without relativistic character and therefore essentially flawed. I favor another possibility. Einstein saw the role of the modified Poisson equation as suggesting a structurally analogous modification of his gravitational field equations of general relativity by a cosmological term. But the consideration that really motivated the introduction of the cosmological term was Einstein's need for the Einstein universe to be a solution of his field equations. And this need was in turn a result of his program of ensuring Machian character for general relativity. Thus, although analogous, the deeper needs for the two modifications were different. One allowed convergence of an integral; the other brought compliance with Einstein's Machian program. So the two differed at this deeper level and Einstein's hesitation was intended to prevent confusion of his Machian interests with Seeliger's problem.

8. Escape by modifying the universality of gravitation

Seeliger, Neumann and Einstein showed that a slight adjustment of the inverse square distance dependence of gravitation force is sufficient to allow Newtonian cosmology. It was also immediately apparent that this is not the only way that Newtonian gravitation may be adjusted to readmit cosmology.

8.1. Föppl's negative masses

At the February 6, 1897 sitting of the Bavarian Academy of Sciences, Seeliger presented a paper by August Föppl (1897). The latter sought a way of escaping

43 "welche der durch Gleichung [(11)] gegebenen Erweiterung der Poissonschen Gleichung vollkommen analog ist."

Seeliger's cosmological problem that, as he put it, did not involve giving up Newton's law of gravitation. His inspiration was the strong analogy between electric fields and gravitational fields. Since there were both positive and negative electric charges, there could be either attraction or repulsion between charges according to their signs. Thus, as long as the amount of positive and negative charge in the universe were equal, the "flux of force" through the surface of a sphere would converge to zero as the sphere grew infinitely large. (This result is seen most vividly using the lines of force model that Föppl did not invoke. Lines of force originating in charges of one type end in charges of the other. So the total number of lines of force emanating from a spherical portion of space is a measure of the difference between the quantities of positive and negative charge enclosed. The number of lines of force would converge to zero if the quantities of positive and negative electric charge suitably approached equality in the limit of large volumes of space.)

One way we saw above that Seeliger found of expressing the cosmological problem was as the infinite build up of field strengths on the surface of a piece of the universe as the piece grows without limit. Einstein's later version of this same formulation involved in infinite accumulation of lines of force on the surface of a spherical piece. Föppl pointed out that the existence of gravitationally negative masses would immediately resolve the cosmological problem, as long as the amounts of masses of both signs were equal in the universe. For then, just as an infinite build up of the flux of electric force did not occur, so also would there be no infinite build up of the flux of gravitational force.[44] Föppl saw this supposition of negative masses as an extension of Newton's law, not a compromise of it. However there is a sense in which it is a violation of Newton's law. Newton proposed a *universality* for gravitation: all bodies attract all others. That, Föppl proposed, should now be given up.

The bulk of Föppl's discussion is given over to more detailed consideration of the proposal. He sought to establish, for example, that gravitational masses of like sign attract and of opposite sign repel. This he did by invoking the analogy to electromagnetism. On the basis of the analogy, he proposed an expression for the energy density of the gravitational field and from it the desired result followed. He did try to address the awkward fact that we have no observational record of bodies that gravitationally repel. That, he explained, was easy to understand. Masses of opposite sign must have been expelled by repulsive forces from our vicinity.

[44] Föppl (1897: 95) made clear that he proposed just the gravitational mass—the measure of the mass as gravitational field source—to take either sign. The inertial mass—the measure of its resistance to acceleration—must remain positive, a result "Mach has already proved in detail in his *Mechanik*." ("hat schon Mach in seiner Mechanik ausführlich nachgewiesen.") We may wonder here precisely what Föppl intended. If *Mechanik* refers to Mach 1883, *Die Mechanik in ihrer Entwicklung Historisch-kritisch dargestellt*, then we would look to Mach's celebrated definition of mass in terms of acceleration induced in interacting bodies. However, far from demonstrating that negative inertial masses are impossible, he assures us (pp. 266–267) :"that there are, by our definition, only positive masses, is a point that experience teaches, and experience alone can teach." ("nach unserer Definition bloß positive Massen gibt, lehrt die Erfahrung und kann nur die Erfahrung lehren.") But Mach allows that the constancy of *vis viva* is an experience and this experience shows through a simple thought experiment the transitivity of equality of mass (p. 269). Perhaps this experience can teach us that there are no negative inertial masses through a similar thought experiment.

Perhaps because of the doubts one must have over positing entities one never expects to encounter, Föppl's concluding paragraph hedged carefully. He was far from asserting that there really are such negative gravitational masses; he merely noted that we could not now deny their possibility and, since their existence could be confirmed observationally, the possibility ought to be made known.

8.2. Kelvin's imponderable ether

Föppl's proposal of negative masses was just one way that a modification of the universality of gravitation could allow escape from the problem. Kelvin invoked a more direct modification that could be applied only to a special form of universe filling matter. He had concluded that his ether would need to sustain infinite forces if it gravitated. So he posited that it did not. The universality of Newtonian gravitation did not extend to the matter of ether:

> If we admit that ether is to some degree condensible and extensible, and believe that it extends through all space, then we must conclude that there is no mutual gravitation between its parts, and cannot believe that it is gravitationally attracted by the sun or the earth or any ponderable matter; that is to say, we must believe ether to be a substance outside the law of universal gravitation. (Kelvin 1901: 167; 1904: 266)

9. Escape by modification of the cosmological assumptions

The escapes considered so far depend on modifying one or other aspect of Newton's law of gravitation. These modifications then allow a cosmology with a uniform matter distribution. A quite different response to the problem was possible. One could leave Newton's law intact and ask just how much the assumptions of cosmological character needed to be modified in order that one may retain an interesting cosmology compatible with Newton's law of gravitation. Escapes of this character emerged rapidly.

9.1. Seeliger's dispute with Wilsing

One of the earliest responses to Seeliger's 1895 original paper was a critique by Wilsing (1895a) in a later issue of *Astronomische Nachrichten*. Seeliger (1895b) responded impatiently to Wilsing's critical remarks and they descended into a cycle of exchanges, Wilsing (1895b), Seeliger (1895c), with the journal's editor calling a halt at the last reply. One significant notion did emerge in Wilsing's opening salvo. He urged that a contradiction between a cosmological model and gravitation theory ought not to be resolved by modifying gravitation theory. Because of their highly speculative character, one ought to give up the cosmological assumptions:

> Herr Seeliger sees in the fact that, according to Newton's law, an infinitely long matter-filled cone would also exert an infinitely great attraction on a point on its axis, an objection against the general validity of the law itself; the author of a cosmogonic hypothesis can only draw the converse conclusion, that that mass distribution, purely imagined by him and inaccessible to observation, could have been possible at no time, for it cannot be explained

> by the forces abstracted from the world of experience accessible to us.[45] (Wilsing 1895a: 388)

It is hard to fault the good sense of this viewpoint. It does presuppose, however, that we retain Newtonian gravitation theory in its unmodified form, which is incompatible with a homogeneous matter distribution. Many then found this incompatibility hard to accept. Wilsing then proceeded to skirt a possibility that, after 1930, would prove to be the key to the modern resolution. Seeliger and most who considered the problem in the decades to follow, including Einstein, did not contemplate seriously how the homogeneous cosmology might develop in time, presuming that it must be static.

> But even if one now wanted to pursue the point of view of Herr Seeliger in relation to the possibility of the mass distribution imagined, then the consideration of the *temporal* development of the system itself, ignored by Herr Seeliger, would not permit a conclusion in relation to the exactness of Newton's law. For the action of the latter would only have the result of the destruction of the mass distribution thought of and its replacement by such a distribution that was compatible with it. Rather, an objection against Newton's law would arise from a mechanical stand point, if one converts the problem, with Herr Seeliger, from a dynamic to a static one by neglecting time, and adds in yet a further demand that the corresponding, imagined mass distribution should represent a state of equilibrium. This augmentation of the conditions satisfied by the imagined distribution appears especially adapted to demonstrate the inadequacy of a conclusion from merely thought up mechanical processes against the laws objectively valid in the world of experience.[46] (Wilsing 1895a: 388–389; emphasis in original)

---

[45] "Herr Seeliger erblickt in der Thatsache, daß nach dem Newton'schen Gesetze die einen unendlich langen Kegel erfüllende Materie auf einen Punkt in seiner Axe auch eine unendlich große Anziehung ausüben würde, einen Einwurf gegen die allgemeine Gültigkeit dieses Gesetzes selbst; der Urheber einer kosmogonischen Hypotheses kann nur den umgekehrten Schluß ziehen, daß jene bloß von ihm vorgestellte, der Beobachtung unzugängliche Massenverteilung zu keiner Zeit möglich gewesen sein könnte, da sie durch die bekannten, aus der uns zugänglichen Erscheinungswelt abstrahirten Kräfte nicht erklärt werden kann."

[46] "Aber selbst, wenn man der Anschauungsweise des Herrn Seeliger bezüglich der Möglichkeit der vorgestellten Massenvertheilung zunächst folgen wollte, so würde selbst dann die Berücksichtigung der von Herrn Seeliger außer Acht gelassenen zeitlichen Entwicklung des Systems einen Schluß bezüglich der Exactheit des Newton'schen Gesetztes nicht zulassen. Denn die Wirkung des letzteren würde dann nur die Zerstörung der gedachten Massenvertheilung zur Folge haben und ihre Ersetzung durch solche Vertheilungen, welche eben mit ihm verträglich sind. Ein Einwurf gegen das Newton'sche Gesetz würde sich dagegen vom mechanischen Standpunkt aus nur ergeben, wenn man mit Herrn Seeliger durch Vernachlässigung der Zeit das Problem aus einem dynamischen zu einem statischen machte, und noch die weitere Forderung hinzufügte, daß die betreffenden vorgestellten Massenvertheilungen Gleichgewichtslagen darstellen sollen. Dieses Vermehrung der von den vorgestellten Massenvertheilungen zu erfüllenden Bedingungen erscheint besonders geeinget, um das Unzulängliche einer Schlußfolgerung von nur gedachten mechanischen Vorgängen auf die in der Erscheinungswelt objectiv gültigen Gesetze darzuthun." This is followed immediately by remarks similar to those of Arrhenius who thought Seeliger had merely proved that one cannot use Seeliger's methods to compute gravitational force in a Newtonian cosmology:

> If Herr Seeliger further assumes, as an example of a mass distribution mentioned, the infinitely extended universe as spherically bounded and filled with mass of uniform density; and he stipulates a midpoint; and then calculates the forces at definite points to have arbitrary values; then this result proves only that the form of the mathematical investigation fails as soon as one leaves the grounding of the imaginable and applies the same to a transcendental problem. ("Wenn Herr Seeliger weiter, als Beispiel einer Massenvertheilung der erwähnten Art, das unendlich ausgedehnte Weltall kugelförmig begrenzt und mit Masse gleicher Dichtigkeit erfüllt annimmt,

Had Wilsing followed his own suggestion and calculated how a homogeneous matter distribution develops in time, he would have been in a position to discover the neo-Newtonian dynamic cosmologies of Milne (1934) and Milne and McCrea (1934) out of which the modern resolution of the problem would come. However he did not and we may well wonder what sense he would have made of their dynamics prior to the discovery of the expansion of the universe. He gives no hint of the form of the distribution that dynamical evolution would provide after the initial instant and we are left to wonder how he would apply Newton's theory to a homogeneous distribution of matter at its initial instant, given the problems Seeliger had identified. Seeliger's (1895b) response set the tone by referring to Wilsing's "irrelevancies" (*Nebensächliche*) and "errors" (*Irrthümern*) and the resulting tension effectively precluded profit from the ensuing exchange.

9.2. Einstein's and Seeliger's false dilemma

Einstein's (1917a) argument was based on a dilemma for Newtonian cosmology: either matter was concentrated in an island or it was homogeneously distributed. Either case brought disaster: the island would evaporate and the homogeneous distribution would suffer Seeliger's divergent forces. Seeliger had explicitly posed essentially this same dilemma without the evaporation argument.

> Therefore it is necessary to choose between two assumptions:
> 1) the total mass of the universe is immeasurably great, then Newton's law cannot hold as a mathematically strict expression of the prevailing forces of attraction,
> 2) Newton's law is absolutely exact, then the total matter of the universe must be finite or, expressed more exactly, there cannot be infinitely great parts of space filled with mass of finite density.[47] (Seeliger 1895a: 132–133)

Seeliger's dilemma depends on two erroneous presuppositions: that an everywhere non-vanishing mass density entails that the total mass of the universe is infinite; and that Newton's law is incompatible with a universe that has any everywhere non-vanishing mass density. The latter, in particular, was not a consequence of his divergent integrals.

Kelvin seemed to have a more acute sense of what these divergences entailed. While he inferred from them that his ether did not gravitate, he could not draw a similar conclusion for ordinary matter, such as constitutes the stars, for the latter clearly does gravitate. For such matter, his gloss of the consequences of the divergences was more cumbersome than Seeliger's, but it avoided the pitfalls. The flux argument provided

---

> einen Mittelpunkt festsetzt und dann bei Berechnung der Kräfte in bestimmten Punkten zu beliebigen Werthen gelangt, so beweist dieses Ergebnis nur, daß die Formen mathematischer Untersuchung versagen, sobald man den Boden des Vorstellbaren verlässt und die selben auf ein transcendentales Problem anwendet.")

But Wilsing, unlike Arrhenius, clearly thought a homogeneous matter distribution untenable.

47 "Es wird deshalb nothwendiger Weise zwischen den beiden Annahmen eine Wahl zu treffen sein: 1) die Gesammtmasse des Weltalls ist unermesslich groß, dann kann das Newton'sche Gesetz nicht als mathematisch strenger Ausdruck für die herrschenden Anziehungskräfte gelten, 2) das Newton'sche Gesetz ist absolut genau, dann muß die Gesammtmaterie des Weltalls endlich sein oder genauer ausgredrückt, es dürften nicht unendlich große Theile des Raumes mit Masse von endlicher Dichtigkeit erfüllt sein." A similar statement of this dilemma is in the introduction to Seeliger 1896 (p. 373).

> decisive proof that the mean density of ponderable matter through any very large spherical volume of space is smaller, the greater the radius; and is infinitely small for an infinitely great radius. (Kelvin 1901: § 11; 1904: § 11)

We shall see below in Section 9.3 that this formulation is still not quite strong enough.[48] To avoid the divergences, one must make a further stipulation on how rapidly the mean density at a given radius diminishes as the radius becomes infinite. We do not know if Kelvin was aware of the need for this further restriction. Rather than pursuing such theoretical issues, the practical Kelvin proceeded to a practical issue: estimating just how much matter might reside in the stars in our observable neighborhood. Taking that neighborhood to be spherical, Kelvin used the forces computed by the flux argument to determine the accelerations on stars in our neighborhood. He delimited the matter density on the assumption that it must be sufficiently small so that the accumulated velocity imparted to the stars over millions of years remains within the small bounds delivered by observation and arrived at a densities he found plausible.

9.3. The hierarchic universe: between the horns of the dilemma

Kelvin did not ask what sorts of cosmologies might be compatible with avoidance of the gravitational problem. That question was asked by Fournier d'Albe (1907), who noticed that Kelvin's condition on the matter density, when suitably strengthened, was compatible with a matter distribution that was homogeneous, in a significant sense. Charlier (1908) developed the idea in the following year. In brief, what they found was a scheme for distributing matter through infinite space that required no preferred center, but led to a vanishing matter density when averaged over all space. They imagined matter grouped locally in clusters; then these clusters were in turn grouped into clusters; and this next level of clusters into a higher level of clusters; and so on indefinitely. The spacing between the clusters could be so contrived that the mean matter density vanished over infinite space and in a way that escaped the gravitational problems. This hierarchic universe[49] passed between the horns of Einstein's and Seeliger's dilemma, for it was a universe with no preferred center, with matter distributed throughout, but free from Seeliger's gravitational divergences.

The work of Fournier d'Albe and Charlier and those who later developed the hierarchic cosmology depended on a small number of results that can be sum-

[48] Nonetheless Seeliger carefully corrected his earlier misstatements of the consequences of his calculations with a similar formulation: "If Newton's law is to hold absolutely exactly, then infinitely great spaces of the universe cannot be filled with mass of finite density on average." (Seeliger 1909: 9). ("Wenn das Newton'sche Gesetz absolut genau sein soll, dann dürfen nicht unendlich große Räume des Weltalls mit Masse von durchschnittlich endlicher Dichtigkeit erfüllt sein.") Oddly Seeliger makes no mention of Charlier's 1908 hierarchic cosmology, which made manifest the need for such reformulation, although he must have been aware of it. For Seeliger (1909: 7) cites Arrhenius 1909 and the latter has an extensive discussion of Charlier's work. And we shall see below in Section 9.4 that Charlier later thanks Seeliger for a letter of 28 April 1909 in which Seeliger corrects a technical result in Charlier 1908. Seeliger 1909 was presented to the Bavarian Academy just three days later on 1 May 1909.

[49] The term 'hierarchic' seems to have been introduced as late as Selety (1922) who spoke of a "Molekularhierarchie" (p. 298) and "die molekularhierarchische Welt" (p. 304).

marized as follows. Assume that the matter distribution is so contrived that the density of matter ρ at a distance r from some point in space diminishes according to[50]

$$\rho(r) = \frac{K}{r^n} \tag{13}$$

for $K > 0$ a constant and $n \geq 0$. Different values for n lead to more or less rapid approaches to zero density as r becomes infinite. Table 1 shows how suitable selection of n can ensure convergence of virtually every quantity involved in the cosmology as the volume of space under consideration grows indefinitely. (See Appendix B for supporting calculations):

<table>
<tr><th></th><th>$n = 0$</th><th>$n = 1$</th><th>$n = 2$</th><th>$n = 3$</th><th>$n = 4$</th></tr>
<tr><td>Total
Mass</td><td colspan="4">Diverges for
$n \leq 3$</td><td>Converges for
$n > 3$</td></tr>
<tr><td>Mean
Density</td><td>K for
$n = 0$</td><td colspan="4">Vanishes for
$n > 0$</td></tr>
<tr><td>Gravitational
Potential</td><td colspan="3">Diverges for
$n \leq 2$</td><td colspan="2">Converges for
$n > 2$</td></tr>
<tr><td>Gravitational
Force</td><td colspan="2">Diverges for
$n \leq 1$</td><td colspan="3">Converges for
$n > 1$</td></tr>
<tr><td>Tidal Force
Z_x</td><td>Diverges for
$n = 0$</td><td colspan="4">Converges for
$n > 0$</td></tr>
</table>

Table 1. Convergence properties of quantities in hierarchic cosmology

In particular, if we set a value of n for which $2 < n \leq 3$, then we have a cosmology with

- total matter of infinite mass
- vanishing mean density
- convergent gravitational potential, force and tidal force.

Such a cosmology passes between the horns of Seeliger's dilemma in so far as it has infinite total mass but there is no need to forgo Newton's law since the relevant gravitational quantities are well defined. We see also that Kelvin's condition of vanishing mean density is not sufficiently restrictive, since this vanishing obtains when $0 < n < 2$ for which not all the gravitational quantities are well defined.

---

[50] More precisely, $\rho(r)$ is the average matter density on the *surface* of a sphere of radius r. It is important not to confuse this density with the mean density of matter over the entire *volume* of the sphere. The latter mean density only cannot diminish faster that $1/r^3$—the case of single centrally located mass.

9.4. How an infinite world may be built up

While it is easy to assume that cosmic matter dilutes with distance according to a rule such as (13), it is another matter to show that such dilution can be achieved within a matter distribution that is, in some significant sense, homogeneous. Modern work on this problem was initiated by an unlikely source. Fournier d'Albe 1907 is a somewhat inflated, semi-popular work intended to develop the notion that the world of the scale of our solar system is but one of an infinite hierarchy of worlds extending into the small and large. Proceeding into the small, the next level down is the atomic level; the atoms are the sun and the electrons the planets (pp. 84–85). Proceeding into the large, the units of the next level are stars clustered into galaxies. Fournier d'Albe's essential purpose was to elaborate this grandiose vision through a seamless fusion of simple technical results and wild speculation. However, the hierarchic structure of his world just happened to allow him to escape two problems of an infinite cosmology. The first was Olbers' paradox of the darkness of the night sky (see below). The second was Kelvin's concern that stellar matter be not so densely distributed that stars falling into our system acquire too great a velocity. To escape both difficulties, Fournier d'Albe assumed his systems so distributed that the matter in a sphere of cosmic size increases only in direct proportion to the radius of the sphere (Part II, chap. II). This rate of dilution corresponds with the setting of $n = 2$ in (13). It escapes Olbers' paradox, but Fournier d'Albe apparently did not notice that it only just fails to ensure the convergent potentials needed to escape high stellar velocities. The latter requires any $n > 2$.

The real value of Fournier d'Albe's work lay in the fact that Carl Charlier read the work and, as he tells us in the opening paragraph of Charlier 1908, saw in the hierarchic proposal a way to escape the problems of an infinite cosmology. Charlier gave the escape a mathematically precise form in his (1908) "Wie ein unendliche Welt aufgebaut sein kann." The essential content of (1908) was repeated, corrected and extended in a different article, Charlier 1922, but with the same title translated into English, " How an Infinite World May Be Built Up." His model proposed a hierarchy of larger galaxies and galaxy clusters, systems $S_1, S_2, \ldots$ such that

N_1 stars form a galaxy S_1,
N_2 galaxies of type S_1 form a second order galaxy S_2,
N_3 galaxies of type S_2 form a third order galaxy S_3,
etc.

The systems $S_1, S_2, \ldots$ are presumed spherical with radii $R_1, R_2, \ldots$ and have masses $M_1, M_2, \ldots$. As galaxies of the $(i-1)$th order are packed to form a galaxy of the ith order, empty space must be introduced in sufficient amount to ensure the dilution of the mean matter density. This is ensured by locating each galaxy of $(i-1)$th order and radius R_{i-1} in a sphere of otherwise empty space of radius ρ_i within a galaxy of ith order, as shown in Figure 5. Thus the mean density of matter will decrease as we proceed up the hierarchy if $R_{i-1}/\rho_i < 1$, for each i, since the ratio of mean density in systems of order $(i-1)$ and i is $(R_{i-1}/\rho_i)^3$.

Charlier identified three problems facing an infinite cosmology. Each could be

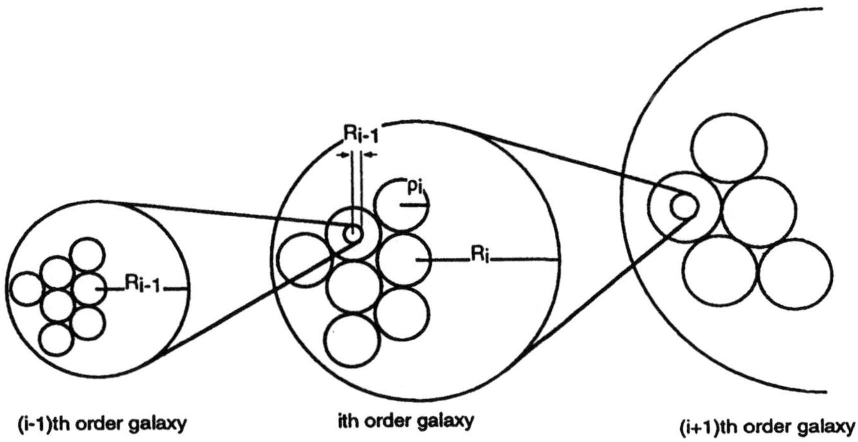

Figure 5. Charlier's hierarchic universe.

solved by requiring that the mean matter density dilute with increasing volumes of space at a suitable rate. Thus each problem could be restated as what he called a criterion that could be translated into a specific rate of dilution. The three criteria are:

> **Olbers' Criterion.** In 1826, Olbers had remarked that, if the universe were filled uniformly with stars, then the sky should glow as brightly as the face of the sun.[51] To see how this arises, in a form relevant to Charlier's work, note that the intensity of light from distant stars diminishes with the inverse square of distance. But if stars are distributed uniformly in space, then the number of stars increases with the square of distance. Thus the total light incident from such stars on the earth is represented by a diverging integral. The intensity of starlight in the sky of such a universe would be infinite.[52] This intensity should be finite.
>
> **Seeliger's Criterion.** The gravitational force in a universe with a uniform matter distribution diverges, as Seeliger pointed out. This force should be finite.
>
> **Velocity Criterion.** As celestial objects fall into the cosmic gravitational field, the depth of the potential well determines how great these velocities are. In an infinite universe with a uniform matter distribution, this well is infinitely deep. Since Charlier knew only

[51] The history of the paradox is a great deal richer than this simple remark suggests. See Jaki 1969.

[52] If the intensity of light due to a star at distance r is A/r^2 and the uniform density of stars is ρ, then the intensity of light due to stars at distance $r > R_0$ is $\int_{R_0}^{\infty}(A/r^2)\rho 4\pi r^2\, dr = \infty$. The occulting of more distant stars by nearer stars reduces the intensity expected to that of the surface of the sun.

of small velocities observed among celestial objects, he required that this well not be so deep as to yield stellar velocities beyond the modest ranges then known. In effect this criterion calls for the convergence of the gravitational potential.

We need not recapitulate Charlier's fairly complicated analysis since the essential results are already implicit in Table 1. Instead we can recover his results rapidly once we have connected the quantities of Charlier's model with the dilution rates of (13). To do this, when $r = R_i$, we will approximate the density $\rho(r)$ of (13) by the mean density $\overline{\rho}_i$ of the ith order galaxy. If the mass of a star is M_0, then the total mass of an ith order galaxy is $M_i = M_0 N_1 N_2 \cdots N_i$ and its mean density is

$$\overline{\rho}_i = \frac{M_0 N_1 N_2 \cdots N_i}{(4/3)\pi {R_i}^3}.$$

It follows that the mean densities of $(i+1)$th and ith order galaxies are related by

$$\frac{\overline{\rho}_{i+1}}{\overline{\rho}_i} = N_{i+1} \left(\frac{R_i}{R_{i+1}}\right)^3.$$

But we require that $\overline{\rho}_i$ dilute with distance as $1/r^n$. That is, that

$$\frac{\overline{\rho}_{i+1}}{\overline{\rho}_i} = \left(\frac{R_i}{R_{i+1}}\right)^n.$$

Comparing the last two equations, it follows that the requirement that the mean density dilute as $1/r^n$ is equivalent to requiring that the number of ith order galaxies in an $(i+1)$th order galaxy be set by[53]

$$N_{i+1} = \left(\frac{R_{i+1}}{R_i}\right)^{3-n}. \qquad (13')$$

Without calculation, we can see that both Olbers' and Seeliger's criterion amount to the same constraint. For Seeliger's criterion amounts to requiring that matter dilute sufficiently rapidly so that the gravitational lines of force emitted by cosmic matter not build to infinite density. These lines of force dilute with the inverse square of distance. Light from distant stars also dilutes with the inverse square of distance. Therefore any star distribution that satisfies Seeliger's criterion will satisfy Olbers' criterion and vice versa. We know from Table 1 that gravitational force converges whenever we set $n > 1$. Thus the two criteria are satisfied by

[53] It follows from (13′) that Charlier's model admits a maximum n of 3. This limit arises since the density $\rho(r)$ of (13), the density at the surface at radial distance r, is being approximated by a mean density over the whole volume of the various orders of galaxies. This mean density dilutes at its fastest when the entire universe has just one star, which is the matter of every order galaxy. It is the case of $n = 3$.

requiring that (13′) with $n = 1$ set an upper limit for N_{i+1}. Written after the form of Charlier 1922: 4, 6, the condition is

$$\frac{R_{i+1}}{R_i} > \sqrt{N_{i+1}}.$$

Unfortunately Charlier did not see the equivalence of the two conditions in 1908. He gave the correct result for Seeliger's criterion (1908: 9). But he gave a result for the Olbers' criterion (1908: 8–9) which he later described as "inexact" (1922: 5) and then published a corrected analysis supplied to him by Seeliger in a letter of 28 April 1909. He also reported (1922: 5) that Franz Selety had informed him through a letter of the correct results, apparently at the time of publication of Charlier 1922. Selety (1922: 299–300) recounts Selety's corrected analysis.

The velocity criterion can be treated similarly. From Table 1 we see that it amounts to the stronger requirement that $n > 2$. Thus, within Charlier's model, the criterion amounts to setting (13′) with $n = 1$ as an upper limit for N_{i+1}:

$$\frac{R_{i+1}}{R_i} > N_{i+1}.$$

Charlier's (1922: 14–15) version of this result replaces the strict inequality $>$ by an inequality $\geq$, apparently in error. Selety (1922: 300–301) again points out the error and in an addendum to proofs (1922: 322–324) criticizes Charlier's revised 1922 treatment.

9.5. The Einstein-Selety debate

The hierarchic cosmologies had already passed squarely between the horns of the dilemma Einstein had presented in his famous cosmology paper (1917a). Someone had to respond. So Franz Selety (1922) took on the task of dismantling Einstein's dichotomy and a great deal more. The paper is fairly long-winded and covers more material than can be reviewed here. However it does lay out quite clearly the options provided by Charlier's hierarchic model. As Selety summarized in his Introduction, this model allowed a Newtonian cosmology to satisfy all of

> 1. Spatial infinity of the universe,
> 2. Infinity of the total quantity of mass,
> 3. Complete filling of space with matter of overall finite local density.
> 4. Vanishing of the average density of the universe.
> 5. Non-existence of a singular center point or central region of the universe.[54] (Selety 1922: 281)

In addition, of course, he recalled how a suitable rate of dilution of matter with distance would escape Olbers' paradox and the divergence of gravitational force

---

[54] "1. Räumliche Unendlichkeit der Welt,
2. Unendlichkeit der Massengröße,
3. Vollkommene Erfüllung des Raumes mit Materie von überall endlicher lokaler Dichte.
4. Verschwinden der mittleren Dichte der Welt.
5. Nichtexistenz eines singulären Mittelpunktes oder Mittelgebietes der Welt."

noted by Seeliger. Finally Selety expended considerable effort to urge that the hierarchic universe provides a way of satisfying the Machian requirements that had influenced Einstein so profoundly in his work on general relativity.

However Selety differed from the standard view of which divergences must be avoided in Newtonian cosmology. He recognized that diverging gravitational potentials could be avoided by allowing the matter density of (13) to dilute with distance faster than $1/r^2$ (p. 287). However he was willing to entertain the divergence of the gravitational potential because he found the case of the $1/r^2$ dilution to have especially interesting properties. In particular, in this case, the mass M_i of an ith order galaxy would be proportional to its radius R_i. The $(i-1)$th order galaxies within it would be in gravitational free fall and the order of magnitude of their velocities would be determined by the depth of the gravitational potential well of the ith order galaxy. The depth of that well was fixed by M_i/R_i. So if M_i was proportional to R_i that depth would be same at all levels. Thus Selety could conclude something very pretty about the velocity of any galaxy with respect to the next order galaxy that contains it: at all levels of the hierarchy, they have the same order of magnitude. Selety had arrived at a similarity between the structures of each order in the hierarchy. As one moved up the hierarchy the lengths, times and masses of processes would preserve their ratios. Selety was sufficiently impressed with this result to want to suggest a "relativity of magnitude" (*Relativität der Größe*) in the hierarchic universe in which matter density diminishes as $1/r^2$ (p. 304).[55]

This relativity of magnitude allowed Selety to address a thorny problem of the hierarchic universe. Einstein had observed that a finite cluster of stars in empty space would scatter. Might not the same fate befall a hierarchic universe? Each galaxy is a region of greater density within a region of lesser density.[56] Selety had to concede that it is of infinitely small probability that normal dynamical processes could bring about a hierarchic structure. But, once it was in existence and if it diluted its matter density as $1/r^2$, then this relativity of magnitude would ensure that it could not lose its hierarchic structure in any finite time. The reasoning was simple. Imagine some process through which a galaxy of some order is destroyed. That same process could also befall higher order galaxies. However the time required for the process to be completed would grow in direct proportion to the size of the galaxies. There is no upper limit to the orders of the galaxies and no upper limit to their size. Therefore, for any such process of destruction and any nominated finite time, there is some order in the hierarchy beyond which it has not had an effect by that time.[57]

---

55 Selety did not give details of the derivation of this necessarily vague result. Presumably we must imagine that the velocity of an $(i-1)$th order galaxy is given as the ratio of some characteristic length and time. If the velocity is to remain roughly constant as we proceed up the hierarchy, these lengths and times must preserve their ratio. The masses of systems would also grow in direct proportion to the lengths and times since the total mass of a galaxy grows, by supposition, in proportion to its radius, as we move up the hierarchy.

56 Arrhenius (1909: 227) worried about exactly such instabilities, including also the possibility of eventual collapse of the galaxies under their own gravity.

57 I cannot resist observing the unhappy corollary. Since Newton's theory of gravitation is time

The case of matter diluting as $1/r^2$ had further special importance. For it enabled Selety to mount a challenge to a major part of Einstein's 1917 argument. There, as we have seen in Section 5 above, Einstein (1917a: § 1) had argued against the possibility of the stars clustering in an island in an otherwise empty universe. If we assumed that the gravitational potential at spatial infinity is finite, then by statistical mechanical arguments, that island would evaporate. Each star would, over time, through random processes acquire sufficient energy to escape the island. Boltzmann's analysis of a kinetic gas in a gravitational field also yielded a relationship between the equilibrium densities ρ and ρ_0 of an isothermal gas in regions, where the potentials are φ and φ_0 respectively:

$$\rho = \rho_0 \exp\left(-\frac{\varphi - \varphi_0}{RT}\right). \tag{14}$$

The gas temperature is T and R is the ideal gas constant.[58] We can now estimate the equilibrium density ρ at spatial infinity that must be supported by a density ρ_0 at the island. The potential difference $(\varphi - \varphi_0)$ between spatial infinity and the stellar island is assumed to be finite. Thus the ratio ρ/ρ_0 would be greater than zero and the density at infinity not zero after all. A stable island is possible only if it is not an island!

The weak point in Einstein's argument was the presumption that the potential at infinity had to be not just finite but quite small. This followed, Einstein urged, because infinite potentials would be reflected in huge stellar velocities and since "the value of the gravitational potential [is] itself necessarily conditioned by heavenly bodies." (1917a: § 1)[59] If the latter remark was intended to indicate that no island matter distribution could generate an infinite potential difference, then Einstein was wrong, as Selety now demonstrated. He assumed (pp. 288–295) a central massing of matter whose density diluted indefinitely with distance. He also assumed that this matter was an ideal gas with pressure p and temperature T obeying the ideal gas law

$$p = \rho RT.$$

He asked how the density ρ might dilute with distance if the gas was assumed to be held in static equilibrium in its central massing by its own gravitational field. He modeled this equilibrium by assuming the gas to be in an isothermal state and that the outward forces of its pressure p exactly balanced the inward gravitational forces. In a brief analysis, he showed that this latter condition of mechanical equilibrium was satisfied by an isothermal, ideal gas whose density diluted with

---

reversible, in so far as Selety's result obtains, it also supplies a proof that a hierarchic universe of the type he favors cannot arise through normal processes from a uniform matter distribution in any finite time.

58 In this discussion, the density ρ is a molar density and the potential φ has units of energy per mole.

59 "der Verlauf des Gravitationspotentials durch die Himmelskörper selbst bedingt sein mußte."

distance r from the center as[60]

$$\rho = \frac{RT}{2\pi G r^2}. \tag{15}$$

That is, the density ρ dilutes as $1/r^2$. The gravitational potential $\varphi(r)$ grew logarithmically with r as

$$\varphi(r) - \varphi(r_0) = G \int_{r_0}^{r} \frac{\rho}{r'} 4\pi r'^2 \, dr' = 2RT \ln \frac{r}{r_0}. \tag{16}$$

It now followed that the density dilution of (15) was fully compatible with the Boltzmann's result (14). Indeed, as Selety noted (p. 293), (15) and (16) entailed the Boltzmann formula (14)! And the potential $\varphi(r)$ would grow infinitely large with r while ρ dropped to zero. Thus, kinetic matter, be it an ideal gas or a cluster of stars, could be held in a stable island by its own gravitational field if its density diluted as $1/r^2$.

Einstein's evaporation argument was almost completely defeated. His last refuge was the remark that large potential differences happen not to obtain in our universe for we do not observe the large velocities they would engender. Here Selety faltered with a weak response. Such high velocities are possible, he urged, but they are just improbable and that is why we do not see them. One argument (p. 293) for this low probability was that stellar velocities in such a universe would be distributed according to the Maxwell velocity distribution. In the case of lower temperatures this accords low probability to high velocities.[61]

Within a few months, Einstein (1922) had published his reply. Selety's assault on Einstein's cosmological work clearly found its mark through Selety's assertion that the hierarchic cosmology would have the Machian character Einstein sought in general relativistic universes. Einstein spent the bulk of his response justifying his Machian requirements and arguing that the hierarchic universe fails to meet them. Einstein seemed rather unmoved by Selety's dismantling of his 1917 dilemma for Newtonian cosmology. He wrote only a paragraph on the matter, essentially conceding his error. It read:

> It is to be admitted that the hypothesis of the "molecular-hierarchic" character of the construction of the universe of stars has much going for it from the standpoint of Newtonian theory, even though we may consider the hypothesis of the equivalence of the spiral nebulae and the Milky Way as refuted by the latest observations. This hypothesis explains naturally the darkness of the heavens and resolves Seeliger's conflict with Newton's law, without conceiving of matter as an island in empty space.

[60] Selety's analysis is given in a footnote on p. 292. While I can reproduce the result, I cannot reproduce his second equation of the footnote and suspect it is a typographical error.

[61] The weakness of the argument lies in its vagueness. How low is low? How high is high? How do they compare with observed velocities? It is interesting to note that with the advent of neo-Newtonian cosmologies of Milne and McCrea in the 1930s, the presumption against gravitational potentials that diverge at spatial infinity disappeared. In those cosmologies, the gravitational potential increases with the square of distance from the origin of coordinates, thus diverging at infinity.

> Also from the standpoint of the general theory of relativity, the hypothesis of the molecular-hierarchic construction of the universe is *possible*. But from the standpoint of this theory the hypothesis is nevertheless to be seen as unsatisfactory.[62] (Einstein 1922: 436; emphasis in original)

Selety pursued the debate vigorously with two rejoinders (1923a, 1924) and also (1923c). Einstein did not reply again.

9.6. The brief celebrity of the hierarchic cosmology

Within the renewed interest in gravitation and cosmology at this time, the hierarchic cosmology enjoyed its moment of greatest celebrity. The mathematician, Émile Borel (1922) reported in April 1922 that Einstein's recent visit to the Collège de France had inspired new attention on the problem of the finitude or infinitude of space. Borel's contribution was not directly to the physics. Rather it was to give an appealing plausibility argument for the possibility of matter distributions such as are proposed by the hierarchic cosmology: Their average matter density is zero. But this does not mean that all their matter is concentrated in a central island. They still exhibit a kind of homogeneity that Borel called "quasi-periodic" [*quasi-périodique*].

Borel's contribution was to display an arithmetic model with these same properties. He considered the sequence 'α' of integers which use only the digits 0 and 1:[63]

$$\begin{cases} \ldots, & -111, & -110, & -101, & -100, & -11, & -10, \\ -1, & 10, & 11, & 100, & 110, & 111, & 1000, & \ldots. \end{cases}$$

The density of integers in α amongst all integers $\ldots, -5, -4, -3, -2, -1, 0, 1, 2, 3, 4, 5, \ldots$ has properties analogous to those of the density of matter in a hierarchic universe. Consider that subset of the integers with fewer than n digits. There are $(2^n - 1)$ integers of α in that subset of $(2/9)(10^n - 1)$ integers overall. So the density of integers of α is $9(2^n - 1)/2(10^n - 1)$ which has an upper bound of $(1/5)^{n-1}$. Thus, as the size of the subset of integers considered grows infinitely large, the density of members of a drops to zero. However, at the same time, the distribution of α exhibits a kind of homogeneity expressible as a quasi periodicity. Consider, for example, the pattern of members of α in the vicinity of 100: $\ldots$, 100, 101, 110, 111, $\ldots$ This same pattern is reproduced arbitrarily often through the integers. We find it in the neighborhood of 1100, of 10100, of 11100, of 10000100, etc. In the last case it arises as $\ldots$,10000100, 10000101,

---

[62] "Es ist zuzugeben, daß die Hypothese vom 'molekularhierarchischen' Charakter des Aufbaues der Sternwelt vom Standpunkt der Newtonschen Theorie manches für sich hat, wenn auch die Hypothese von der Gleichwertigkeit der Spiralnebel mit der Milchstraße durch die letzten Beobachtungen als widerlegt zu betrachten sein dürfte. Diese Hypothese erklärt ungezwungen das Nichtbuchten des Himmelsgrundes und vermiedet den Seeligerschen Konflikt mit dem Newtonschen Gesetz, ohne die Materie als Insel im leeren Raum aufzufassen.

Auch vom Standpunkte der allgemeinen Relativitätstheorie ist die Hypothese vom molekular-hierarchischen Bau des Weltalls möglich. Aber vom Standpunkt dieser Theorie ist die Hypothese dennoch als unbefriedigend anzusehen."

[63] This is the series as given by Borel. Presumably the omission of 0 and 1 from the series is an error.

10000110, 10000111, ... This homogeneity, as Borel pointed out, exhibits one defect. The pattern at 0 of three successive members of α, $(\ldots, -1, 0, 1, \ldots)$ is reproduced nowhere else in the integers. Also the distribution of members of a exhibits reflection symmetry only when reflected about 0.

Borel's work attracted immediate response. Costa (1922) pointed out that Borel's pattern of distribution of masses would not allow one to escape infinite gravitational potentials. Selety (1923b) complained that Borel's construction contained a center of gravity and provided a multidimensional alternative free from the problem, based on a construction proposed by Fournier d'Albe.

This activity surrounding Borel's work reflected the growth of interest in the hierarchic cosmology. In a last minute addition (27 June 1922) to his original paper, Selety (1922) had reported Borel's efforts. In his reply to Einstein, Selety (1923a: 58) was pleased to note that he had become aware of even more work on the hierarchic cosmology, listing four recent papers, including Costa 1922. The cosmology also started to enter the relativity textbooks. Silberstein (1924: 540–543) included a short appendix reviewing the cosmology. However the hierarchic cosmology shone only briefly. In later decades it enjoyed no serious presence in relativity or cosmology texts. A major exception is the cosmology article written for the *International Encyclopedia of Unified Science* by Erwin Findlay-Freundlich, an astronomer and former associate of Einstein. Freundlich (1951: 5, 23–27) considered the hierarchic hypothesis within the context of the Milne-McCrea neo-Newtonian cosmologies and argued that it provides an escape from the diverging gravitational potential at spatial infinity of these cosmologies. Freundlich seemed to prefer, however, that Newton's law be modified so that a uniform matter distribution becomes admissible and a big bang singularity avoidable.

10. **Escape through the finitude of space**

We have seen how one can escape the gravitational problem for Newtonian cosmology by reducing the amount of matter the cosmology presumes. The attraction of the hierarchical universe is that this reduction is effected with minimal compromise to the homogeneity of the matter distribution. There is a more direct way of achieving this end. We need only presume that the geometry of space is such that it is closed so that a uniform matter distribution corresponds with a finite total mass. Such was the supposition explored by Lense (1917: 1050–1054), as part of a broader analysis of Newton's law of gravitation in non-Euclidean spaces. Lense assumed a spherical-elliptical space and asked after the gravitational potential, force and tidal force given by Seeliger's formulae (1).[64]

It was not so clear which was the appropriate form of Newton's law to employ in such a space. The simplest choice was to continue to allow gravitational force

[64] The volume element of these formula, $r^2 \sin\gamma\, dr\, d\gamma\, d\theta$, needed to be altered to $\sin^2 r\, \sin\gamma\, dr\, d\gamma\, d\theta$ to accommodate the non-Euclidean geometry.

to vary with the source mass μ and distance r as

$$\frac{\mu}{r^2}.$$

However this dependence is better adapted to a Euclidean space where it entails a conservation of lines of force. Since the areas of spherical shells containing the mass μ grow in proportion to r^2 in a Euclidean space, this μ/r^2 force law ensures that the total flux of lines of force penetrating a sphere remains constant, proportional to μ, on each sphere. In a spherical space, the areas of the corresponding spheres grow with radius r as $4\pi R^2 \sin^2(r/R)$, where R is the radius of curvature of the space. So an analogous gravitation law that conserves lines of force would have a dependence

$$\frac{\mu}{R^2 \sin^2(r/R)}.$$

Lense's results were convergent integrals. The first dependence gives constant values for the potential and tidal force and a vanishing gravitational force. The second law gives a constant value for the potential and vanishing force and tidal force.

11. Conclusion

With the discovery of the recession of the galaxies, cosmology changed and with it Newtonian cosmology. Milne (1934) and Milne and McCrea(1934) proposed neo-Newtonian cosmologies that mimicked the dynamics of the relativistic Friedman universes. They still presumed a uniform matter distribution and retained Newton's inverse square law of gravitation. So Seeliger's original problem still held and had to be addressed. In the decades that followed a new resolution of the problems took the place of the many escapes that had been tried prior to 1930. That new resolution left untouched the 'cosmological' and 'gravitational' commitments of Newtonian cosmology, introduced at the beginning of this paper. Instead it sought to modify the 'kinematical' commitments of Newtonian cosmology. The idea was perhaps implicit in Milne and McCrea's original work; it is stated more explicitly in Milne (1942); and it is developed through the work of a number of researchers over the decades that follow, finding its most complete, modern statement in Malament (1995). That escape amounts to the recognition that there is a relativity of acceleration in Newtonian cosmology, so that the specification of which are the inertial frames becomes as much a matter of convention as does the specification of a rest frame in special relativity. Since gravitational force in Newtonian theory is only defined once the inertial frames are specified, the indeterminacy of gravitational force that Seeliger revealed reflects only this indeterminacy of inertial frames (see Norton 1995).

Perhaps the most astonishing part of our story is that both of the greatest figures of cosmology and gravitation, Newton and Einstein, stumbled on the same problem. When presented with the problem, Newton seemed so sure that his cosmology would be well behaved that he saw no need to think the problem through. Einstein also was overly hasty, seeing in the problem a dilemma for Newtonian cosmology that others showed to be a false dilemma. There is a final irony in Einstein's association with the problem. He thought it a useful foil for his 1917 work in cosmology. The deeper purpose of that work was to show how his new general theory of relativity could be made compatible with the demands of a generalized principle of relativity—something that he believed impossible in Newtonian theory. Yet we now see in Einstein's foil the strongest evidence of a relativity of acceleration in Newtonian cosmology. However debate still rages over whether Einstein has had any success in his efforts to find a relativity of acceleration through his general theory of relativity.[65]

---

65 For a survey of this long standing debate see Norton 1993b.

Appendix A: **Seeliger's derivation of his expressions for gravitational potential, force and tidal force**

Seeliger's (1895a) derivation of his formulae (1) is recapitulated in Seeliger 1896, 1909. It proceeded as follows. He considered a point A in space surrounded by the masses of the universe and sought the gravitational effect of these masses at A. He assumed that the matter in each mass is arranged in a sphere. Within each sphere, the mass is laid out in concentric spherical shells of constant density. (This is a close approximation of how the matter of the stars and planets are actually arranged.) This matter distribution will be discontinuous. Seeliger proposed to replace it by a piecewise continuous distribution with the same gravitational effects. To do this, he expanded each sphere into a much large sphere of the same mass. As long as the matter in the sphere remains in concentric shells of constant density and as long as the spheres do not grow so large as to engulf A, a well-known theorem of Newtonian gravitation theory tells us that the gravitational effect at A of the spheres will remain the same. Choosing any suitable one of the many ways of expanding the spheres that remain within these limits, Seeliger replaced the original matter distribution with a continuous distribution with an everywhere finite mass density ρ.

Seeliger set up a spherical coordinate system (r, θ, γ) centered on a point O *other* than A, with A lying at a distance x from O on the $\gamma = 0$ axis, as shown in Figure A1. Consider a volume element at (r, θ, γ) bounded by coordinate differentials dr, $d\theta$, $d\gamma$. Its mass dm is given by $\rho r^2 \sin\gamma\, dr\, d\theta\, d\gamma$. Consider all such mass elements in the volume V between a surface $r = R_0$ and $r = R_1$, where R_0 and R_1 will vary with the angular coordinates θ and γ. The gravitational potential due to all these elements at A is

$$\varphi(x) = -G \int_V \frac{dm}{s} = -G \int_{\theta=0}^{2\pi} \int_{\gamma=0}^{\pi} \int_{r=R_0}^{r=R_1} \frac{\rho\, r^2 \sin\gamma\, dr\, d\gamma\, d\theta}{s}, \tag{A1}$$

where s is the distance from A to the mass element dm. The function $\varphi(x)$ shows how the gravitational potential φ varies along the line OA as the distance from O to A changes. This functional dependence encodes both the gravitational and tidal force at O, as we see if $\varphi(x)$ is expanded as a Taylor series at O

$$\varphi(x) = \varphi(0) + x \left.\frac{d\varphi}{dx}\right|_{x=0} + \frac{1}{2} x^2 \left.\frac{d^2\varphi}{dx^2}\right|_{x=0} + \ldots . \tag{A2}$$

The term in x contains the gravitational force at O, and the term in x^2, the tidal force at O, $Z_x = -d^2\varphi/dx^2$. To recover F_x and Z_x, Seeliger needed to write (A1) as a power series expansion. This he was able to do by means of a standard result in function theory. In the spherical coordinate system given, it turns out that $1/s$ can be written as an infinite series of Legendre polynomials

$$\frac{1}{s} = \frac{1}{r} \left(P_0(\cos\gamma) + \frac{x}{r} P_1(\cos\gamma) + \frac{x^2}{r^2} P_2(\cos\gamma) + \ldots \right),$$

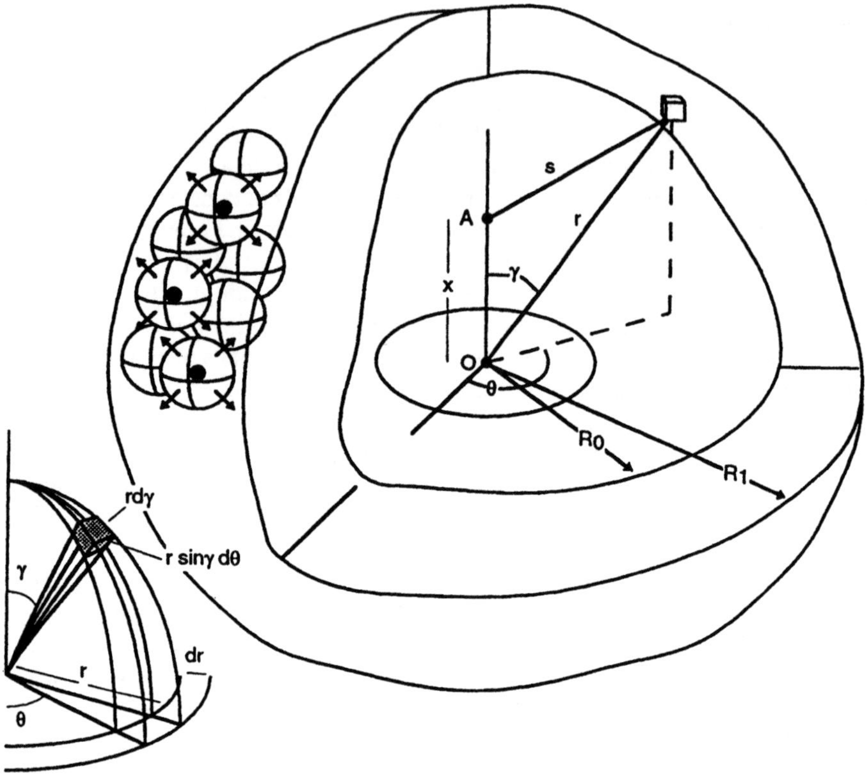

Figure A1. Seeliger's derivation.

where the Legendre polynomials are

$$P_0(\cos\gamma) = 1, \quad P_1(\cos\gamma) = \cos\gamma, \quad P_2(\cos\gamma) = \frac{1}{2}(3\cos^2\gamma - 1), \ \ldots$$

Substituting for $1/s$ in (A1) we recover

$$\begin{aligned} \varphi(x) = &- G \int_{\theta=0}^{2\pi} \int_{\gamma=0}^{\pi} \int_{r=R_0}^{r=R_1} \rho\, r\, P_0(\cos\gamma) \sin\gamma\, dr\, d\gamma\, d\theta \\ &- xG \int_{\theta=0}^{2\pi} \int_{\gamma=0}^{\pi} \int_{r=R_0}^{r=R_1} \rho\, P_1(\cos\gamma) \sin\gamma\, dr\, d\gamma\, d\theta \\ &- x^2 G \int_{\theta=0}^{2\pi} \int_{\gamma=0}^{\pi} \int_{r=R_0}^{r=R_1} \frac{\rho}{r} P_1(\cos\gamma) \sin\gamma\, dr\, d\gamma\, d\theta \\ &- \ldots \end{aligned} \tag{A3}$$

Comparing the terms of (A2) with (A3), we can read off directly Seeliger's expressions (1) for φ, F_x and Z_x at the origin of coordinates O.

For uniformity of exposition I have altered Seeliger's notation. I have restored G which Seeliger set to unity by a choice of units. His density δ, potential V, force X, tidal force Z, displacements a and ρ and his coordinate system (r, φ, γ) have been replaced by my density ρ, potential $-\varphi$, force F_x, tidal force Z_x, displacements x and s and my coordinate system (r, θ, γ). (Note the sign change in V.) Seeliger also employs the convention of writing the integral $\int f(x)\,dx$ as $\int dx\, f(x)$ so that it is easy to mistake what lies within the scope of an integration operation in a multiple integral.

Appendix B: **Convergence of quantities in hierarchic cosmologies**

We assume that the density ρ of matter dilutes with distance r as K/r^n. Hence we compute the entries in Table 1 as follows. The total mass M enclosed between a sphere of radius R and a smaller sphere of radius R_0 is given as

$$M = \int_{R_0}^{R} \rho\, dV = 4\pi K \int_{R_0}^{R} r^{2-n}\, dr$$

$$= \begin{cases} \dfrac{4\pi K}{3-n}\left[R^{3-n} - {R_0}^{3-n}\right] & \text{for } n \neq 3, \\ 4\pi K \log \dfrac{R}{R_0} & \text{for } n = 3. \end{cases}$$

(The volume element dV is $4\pi r^2\, dr$.) Hence M converges as $R \longrightarrow \infty$ just in case $n > 3$. The mean density $\overline{\rho}$ for this shell is likewise

$$\overline{\rho} = \frac{M}{(4\pi/3)\left(R^3 - {R_0}^3\right)} = \begin{cases} \dfrac{3K\left(R^{3-n} - {R_0}^{3-n}\right)}{(3-n)\left(R^3 - {R_0}^3\right)} & \text{for } n \neq 3, \\ \dfrac{3K \log(R/R_0)}{\left(R^3 - {R_0}^3\right)} & \text{for } n = 3. \end{cases}$$

Hence $\overline{\rho} \longrightarrow 0$ just in case $n > 0$. The convergence of the potential φ, gravitational force F_x and tidal force Z_x due to masses in this shell is decided by Seeliger's three integrals (6). For the case of the gravitational potential, we have

$$\int_{R_0}^{R} \rho r\, dr = \int_{R_0}^{R} K r^{1-n}\, dr = \begin{cases} \dfrac{K}{2-n}\left(R^{2-n} - {R_0}^{2-n}\right) & \text{for } n \neq 2, \\ K \log \dfrac{R}{R_0} & \text{for } n = 2. \end{cases}$$

Hence the potential φ converges as $R \longrightarrow \infty$ just in case $n > 2$. For the case of the force F_x we have

$$\int_{R_0}^{R} \rho\, dr = \int_{R_0}^{R} K r^{-n}\, dr = \begin{cases} \dfrac{K}{1-n}\left(R^{1-n} - {R_0}^{1-n}\right) & \text{for } n \neq 1, \\ K \log \dfrac{R}{R_0} & \text{for } n = 1. \end{cases}$$

Hence the force F_x converges as $R \longrightarrow \infty$ just in case $n > 1$. For the case of the tidal force Z_x we have

$$\int_{R_0}^{R} \rho\, r^{-1}\, dr = \int_{R_0}^{R} K r^{-n-1}\, dr = \begin{cases} -\dfrac{K}{n}\left(R^{-n} - {R_0}^{-n}\right) & \text{for } n \neq 0, \\ K \log \dfrac{R}{R_0} & \text{for } n = 0. \end{cases}$$

Hence the tidal force Z_x converges as $R \longrightarrow \infty$ just in case $n > 0$.

References

Arrhenius, Svante. (1909). "Die Unendlichkeit der Welt." *Rivista di Scienza* **5**: 217–229.

'Bach, Rudolph' [Rudolf Förster]. (1918). "Die Anziehung eines unendlichen Sternsystems", *Astronomische Nachrichten* **206** (Nr. 4939): 165–172.

Bentley, Richard. (1756). "Four Letters from Sir Issac Newton to Doctor Bentley: Containing Some Arguments in Proof of a Deity." Printed in *Sermons Preached at Boyle's Lecture*. 203–215. London: Francis MacPherson (1838). [Reprinted: New York: AMS Press, 1966].

Borel, Émile. (1922). "Définition arithmétique d'une distribution de masses s'étendant à l'infini et quasi-périodique, avec une densité moyenne nulle." *Comptes Rendus des séances de l'Académie des sciences* (Paris) **174**: 977–979 [Reprinted: *Œuvres de Émile Borel*. Vol. 2: 811–812. Paris: Éditions du CNRS (1972)].

Charlier, Carl V. L. (1908). "Wie eine unendliche Welt aufgebaut sein kann." *Arkiv för Matematik, Astronomi och Fysik* **4**: 1–15.

——— (1922). "How an Infinite World May Be Built Up." *Arkiv för Matematik, Astronomi och Fysik* **16**: 1–34.

Costa, Amoroso. (1922). "A propos d'une Note de M. Borel." *Comptes Rendus des séances de l'Académie des sciences* (Paris). **175**: 1190–1191.

Einstein, Albert. (1917a). "Kosmologische Betrachtungen zur allgemeinen Relativitätstheorie." *Sitzungsberichte der Königlichen Preussischen Akademie der Wissenschaften* (Berlin): 142–152; [English translation: "Cosmological Considerations on the General Theory of Relativity." In H. A. Lorentz et al. *The Principle of Relativity*. 175–188. London: Methuen (1923). Reprinted: New York: Dover (1952)].

——— (1917b). *Über die spezielle und die allgemeine Relativitätstheorie*. Braunschweig: Vieweg. [Cited after the English translation: *Relativity: the Special and the General Theory*. 15th ed. R. W. Lawson, trans. London: Methuen (1954)].

——— (1922). "Bemerkung zu der Franz Seletyschen Arbeit 'Beiträge zum kosmolgischen System'." *Annalen der Physik* **69**: 436–438.

Finlay-Freundlich, Erwin. (1951). "Cosmology." In *International Encyclopedia of Unified Science*. O. Neurath, R. Carnap and C. Morris, eds. Vol. 1: 505–565. Chicago: University of Chicago Press.

Föppl, August. (1897). "Über eine mögliche Erweiterung des Newton'schen Gravitations-Gesetzes." *Sitzungsberichte der mathematische-physikalischen Classe der K. B. Akademie der Wissenschaften zu München* **27**: 93–99.

Fournier d'Albe, Edmund Edward. (1907). *Two New Worlds: I. The Infra-World; II. The Supra-World*. London: Longmans, Green & Co.

Jaki, Stanly L. (1969). *The Paradox of Olbers' Paradox*. New York: Herder and Herder.

——— (1979). "Das Gravitations-Paradoxon des unendlichen Universums." *Sudhoffs Archiv* **63**: 105–122.

Kargon, Robert & Achinstein, Peter, eds. (1987). *Kelvin's Baltimore Lectures and Modern Theoretical Physics: Historical and Philosophical Perspectives*. Cambridge (MA): MIT Press.

Kelvin, Lord (William Thomson). (1901). "On Ether and Gravitational Matter through Infinite Space." *Philosophical Magazine* **2**: 161–177.

——— (1904). *Baltimore Lectures on Molecular Dynamics and the Wave Theory of Light*. London: C. J. Clay & Sons.

Koyré, Alexandre. (1957). *From the Closed World to the Infinite Universe*. Baltimore: Johns Hopkins University Press.

——— (1965). *Newtonian Studies.* Chicago: University of Chicago Press.

Jammer, Max. (1961). *Concepts of Space.* Cambridge (MA): Harvard University Press.

Lense, Josef. (1917). "Das Newtonsche Gesetz in nichteuklidischen Räumen." *Sitzungsberichte der Akademie der Wissenschaften in Wien, Mathematisch-naturwissenschaftliche Klasse, Abteilung IIa* **126**: 1037–1063.

Mach, Ernst. (1883). *Die Mechanik in ihrer Entwicklung Historisch-kritisch dargestellt.* Leipzig: Brockhaus. [English translation: *The Science of Mechanics: A Critical and Historical Account of Its Development*, 6th ed. T. J. McCormack, trans. La Salle (IL): Open Court (1960)].

Malament, David. (1995). "Is Newtonian Cosmology Really Inconsistent?" *Philosophy of Science* **62**: 489–510.

McCrea, William H. & Milne, Edward A. (1934). "Newtonian Universes and the Curvature of Space." *Quarterly Journal of Mathematics.* **5**: 73–80.

Milne, Edward A. (1934). "A Newtonian Expanding Universe." *Quarterly Journal of Mathematics* **5**: 64–72.

Neumann, Carl. (1873). "Über die den Kräften elektrodynamischen Ursprungs zuzuschreibenden Elementargesetze." *Abhandlungen der mathematisch-physikalischen Classe der Königlichen Sächsichen Gesellschaft der Wissenschaften* **10** (No. 6): 417–524. [Submitted March 31, 1873, printed July 12, 1873, bound in vol. 10, dated 1874].

——— (1896). *Allgemeine Untersuchungen über das Newton'sche Princip der Fernwirkungen mit besonderer Rücksicht auf die elektrische Wirkungen.* Leipzig: Teubner.

Newcomb, Simon. (1895). *The Elements of the Four Inner Planets and the Fundamental Constants of Astronomy.* Washington: U. S. Government Printing Office.

Newton, Isaac. (1687). *Philosophiae Naturalis Principia Mathematica.* London: Joseph Streater for the Royal Society. [English translation: A. Motte, trans. *Mathematical Principles of Natural Philosophy.* (1729). Revised translation: F. Cajori, trans. Berkely: University of Los Angeles Press (1934)].

North, John D. (1965). *The Measure of the Universe: A History of Modern Cosmology.* Oxford: Clarendon.

Norton, John D. (1993a). "A Paradox in Newtonian Cosmology." In *PSA 1992: Proceedings of the 1992 Biennial Meeting of the Philosophy of Science Association.* M. Forbes, D. Hull and K. Okruhlik, eds. Vol. 2: 412–420. East Lansing (MI): Philosophy of Science Association.

——— (1993b). "General Covariance and the Foundations of General Relativity: Eight Decades of Disute." *Reports on Progress in Physics* **56**: 791–858.

——— (1995). "The Force of Newtonian Cosmology: Acceleration Is Relative." *Philosophy of Science* **62**: 511–522.

Oppenheim, S. (1920). "Kritik des Newtonschen Gravitationsgesetzes." In *Encyklopädie der mathematischen Wissenschaften, mit Einschluß an ihrer Anwendung.* Vol. 5.1, *Physik.* Arnold Sommerfeld, ed. 25–67. Leipzig: Teubner.

von Seeliger, Hugo. (1895a). "Über das Newton'sche Gravitationsgesetz." *Astronomische Nachrichten* **137** (No. 3273): 129–136.

——— (1895b). "Über eine Kritik meines Aufsatzes: 'Über das Newton'sche Gravitationsgesetz'." *Astronomische Nachrichten* **138** (No. 3292): 129–136.

——— (1895c). "Bemerkung zur vorstehende Erwiderung." *Astronomische Nachrichten* **138** (No. 3304): 255–258.

——— (1896). "Über das Newton'sche Gravitationsgesetz." *Bayerische Academie der Wissenschaften. Mathematische-Naturwissenschaftliche Klasse* **126**: 373–400.

——— (1909). "Über die Anwendug der Naturgesetze auf das Universum." *Sitzungsberichte der Königlichen Bayerischen Akademie der Wissenschaften, Mathematisch-physikalische Klasse* (1909): 3–25.

SELETY, Franz. (1922). "Beitrage zum kosmologischen Problem." *Annalen der Physik* **68**: 281–334.

——— (1923a). "Erwiderung auf die Bemerkungen Einsteins über meine Arbeit 'Beitrage zum kosmologischen Problem'." *Annalen der Physik* **72**: 58–66.

——— (1923b). "Une distribution des masses avec une densité moyenne nulle, sans centre de gravité." *Comptes Rendus des Séances de l'Académie des Sciences, Paris* **177**: 104–106.

——— (1923c). "Possibilité d'un potentiel infini, et d'une vitesse moyenne de toutes les etoiles égale à celle de la lumière." *Comptes Rendus des Séances de l'Académie des Sciences, Paris* **177**: 250–252.

——— (1924). "Undendlichkeit des Raumes und allgemeine Relativitätstheorie." *Annalen der Physik* **73**: 291–325.

SILBERSTEIN, Ludwik. (1924). *The Theory of Relativity*. London: MacMillan.

SKLAR, Lawrence. (1976). "Inertia, Gravitation and Metaphysics." *Philosophy of Science* **43**: 1–23 [Reprinted in L. Sklar. *Philosophy and Space-Time Physics*. 189–214. Berkeley (CA): University of California Press (1985)].

STRUIK, Dirk J., ed. (1986). *A Source Book in Mathematics, 1200–1800*. Princeton: Princeton University Press.

TOLMAN, Richard C. (1934). *Relativity, Thermodynamics and Cosmology*. Oxford: Oxford University Press [Reprinted: New York: Dover (1987)].

TRAUTMAN, Andrzej. (1965). "Foundations and Current Problems of General Relativity." In A. Trautman, F. A. E. Pirani & H. Bondi. *Lectures on General Relativity* (Brandeis Summer Institute in Theoretical Physics 1964, vol. 1). 1–248. Englewood Cliffs (NJ): Prentice Hall.

WILSING, Johannes. (1895a). "Bemerkung zu dem Aufsatz von Herrn Prof. Seeliger: 'Ueber das Newton'sche Gravitationsgesetz'." *Astronomische Nachrichten* **137** (No. 3287): 387–390.

——— (1895b). "Ueber das Newton'sche Gravitationsgesetz (Erwiderung)." *Astronomische Nachrichten* **138** (No. 3304): 253–254.

ZENNECK, J. (1901). "Gravitation." In *Encyklopädie der mathematischen Wissenschaften, mit Einschluß an ihrer Anwendung*. Band V.1. *Physik*. A. Sommerfeld, ed. 25–67. Leipzig: Teubner.

Genesis and Evolution of Weyl's Reflections on De Sitter's Universe

Silvio Bergia and Lucia Mazzoni

Daß die beiden Sterne ... einem gemeinsamen System angehören, vom Ursprung her miteinander kausal verbunden sind, bedeutet, daß ihre Weltlinien den gleichen Wirkungsbereich ... haben.
Hermann WEYL (1923)

HERMANN WEYL'S CONTRIBUTION to the development of theoretical cosmology in the years from the formulation of Willem De Sitter's model to the systematization of the nineteen thirties has been analysed by various authors. Let us mention, among others, Jacques Merleau-Ponty (1965), John North (1965) and Pierre Kerszberg (1986, 1989a, 1989b). In particular, Kerszberg's 1986 monographic essay provides an extensive and most interesting reconstruction of Weyl's activity in the field of cosmology.

This reconstruction is taken up again in his book, in which Kerszberg stresses that "the years from 1918 ... to 1923 ... constitute a period of extraordinarily intense intellectual activity for Weyl" (Kerszberg 1989b: 235). Within this activity, Kerszberg highlights "the more or less complete clarification [Weyl gave] of the technical difficulties with cosmic time in the De Sitter model of a static universe" (Kerszberg 1989b: 236–237). It is within this context that Weyl formulated a basic principle of cosmology: Weyl is in fact "the creator of a 'cosmological principle,' ... the all-encompassing principle that laid down what is now the almost universally accepted form of the theory" (Kerszberg 1989b: 8). Kerszberg is thus quite definite about the relevance and the meaning of Weyl's main contribution to cosmology. Here we are in fact particularly interested in grasping the logical and physical meaning of Weyl's principle and its relevance for the subsequent development of cosmology. In this respect, the approach of both Kerszberg's essay and book on the subject tends to accept "thoughts and language of a time past," as is perhaps "imperative for the historian,"[1] an attitude which does not always contribute to identify or elucidate the points of greatest contemporary scientific

[1] The quotations are from Hubert Goenner's review (1991) of Kerszberg 1989b.

H. Goenner, J. Renn, J. Ritter, T. Sauer (eds.)
The Expanding Worlds of General Relativity
(Einstein Studies, volume 7) pp. 325–342

interest. Here we will try to provide a reconstruction that may prove interesting from the point of view of modern cosmology. Our wish to achieve a personal understanding of Weyl's achievements in this field first prompted a study which was presented at a conference held on the occasion of Edwin Hubble's centenary,[2] and subsequently led to an investigation of wider scope which formed the object of a thesis (Mazzoni 1991). This analysis represents a natural development of the former contribution in the light of further evidence collected in the latter.

The essential meaning and relevance of Weyl's assumption for the foundations of modern cosmology have been synthetically expressed by Jürgen Ehlers (1990: 29): "Concerning the history of the theory underlying Hubble's law I would mention that H. Weyl in 1923 [***Raum-Zeit-Materie***, 5th ed. (Weyl 1923a)] points out that to have a cosmological model one has to specify, besides a space-time (M, g) a congruence of time-like curves to represent the mean motion of matter." Ehlers adds a further comment: "He then goes on to construct such a congruence on the De Sitter hyperboloid and requires, not only that all these curves emerge from a common origin in the infinite past, but that the causal features of all these curves should be identical. This appears to be the first use of causal structure in relativistic cosmology."

The first purpose of this paper is to analyse the process through which Weyl gradually arrived at the formulation of his assumption. Here we use the word *process* to refer both to the genesis of the idea and to the gradual emergence of its nature: that of an independent principle, or working hypothesis, determining a specific interpretation of De Sitter's universe.

We suggest that the idea eventually stemmed from Weyl's concern for global causality in general relativity; and that Weyl was not quite clear from the beginning as to the status of what he called, at different times, hypothesis or assumption. In particular, he did not make it clear that it was an independent proposition that had to be added to, or implemented in, De Sitter's mathematical model of the universe. He was, in fact, completely clear about these issues only in a paper published in 1930, i.e., some seven years after he first formulated his hypothesis. Ehlers's assessment is thus basically confirmed, but we are suggesting that Weyl was not quite definite about the issue from the very beginning in 1923. It is also pointed out that this might be the reason for the delayed reception of Weyl's ideas by the small community of cosmologists. These aspects are discussed in the following Sections. The purpose of grasping the foundational meaning of Weyl's assumption may, however, justify, in the eyes of the historian, the prefacing of the historical reconstruction by an assessment of De Sitter's universe and of Weyl's principle in modern terms.

The paper is therefore structured as follows: in Section 1, De Sitter's mathematical model of the universe is illustrated; after briefly recalling the riddles it presented to its readers, the way Weyl eventually transformed it into a model universe is anticipated; in Section 2, Weyl's line of thought on De Sitter's universe

[2] Bergia 1990. The presentation there focused on some specific geometric properties of the De Sitter universe; on the other hand, the historical analysis was carried out only at a rather preliminary level.

is traced back to his concern for global causality in general relativity already expressed in the fourth edition of *Raum-Zeit-Materie*; in Section 3, Weyl's shift from Albert Einstein's to De Sitter's cosmology in the fifth edition of *Raum-Zeit-Materie* is analysed; Section 4 deals with Weyl's "wonderful discovery" that De Sitter's universe (DSU) admits ∞^3 classes of world lines each of which had a common origin in the past and his "very natural [*naheliegend*] hypothesis" about celestial bodies, that is the first (implicit) formulation of Weyl's principle; in Section 5, Weyl's basic paper of 1923 is analysed, with an emphasis on some obscurities in the presentation; in Section 6, Ludwik Silberstein's criticism of 1924 is presented; in Section 7, Weyl's reply to criticism and the assessment of his assumption are reported; in Section 8, further reactions to Weyl's contributions in the field of cosmology between 1923 and 1930 are briefly discussed; Section 9 refers to the 1930 paper, in which Weyl presented his hypothesis, with reference to observations put forward by various authors; in Section 10, the impact of Weyl's assumption on authors such as Howard P. Robertson is discussed, in connection with the formulation of the cosmological principle proper; finally, in Section 11, we outline the background for the reconstruction attempted here that can be found in the secondary literature.

1. De Sitter's model: mathematical objectivity and interpretative ambiguities; how to transform it into a model universe

De Sitter's mathematical model of the universe—'De Sitter's universe,' (DSU) as it is usually referred to—has accompanied the development of cosmology through our century. It was, with Einstein's cylindrical model, one of the two mathematical models of the universe that were considered viable by astronomers and cosmologists in the twenties; it later provided the mathematical background for the steady-state cosmology;[3] and, more recently, for inflationary cosmology.[4]

We shall first synthetically present DSU. Its specification implies, first of all, choosing a global geometry for space-time, considered as a four-dimensional maximally symmetrical space with a metric of Lorentzian signature. These require-

[3] In the case of a four-dimensional maximally symmetric manifold with three positive and one negative eigenvalues—the case of DSU, see below—for a curvature constant $K > 0$, one can choose coordinates in which the line element is that of a *spatially* flat space-time (see, e.g., Weinberg 1972: 392):

$$d\tau^2 = dt^2 - e^{(2K^{1/2}t)}(dr^2 + r^2 d\theta^2 + r^2 sin^2\theta d\phi^2), \qquad (*)$$

with K the space-time curvature constant. The line element of the steady-state cosmology, on the other hand, is given by the expression:

$$d\tau^2 = dt^2 - R^2(t_0)e^{2H(t-t_0)}(dr^2 + r^2 d\theta^2 + r^2 sin^2\theta d\phi^2), \qquad (**)$$

where H is the value of the Hubble constant, a fixed value in the steady-state cosmology. The two line elements can be identified if the factor $R(t_0)\exp(-Ht_0)$ is absorbed into the radial coordinate. For Equations (*) and (**) to coincide, one must make the identification: $K = H^2$ (see, e.g., Weinberg 1972: 459, 460).

[4] See, for instance, Guth 1981: 351. In the context of inflationary cosmology, Equation (**) of the previous note expresses the exponential expansion of the universe in a false vacuum state.

ments specify the space uniquely (apart from isometries) as a manifold of negative (if $\Lambda > 0$) constant curvature. The manifold can always be thought of as immersed in a five-dimensional Minkowskian space-time with line element

$$ds^2 = dT^2 - (dX^2 + dY^2 + dZ^2 + dW^2),$$

where it is represented by the equation:

$$X^2 + Y^2 + Z^2 + W^2 - T^2 = a^2 \tag{1}$$

(see e.g., Rindler 1977: 185). Once two space dimensions have been suppressed to allow visualization (see Figure 1), its image is that of a one-sheeted equilateral hyperboloid of revolution about an axis corresponding to the time-like coordinate; described, if X and Y are the suppressed coordinates, by the equation

$$Z^2 + W^2 - T^2 = a^2.$$

In his 1923 paper (Weyl 1923b) Weyl referred to the above as "a metrically faithful image" [*ein metrisch treues Abbild*] of the world, i.e., perhaps something more than just a *solution* of the field equations.

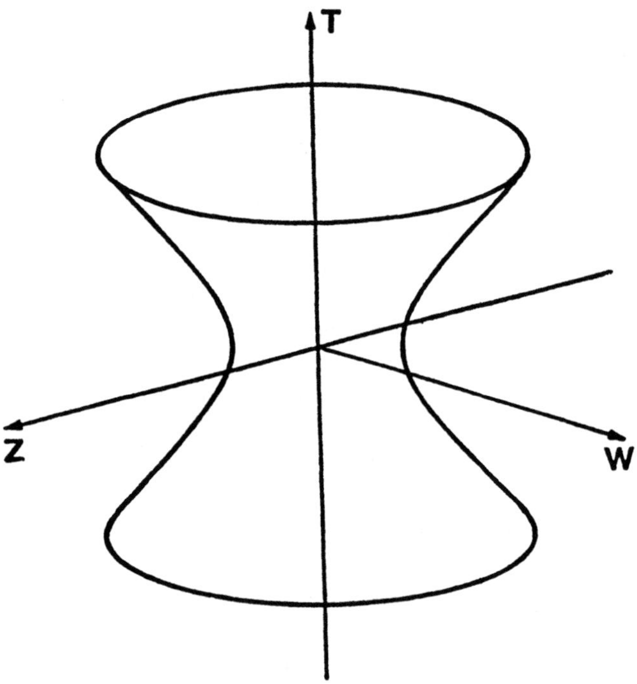

Figure 1. Two-dimensional visualization of the De Sitter universe.

However, it did not provide what we would call a model universe. According to George Ellis and Ruth Williams, in order to obtain a model universe, one must in fact specify both the space-time itself *and* the average motion of matter: "We will call a space-time a *model universe* when a family of preferred world-lines is specified in it" (Ellis & Williams 1988: 172). The family identifies the so-called 'substratum'. It should be noted that this definition only considers kinematic aspects: the notion of a Friedmann model is more specific, but will not be dealt with here. In standard cosmology, the notion of a model universe in this sense is intertwined with the requirement of the validity of the cosmological principle, which implies the foliation of space-time in terms of space-like hypersurfaces, which are hypersurfaces of homogeneity, and the perpendicularity of these to the world lines of the substratum. Read in this way, the cosmological principle necessarily introduces a cosmic time, as the time-like coordinate parametrizing the one-parameter family of space-like hypersurfaces and, at the same time, as the physical time measured by observers following a family of preferred geodesics, those perpendicular to the hypersurfaces.

In the original formulation to which Weyl refers, De Sitter's model was not a model universe in this sense, but only a geometrically defined space-time. It is mainly, if not only,[5] for this reason that for a long time it seemed hard to decipher.[6] Weyl made it a model universe by introducing a family of preferred geodesics into it. This is all very clear today, but it was much less obvious in the course of the long process that led to this clarification.

In order to identify families of geodesics on the manifold, one should note that one can go from the equation of the hyperboloid to that of a (pseudo)sphere by means of a linear and homogeneous transformation, i.e., by replacing T with iT. This makes it possible to associate the geodesics of the hyperboloid with the great

---

[5] There were in fact other difficulties: De Sitter had presented his solution in the form

$$ds^2 = -R^2\left(d\chi^2 + sin^2\chi(d\phi^2 + sin^2\phi d\theta^2)\right) + cos^2\chi c^2 dt^2, \quad r = R\chi.$$

As is well-known, it can also be obtained as a modified Schwarzschild solution with vanishing mass and cosmological constant Λ:

$$\begin{cases} ds^2 = \gamma dt - \gamma^{-1}dr^2 - r^2(d\theta^2 + sin^2\theta d\phi^2), \\ \gamma = 1 - \dfrac{2M}{r} - \dfrac{1}{3}\Lambda r^2. \end{cases}$$

In both cases it exhibits a coordinate singularity, respectively at $\chi = \pi/2$ and at $r = (3/\Lambda)^{1/2}$. At the time, it was not clear at all how the manifestation of such a singularity should be dealt with. Of course, a further difficulty arose in connection with the emptiness of DSU. However, one could consider it, together with Einstein's static universe, as one of the limit cases, respectively for low and high matter density.

[6] "De Sitter's universe appeared for a long time as a veritable mathematical unicorn. It concealed so many riddles and raised so many mirages that even Einstein and Weyl were allured, and for a time attributed imaginary properties to it" (Merleau-Ponty 1965: 53); Merleau-Ponty calls it a "boîte à secrets" (p. 61). As pointed out by Ellis, the origin of these difficulties is to be traced back, in particular, to the fact that DSU can be written in many different ways as a cosmological model. It can in fact be written as a *static universe*, as it was initially generally interpreted, or as a *FLRW universe with* $k = 0, +1, -1$ (Ellis 1990: 100; emphasis in the original).

circles of the sphere. Geodesics are thus given by the hyperboloid's conic sections obtained by cutting it with (two-dimensional) planes through the origin. On the hyperboloid one may consider various families of time-like geodesics, such as, for instance,

i) those obtained by cutting the sphere with a bundle of planes through the T-axis;
ii) those obtained by cutting the sphere with a bundle of planes through a generator of the asymptotic cone; as the generic plane of the bundle rotates about the generator, curves like those shown in Figure 2 are specified on the hyperboloid; the time parameter increases from bottom to top (given the construction, it follows that the geodesics tend to the generator in the distant past).

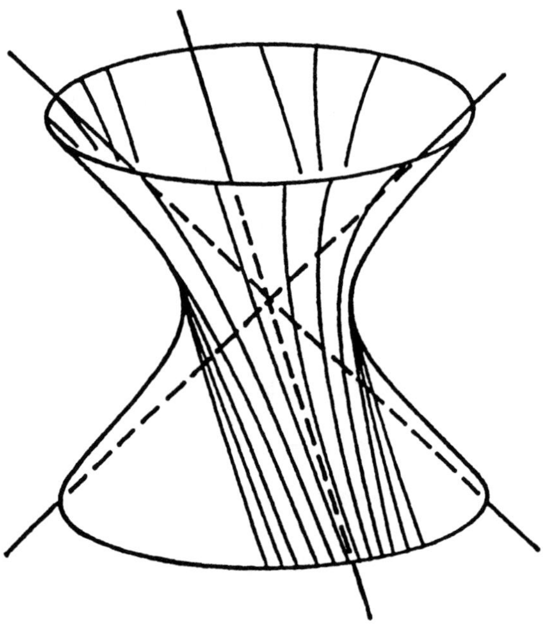

Figure 2. Generators of the asymptotic cone with the family of preferred geodesics singled out by a pencil of planes through a generator.

Another important factor is linked to the quadric in question being a ruled surface. It can be easily shown that the pair of generating lines (one for each family) at each point coincides with the null geodesics through the point.[7]

[7] To show it is an easy exercise; see, for instance, the appendix in Bergia 1990.

Choosing between the two families of geodesics means, as we know, choosing between cosmological models based on the same space-time; the cosmological model shall then be compared with a full set of observations, and it is from this comparison that we shall be able to establish the most suitable system (if any).

This provides an external reference on which to rely. In its absence, no reason intrinsic to space-time can determine what its 'natural motions' will be; nor is there any reason why, in particular, these natural motions should follow the geodesics of a certain family, nor those of one family rather than those of the other. This basic arbitrariness is the problem Weyl was confronted with.

2. Weyl's concern for causality in the fourth edition of *Raum-Zeit-Materie*

The origin of Weyl's line of thought on DSU, and of his basic assumption about it, can be traced back to his concern for global causality in general relativity. Section 34 of the fourth edition of *Raum-Zeit-Materie* was devoted to "the inter-connection of the world as a whole." In it he observed that general relativity limits itself to assigning a local topology to the space-time manifold, whereas "it makes no assumptions at the outset about the interconnection of the world." It thus becomes possible "for the cone of the active future to overlap with that of the passive past." In other words, closed time-like world lines cannot in general be avoided. Although "considerable fluctuations of the g_{ik}'s that would be necessary to produce this effect do not occur in the region of world in which we live," he went on, "there is a certain amount of interest in speculating on these possibilities inasmuch as they shed light on the philosophical problem of cosmic and phenomenal time." The rest of the section is devoted to a demonstration that the laws of gravitation, in Einstein's formulation, have a form such as is demanded by the *principle of causality*, meaning that physical quantities are expressed in terms of these quantities themselves and their spatial differential coefficients. According to what was previously said, the demonstration would only be valid locally. This is interesting insofar as, transferring the systems of coordinates introduced by Gauss in the theory of surfaces to the context of the space-time continuum, it introduces, in the neighborhood of a given world-point, a foliation of the space-time manifold in terms of space-like hypersurfaces of fixed x^0 and a family of geodesics orthogonal to them, such that the proper time of the arc between the corresponding points of two surfaces is x^0; just the kind of foliation that would later be considered in standard cosmology.

3. Weyl's shift from Einstein's to De Sitter's cosmology in the fifth edition of *Raum-Zeit-Materie*

Section 39 of the fifth edition of *Raum-Zeit-Materie* (Weyl 1923a), "Über die Zusammenverhältnisse der Welt in Grosse" [On the interconnection of the world as a whole], is characterized by Weyl's tendency to withdraw his support for Einstein's cosmology: "As fascinating [*verlockend*] as it may appear, it meets with serious objections." First of all, from *facts* (Weyl's emphasis). Among them, the circumstance that "*the present state of the firmament has nothing to do with 'a final statistical state*'" (emphasis in the original). There follows an apodictical, and—at this stage, and out of any context—rather surprising, statement: "The low velocities of the stars are due to a common origin [*einem gemeinsamen Ursprung*] rather than to a compensation"; furthermore, "it also appears that the farther away the structures involved, the higher, on average, the velocities found."[8] Thus DSU seems more promising, and Weyl devotes renewed attention to it (he had elaborated on DSU in the fourth edition of *Raum-Zeit-Materie*, Weyl 1921: 281–282). First, he observes that it can be put in a strict relationship with Einstein's cylindrical universe in terms of static coordinates. However, these provide a representation of a limited portion of De Sitter's hyperboloid. He then concludes that De Sitter's solution, on the whole, is not static (Weyl 1923a: 294). This circumstance invites us, once again, to think over the issues of the global topology of the space-time manifold and causality. The representation in terms of an hyperboloid now highlights an element of DSU that Weyl had overlooked in his previous work, an element which, in his eyes, privileged it over Einstein's universe. While "the four-dimensional numerical space" ($\mathbf{R}^4$) has a unique (connected) boundary at infinity, this view of DSU implies that it has two separate boundaries, "the infinitely distant past ($-\infty$) and the infinitely distant future ($+\infty$)." If the universe has two boundaries, the simple postulate that all the null geodesics are oriented, in one direction towards the boundary at $+\infty$, in the other direction towards the boundary at $-\infty$, can avoid the overlapping of the past and future light cones.

The cylindrical and hyperbolic universes share this feature. However, as is easily visualized by suppressing two spatial dimensions, in the cylindrical world the past light cone, indefinitely prolonged backwards in time, overlaps with itself infinitely many times. It may thus happen that we see multiple images of the same star, showing it in different epochs, separated by the huge time intervals taken by light to complete a round-trip of the universe. What is worse, the diffuse light generated would be so strong that stars would on average absorb as much light as they emit, and a thermal equilibrium would arise, just as in a cavity radiator.

De Sitter's hyperbolic universe has a past and a future boundaries as well, but, contrary to what happens in the cylindrical case, the mutual overlapping of the null cones does not occur.

4. A "wonderful discovery" about De Sitter's universe

At this stage Weyl introduced what he described as "another wonderful property of the hyperboloid," associated with one possible choice of geodesic lines on it: those which are cut on the hyperboloid by planes through a generator of the asymptotic cone. The null cones emitted, so to speak, by a point on one such geodesic cover

[8] No observational work is quoted. It may be noted that De Sitter had made the first prediction of a cosmological redshift already in his 1917 paper (De Sitter 1917) and that Slipher had started measuring redshifts of the spirals as early as 1914 (see, for instance, Kerszberg 1989b: 40).

only a part of the hyperboloid, limited by two generators. In the past direction, the zone delimited shrinks asymptotically to a point; in the future direction, instead, it covers almost the entire circumference. This "domain of action" [*Wirkungsgebiet*] is common to ∞^3 geodesics forming a bundle diverging in the future direction and converging in the past direction. There exist ∞^3 such bundles, differing from each other in their direction of convergence, just as in ordinary four-dimensional space there exist ∞^3 bundles with ∞^3 parallel lines each. "Here, however, only those elements of matter whose world-lines belong to the same bundle are in mutual connection [*Wirkungszusammenhang*] from the beginning [*von Anfang*]"; two different systems establish a reciprocal causal connection only in the course of their history.

"It is a very natural hypothesis," concluded Weyl, "that all the celestial bodies that we know belong to a system of this type." This is in fact a conclusion, since the hypothesis is not further discussed.

5. Explicit hypothesis versus implicit assumption

These matters were further discussed by Weyl in a paper published later that year (Weyl 1923b).[9] Again he proposed a comparison between Einstein's and De Sitter's universes.[10] He regarded the last as "by far the more satisfactory" since, according to it, "for two stars which, from the origin [*von Ursprung*], belong to the same connected world of action [*Wirkungswelt*], one obtains a reciprocal shift of the spectral lines." The terms used would have had a well-defined meaning (although here *Ursprung* and *Wirkungswelt* replace *Anfang* and *Wirkungsgebiet* respectively) in the context of the discussion contained in the fifth edition of *Raum-Zeit-Materie*. However, they sound out of context in this paper, where they are introduced before their meaning is clarified in similar terms to those used there.

The spectrum from a star A observed by an observer B is redshifted according to the formula

$$\frac{\Delta\lambda}{\lambda} = \tan\frac{r}{a},$$

where a indicates the constant radius of curvature of the world (see Equation (1) in Section 1 of this paper) and r the distance between the star and the observer. Weyl's intent in the paper was to "carry out the little calculation" of the effect.

After describing De Sitter's hyperboloid and recalling that its "geodesics are cut out [of the hyperboloid] by the two-dimensional planes passing through the origin of the five-dimensional space," (translation: Kerszberg 1989b: 317), Weyl defined what he now called the "range of action" [*Wirkugsbereich*] of an event: "The null cone, opening into the future, which issues from the world point P of

[9] In his paper, Weyl refers the reader to the fifth, 1923 edition of *Raum-Zeit-Materie* (Weyl 1923a). The preface to it is dated "Zürich, Herbst 1922"; the paper was received on 17 April 1923, the same year.

[10] He actually reminded the reader that he had previously analysed three representations of the world for comparative purposes: the "elementary" model, Einstein's and De Sitter's.

such a geodesic g with time-like direction, fills the range of action of g."[11] And continues:

> Dass die beiden Sterne A und B einem gemeinsamen System angehören, vom Ursprung her miteinander kausal verbunden sind, bedeutet, dass ihre Weltlinien den gleichen Wirkungsbereich Σ haben. (Weyl 1923b: 231)

Kerszberg has translated his sentence into English as:

> That both stars A and B belong to a common system, i.e., that they are causally linked together from the origin, means that their world lines have the same range of action Σ. (Kerszberg 1989b: 316)

The clause beginning with "vom Ursprung" is thus intended to define the notion of a "common system of stars"; two stars belong to a common system if they are causally linked together "from the origin"; and this "means" [*bedeutet*] that their world lines have the same range of action Σ. The sentence that follows reads: "The world-lines of all stars of the system Σ are cut by planes which pass through a common axis, i.e., a generator of the asymptotic cone" (translation: Kerszberg 1989b: 319). This seems intended to provide an answer to the question: How is it possible to introduce a common system of stars in DSU? The answer is to put them on the geodesics of option (ii) of Section 1. It seems that the discussion, up to this point, could be summarized in these terms: in De Sitter's hyperboloid *there exists the possibility* of selecting a family of geodesics such that stars sitting on them would form a common system; this means that they would be causally linked from the origin, in the sense that they would have the same range of action. Surprisingly enough, at the end of the sentence I have just analysed in detail, Weyl refers to the note, which we also reproduce in its original German version:

> Bei De Sitter sowohl (Monthly Notices of the Roy. Astronom. Soc., Nov. 1917) wie bei Eddington (Math. Theory of Relativity, Cambridge 1923, S. 161ff.) fehlt noch diese *Annahme* über den "Ruhzustand" der Sterne—die einzig möglich übrigens, welche mit sich der Homogenität von Raum und Zeit verträgt. Ohne eine solche Annahme läßt sich aber natürlich nichts über die Rotveschiebung ausmachen.[12] (Weyl 1923b: 231; emphasis added)

We say "surprisingly" since, as we have tried to argue, the previous text points to a mere possibility and the author does not state that he will assume it as a working hypothesis. The reader is invited to insert, between the last two sentences, the following two:

1) *It is a very natural hypothesis that all the celestial bodies that we know belong to a system of this type,*
(this is the sentence that closed the discussion in the relevant section of the 1923 edition of *Raum-Zeit-Materie*), and

[11] Translation: Kerszberg 1989b: 317. Above, in Section 4, we have used 'domain of action' to translate *Wirkungsgebiet*. The two terms appear to have exactly the same meaning.

[12] "In De Sitter ... as well as in Eddington ... this *assumption* on the 'state of rest' of the stars—the only one possible, on the one hand, that carries with it the homogeneity of space and time—is still lacking. Without such an assumption, on the other hand, one cannot establish anything about the redshift in a natural way." (our translation).

2) *Accordingly, we shall convert this hypothesis in an assumption to be implemented in De Sitter's universe.*

It seems to us that an insertion of this kind would have made much clearer what Weyl was proposing.[13] It should therefore not come as a surprise that some of his readers should find his presentation obscure and deceiving.

Weyl came back to the subject in a paper published in *Naturwissenschaften* in 1924 (Weyl 1924a).[14] The paper is conceived in the form of a dialog. It presents once more Einstein's and De Sitter's versions of the universe, contrasted with an elementary (Newtonian) cosmology. It is worth stressing that Weyl's favored family of geodesics is introduced with no emphasis on the fact that it is to be added to, or implemented in, De Sitter's mathematical model, hence on its arbitrariness. It is unlikely that the paper was much read by people active in the field; but, in any case, its readers would have found hard to realize from it that a new principle had been introduced in cosmology.

6. An early reception: Silberstein

In 1924, Ludwik Silberstein devoted a paper (Silberstein 1924) to the study of DSU. According to him, De Sitter's formula leading to a "spectrum shift" "was based on the artificial assumption of a star coerced to remain at a fixed distance from the observer." As a consequence, it cannot be applied to actual stars "if these, as well as the observing station, are to behave as free particles, describing world geodesics (...) To be of any use at all, the Doppler effect formula [De Sitter's formula] ought to be constructed for a star and an observer in free inertial motion, each describing some geodesic $\delta \int ds = 0$ of De Sitter's space-time ... or, in view of the practically evanescent transversal effect, a geodesic of its two-dimensional section ... covering the sub-case of all radial motions." "This broader problem," Silberstein continued, "was recently taken up by Weyl,[15] whose merit is to have formulated a general principle for computing intrinsically any such Doppler effect (...) Unfortunately, applying this general principle ... to the problem in hand, Dr. Weyl introduced the perfectly gratuitous assumption, more or less disguised as a necessary feature of De Sitter's world, according to which the world-lines of all the stars belong to a 'pencil of geodesics diverging towards the future'— which, in plain English, amounts to assuming that some time ago the stars were all assembled and rather crowded around some place and had, there and then, evanescent relative velocities to grow later on into huge and (apart from gravitation) ever increasing positive radial motions." Before going any further with Silberstein's criticism, we would like to stress that Silberstein did in fact find an assumption

[13] Elsewhere one of us has suggested the possibility of a mere oversight on Weyl's part (Bergia 1990). As we shall see, there is good contextual evidence that it was not so.

[14] The paper is wrongly referred to by Weyl in his 1930 paper under the title of another of his 1924 articles: "Was ist Materie?"

[15] Silberstein referred the reader both to the fifth edition of *Raum-Zeit-Materie* and to the 1923 paper.

where Weyl had written there was one, and, secondly, he understood perfectly well what it meant, to the point of making the comment, that sounds ironical, about the crowding of stars in the remote past (his interpretation of "the state of rest" as a state of evanescent relative velocities is also perfectly appropriate). What he complained about was the *gratuitousness* of the assumption and that it should have been more or less *disguised as a necessary feature of De Sitter's world.* These two criticisms might be justified to some extent by Weyl's presentation. According to Silberstein, "with this arbitrary hypothesis, and through a number of rather obscure technicalities,"[16] Weyl had obtained his redshift formula, "yielding always a redshift or what the astronomers call a *positive* 'radial velocity'."[17] Here came the last of Silberstein's criticisms. "No matter how unlikely and arbitrary, Weyl's hypothesis is entirely undesirable, as it is flatly contradicted by the famous nebula in Andromeda, ... by two or more spirals" and by a host of globular clusters. "Under these circumstances," wrote Silberstein, "it has seemed worthwhile to investigate the problem from the outset, basing its solution on De Sitter's theory unsupplemented by any such arbitrary assumptions by which it was unnecessarily limited or actually mutilated."

7. Weyl's reply to criticism and the assessment of the assumption

Weyl answered Silberstein's criticisms in a short note which appeared five months later in the same journal (Weyl 1924b). He wrote that he had "by no means 'more or less disguised as a necessary feature of De Sitter's world' the assumption regarding the world lines of the stars." "On the contrary," he went on, "going further than De Sitter and Eddington, [he had] strongly emphasized the necessity for *adding* an assumption regarding the 'undisturbed state' of [the] stars, if anything in the theoretical line regarding the displacement to red *had* to be formulated." His hypothesis, he went on, " [was] not 'perfectly gratuitous,' but simply [meant] that the stars of the system are able to act upon one another from eternity" (this is evidently the way Weyl preferred to translate his term *Ursprung*).

We would like to stress three points which characterize this reply. The first, and fairly obvious one, is that this is the first time Weyl made clear that the assumption was something that had to be added to DSU; for the reasons we have discussed, neither Weyl 1923a nor Weyl 1923b complied with this requirement.. Secondly, the rest state of the stars has now become their *undisturbed* state. This amounts to saying: except for local motions, due to the gravitational attraction between nearby bodies, celestial bodies (we would replace Weyl's stars with galaxies) follow the geodesics of a preferred family; therefore in our terminology they specify a substratum. Thirdly, now Weyl finds it necessary to justify his assumption: the

[16] Some of them, according to Silberstein, were responsible for a replacement of $\tan\sigma$ for $\sin\sigma$. Weyl, in his answer, pointed out that Silberstein had unduly mixed concepts that were distinct in the two approaches.

[17] Silberstein wanted to stress that Weyl himself had insisted on this feature "as a result of a sublime guess about the remote past of all the stars."

point is that one wants to deal with a system of stars "causally linked fron eternity," to use one of his previous expressions. Still, he does not yet tell us why.

In the final paragraph of his short note, Weyl mentions a curious contradiction in Silberstein's paper, namely, that at the end of his paper "he uses exactly the same assumptions as a basis, the only difference being that he adds to my group of world-lines, which diverges into the future, that which results from it through the interchange of past and future ... "[18]

8. Further reference to Weyl before 1930

One of us (Mazzoni 1991) has systematically analysed the literature in order to verify the possible influence of Weyl's work on the development of cosmology. References to Weyl's work between 1923 and 1930 are scanty. The list of the authors who do quote it does not extend beyond Georges Lemaître, Richard Tolman and Howard Robertson. Weyl's 1923 paper is quoted in the bibliography of Lemaître's 1927 paper with no explanation as to the reason for its appearing there. Since Lemaître proposed a calculation of the redshift, he perhaps quoted Weyl just to indicate that he was aware of other proposals that had been put forward. Tolman suggested comparison with Weyl 1923b in two footnotes in a paper published in 1929. Incidentally, this paper is the only one that Tolman devoted to DSU. In one of the notes, Tolman wrote that he "had not been able to follow the derivation of Weyl's formula" (for the redshift), and stressed its lack of generality (Tolman 1929: 261). We will discuss Weyl's reply to Tolman in the next Section. We were able to find a total of seven references to Weyl's work in Robertson; however, prior to 1930, the only reference concerns the calculation of z in DSU (Robertson 1928).

It would perhaps not be worthwhile to go through this survey were it not for the fact that it allows the conclusion that, except for Silberstein, no notice was taken of a potentially new principle in cosmology before the publication in 1930 of Weyl's second specific contribution on the subject. We stress the absence, in the period examined, of any pertinent reference to the fifth edition of *Raum-Zeit-Materie*.[19] Given the obscurities in the presentation of 1923b, and the relative perspicuousness of that given in 1923a, we venture to say that the apparently scarce acquaintance with Weyl's book on the part of the small community of cosmologists might be responsible for the delayed answer to the novelty that Weyl had introduced.

9. Further progress: the 1930 paper

Weyl returned to the subject seven years later. Meanwhile, cosmology had progressed at a terrific pace. The Great Astronomy Debate had seen its conclusion

---

[18] The Weyl-Silberstein controversy is discussed in some detail in Kerszberg 1986: 72; 1989b: 324–325.

[19] We refer to Mazzoni 1991 for an analysis of the possible reasons for this neglect.

in 1924 with the assessment of the extragalactic nature of the spiral nebulae; Edwin Hubble had published the results of his and Milton Humason's investigations over the redshift-distance correlation, Lemaître's (not to mention Aleksandr Friedmann's) work had been finally rediscovered, to mention but the most spectacular events. Hubble's paper is in fact the first to be quoted by Weyl in his paper. The systematic character of the redshifts of the galaxies, and Hubble's results showing that they increased with distance indicated, Weyl argued, that questions about the structure of the world as a whole had acquired a wider empirical interest. It could then well be that one had to interpret the data "in a more physical manner," but "the cosmologic-geometric conception must on any account be examined seriously as a possibility." He was thus led to re-examine DSU and his previous speculations on the subject.

The paper was actually conceived as an answer to observations made by various authors, among them Robertson (1928) and Tolman (1929). "In particular," Weyl wrote, "Tolman's careful investigation shows anew that nothing like a systematic redshift can be derived solely on the basis of the constitution of the metric field in its undisturbed state, i.e., from the De Sitter solution of the gravitational equations."[20] "In a complete cosmology," Weyl went on, "supplementary assumptions must be added:—(1) assumptions of topological character, which determine whether the entire De Sitter sphere or which part of it corresponds to the real world, and (2), closely connected with these, *an assumption about the 'undisturbed' motion of the stars* by which ∞^3 geodesic world-lines are set off from the manifold of all these lines" (Weyl 1930: 937–938; emphasis added).

"As regards the first point," explained Weyl, "the static coordinates ... with which Mr. Tolman works by preference, represent only a cuneiform sector of the entire sphere." As far as the second, crucial, point was concerned, Weyl recalled once more his own choice of a family of geodesics, corresponding to those shown in Figure 2 in Section 1 above, and, in connection with it, his definition of the "range of influence" of a star; he also reminded the reader that there are ∞^3 geodesics with the same range of influence as the chosen star A, and that they form a "system that has been causally interconnected since eternity." The stars of such a system may be described as stars of common origin (a common origin that lies in an infinitely distant past). And finally, in the right place, and with due emphasis, he added:

> *Our assumption is that in the undisturbed state the stars form such a system of common origin.* (Weyl 1930: 939)

That is, just the sentence that one would have liked to find inserted at its proper place in the 1923 paper.

Weyl went on to discuss the topological aspects alluded to at the beginning of the paper, together with some more specific topics. The point we wish to stress here is that, finally, a clear self-consistent introduction of Weyl's basic assumption had been achieved; one that could hardly be ignored by people active in the field, due to its being conceived as an answer to issues that had come to the foreground.

---

20 As we mentioned in Section 8, Tolman's paper dealt with redshifts in De Sitter's universe.

10. From Weyl's principle to the cosmological principle

The cosmological principle states that the space-time continuum is a foliation of space-like homogeneous hypersurfaces parametrized by a time coordinate x^0, whose tangent vector is orthogonal to the hypersurfaces. When a substratum is identified as a family of world lines orthogonal to the hypersurfaces, the time coordinate becomes a physical cosmic time measured by the preferred observers of the substratum. By identifying a substratum in DSU, Weyl endowed it with a cosmic time. In this sense, Weyl's principle may be said to anticipate the cosmological principle in the specific case of DSU. It is interesting to ask whether the authors who, in one way or another, introduced the cosmological principle in standard cosmology acknowledged Weyl's priority and/or were to some extent influenced by him.

The survey recalled in Section 8 has been extended to the authors who contributed to the introduction of the cosmological principle. A relevant quotation is in Edward A. Milne 1933. One should recall that Milne had set up a cosmology of his own, his "kinematic relativity," in which what he called Einstein's postulate ("All points in the universe are equivalent") was, in his opinion, "completely satisfied"; he further stated that Weyl had developed "an approach towards a similar result." Milne thus showed he had understood that Weyl's assumption amounted to selecting a substratum in DSU. An even more relevant quotation is to be found in Robertson 1933, where the author introduced the line element in the form

$$ds^2 = c^2 dt^2 + g_{ij}\, dx^i\, dx^j, \quad i, j = 1, 2, 3,$$

with $g_{ij} = g_{ij}(t, x^i)$, where t is the cosmic time and the matter of the substratum has the geodesics x^i = const as world lines. He then stated explicitly: "The possibility of thus introducing in a natural and significant way this *cosmic time* t we consider as guaranteed by Weyl's postulate, which is in turn a permissible extrapolation from the astronomical observations." (Robertson 1933: 65). In his 1929 paper, Robertson had given no justification for his introduction of a cosmic time. As we have just seen, he did offer some in 1933, guided by Weyl's principle. Therefore the continuity between Weyl's and the cosmological principle seems fairly well established.[21]

11. The evolution of Weyl's reflections and the impact of Weyl's speculations on later developments in secondary sources

Merleau-Ponty devotes a good deal of attention to Weyl, who is mentioned throughout the first part of his book (Merleau-Ponty 1965). Weyl is quoted for understanding that in general a Riemannian metric implies a spectral shift, which turns out to be a redshift in the case of DSU, and for his definition of a substratum and of cosmic time. It is interesting to note that Merleau-Ponty stresses the a priori choice

---

[21] We have reported here only two significant episodes of a wider process. A detailed analysis has been carried out in Mazzoni 1991.

of a causal connection as the central element characterizing Weyl's approach to cosmology. As far as Weyl's 1923 paper is concerned, Merleau-Ponty expresses the opinion that it must have been difficult to read for all except very skilled geometricians: "In order to understand it correctly," he states, "even Tolman and Robertson needed some explanation which the author gave them both orally and in written form."[22] There is very little reference to Weyl's work in Merleau-Ponty's historical reconstruction following the discussion of DSU. However, it is stressed that Robertson founded his hypothesis on the existence of a cosmic time on a generalization of the postulate introduced by Weyl in his discussion of DSU.

North recalls Weyl's supposition that "the stars (or nebulae, in a later context) lie on a pencil of geodesics diverging from a common event in the past," a supposition which "has often been called 'Weyl's Principle'" and that "played an important part in later cosmology" (North 1965: 100). Concerning the meaning of the principle, North emphasizes the fact that it selects stars in an undisturbed state. He attributes a certain relevance to Silberstein's contributions, and stresses that he "did not consider Weyl's principle as such" (North 1965: 102). Due relevance is given to the circumstance that Weyl's principle "is a restriction on most accounts of relativistic cosmology which is accepted at the outset" (North 1965: 137). North does not insist specifically on the evolution of Weyl's approach nor on the reception of his ideas.

As previously recalled, Kerszberg's book and essay dwell extensively on Weyl's involvement in cosmology. However, both are mainly concerned with the period 1917–1923 and contain scanty references to the subsequent period, which, as we have tried to show, is important both for a reconstruction of the evolution of Weyl's reflections and the impact they had on the development of cosmology. Kerszberg outlines the background on which our analysis has been carried out, but leaves somewhat open the question as to where the actual formulation of Weyl's principle should be located. In the essay, he writes that what we agree to call Weyl's principle "expressed Weyl's thought in 1923, although Weyl did not state it then in the form of a principle proper" (Kerszberg 1986: 5). Elsewhere he indicates Weyl's paper of 1930 as the one "in which he proposed something wholly distinct and novel, a *principle* capable of relating all the parts of the universe to one another" (Kerszberg 1989b: 4), although it "reflects Weyl's thought from as early as 1923, long before others saw its full significance as a foundation principle." (Kerszberg 89b: 17). As far as subsequent events are concerned, Kerszberg correctly states that "only the non-statical cosmology was capable of adopting Weyl's intuition so swiftly as a presiding principle; and cosmology *became* evolutionary when the shift was performed" (Kerszberg 1989b: 17). However, he does not analyse in detail the way people such as Robertson or Milne quoted Weyl. As far as Weyl's motivations are concerned, Kerszberg suggests an interplay between various philosophical motivations. In particular, he considers the demand for a cosmic time as the main

[22] All references are to Merleau-Ponty 1965: 61–62. According to the same author, Weyl's 1930 paper was in fact the written answer to Tolman and Robertson; as to the oral answer, Merleau-Ponty alludes probably to what Weyl, in that same paper, referred to as "a conference with Professor Robertson at Princeton" (Weyl 1930: 937).

factor in determining Weyl's approach to cosmology, while we would rather stress, in agreement with Merleau-Ponty, the search for global causality.

Recent references to Weyl are to be found in papers by Ellis. In the framework of a survey on the non-static coordinate systems in which DSU can be described (Ellis 1989), Ellis attributes a significant role to Weyl's paper of 1930, and favors the causal motivation for Weyl's choices; he also points out that "its statement in causal terms was not well understood by [his] contemporaries." (p. 375). Ellis indicates a link between Weyl and Robertson: Robertson considered the existence of a flow of matter converging toward the past as the cause for the existence of a cosmic time.[23] No detailed analysis of the evolution of Weyl's views and of the reasons for their reception is undertaken.

Acknowledgements

One of us (S. B.) is indebted to M. Pauri for first pointing out the importance of Weyl's principle in the history of cosmology. We wish to thank Barbara Frentzel-Beyme and G. Gottardi for linguistic advice.

References

Bergia, Silvio. (1990). "Il Principio di Weyl, passo essenziale verso la precisazione della nozione di modello di universo." Paper presented at the conference "La Cosmologia nella cultura del Novecento—Conferenza internazionale nel centenario della nascita di Hubble," Turin, 21–23 November 1990. Published in *Giornale di Astronomia* **1/2** (1991): 48–53.

Bernstein, Jeremy & Feinberg, Gerald, eds. (1986). *Cosmological Constants. Papers in Modern Cosmology.* New York: Columbia University Press.

Bertotti, Bruno, Balbinot, Roberto, Bergia, Silvio & Messina, Antonio, eds. (1990). *Modern Cosmology in Retrospect.* Cambridge & New York: Cambridge University Press.

De Sitter, Willem. (1917). "On Einstein's Theory of Gravitation and its Astronomical Consequences III." *Monthly Notices of the Royal Astronomical Society* **78**: 3–28 [Reprinted in Bernstein & Feinberg 1986: 27–48].

Ehlers, Jürgen. (1990). [Discussion]. In Bertotti et al. 1990: 29–30.

Ellis, George F. R. (1989). "The Expanding Universe: A History of Cosmology from 1917 to 1960." In Howard & Stachel 1989: 367–401.

——— (1990). "Innovation, Resistance and Change: the Transition to the Expanding Universe." In Bertotti *et al.* 1990: 97–113.

Ellis, George F. R. & Williams, Ruth M. (1988). *Flat and Curved Space-Times.* Oxford: Clarendon Press.

Goenner, Hubert. (1991). [Review of Kerszberg 1989b]. *General Relativity and Gravitation* **23**: 615–616.

Guth, Alan. (1981). "Inflationary Universe: A Possible Solution to the Horizon and Flatness Problems." *Physical Review D* **23**: 347–356 [Reprinted: Bernstein & Feinberg 1986: 299–320 and in *Quantum Cosmology.* L. Fang & R. Ruffini, eds. 139–148. Singapore: World Scientific (1987)].

---

[23] Ellis stresses that actually "this only follows because the flow is vorticity-free, which happens to be true for the family of geodesics picked out in De Sitter space-time by Weyl's causal postulate, but does not follow in general from Robertson's statement." (Ellis 1989: 375).

HOWARD, Don & STACHEL, John, eds. (1989). *Einstein and the History of General Relativity* (Einstein Studies, vol. 1). Boston: Birkhäuser.

KERSZBERG, Pierre. (1986). "Le Principe de Weyl et l'invention d'une cosmologie non-statique." *Archive for History of Exact Sciences* **35**: 1–89.

——— (1989a). "The Einstein-De Sitter Controversy of 1916–1917 and the Rise of Relativistic Cosmology." In HOWARD & STACHEL 1989: 325–366.

——— (1989b). *The Invented Universe. The Einstein-De Sitter Controversy (1916–17) and the Rise of Relativistic Cosmology.* Oxford: Clarendon Press.

LEMAÎTRE, Georges. (1927). "Un univers homogène de masse constante et de rayon croissant, rendant compte de la vitesse radiale des nebuleuses extragalactiques." *Annales de la Societé scientifique de Bruxelles A1* **47**: 49–59. [English translation: "A Homogeneous Universe of Constant Mass and Increasing Radius Accounting for the Radial Velocity of Extragalactic Nebulae." *Monthly Notices of the Royal Astronomical Society* **91**: 483–490. Reprinted: BERNSTEIN & FEINBERG 1986: 92–101].

MAZZONI, Lucia. (1991). "Dal postulato di Weyl al principio cosmologico." Tesi di Laurea in Fisica, University of Bologna.

MERLEAU-PONTY, Jacques. (1965). *Cosmologie du XXe siècle. Étude épistémologique et historique des théories de la cosmologie contemporaine.* Paris: Gallimard.

MILNE, Edward A. (1933). "World-Structure and the Expansion of the Universe." *Zeitschrift für Astrophysik* **6**: 1–94.

NORTH, John D. (1965). *The Measure of the Universe. A History of Modern Cosmology.* Oxford: Clarendon Press.

RINDLER, Wolfgang. (1977). *Essential Relativity: Special, General, and Cosmological.* 2nd edition. New York: Springer.

ROBERTSON, Howard P. (1928). "On Relativistic Cosmology." *Philosophical Magazine* **5**: 835–848.

——— (1933). "Relativistic Cosmology." *Reviews of Modern Physics* **5**: 62–90.

SILBERSTEIN, Ludwik. (1924). "Determination of the Curvature Invariant of Space-Time." *Philosophical Magazine* **47**: 907–917.

TOLMAN, Richard C. (1929). "On the Astronomical Implications of the De Sitter Line Element for the Universe." *Astrophysical Journal* **69**: 245–274.

WEINBERG, Steven. (1972). *Gravitation and Cosmology: Principles and Applications of the General Theory of Relativity.* New York: Wiley.

WEYL, Hermann.

——— (1921). *Raum-Zeit-Materie.* 4th ed. Berlin: Springer. [References are to the English translation: *Space-Time-Matter.* H. L. Brose, trans. London: Methuen (1922). Reprinted: New York: Dover (1952)].

——— (1923a). *Raum-Zeit-Materie.* 5th ed. Berlin: Springer.

——— (1923b). "Zur allgemeinen Relativitätstheorie." *Physikalische Zeitschrift* **25**: 230–232.

——— (1924a). "Massenträgheit und Kosmos. Ein Dialog." *Die Naturwissenschaften* **12**: 197–204.

——— (1924b). "Observations on the Note of Dr. L. Silberstein: Determination of the Curvature Invariant of Space-Time." *Philosophical Magazine* **47**: 348–349.

——— (1930). "Redshift and Relativistic Cosmology." *Philosophical Magazine* **9**: 936–943.

Milne, Bondi and the 'Second Way' to Cosmology

George Gale and John Urani

> *But then, we all three of us, were pretty well convinced it's right, Tommy and I perhaps slightly more on philosophical grounds, Fred more on the—sort of very general grounds.*
>
> Hermann Bondi (1975)

Hermann Bondi's steady-state cosmology exhibited a remarkable philosophical purity.[1] This is not an accident: Bondi explicitly developed the theory along the lines of a deeply-held, clearly-expressed philosophical position. Much is already known about Bondi's philosophical views; after all, he has never been the least bit reticent about expressing them. But what is not already, or widely, known about Bondi's philosophical views is where they came from. There are some surprises here. As we will show, Bondi's views emerged from the context of a twenty-year-long debate among British and American scientists and philosophers about method in the newly founded science of physical cosmology. Although the very existence of this debate will surprise some historians and philosophers of science, even more surprising will be the name of the participant who most influenced Bondi. This was none other than Edward A. Milne, who, today, is most often remembered—if he is remembered at all—as the founder of a strange and ill-fated cosmological theory called *kinematic relativity* (KR). As we show, Bondi's philosophy in the steady-state theory owes much to Milne's theory.[2]

[1] As the introductory quote from Bondi itself clearly asserts, Bondi developed his version of steady-state cosmology in conjunction with Thomas Gold. In what follows, however, we will refer to this version of the theory as though Bondi developed it alone by himself. This would not seem to be entirely unfair, since, in the first place, Bondi is most likely most responsible for the main development of the theory. Moreover, and most germane to our project here, the philosophical and methodological bases of the theory are most certainly due to Bondi.

[2] Bondi also owes a clear debt to Popper. Although this debt is not our focus here, it will come up for some mention.

H. Goenner, J. Renn, J. Ritter, T. Sauer (eds.)
The Expanding Worlds of General Relativity
(Einstein Studies, volume 7) pp. 343–375

As Milne eventually developed it, KR included not only a proposal for a scientific cosmological theory, a theory rival to that based upon general relativity, but also a proposal for philosophical and methodological programs sharply in contrast to those prevailing at the time. In our earlier work, we have focused for the most part upon the scientific and philosophical aspects of KR and its reception in the typically general relativity-based cosmological community of the thirties.[3] We now turn our attention to the methodological aspects of KR and their reception by that same community. As we shall demonstrate, there was vigorous initial opposition to Milne's methodological proposals. Cosmology, according to Milne, should be done in an axiomatic fashion, beginning from a secure foundation in the most general notions we could obtain and affirm as true on the basis of an examination of our intuitive reason. Tolman called this method of doing cosmology "the second way", a term often used by Milne himself during later conflicts with Dingle, McVittie, and others. Following a discussion of exactly what Milne's methodological proposal entailed, and its consequences for cosmological practice, we will examine the details of the debate over Milne's Methods, focusing especially upon the events surrounding the 12 June 1937 special supplement to *Nature*, published solely in order to air the controversy publicly. Following this discussion, we begin our final project, investigating the close intellectual relations between Milne, his Methods and the philosophy adopted by Bondi in his steady-state cosmologizing. Analyses of published works, personal correspondence, and interviews with participants will provide background and evidence for our examination of these crucial events in the origin of modern cosmology.

1. The 'Second Way' to do cosmology: a brief overview

British and American physical scientists and applied mathematicians[4] in the twenties and thirties for the most part subscribed to a rough-and-ready *empirico-inductivist* methodology. This methodology was the methodology of Galileo and Newton, or so most of them believed; it was thereby sanctioned from the very beginnings of modern science. Moreover, according to a leading spokesman for this view, through the centuries the Royal Society itself had functioned as the protector—indeed, guarantor—of this way of doing science. (Dingle 1937: 786) Put most simply, the view was that genuine science began in observation, most especially observation in experimental contexts, and proceeded via inductive generalizations of the observations. Theories were the ultimate products of this process. When physical cosmology became acceptable in the very early 30s, this view of methodology was imported into it in a straightforward, uncritical fashion, from the other physical sciences. (Gale 1991) Yet, not all Anglo-American physical scientists limited their views to this single 'way' of doing science.

---

[3] We have provided the historical background in a series of publications over the past several years: Gale 1991, 1992, Gale & Urani 1994, Urani & Gale 1994. Another paper touching briefly, albeit differently, on some of the issues herein is Gale & Shanks 1996.

[4] Much of what goes on in an American physics department, goes on in a British applied mathematics department.

Caltech's R. C. Tolman, a theorist who frequently collaborated with Hubble, was particularly sensitive to methodological possibilities other than the dominant empirico-inductivist one. In a 1932 address to UCLA's Philosophical Union he described the possibilities at some length:

> Principles [of physical theory] are actually suggested to the theoretical physicist in two ways—not only as immediate generalizations of experimental findings, but also as desiderata for the inner harmony and simplicity of the theoretical structure he is attempting to build. We must admire Galileo for insisting on observational fact as the ultimate arbiter. ... But we must not let this just admiration blind us to the power and skill of those other theoretical physicists who obtain the suggestion for physical principles from the inner workings of the mind and then present their conclusions to the arbitrament of experimental test. (Tolman 1932: 373)

In this passage, Tolman obviously takes the 'Galilean' empirico-inductivist way of discovering physical principles ("generalizations") as the *first way*. But, he notes, there are "*actually*" two ways. In the *second way*, principles are desired insofar as they aid in the "inner harmony" and "simplicity" of the theory under construction. These principles, however, are not suggested by observations based in experiment, but rather are suggested somehow by "the inner workings of the mind." Following the initial, or *suggestion*, stage, the principles are then presented in a second stage "to the arbitrament of experimental test." We shall call this two-stage 'Second Way', the *cognitive-hypothetico-deductivist* (CHD) way.[5]

With one significant exception, Milne's description of the second way to do cosmology didn't differ appreciably from Tolman's. In an address—his first major address to the cosmological community after a long period of silence—given at the dedication of the McDonald Observatory in Texas, Milne described the Second Way like this:

> There are, fundamentally, two different possible starting points for an investigation of this kind—two possible modes of attack. ... The second main approach to a solution of the cosmological problem ... is to study the universe as it actually presents itself to us, without appealing to any small-scale experiments or to knowledge derived from small-scale experiments ... , without supposing ourselves to know anything, to start with, of the facts used in the first method. In the second method we attempt a complete reconstruction of physics from the bottom upward, on an axiomatic basis. (Milne 1940: 132–133)

According to Milne's version of CHD, the suggestion of theoretical principles in the Second Way is to occur as if in ignorance of facts or knowledge gained by the First Way. Although Tolman had not called for anything so restrictive, Milne's restriction is not inconsistent with Tolman's version of CHD. But the major difference between Tolman's CHD and Milne's is the latter's insistence

[5] It must be specifically remarked how this method differs from the more general *hypothetico-deductivist* (HD) method. From Tolman on, through Milne, Eddington, Dirac, and ultimately to Bondi, the CHD method, unlike HD—which is indifferent to the source of the suggestion—relies strictly upon just one specific source for the suggestion of hypotheses: the "inner workings of the mind", to use Tolman's own words. How this reliance plays out in particular cases will be clear as we consider each case in turn.

upon the axiomatization of cosmological theory's physics. This addition to CHD will have no small effect upon cosmological practice, as we shall soon see.

CHD, as originally described by Tolman, and promulgated, first, by Milne, and soon thereafter by Eddington and Dirac, was not allowed by proponents of the First Way to languish unaffected. It soon came under attack, as duly noted by Edmund Whittaker, at that point certainly the dean of applied mathematicians in England:

> A lively debate is in progress at the present moment between Sir Arthur Eddington and Dr Harold Jeffreys of Cambridge, Professor Milne of Oxford, Sir James Jeans, and Professor Dingle of the Imperial College, the subject being the respective shares of reason and observation in the discovery of the laws of nature. (Whittaker 1941: 160)

Whittaker's allusion here to "reason" is an important one: it is quite plausible to understand the 'C' (cognitive) aspects of CHD as implying a particular sort of *rationalism*, in that the original sources of hypotheses are limited to those accessible by reason alone. This is a classic interpretation of rationalism *pace* e.g., Kant.[6] It must also be noted here that Whittaker's intent is to refer to the *discovery* of laws of nature. This is precisely in line with Tolman's original proposal that the disagreement between proponents of the two ways focused upon the initial, or suggestion, stage of scientific development, and not on the second, or verification, stage.[7] A final point made by Whittaker's comments here is seen in his reference to how "lively" the debate is between proponents of the two ways. And indeed the debate was lively! At times, particularly when it was Dingle espousing the First Way, the debate got not just lively, but vituperative.[8]

So far in this general overview of the Two Ways we have seen comments from Tolman, Milne, and Whittaker. We now conclude our brief introduction with two comments from Bondi, which exhibit his understanding of both the opposition between the Two Ways, and Milne's allegiance to the Second Way. The first comment is taken from Bondi's 1948 article "Review of Cosmology:"[9]

> One can approach the problem either by extrapolating from laboratory physics to form a comprehensive mathematical theory of relativity first and consider its cosmological implications only afterwards or, alternatively, one may try to postulate a rigorous cosmological principle and attempt to derive the corresponding theory of relativity and gravitation afterwards. The first procedure was adopted by the school of general relativity, the second by Milne and his followers in kinematical relativity. (Bondi, 1948: 107)

---

[6] See Smith 1929: 666. Milne's methodology is explicitly claimed to be "rationalist" in Singh 1970. For a fuller discussion of this point see Gale 1992.

[7] That is, both sides relied upon observation at some stage in the theory-construction process.

[8] As Bondi remarks, with respect to KR: "When the theory was first developed it met with great hostility and was criticized very severely, often unjustly, and sometimes frivolously." (Bondi 1960: 123)

[9] The first issue of *Monthly Notices* each year was marked by one (or sometimes more) "Reports on the Progress of Astronomy" devoted to some particular topic, e.g., solar astronomy, planetary astronomy, etc. Bondi's article here cited is one of these reports. In a straightforward fashion this article provided a dry run for Bondi's marvelous book *Cosmology*, which appeared two years later.

In this passage it is clear that Bondi's understanding of the Second Way is precisely in line with that of Tolman and Milne. What is especially clear is the fact that the initial *formal* move in the Second Way is to *postulate* the first principle. All subsequent formal movement consists of deductions therefrom. Obviously, what Bondi alludes to here is a rough sort of axiomatic method; we shall present below Milne's (and a later-Bondi's) much more polished version.

As our final example, we present a passage from Bondi's 1950 *Cosmology*. Most important to remark in this passage is the new point made in the last sentence:

> In particular, there are two important approaches to the subject [cosmology] so different from each other that it is hardly surprising that they lead to different answers. ... The contrast between the 'extrapolating' and the 'deductive' attitudes to cosmology is very great indeed, and most of the workers in the field adopt an intermediate position combining parts of both. ... There is, however, a widespread lack of understanding and of appreciation of the differing points of view that makes the disagreements less fruitful for advancing the subject than they would otherwise be. (Bondi 1960: 3–5)

As Whittaker noted above, there was a lively debate over the status of the Two Ways. Bondi minces no words; for him, "lively debate" in this case simply means "widespread lack of understanding" and "disagreements." More importantly, because of the character of the disagreements, they do not advance cosmology as well they might. Moreover, as we shall see below, the character of the scientists is also relevant to the lack of progress manifested by the disagreements.

One final point: Bondi here asserts that "*most*" working cosmologists adopt an intermediate position combining parts of both Ways. We're not at all sure whom Bondi has in mind. Participants in the debate—which comprises most cosmologists active in Britain at the time—certainly came down on one side or another. This is also true for Bondi himself, who, as we shall see, is a clear proponent of the Second Way.[10] It's not clear that there was anyone left who could adopt the intermediate position limned by Bondi.

The passages we have seen make several things clear. First, there are two distinct methodological positions on offer during the early days of modern cosmology. The first—the First Way—is centered on an empirico-inductivist method, wherein general principles of cosmological theory are to be induced from observations made on the small-scale, in laboratories, on workbenches, etc. The Second Way is the cognitive hypothetico-deductive method, wherein general principles of cosmological theory are to be suggested (somehow) by the "inner workings of the mind", only subsequently to be tested, via their deductive consequences, against the observable world. Secondly, various cosmologists, in particular Milne, Eddington, Jeans, Dingle and Bondi explicitly and publicly adopt one or the other of these two methodologies. Finally, a "lively" debate ensues between proponents respectively of the First or Second Way.

---

[10] And just as clear an antagonist of the First Way! See, for example, Bondi 1955.

To conclude this overview, we should like to indicate what to us seem to be the three main points of historical and philosophical interest which will emerge in our discussion of this debate over methodology in cosmology.

First, even though the debate is undeniably philosophical in nature, the participants are overwhelmingly drawn from the scientists' camp, and not from the philosophers' camp. On its face, then, the philosophical issues of the debate are significant and meaningful for the scientists. Moreover, they are consequential for scientific practice and development, as we shall demonstrate by pointing later to Bondi's reliance upon Milne's methods in the construction of his steady-state theory.

Secondly, because cosmology is so young a science at this point, the debate plays a significant role in shaping the future of the science. Again, this will be demonstrated by reference to the construction of steady-state theory.

Finally, the significance of the debate is great in part because of the peculiar nature of cosmology *as a physical science*. That is, because cosmology is and must ever be theory-rich and data-poor, philosophical controversy, such as this debate over method, will always have an effect disproportionately large as compared to, say, areas of physical science with large amounts of data.[11]

This completes our overview of the issues involved in the controversy over the Second Way. We turn now to filling in some of the details, beginning with Milne's contributions to CHD.

2. Milne's version of the Second Way

Milne's training was in mathematics. But before his time at Trinity College, Cambridge, was completed, he joined the war effort as a mathematician working with A. V. Hill's Anti-Aircraft Experimental Section. This section "in due course came to include a big proportion of the most able mathematicians in the country", all working on problems in ballistics and sound ranging. (McCrea 1950: 422) Approaches that Milne learned during this period, in particular a strong commitment to algebraic rather than geometrical analysis, stayed with him the remainder of his life. At this time, Milne also closely studied Whitehead and Russell's *Principia Mathematica*, of which he could recite "chapter and verse." (Crowther 1970: 12) His familiarity with, and, indeed, devotion to the axiomatic methodology of Whitehead and Russell shaped his approach to scientific theorizing.

Milne's initial scientific work was in theoretical astrophysics, specifically, stellar atmospheres. During the 20s, in addition to his Fellowship at Trinity, Milne worked variously as an " 'astronomer', in a very broad sense of that term", serving, first, as assistant director of Cambridge's Solar Physics Observatory, and later as University Lecturer in Astrophysics. (McCrea 1950: 422) In 1928 Milne was elected Rouse Ball professor of mathematics in Oxford, the position he held for the rest of his

[11] This point is related to one made to us by Dennis Sciama. Sciama noted that cosmology, especially in its early days, was, because of the size of its participant-base, unstable: individual participants—Eddington, for example—could lead it astray singlehandedly. This was especially dangerous, he added, because of the lack of tight control exerted by a large observational database. (Sciama 1990)

life. His inaugural lecture "The Aims of Mathematical Physics" (Milne 1929) laid out his philosophical views, especially his methodological commitments, in clear, unambiguous fashion.

According to Milne, although the mathematical physicist might very well find his "problems suggested by experiment and observation",

> The greatest role of the mathematical physicist is, however, the conjecturing of laws of nature, where he rises superior to the hypotheses with which he has been presented, and to satisfy some mathematical yearnings, suggests a new hypothesis as one worth testing. (Milne 1929: 6)

All the elements of the Second Way are present in this thought. First, the mathematical physicist finds a hypothesis suggested by "some mathematical yearnings"—for example, "the new hypothesis may be directly inspired by mathematical symbolism." (Milne 1929: 6) Then, after consideration, the new hypothesis may be found to be "one worth testing." It is this second stage, the testing, which differentiates the mathematical physicist from the pure mathematician. Although both begin with "*insight*", it is the peculiar province of the mathematical physicist's insight[12] which leads to testing:

> The pure mathematician exercises his insight into the possibilities of mathematical form—discovers Riemannian geometry for instance. The mathematical physicist by the exercise of a different kind of insight recognizes this as a possibility concerning the nature of the world, and proceeds to test it . . . (Milne 1929: 17)

The aim of the theoretical astrophysicist is part and parcel of this same notion. According to Milne, the "peculiar contribution" made by the theoretical astrophysicist is a "set of models constructed on as many different plans as he can conceive, with a corresponding set of consequences." (Milne 1929: 26) Whether or not the models "reproduce nature"—have consequences which agree with observation—is of "little importance". This, because, in the end:

> A model which fails to reproduce nature is really more valuable than one which does—it at least shows what nature is not like, whilst a successful model leaves it an open question which of its characteristics is responsible for its success. (Milne 1929: 26)

Milne's commitment to falsificationism is quite explicit. In a later passage, he speaks again of the theoretical astrophysicist's aim to "infer something of the real nature of the bodies observed." But, again, even if the predictions inferred from the astrophysicist's hypotheses "predict results in agreement with observations", that is of little scientific benefit. "I have already said," Milne remarks in a reference to the passage above, "that observation can never confirm our hypotheses, it can only disprove them." (Milne 1929: 26)

---

[12] An extremely interesting issue—one which we cannot discuss at any length here—is the question what distinguishes the pure mathematician from the mathematical physicist. Put very briefly, physics involves time: "An essential difference between a theorem of kinematics and a theorem of geometry is that the former involves mention of temporal changes. . . . We can go on proving theorems in geometry ad libitum without encroaching on the traditional domain of physics; to take the step . . . it is necessary to pass from static theorems to theorems about motion." (Milne 1941: 356

As these comments make clear, Milne was already committed to a falsificationist hypothetico-deductive method by 1929.[13] Although this position was not yet well-known among British scientists, it at least was slightly familiar. Had Milne's methodological commitments been limited to this sort of HD position, the ensuing debate most likely would not have occurred. But then, to this somewhat unfamiliar position, Milne added something not just unfamiliar, but downright repugnant, to the majority, to the practicing empirico-inductivists. That is, he added his cognitive-source requirement for hypotheses.

Although, as we noted earlier, Milne was agreeable in 1929 to the notion that hypotheses might very well be suggested cognitively, e.g., by "mathematical symbolism", his understanding of the full possibilities of this mode of discovery developed only slowly during the ensuing years. The first significant realization is reported in a September 1934 letter to his student, Walker:

> I begin now to see that my theory is an approach to the ideal of deducing so-called laws of nature without any appeal to experience whatever—merely the embodiement of the compatibility or self-consistency of different observations by different observers of the same phenomena. I shall develop this idea later in some lecture or another. (Milne *Corr. W*: 12 September 1934)

A letter to Subrahmanyan Chandrasekhar written fifteen days later discusses many of the same points as the letter to Walker, and ends with the remark that "I have not developed these ideas in the book—they have only occurred to me since I finished it."[14] Two ideas mentioned in the letter are most relevant to us here. First, there is the question of the *nature* of the methodology, which Milne in the letter to Walker had described as "*deducing* the so-called laws of nature without any appeal to experience." Secondly, there is the question of the *content* of the starting cognitive suggestion. We examine the second idea first.

In the letter to Walker, Milne had described the content of his starting point as "merely the embodiement of the compatibility or self-consistency of different observations by different observers of the same phenomena." In the letter two weeks later to Chandrasekhar he asserts that he has

> the expectation that laws of nature are not idealities, or projections of the mind, but inevitable general relations following from the condition of the compatible observation of phenomena by different observers. (Milne *Corr. C*: 27 September 1934)

What Milne is getting at here is his belief, which he soon affirms with great strength (Milne 1934), that every observer in the universe must see exactly the

---

[13] It is not at all clear from whence came Milne's position. Menachem Fish has suggested that Milne's position came, via Trinity, from Whewell. This is an extremely plausible suggestion. (Fish 1996) Milne heartily approved De Morgan's "devastating analysis" of Bacon's inductivism, and the rough-and-ready HD (but not falsificationism) offered as replacement. (Milne 1935b: 125)

[14] The reference is to Milne 1935a, *Relativity, Gravitation, and World-structure*. Milne's own appraisal of the book is prescient: "I got my book off, but have not had any proofs yet. Unless the book is wholly ignored as the work of a 'crank', it is bound to arouse hostile criticism." (Milne *Corr. C*: 27 September 1934)

same universe, or, as he puts it, "get the same world-picture." That is, each observer must see himself as observing the entire rest of the universe expanding away from him, with expansion rate increasing as distance. Moreover, each observer must get exactly the same physical laws as every other. As it turns out, Milne is able to satisfy these desiderata[15] in an incredibly ingenious and economical fashion. To put it most simply, Milne devises a physics—an axiomatized conceptual system—based upon one and only one phenomenon, one and only one appeal to experience: "The only appeal to experience is the existence of a temporal sequence for the individual, necessary in order to introduce time." (Milne *Corr. C*: 27 September 1934) Milne asserts that an observer, conscious of his internal time sense, may apply the simple predicates 'before', 'after', and 'at the sime time' to events that he observes. Using this qualitative scale, the observer can construct some sort of a primitive 'clock', and thereby mark his observations. Then, he need only have some sort of signalling device, and a fellow observer similarly equipped—and willing and able to communicate with him, of course. Thus, holds Milne, given only an observer's primitive sense of before-and-after, and a means of signalling some other similar observer, between the two of them, physics can be constructed. Add a third observer, and cosmology can be constructed (Milne 1940: 136).

Milne's development of his axiomatized cosmology—KR—based upon nothing other than an appeal to an observer's internal sense of before-and-after must rank as one of the more interesting developments in mathematical physics during the first part of the century.[16] However, since we have discussed the nature and role of Milne's development of KR in another place, we need not do so here. (Urani & Gale 1994) Rather, we here focus our gaze instead upon the methodological point raised above in the letter to Walker, what Milne (and others) call the "*deductive*" character of his methodology. The letter to Chandrasekhar provides a useful starting point.

Just as in the letter to Walker, Milne's letter to Chandrasekhar describes his methodology as a 'deductive' one, and then goes on to raise an associated issue:

> The tendency of all scientific theory is to reduce the axiomatic basis, to deduce more and more phenomena from fewer and fewer statements of general principles. When will this process stop? Can we reduce the axiomatic basis to zero? My work strongly suggests that we can ... (Milne *Corr. C*: 27 September 1934)

To begin with, the method Milne has in mind is fairly straightforward and can be summarized simply. One or more general principles are conjectured or hypothesized to serve as axioms; these form the starting point for the deductive process, which process culminates in theorems. Frequently, the axiom set consists

[15] Which desiderata Milne embodied in a principle most came to call 'The Cosmological Principle'. The name, so far as we can tell, is Robertson's. This is the principle that Bondi later generalized as the content of the starting point of his steady-state cosmology.

[16] Bondi argues that Milne's KR definition of distance—the primitive dimension in cosmological theory—"is probably the best yet devised." (Bondi 1960: 128) Bondi's comparison focuses especially upon the general relativistic definition.

mainly of definitions.[17] Typically, Milne uses as his model of this axiomatic-deductive process arithmetic and geometry à la Russell and Whitehead. (Milne 1941: 358) But these subjects provide only *mathematical* examples of his method; ultimately, Milne is concerned to construct an axiomatic physics and cosmology. How would *this* project proceed—how would it differ from the construction of purely mathematical axiomatic systems? Moreover, in the passage, Milne raises an associated issue: he seems to proffer hope that he might reduce the axiomatic basis of cosmology to zero. How would this process of reduction be carried out? Over the next few years, as Milne's commitment to his 'deductive' methodology got stronger, he came to believe the reduction *could* continue to zero, at least in one sense.

Milne's descriptions of the process typically relied upon a putative history of those disciplines based upon mathematics. Although these disciplines—for example number theory, geometry, or even physics and cosmology—might originate in empirical contexts, their whole course of evolutionary development was away from their empirical origins. For example, regardless of its empirical origins, some theorem in number theory

> cannot be held to be connected with the discipline which is the theory of numbers so long as the evidence for it is wholly or partly empirical. The mathematician must, at some stage, perform an act of renunciation and proceed without reference to empirical facts. (Milne 1940: 133)

In the end, the discipline in question must stand by itself alone, as a purely cognitive object, unconnected to the empirical world. Indeed, even in the physical sciences

> The more advanced a branch of science, the more it relies on inference and the fewer the independent appeals to experience it contains. As a science progresses, the most diverse phenomenal laws are seen to be deducible from a few general principles, and the number of these, in turn, tends to become smaller and smaller. We deduce the more and more from the less and less, or at least from the simpler and simpler. (Milne 1943: 11)

And here the question arises again "as to whether this process of inferring can come to a stop, and if so, where?" That is to say,

> Is there an irreducible number of brute facts derived from observation? Or can the complete science be fully constructed by processes of logical inference, without specific appeals to laws based on experience? And, if the latter, what is the real basis of argument? (Milne 1943: 11)

As he later answers, Milne believes the answer is affirmative: "The answer seems to me to be that we can reduce the appeals to *quantitative* experience to zero." (Milne 1943: 11) By quantitative, Milne here at bottom means *perceptual*, that is, gotten from observation. What one starts from, according to Milne, are purely *qualitative*—prior to observation—"propositions which assert how I, as an observer, can become aware both of an external world and of other observers in

[17] In particular, operational definitions: "We require operational definitions, not definitions by enumeration [ostension]." (Milne 1941: 358)

it." (Milne 1943: 12) These propositions take as their subject matter the most fundamental cognitive objects—primitive intuitions:

> That is, in short, to use only such brute facts, such irreducible facts, as are of the intuitive sort or do not rest on the questionable principle of induction, and thus to appeal to no empirical 'laws of nature' of a quantitative kind. (Milne 1940: 132)

These "brute facts" are those relating ultimately to the observer's experiences of before-and-after. In a fashion strongly reminiscent of Kant's construction of the a priori spatio-temporal framework from the internal sense of time, Milne, after rejecting Descartes' *cogito* (again much like Kant), bases everything on the intuition of time:

> First things first. Before I look out on the world, I am conscious of myself, as thinker and as capable of being a percipient. But "cogito, ergo sum" is mere metaphysics, and takes us nowhere in Natural Philosophy[18] What is important is that I possess a consciousness, and that my sensations—the events in my consciousness—arrange themselves in a single linear order, which we call a time-order. This is an appeal to a contingent proposition, for sensations might have been capable of a multi-dimensional arrangement. But apart from this initial appeal, we make at present no appeal to our empirical knowledge of the laws obeyed in nature. (Milne 1943: 18)

This, then, is Milne's 'bottom line': he holds that it is possible to construct cosmology on the basis of hypotheses deduced from the brute fact of a percipient's internal experience of the before-and-after of his perceptions. Nothing else is needed.

Did Milne's Methods succeed? His critics, of course, said no.[19] His supporters, of whom there were several, said yes.[20] Two fellow-travellers with Milne were quite famous in their own right; I speak, obviously, of Eddington and Dirac, to whom we now turn.

3. **Eddington and Dirac: Milne's 'fellow travellers'**

Milne called his method "deductive", which, as we have seen, is a synechdoche: the whole scheme involves not just deduction, but an earlier hypothesis based, in Milne's case, upon the fundamental intuition of time. Although both Dirac and Eddington also use "deductive" and related terms to describe their methods, there are differences between their deductive methodologies and Milne's. Eddington's methodology is the most developed in this regard.

Eddington first introduced his method—later called "pure," or "fundamental theory"—during the cosmology session of the 1931 British Association meeting

---

[18] The term is here used in its historical, technical sense; after all, Milne's paper is directed to faculty of Scottish Universities, as he himself notes! (Milne 1943: 10)

[19] McVittie's attack (McVittie 1940) is specifically focused upon the places where Milne has appealed "unconsciously no doubt, to an empirical fact." (McVittie 1940: 275)

[20] Bondi, as we shall shortly see, thought highly of Milne's Methods, calling them, in various places, "beautiful", "lucid", "remarkable", etc. (Bondi 1960: 126–133)

(Eddington 1931). In this meeting he proposed 'deductive' reasons for the expansion of the universe, and for its velocity. Needless to say, the audience was astounded (Dingle 1931). A year later, in lectures published as *The Expanding Universe*, Eddington was more definite in what his pure theory involved: "According to the argument developed here, we can calculate by pure theory what ought to be the speed of recession of the spiral nebulae. ... No astronomical observations of any kind are used in this calculation." (Eddington 1933: 94) The parallel between Eddington's notion here of the starting point for pure theory, and Milne's rejection of empirical observations is obvious. Unfortunately, Eddington himself was never entirely clear about what the starting point for his deductions would be, beyond the claim that it wasn't astronomical observations. Luckily enough, Whittaker was quite explicit in his description of Eddington's starting point:

> All the quantitative propositions of physics, that is, the exact values of the pure numbers that are constants of science, may be deduced by logical reasoning from qualitative assertions, without making any use of quantitative data derived from observation. (Whittaker 1951: 3)

A *qualitative* assertion is one which does not mention any particular number.[21] Examples would include "The velocity of light is independent of the motion of its source"; "A material body which occupies a certain space can also occupy a different space without being changed in its properties"; and "Inside a hollow electrified conducting shell there is no electric field." (Whittaker 1951: 2)[22] In his rejection of quantitative statements, Eddington resembled Milne; however, as these examples of qualitative statements show, Eddington was willing to accept—at least in the beginning—much more empirical-knowledge-laden propositions than Milne found acceptable. In the end, however, Eddington hoped just as strongly as Milne to reduce the deductive starting point to nil. His efforts revolved around something Whittaker mentions, above: the constants of nature.

In *The Expanding Universe*, Eddington had argued strongly against removing λ, the "cosmic constant", from Relativistic Cosmology. This argument apparently interested him in the role of the other constants of nature. Within a very short time, the constants appeared in the very foundations of fundamental theory. In a passage sounding marvelously Pythagorean, Eddington lays out his view of the role to be played by the constants in his deductions:

> The idea is that we ought to be able to calculate out of these every other constant displayed in natural phenomena. ... We may thus look on the universe as a symphony played on seven primitive constants[23] as music is played on the seven notes of the scale and when he has supplied us with

[21] Although Milne and Eddington both use the term "qualitative propositions", their meanings are not exactly the same. Qualitative propositions, for Milne, refer to primitive intuitions; for Eddington, this class of propositions merely mentions no number.

[22] Whittaker himself had expressed some interest in the use of a restricted class of such statements—which he called principles of impotence—in codifying a field (Whittaker 1941).

[23] The seven constants are: e, the charge of an electron; m, the mass of an electron; M, the mass of a proton; h, Planck's constant; c, the velocity of light; G, the constant of gravitation; and λ, the cosmical constant. (Eddington 1935: 230)

> these numbers, the experimental physicist might retire and leave all the rest of physics to the mathematician. (Eddington 1935: 231)

Deduction—carried out by the mathematician—proceeds from the constants given by the physicists; the end result is a calculation of *every* other constant in physics, and, thereby, of all of physics. But Eddington didn't rest content here. He soon came to believe that there was a starting point even more primitive than the constants. That point was none other than the nature of the human knowing system itself:

> But in the more far-reaching investigations to which I am now referring, the constants as well as the algebraic forms are included. My conclusion is that not only the laws of nature but the constants of nature can be deduced from epistemological considerations, so that we can have a priori knowledge of them. (Eddington 1939)

Since we do not have the time or space here to examine Eddington's arguments, we will simply note that by "epistemological considerations" Eddington is referring to the fact that quantum theory—the basic theory of matter—is a probabilistic theory. Moreover, since Eddington gives a straight-forward subjectivist interpretation to probability theory, it follows for him that quantum theory is shaped and formed by the operations of the human subjectivity, that is, the knowing mind.[24]

No matter how we judge its success or failure, there simply can be no denying the similarities between Milne and Eddington vis a vis their commitment to the method they both call "deduction". What differences there are between the two men lie mostly in the area of their choice of starting points for the 'deductive' method. Milne favored the intuition of time, Eddington, the subjective features of probabilistic reasoning.

Dirac was the third member of the group of 'deductivists.' But Dirac's allegiance to the method was much more restricted than either Milne's or Eddington's. Dirac allows that the arguments used by Eddington to "purely deductively" calculate the constants of nature, "while they give one the feeling that they are probably substantially correct in the case of the smaller numbers [e.g., hc/e^2], the larger numbers [e.g., N] are so enormous as to make one think that some entirely different type of explanation is needed for them." (Dirac 1937: 323). Dirac's "different type of explanation", one which has been so influential in contemporary anthropic principle cosmologizing, relies upon the larger constants being functions of the epoch of the universe. This implies that, at different epoches, the 'constants' have different values. Interestingly enough, this particular feature of Dirac's position did not arouse much criticism.[25] What triggered a reaction was his throwing in with Eddington, and thereby Milne, on the 'deductive' method. The attack on all three came soon after publication of Dirac's 1937 paper cited just above.

---

[24] Broad's essay review of Eddington's work presents what is most likely the most sympathetic treatment of this analysis of "epistemological considerations" (Broad 1940). It is brutally critical, nonetheless.

[25] Indeed, as Barrow and Tipler demonstrate, a small-scale 'school' of followers developed, whose main focus was examining the implications of varying constants for physical theory (Barrow & Tipler 1986: 235).

4. Stage Two: Dingle's inductivo-empiricist counterattack

In March 1937, Herbert Dingle launched a vigorous frontal assault upon the deductive methodology of the Second Way. (Dingle 1937) In a *Nature* article entitled "Modern Aristotelianism", Dingle took to task Dirac, Eddington, and, especially, Milne, for their "treason" to the Galilean heritage of empirico-inductivism as preserved by the Royal Society.[26] (Dingle, 1937: 786) According to Dingle, their 'Aristotelianism'[27] consisted in the belief that "Nature is the visible working-out of general principles known to the human mind apart from sense perception." The issue, Dingle believed, concerned what it was methodologically "proper to do: Should we deduce particular conclusions from a priori general principles [Second Way] or derive general principles from observations [First Way]?" Given the choice between the Second Way or the First Way, Dingle believed that the First Way was the only way. He stated his own preferred version of the First Way in clear and succinct fashion. The "experimental philosophy", Dingle asserted, required that "the first step in the study of Nature should be sense observation"; moreover, with respect to subsequent steps in the method, "no general principles being admitted which are not derived by induction therefrom." Both elements of this classic empiricist epistemology plus inductivist logic were violated by the proponents of the Second Way's deductive methodology. Yet not all elements of the Second Way rated equal condemnation by Dingle. Perhaps because there is no gainsaying the efficacy of deductive logic, Dingle focused especially upon the epistemological aspects of the CHD method. In regard to the starting points of the two methods, as Dingle saw it, "the question presented to us now is whether the foundation of science shall be observation or invention." Invention, as Dingle made clear, is simply the free creation of hypotheses, independent of prior observation. Thus, whether the issue concerned Eddington's "epistemological considerations" or Milne's internal sense of time made no difference. Either was equally reprehensible, simply because it was not a sensory observation.

One other aspect of "invention" should be mentioned briefly here; it comes up later for fuller discussion. Dingle's use of the term "invention" allows of a stronger interpretation than that just given it. Rather than the weaker, epistemological-methodological character involved in merely *inventing hypotheses*, the stronger sense is purely metaphysical: what goes on is *inventing Nature itself.* This position is traditionally called *metaphysical idealism*; it was propounded by various philosophers, especially post-Kantians such as Fichte and Schelling. According to this interpretation of Dingle's "invention"—an interpretation explicitly affirmed by some of his First Way allies in the ensuing debate—what Milne and the Second Way proponents were engaged in was "inventing Nature" according to their own ideas. Nothing, according to the empirico-inductivists, could be more *un*scientific.

[26] One surprising but heretofore little-known fact about this controversial article by Dingle is that it was solicited by the Editor of *Nature*! The original is in the Dingle archives at Imperial College, London.

[27] Dingle has of course confused Aristotle's scientific method with Descartes', a point not lost on some of the critics who answered Dingle almost immediately.

Dingle himself ascribed the cause of this heresy to one root: If cosmology "may be described in broad terms as an idolatry of which 'The Universe' is the god", the ultimate source of the false worship is clearly evident: "This cosmolatry, as might be expected, came ... out of *mathematics* [emphasis added]." In other words, the mathematicians were at fault.

Dingle had always exhibited a healthy distrust of mathematics. For him, the subject existed solely to solve practical problems (Whitrow 1990). In his own work in astrophysics, Dingle typically concentrated upon the experimental; his mathematical activities were limited to those techniques required for spectral analysis. Obviously, Milne's and Eddington's importing highly speculative mathematics into what should have been a purely observation-based discipline grossly violated Dingle's methodological sensibilities.

But for its own part, Dingle's vituperative attack grossly violated more than a few sensibilities as well. Indeed, the reaction was so strong and so wide-spread, that *Nature* was forced to issue a supplemental number to accommodate it all.

5. The great methodological debate of 1937

On 12 June 1937 *Nature* published a special supplement entitled "Physical Science and Philosophy." The issue was entirely devoted to reactions—pro and con—to Dingle's attack upon the deductive methodology of Milne, Eddington, and Dirac. Participants in the debate included some of the most well regarded scientists and thinkers in Great Britain (see Table 1). The deductivist principals generally contented themselves with trenchant restatements of their positions, although Dirac's defense presented a mild sop to Dingle and his allies.

5.1. Responses of the Second-Way principals

Milne's response, other than its mention of a heretofore unmentioned "principle of economy of thought", provided nothing new. His methods, he claimed, were innocent enough:

> The general object of the investigations in question was to determine the consequences of the assumption that the universe is, on the average, homogeneous. ... [28] By an extreme application of the principle of the economy of thought I investigated the consequences without appealing to any empirical quantitative 'laws of Nature' whatsoever. ... I did so because it early appeared that very much more could be deduced from it than was commonly recognized. (Milne et al. 1937: 997)

Milne's response went on in this vein, ending with the rather righteous claim that "the individual investigator must be left in peace to state his own problems and solve them by his own methods." And, in a direct poke at Dingle, Milne argued that "criticism of these [methods] on the ground that they go beyond Renaissance science is merely a form of authoritarianism." (Milne et al. 1937: 999)

Eddington's response provided not a little humor—in the deductivist vein, of course. First, Eddington tweaked Dingle and his ilk:

[28] This is an alternate way to describe the Cosmological Principle.

"PHYSICAL SCIENCE AND PHILOSOPHY" *Nature* Supplement (12 June 1937)	
Participants	
Campbell, N. R. [Gen. Elec. Co.]	Haldane, J. B. S. [FRS, Univ. College-London]
Darwin, C. G. [FRS, Cambridge]	Jeffreys, H. [FRS, Cambridge]
Dingle, H. [Imperial College]	McCrea, W. H. [Queen's, Belfast]
Dirac, P. A. M. [FRS, Cambridge]	McEntegart, W. [Heythrop College]
Eddington, A. S. [FRS, Cambridge]	Milne, E. A. [MBE, FRS, Oxford]
Filon, L. N. G. [CBE, FRS, Univ. College, London]	Peddie, W. [Univ. College, Dundee]
Hatfield, H.	Sampson, R. A. [FRS, Roy. Observatory. Edinburgh]
Whitrow, G. J. [Oxford]	

Table 1. The 'Great Epistemological Debate'.

In the debate, Filon, Hicks, Jeffreys and Peddie backed Dingle's empirico-inductivism. Campbell, Haldane, Hatfield, McCrea, Sampson and Whitrow came down on the side of the deductivists. Darwin was skeptical of the whole contretemps, and McEntegart pointed out that Dingle's use of "Aristotelianism" was historically incorrect; but went on to add that the genuine Aristotelians would be on Dingle's side!

> We are given—
> (a) It is impossible to have a priori knowledge of an objective universe.
> (b) The mass-ratio [e^-/p^+] has been found by an a priori method.
> Therefore: Knowledge of the mass-ratio is not knowledge of an objective universe. (Milne et al. 1937: 1000)

The barb here is Eddington's claim to have found the mass-ratio by a priori—deductive—methods. Dingle and the others naturally would deny the claim. But the argument is valid, and at least serves to modestly mock the empirico-inductivists. But the real force of Eddington's rejoinder is his concluding note:

> After a rather extensive series of researches, I have found that a great part of the current scheme of physics is deducible by a priori argument and therefore does not constitute knowledge of an objective universe. (Milne et al. 1937: 1000)

Eddington's conclusion, unstated but validly deducible from his premises, is that "a great part of the current scheme of physics" is not "knowledge of an objective universe." This conclusion squares precisely with what we saw earlier about the basis of his views: since quantum mechanics—the basic theory of physics—is inherently subjective, it is deducible from "epistemological considerations."

Further, any physics deducible from quantum mechanics preserves the subjective nature of its starting point. It is no wonder that Dingle et al abhorred Eddington's deductive methodology.

Unlike Milne's and Eddington's defenses, Dirac's rejoinder to Dingle was not nearly so sharp nor so direct.

Dingle, Dirac notes, criticizes his, Dirac's, acceptance of Eddington's deduction of the smaller constants because "it departs from the Galilean scientific method of building up theory to fit observations." Yet, Dirac responds, there is more than one method useful to science:

> The successful development of science requires a proper balance to be maintained between the method of building up from observations and the method of deducing by pure reasoning from speculative assumptions, and I think my [work] satisfies this condition; but as it was written rather concisely, I would like here to restate the main points ... (Milne et al. 1937: 1001)

And here Dirac goes on to restate, in rather conciliatory fashion, the main line of his argument in favor of the hypothesized relation between the age of the universe and the values of the large constants.[29]

So much for the responses of the principal adherents of the Second Way.

5.2. Responses by the Deductivists' Allies

But the principals were not the only adherents to deductivism. Several other prominent scientists and philosophers supported the method as well. A brief survey of some of their remarks would not be out of place here. What is most interesting about the remarks is the range of reasons their authors offer for their support of deductivism. Reasons stretch all the way from the pragmatically concrete to the methodologically abstract. As an example of the former, we need only look to Hatfield. He has practical worries about attracting young minds into a Dinglian science:

> Dr. H. Dingle's rather drastic remarks concerning present-day physical speculation seem to overlook one main factor in the progress of science—the capture of young minds of the highest order of genius. The genius of every age turns mainly in that direction which offers the greatest scope for the creative imagination. ... A science kept strictly to facts and formulæ by 'pale Galileans' would recruit very few geniuses, fewer rank-and-file workers, and—fewer endowments! (Milne et al. 1937: 1009)

Although Hatfield's remarks have a definite point, his recommendations, however, say little about the methodological benefits (or otherwise) of the Second Way. Haldane's response, on the other hand, albeit somewhat pragmatic in part, is measured and directed precisely at Dingle's methodological argument. In the first place, Haldane claims, biologists and geologists, since they are used to a science in which "chains of reasoning are shorter than in cosmology", might "inevitably" have some sympathy for Dingle's claims. This would be unfortunate, Haldane

[29] It should not go unremarked that Dirac here specifically recommends a proper balance of the First and Second Ways. Perhaps he is the cosmologist Bondi referred to in Bondi 1960: 3–5! See our discussion, Section 1 above.

believes, because "Milne's cosmology . . . , if accepted as a working hypothesis, will profoundly affect their sciences." Here Haldane has in mind the potential usefulness of Milne's two timescales[30] for squaring the apparent discrepancies between then-extent physical and biological estimates of the duration of the past. "If the results serve to illuminate the history of geological and organic evolution," Haldane says of Milne's work, "we shall be kept too busy to find much time to blame him for a perhaps unduly idealistic[31] account of their origin." (Milne et al. 1937: 1003-1004)

A second, even more measured defense of the deductivists is found in N. R. Campbell's rebuttal to Dingle's claims.[32] .Campbell's response uses both logical and historical analyses to argue for acceptance of the new methodology. After a fast but fair summary of the logical relationships between, on the one hand, observational laws and inductive logic, and, on the other, theoretical explanations and deductive logic, Campbell goes on to argue that beginning in the Twentieth Century the terms found in theoretical explanations became more and more difficult to correlate with the referents of terms in observations. As he puts it, "laws were discovered for the explanation of which no 'mechanical' theory . . . could be found." That is, the laws had no direct reference to any *demonstrable experiment*.[33] Explanations thus were required to be embedded in large sets of hypotheses, with the hope that ultimately some sort of observational consequence would be found. With this history in mind, Campbell argues that it is entirely understandable that the mathematician-deductivists emerged, especially in cosmology. Speaking directly to Dingle and the other empirico-inductivists' animosity toward mathematicians, Campbell makes the following clear statement of the criteria of the deductivists' method:

> Since the propositions [hypotheses] were mathematical and invented by mathematicians, it was inevitable that the hypothetical ideas should be highly abstract and that the acceptability of the resultant theory should lie in a conformity to mathematical ideals, that is to say, in the rigid deduction of elegant conclusions from the minimum of self evident axioms. The work that makes Dr. H. Dingle so indignant appears to me a laudable attempt to solve this problem . . . (Milne et al. 1937: 1006)

[30] The two timescales are kinematical and dynamical. Kinematical time is appropriate for describing chemical (and biological) events; it is finite in the past. Dynamical time, which measures physical events such as the earth's rotation, is apparently infinite in the past.

[31] Haldane's intended meaning of the term "idealistic" is unclear. Certainly it is at least epistemological, in that the origin of Milne's theory is in his (Milne's) own cognitive activities. But it is also possible that Haldane might intend us to take a certain flavor of metaphysical idealism as well from Milne's work. It is well known that Haldane was, at this time, a devoted Marxist; hence, his philosophical inclinations would be to second Marxism's condemnation of 'idealist' metaphysics.

[32] Campbell was a member of Trinity, and graduate of Cambridge in Physics. He worked under J. J. Thompson at the Cavendish 1903–1910. He later worked as a researcher at General Electric for nearly twenty-five years. His well-known efforts in philosophy of science were much influenced by Mach and Poincaré

[33] Campbell uses "demonstrable experiment" to refer to observational situations associated with "mechanical" theories. Hence, a "mechanical" theory is what physicists today would call a 'phenomenological' theory.

Yet, Campbell concludes, even given the abstract mathematical nature of the procedures described, it is clear that they are not completely divergent from earlier standards of observational situations. Both Dirac and Milne, he argues, introduce "as fundamental an idea closely analogous to the agreement between observers, which is implied in the conception of a demonstrable experiment." And this is enough to refute Dingle's claim of treason to the founding ideals of the Royal Society. After all, if Dingle

> does not deem it important to observe the distinction between what is and what is not demonstrable experiment, surely he should welcome a movement to amalgamate the Royal with the Aristotelian Society. (Milne et al. 1937: 1006)

Of the several other backers of the deductivists, the most important response was that made by McCrea. By that time already professor of mathematics at Queen's University, Belfast, McCrea was one of the first workers (along with McVittie) who had had explicit training in the cosmological trade.[34] He had made his mark early, and was well respected. That he came in on the side of Milne and the other deductivists was significant, since it represented the support of both youth and talent.

McCrea begins with a delightful *ad hominem*. Dingle's objection, he points out, "seems to be itself what he would call Aristotelian rather than Galilean." After all,

> a truly Galilean point of view would be to observe whether Eddington, or Milne, or anyone else, can in fact deduce properties of the physical world from a knowledge of "the system of thought [of] the human mind ... " Should it turn out that such a deduction is possible, the fact would be a new and exceedingly important experimental result ... worthy of the attention of all true Galileans. (Milne et al. 1937: 1002)

Of course *ad homina* do not advance the state of the argument, no matter how delightful they might be! Hence, McCrea moves immediately to the heart of the matter. "What Dr. Dingle has done" McCrea says, "is to reopen the question of the relation of mathematical physics to experimental physics." This, especially, because Dingle claims "to detect a new and perverted point of view in the former [mathematical physics]." To address this newly reopened question, McCrea delivers a succinct description of what he takes this relation to be: apart "from the alleged perversion", "a system of mathematical physics ... is the working out of the mathematical consequences of certain hypotheses." This initial claim is agreeably restrained—essentially everyone party to the debate could affirm it; moreover, it tactfully avoids any reference to the origin of the hypotheses in question, which, of course, is a major sticking point for Dingle and his allies.

According to McCrea, the system of mathematical physics (thus defined) is evaluated according to two sorts of criteria. First, the internal ones: "The worth of the theory is judged on one hand by the fewness and simplicity of its hypotheses."

---

[34] While McCrea was an undergraduate in Cambridge (1923–26), he followed the received advice and avoided astronomy courses. But during his graduate work under Whittaker at Edinburgh he was introduce to relativity; the rest, as the saying goes, is history. (McCrea 1987) For example, McCrea's 25 October 1936 Inaugural Lecture in Queen's was entitled "Mathematics and Cosmology."

These are precisely the elements mentioned by Milne initially, and by Campbell in his defense. The second criteria are exactly what we should expect: theory is judged "by the closeness of the agreement of its predictions with the results of observation, and also the number of phenomena which it can so predict from the one set of hypotheses." (Milne et al. 1937: 1006)

Agreement with observation is the hallmark of the "D" aspect of the HD methodology. Here McCrea affirms that explicitly and precisely.

McCrea's additional mention of the richness criterion—how rich in predictions the hypotheses-set is—is not part of standard HD methodology. Yet, of course, it is not inconsistent with it.

McCrea ends his defense of Milne by alluding to what we have earlier seen in Milne's letters to Walker and Chandrasekhar. "Mathematical physics," McCrea claims, "advances then, in one direction, by reducing the number of its hypotheses."[35] As an historical example of this sort of progress in reducing the axiom-set, McCrea argues that there is a continuous series beginning in Newton's theory, going on to general relativity, and progressing to Milne's theory.[36]

Thus, although Milne's work, as well as Eddington's, according to McCrea, "is still too new for a consensus of scientific opinion to have been expressed; it is [my] purpose ... merely to suggest that it is in harmony with the historic evolution of mathematical physics." (Milne, et al. 1937: 1003) In the end, McCrea concludes, "the scientific attitude is, not to cavil at the attempt, but to see if it is successful." That, after all, is the Galilean way!

One other defense of Milne and the deductivists worth mentioning is that of Milne's student, G. J. Whitrow. Whitrow's counter-argument contains one quite wicked *ad hominem*, and one excellent methodological point. The *ad hominem* hoists Dingle on his own petard. Dingle's claim, Whitrow notes, is that the deductivists are traitors to the methods of modern science. If this is the case, then:

> Since modern scientific method originated with the Copernican theory of the universe, let us consider the attitude which Dingle, as an empiricist, would have adopted towards the then new cosmology. (Milne et al. 1937: 1008)

Of course, as Whitrow notes, exploding the petard, since "there were no known celestial phenomena which were not accounted for by the Ptolemaic system", Copernicus' system must necessarily be branded by Dingle a piece of "cosmomythology." In other words, "Dingle himself would have been in the 'Aristotelian' camp."

Wicked though this *ad hominem* is, Whitrow's methodological point is of much heavier weight. To begin, he notes—correctly—that Dingle's point of view "is based upon a pure myth—in this case a mythical interpretation of the history of

---

35 In one sense, this principle is the inverse statement of the richness principle just seen. The richness principle states that Hypothesis H_1 is better than H_2 just in case the former predicts more observable consequences than the latter. The inverse of this would be, if Hypothesis Set H_1 consisting of hypotheses $\{H_a+H_b+H_c\}$ predicts consequence set C, but Hypothesis Set $H2$ consisting of hypotheses $\{H_a+H_b'\}$ also predicts consequence set C, then $H2$ is better than $H1$.

36 This theme is developed at length in McCrea's 1939 *Philosophy of Science* paper "The Evolution of Physical Theories." (McCrea 1939)

scientific method." In particular, Dingle has got Galileo precisely wrong, or so Whitrow believes.

As we have seen, Dingle (and, as we shall soon see, his allies) is particularly exercised by the role mathematics plays in the new Aristotelians' "cosmomythology": the abstract entities and unobservable relations populating the deductivists' theories are particularly galling to the empiricist sentiments of Dingle and his allies. Galileo, they claim, exhibits the empirico-inductivist antidote to the modern mathematical sinners. Whitrow attacks this view of Galileo directly, relying on two quotes from Galileo himself, and one from Galileo scholar J. J. Fahie. One of the Galileo quotes is the famous one ending with the claim about the book of the universe, namely, "This book is written in the mathematical language." The second Galileo quote is not so well known and thus deserves full citation: "We do not learn to demonstrate from the manuals of logic but from the books of demonstration which are the mathematical."[37] Again, the force of the passage is that mathematics provides the books which must be read in order to carry out science.

Fahie's point is an interesting and trenchant one, namely, that as far as his private beliefs were concerned, Galileo was a *deductivist*, exactly the opposite of what Dingle and his allies have claimed! According to Fahie, Galileo conducted experiments solely "in order to demonstrate to his opponents the truth of his conclusions." But "to satisfy his own mind alone he had never felt it necessary to make any."[38] So much for Galileo the Great Experimenter![39]

As this short review has shown, the defense of the Second Way is broadly based, including an array of arguments, *ad homina*, and reasons from all across the philosophical spectrum. Let us now turn from supporters of Milne and the Second Way, to the points made by the supporters of Dingle and the First Way.

5.3. Responses by Dingle's First-Way allies

Dingle was not without allies in his attack upon the Second Way's 'mathematicians.'[40] Most followed the line offered by L. N. G. Filon.[41] Filon admitted that, while he didn't agree with all that Dingle had claimed, "one may be permitted to welcome a protest against the modern tendency in physics to proceed from abstract and universal a priori mathematical theories." (Milne et al. 1937: 1006)

---

37 Quoted in Milne et al. 1937: 1008.

38 Quoted in Singer 1921: 251.

39 My point here does not deny that Galileo carried out some great experiments, since it's obvious that he did. However, if Fahie is right (and I see no reason to doubt it) Galileo's impetus to carry out experiments was not based in his own concept of methodology; apparently, mathematical demonstration alone would suffice for his own conviction.

40 In addition to those taking part in the public debate, Dingle's Nachlaß in the Imperial College archives contains letters of support from many other scientists, among whom the most surprising is found to be H. P. Robertson, who, by the time of the debate, had already published his remarkable series of three papers based upon the work of none other than Dingle's arch villain, E. A. Milne! (Robertson 1935, 1936a, 1936b)

41 Filon was C. B. E. and F. R. S. More importantly, he was a highly successful applied mathematician who had achieved high rank (Dean of the Faculty of Science, and Vice-Chancellor) in University College, London.

More explicitly, Filon claims,

> The real trouble seems to be that, instead of starting from the facts of observation and gradually building up by induction particular laws, which may or may not eventually be linked up, some men of science appear to think that they can solve the whole problem of Nature by some all-inclusive mathematical intuition. (Milne et al. 1937: 1006)

Here we find stated the crux of the controversy. Dingle and his allies, Filon included, believe that observation and induction, together, constitute the sole acceptable methodology for science in general, and physics in particular. This, of course, is the First Way. In opposition they find the "mathematical intuition" of the Second Way. But what is curious about Filon's description of the distasteful 'methodology' is its incompleteness: he refers entirely and only to the *source* of the CHD methodology—the cognitive element—and not at all to its *sink*, namely, the deduction of observational consequences. Only through this unfairly incomplete characterization of his opponents' methodology can Filon go on to conclude that "what they are *really* doing is not to explain Nature, but to explore the possibilities of the human mind." (Milne et al. 1937: 1006; emphasis added) Apparently, then, Filon thinks that the proponents of the Second Way are practicing some brand of Psychology, and not Physics at all.

A similar attack is mounted by W. Peddie, professor at University College, Dundee. Following his approval of "the recent timely strictures made by Dr. Dingle on modern Aristotelian tendencies in some theoretical investigations in physics", Peddie proposes a set of four postulates which define an observational basis for space-time and motion. His point is that by defining the postulates in the way that he does, physics—relativity theory in particular—is guaranteed an empiricist foundation. This, he claims, will put physics back on the right track:

> In effect we can [now] blame the objective physical universe for the formal laws of thought, instead of the laws of thought for the structure of the universe. We can escape from Aristotle and get back to Galileo and Newton. (Milne et al. 1937: 1006)

Clearly, what we see here (again) is the usual charge of *metaphysical idealism*—the notion that the mind creates reality in its own image (instead of vice-versa)—that Dingle and his ilk frequently level against the deductivists. But as was true with Filon, such a charge may be brought against the deductivists only by ignoring the fact that their method requires checking deduced conclusions against observations of Nature.

The most coherent and salient defense of Dingle is that of Jeffreys.[42] Jeffreys' defense depends upon a good offense—he attacks Milne's rejection of the principle of induction, both directly and by *ad hominem*. Moreover, in a parallel assault, he denies the deductivists' claims of non-empirical, a priori origins for their hypotheses. Jeffreys begins by noting that Dingle has presented "an excellent statement of the position, and I am in full agreement with it." (Milne et al. 1937: 1004) He then

---

[42] Jeffreys was an important ally for Dingle. He was FRS, and had written highly regarded texts in both geophysics (*The Earth*, 1924) and scientific method (*Scientific Inference*, 1937).

launches into his first attack, the sally against Milne's anti-inductivism. The problem with CHD, he says, is not the giving of the definitions based upon "intuitive suggestions", nor is "working out the consequences deductively, and then testing the results by comparison with observation" anything but 'legitimate.' What *is* problematical about the method, though, is that, as Russell and Whitehead themselves argue, definitions can be shown to be *valid*[43] only by showing the existence of the thing defined, which "therefore requires either a postulate or an existence theorem." Moreover, since both of these latter rely upon experience, which in turn is dependent upon the principle of induction, it follows finally that CHD cannot be *validated* as a method without the principle of induction. If Jeffreys is right, then Milne cannot attack the principle of induction without at the same time attacking his own deductivism.

And just to complete the argument, Jeffreys slips in the *ad hominem* that "without using induction, Milne and Eddington could not order their lives for a day." Presumably, since Milne and Eddington *do* order their lives, even if only for a day, they do use the principle of induction. Jeffreys wraps up this stage of his overall argument by reference to the inductive/probabilistic methods of Pearson, and Fisher and other statisticians.

As a second stage of his argument, Jeffreys makes several references to occurrences where, he believes, the deductivists rely upon undisclosed, hidden experiential assumptions for propositions they purport to be based upon pure a priori grounds. His most interesting case concerns some work he himself did on the bending of starlight, most famous of the tests for general relativity, and cause célèbre for the deductivists. Unsatisfied with the deductivists' derivation of the results of the bending experiment from first principles, Jeffreys himself tests a number of alternate hypotheses, for example, that the bending is due to refraction in the diffuse gasses of the sun's photosphere. But he finds that the probability of this alternative explanation was such that it could not occur more than "once in 170 times." In other words, here was perfectly good empirically-based information arguing in favor of the relativistic result. But did the deductivists embrace Jeffreys' results? Not at all:

> Now the remarkable thing is that, so far as I am aware, no exponent of relativity has ever mentioned this work or any improvement on it. ... It is apparently considered satisfactory to give a purely deductive theory ... while the fact that the theory is the *only* suggested one that is observationally tenable is not thought worth mentioning. (Milne et al. 1937: 1005)

Indeed, Jeffreys asks with regard to the non-mention of his result, "is it not mentioned *because* mention of it would admit that the postulates of the theory are not a priori certain?" (Milne et al. 1937: 1005) In the end, the point tells against the root cause of the whole deductivist methodology, namely, its mathematicism:

> I think the source of the trouble is the belief that there is some special virtue in mathematics. Instead of being regarded as what it is, a tool for dealing with arguments too complicated to be presented without it, it has become

---

[43] Jeffreys' use of the term 'valid' here would certainly be unacceptable today.

> emotionalized to such an extent that many people think that nothing but mathematics has any meaning ... (Milne et al. 1937: 1005)

Jeffreys' parting shot contains a sly poke at the very basis of the CHD method, deductive logic: "The utility of deduction is that in investigating the consequences of a well-supported law it is a convenient approximation to induction." But by far the worst canard Jeffreys lays upon the poor methodologist is contained in his rhetorical query that "Is not the salient fact about the philosophy of science that no professional philosopher can write a book that a man of science wants to read?" (Milne et al. 1937: 1008)

A final party to the debate should be mentioned in conclusion to our review of the positions pro and con. This is C. G. Darwin, whose position is a truly balanced one: he thinks the whole debate is fruitless, indeed, pointless, and a pox on both their houses![44]

Darwin makes his argument quite forcefully:

> The noticeable point is that these contradictious views do not seem to matter in the very least; we none of us doubt that both Dingle and Milne will help on the progress of science, and that the advances they make will have scarcely anything to do with their opposite metaphysical ideas ... ; the fact remains that it is the science and not the philosophy that matters, and that most men of science do not find it worth their while to read the works of metaphysicians. (Milne et al. 1937: 1008)

Of course, even Darwin admits that everyone must necessarily have **some** philosophy or other, one which may be "extracted from us with the methods of Socratic dialogue by some more expert philosopher." After going through this process, however, "we shall end by wanting to do to our inquirer what the Athenians did to Socrates."

Obviously, however, the very fact that Darwin took part in the debate refutes his position, since, if philosophical debates about cosmological method were *genuinely* pointless, there would be no point in his, Darwin's, taking part so vociferously in the debate in question!

This completes our review of the great methodological debate of 1937. It is clear and evident on its face that the debate was vigorous, hotly contested, and deeply philosophical. Moreover, as we shall now show, its consequences were significant to the conduct of cosmology itself: we turn to our last effort, namely, the investigation of the consequences of the First vs. Second Way debate in the work of Hermann Bondi.

6. Methodology Redux: Bondi and the Second Way

Bondi's involvement with the Second Way proceeds under several guises. First, he specifically remarks the excellence of Milne's Methods, and the theory, KR,

---

[44] Darwin was another Trinity physicist. His work was in atomic physics and x-ray diffraction. Darwin was FRS in 1922, and received the Royal Medal in 1935. His *The New Conceptions of Matter* (1931) was highly regarded.

created therefrom. Secondly, he argues in favor of hypothesis and deduction, that is, in favor of the Second Way. Finally, he argues against induction and extrapolating from small-scale experiment, that is, against the First Way. Taken together, these three categories of support provide strong evidence both of Bondi's commitment to the Second Way, and to Milne's influence upon Bondi's work.[45] Let us look at each in turn, beginning with Bondi's more general comments about the Two Ways.

First, we re-iterate a comment from Bondi's 1950 *Cosmology* noted earlier:

> In particular, there are two important approaches to the subject [cosmology] so different from each other that it is hardly surprising that they lead to different answers. ... The contrast between the 'extrapolating' and the 'deductive' attitudes to cosmology is very great indeed. (Bondi 1960: 3–5)

Proponents of the *extrapolating* approach (whom Bondi sometimes calls the *empirical school*), emboldened by their continuing successes in applying laws of dynamics discovered in terrestrial labs to ever larger, ever more distant physical systems, have a straightforward view of method in cosmology:

> According to this school of thought, then, cosmology is the largest workshop in which we may assemble equipment, the elements of which are entirely composed of terrestrially verified laws of physics. (Bondi 1960: 4)

Opposed to the extrapolative approach is the *deductive* approach, which "is reached from investigations in the borderland between physics and philosophy." According to one major version of this approach in cosmology, philosophical analyses reveals that either the fundamental methodological assumptions of physical science are illogical, or space and time must be presumed to possess certain specific properties. Since the former disjunct is difficult to face, the latter must be accepted. It then directly follows that:

> the fundamental laws of cosmology are immediate consequences of the a priori[46] assumptions involved in any physical science. These fundamental laws play the part of axioms, and the main body of cosmology may be deduced from them in the manner of a discipline of pure mathematics. (Bondi 1960: 4–5)

Although Bondi finds good points in both approaches, he also finds problems in both approaches. In the end, cosmology is the worse for excesses from *either* end of the spectrum:

> Just as some adherents of the 'empirical' school tend to regard cosmology as a testing ground for their extrapolations and as a legitimate playground for the geometers, so some adherents of the deductive approach appear to regard cosmology as a purely logical subject. (Bondi 1960: 7)

---

[45] When he heard the following evidence, Clive Kilmister, a longtime Bondi collaborator, claimed that the author (GG) "certainly is correct—Bondi was strongly influenced by Milne." (Remarks made at the HGR4 Conference, Berlin, 1995.)

[46] The term 'a priori' here is not to be taken in its Kantian sense; rather, it is evidently intended that something more like Collingwood's sense of 'relative presupposition' is involved (Robinson 1958).

In this latter case, the deductive extremists, in their mathematical zeal, seem to forget that cosmology, after all, *must* have some relation to observation: "To them all that is of interest in a theory is its logical character, not its relevance to the interpretation of observational data." This leads to a dangerous state, one in which cosmology risks losing its empirical character:

> They are willing to discard a theory only if a logical internal inconsistency is found, and they consider any disagreement with observation as of minor importance only, relating merely to the applicability of the theory. (Bondi 1960: 7)

The dangers of this deductivist extreme may be seen in a comment on the differences between the Two Ways which Bondi makes seven years after the above. There, Bondi's remarks reveal that he has added something new to the Second Way precisely as an antidote to the logical extremes of some deductivists:

> There are two distinct ways of constructing a model of the universe:
> (i) One can take the laws of physics as found locally and then try to apply them to the whole universe;
> (ii) One can postulate (as a kind of working hypothesis) some general properties of the entire universe and then, by applying the known laws of physics as far as is possible, one can try to infer local observable consequences of the original postulates. The observational comparison that can now be made makes a disproof of the postulates possible and so gives them *scientific status*. (Bondi 1957: 196; emphasis added)

According to Bondi here, deductivism can be a *scientific* approach in cosmology only if its postulates (or axioms) are candidates for disproof. Given this criterial addition, Bondi now holds that when certain deductivists go too far, and retreat into solely internal, logical/mathematical criteria for evaluating their models, they have removed their efforts from the domain of science.

Clearly Bondi has added a distinctly Popperian element to the deductivist methodology, one which had not previously appeared in the works of any of the earlier members of the school. Much later Bondi was to make explicit his debt to Popper:

> I think the person from whom we had most help on the philosophical side was Popper. His analysis of science encouraged one to be imaginative, and encouraged one to go for something that was very rigid and therefore empirically disprovable. (Bondi 1990: 194)

However, with the exception of the addition of the Popperian element, Bondi's version of the Second Way is precisely that of the deductivists, especially Milne. To see this, we need only look to some of Bondi's comments about Milne and KR.

6.1. Bondi on Milne's methods and kinematic relativity

According to Bondi, Milne's theory was through and through deductive, which was reason enough for some of his colleagues to condemn it:

> The aim of this discipline [KR] is to deduce as much as possible merely from the cosmological principle and the basic properties of space, time and the propagation of light. The beauty of this, as indeed of any deductive

> theory, rests on the rigour of the arguments and the small number of the axioms required. ... When the theory was first developed it met with great hostility and was criticized very severely, often unjustly, and sometimes frivolously. (Bondi 1960: 123)

Bondi's admiration for Milne's deductive methods is obvious; moreover, his displeasure about the reception accorded Milne is equally obvious.[47]

Bondi's response here is by no means unique. In his extended discussion of KR in *Cosmology*, Bondi frequently speaks of Milne's theory in glowing terms. For example, in speaking about Milne's "searching analysis of the time concept", Bondi judges that this analysis "is possibly the greatest achievement of the theory."[48] Moreover, when Milne uses his concept of time as a basis to then construct a measure of distance,[49] Bondi again applauds the effort:

> Imperfect as Milne's definition of distance may be, it is very much better than the 'rigid ruler' one used in most other theories. ... Milne's definition of distance, by no means perfect as it is, is probably the best yet devised. (Bondi 1960: 126–129)

In other places, Bondi refers to elements of Milne's work as "beautiful" (p. 129), and "remarkable" (p. 133); he finds KR's central concept "particularly simple and lucid" (p. 130). In the end, Bondi sums up Milne's deductivist contributions with no uncertain praise:

> The foregoing brief description will have indicated the remarkable success of kinematic relativity in attempting to use the cosmological principle not only for the construction of the substratum but as chief guide in formulating ordinary physics. In this respect it differs greatly from all other cosmologies which either rely on a conventionally obtained body of physics or have not yet in succeeded in drawing conclusions of local interest from the cosmological principle. (Bondi 1960: 136)

In this passage, Bondi not only explicitly lauds Milne and his successes with KR, he also implicitly supports Milne's deductive methodology in its "drawing conclusions" from the extremely general postulate of the *Cosmological Principle* (CP). Reference here to the CP is significant. Bondi is unhesitating in placing his own work with the CP in his *steady-state theory* (SS) in direct line of descent from Milne's work with the CP in KR:

---

[47] More recently, Bondi has also praised the deductivists, in particular Milne, whose lack of due credit today is most likely his own fault: "At that time [1930s and 1940s] the work of Eddington, of Milne, and of Dirac showed that cosmology was ripe for imaginative ideas, ripe for something to bring order into it. I think it was Milne's pugnacious unwillingness to accept that people might swallow only part rather than the whole of his theory which makes us all give him less credit than he really deserves ... " (Bondi 1990: 190)

[48] Bondi was never reticent about his admiration for Milne on time. Speaking about Milne's notion of regraduation of clocks, McVittie remarks "That contains things which H. Bondi and postwar people made a great song and dance about." (McVittie 1990: 41) More recently, Bondi himself, speaking about Milne and the clock regraduation problem, remarks "Milne should be praised for his idea that there might be different time scales for different clocks. This still seems to me one of the most imaginative ideas of my time." (Bondi 1990: 190)

[49] Milne uses what Bondi calls "the radar method", namely, timing the transit of a light signal out to a point and then back, in order to calculate $d = ct/2$. According to Bondi, those of us caught speeding by police using radar guns have Milne to thank for the method! (Bondi 1990: 190)

> The alternative type [Second Way] postulates certain general properties of the universe in the form of a cosmological principle, possibly together with some other axioms as a basis, and then attempts to deduce observable properties of the universe as a whole and of its parts from these postulates. ... Kinematic relativity was historically the first representative of the second type of theory, which now also includes the steady-state theory. (Bondi 1960: 65)

According to Bondi's understanding of the structure of both KR and SS, the major axiomatic role is played by the CP. As we have earlier seen (see footnote 15), Milne had originally identified the content of the principle, and Robertson named it. According to Milne's version of the principle, every observer in the universe should get the same world picture, that is, should make precisely the same observations of the universe at the same moment as any other observer. (Milne 1934) Milne's invoking of the principle guarantees that the Universe is uniform throughout its vast spatial expanses. Yet Milne's Universe evolves, it changes its form over time; in particular, it expands from a dense to a less dense state. Hence Milne's universe has no temporal uniformity. Bondi felt that this raised the possibility that physics itself might change over time. And, if physics changed over time, then there would be an unacceptable risk in extrapolating the physics of *our* epoch into any other. Because of this risk, Bondi generalized Milne's cosmological principle into what he called the *Perfect Cosmological Principle* (PCP).[50] PCP states that the Universe presents the same picture to all observers *at all epoches*.[51] Holding to this version of Milne's original principle severely constrains Bondi's choices about the form and content of SS. The resulting theory, as McVittie remarked, is "much more restrictive than general relativity", (McVittie 1990: 45) But it is this very restrictiveness which satisfies Bondi's Popperian wishes:

> For the correct argument has always been that the steady-state model [rather than, e.g., the Big Bang model] was the one that could be disproved most easily by observation, Therefore, it should take precedence over other less disprovable ones until it has been disproved. (Bondi et al. 1959: 55–56)

In another place, Bondi makes a similar point: "Comparison with observation becomes then possible and renders the PCP liable to observational disproof. This possibility of a clear-cut disproof establishes the scientific status of PCP." (Bondi 1957: 198) Comments such as this make clear Bondi's commitment to a Popperian addition to the basic deductive methodology he inherited from Milne.

In the end, what we find is that Bondi has made two modifications to his Milnean inheritance. First, he methodologically modifies Milne's deductivism by adding the Popperian falsifiability requirement. Bondi does so specifically to limit the damage created by those critics who claimed that the deductive methodology as practiced by Milne and his ilk was *unscientific*. Secondly, Bondi strengthens Milne's Cosmological Principle, requiring that it legislate uniformity not only over

---

[50] Dingle, in his Presidential Address to the Royal Astronomical Society (Dingle 1953), witheringly satirizes Bondi's principle, as well as inditing him for his rationalism à la Milne.

[51] Bondi's PCP explicitly rules out systems of physics based upon changing 'constants' and laws, most prominent of which are those rooted in Dirac's notions. (Barrow & Tipler 1986: 235)

space, but over time as well. Bondi's second move obviously ties in with his first: the strengthened CP more easily satisfies the Popperian falsifiability requirement. Yet, even after noting these two additions Bondi makes to Milne's Methods, it remains obvious that Bondi's *fundamental* commitment is to Milne's deductivism and the Second Way. Let us look briefly at this commitment.

6.2. Bondi's commitment to the Second Way

One of the clearest expressions of Bondi's commitment to the deductive approach comes in a discussion of a problem we saw earlier, namely, the accusation by McVittie and others that Milne and the deductivists had somehow appealed "unconsciously no doubt, to an empirical fact" during their supposedly 'pure' deductions. (McVittie 1940: 275) While he agrees that such a risk genuinely lurks here, Bondi nevertheless strongly believes that a pure deduction is possible:

> When an attempt is made to deduce from first principles a fact already known from ordinary physics, only the frankest discussion between proponents and critics can lead to a formulation in which it is clear that the axiomatic deduction is strict and no empirical argument has been slipped in subconsciously. (Bondi 1960: 123)

For his part, Bondi quite clearly believed that he had produced just such a *strict* axiomatic deduction—SS. The theory had one and only one first principle: PCP, the perfect cosmological principle. Unlike Relativistic Cosmology, which depended upon scaling up local laws, thereby raising difficulties at long ranges and distant times, SS was complete in itself:

> If the uniformity of the universe is sufficiently great none of these difficulties arises. The assumption that this is so is known as the perfect cosmological principle. ... It was shown by Bondi and Gold that this single principle forms a sufficient basis for developing without ambiguity a cosmological theory capable of making definite and far-reaching physical statements agreeing with observation. (Bondi 1960: 13)

As we have seen earlier, Bondi was well aware of the reception afforded the deductivist methodology of his predecessors; in particular he knew how antagonistic the empirico-inductivists were toward the starting points of the deductivists, the cognitively-based hypotheses. But in the face of this antagonism, Bondi didn't flinch one bit:

> The initial emotional judgement of the likelihood of the truth of the basic hypothesis is completely irrelevant; what is important is whether its consequences form a self-consistent scheme agreeing with the actual observations, though not necessarily with commonly accepted extrapolations from the observations. (Bondi 1960: 7)

Although, in the discussion surrounding this passage, Bondi evidently is referring merely to "*A theory* ... based upon a powerful hypothesis which for various reasons may be widely regarded as implausible", it should be obvious that a primary instance of such a theory is none other than SS; and the powerful hypothesis is none other than PCP. Forty years after the fact, Bondi described the "philosophical attitude" which underlay his "implausible" PCP:

> But the essential point of the philosophy was and is that if the universe was evolving and changing, then there is no reason to trust what we call the laws of physics, established by experiments performed here and now, to have permanent validity. (Bondi 1990: 192)

Hence, or so Bondi's argument goes, since there is reason *not* to trust the laws of physics, let us presume that the universe is *not* evolving and changing; that is, let us presume PCP. Obviously, the linchpin of the argument here involves what Bondi takes to be the reasons against trusting the laws of physics. For those reasons, he is again indebted in no small measure to Milne. As a conclusion of this Section, let us review Bondi's arguments against observation and induction, against fact and extrapolation. As will be immediately obvious, the parallel between Bondi's stance and Milne's stance is overwhelming.

6.3. Bondi and the problems with observation and induction

First, consider observation and 'facts'. Milne's methodological starting point was the most general observation he could imagine: every observer's felt internal sense of the before-and-after of time. Other 'facts', Milne thought, were far too complicated and removed from direct observation to serve as secure starting points. Although Bondi's position is essentially identical to Milne's, he adopts it even more vehemently:

> The purely factual part of the vast majority of observational papers is small. It is also important to realise that these basic facts are frequently obtained at the very limit of the power of the instruments used, and hence are of considerable uncertainty. To refer to observational results as 'facts' is an insult to the labours of the observer, a mistaken attempt to discredit theorists, a disservice to astronomy in general and exhibits a complete lack of critical sense. Indeed I would go so far as to say that this sort of irresponsible misuse of terminology is the curse of modern astronomy. (Bondi 1955: 158)

An observational 'fact' for Bondi—as exactly for Milne before him—is extremely low-level: "At most it is a smudge on a photographic plate." (Bondi 1955: 158)[52] To get from the smudge to an observational 'fact', "corrections may have to be applied, calculations and reductions may have to be carried out, and above all, interpretations requiring a great deal of theoretical background may have to be made." (Bondi 1955: 157) It is this 'cooking' of the data—the smudges—which brings in the fundamental problem with observational astronomy. Ultimately, in observational work "*long chains of inferences are based on frequently somewhat uncertain data.*" (Bondi 1955: 158; emphasis in original) Because of this, Bondi believes, it is most likely that "theoretical results are of greater reliability." (Bondi 1955: 158)[53]

But if the starting point of the First Way—observation—is no better than uncertain, what can be said about the subsequent inductive processes, especially

[52] Compare Milne: "A certain area on a photographic plate is taken ... and attention is fixed on a number of small nebulous patches" (Milne 1934: 25).

[53] One is reminded here of the perhaps apocryphal statement attributed to Eddington: "No observation should be accepted until it has been verified by theory."

extrapolation? Just as Milne before him, Bondi eschews inductive logic: "it is this impossibility of direct abstraction from observations that rules out the usual inductive approach" from use in cosmology. (Bondi 1960: 10) Indeed, in the cosmological situation, no matter how 'natural' is the inductive impulse, it is far more dangerous than is theoretical 'speculation':

> There seems to be a widespread tendency to consider any extrapolation from observational data, however great, to be 'self-evident.' ... It is a dangerous habit of the human mind to generalize and to extrapolate without noticing that it is doing so ... ; extrapolations constitute a far greater danger to the progress of physics than so-called speculation. (Bondi 1960: 6)

It is clear that Bondi, just as Milne before him, mistrusted the empirico-inductivist First Way in both its epistemology and its logic: observational 'facts' are too uncertain, and inductive extrapolation is too dangerous.

As the evidence of this Section exhibits, there is a strong coupling between Bondi and Milne. First, Bondi much admired both Milne's work and his methods. Secondly, Bondi not only applauded the Second Way, the deductive way, he used it himself to create the steady-state theory, a theory which, by his own admission, is a descendent of Milne's KR. Finally, Bondi eschewed both the empiricism and the inductive logic of the adherents to the First Way, and this for the same reasons as Milne. If, as Dingle claimed, Milne is a Modern Aristotelian, then so also is Bondi!

7. Conclusion

Our goal in this paper was to reveal the connections between Milne and Bondi via a description and analysis of, first, Milne's commitment to the Second Way—the deductivist method—of doing cosmology; and secondly, the debate over this method among, on the one hand, Milne and his supporters, and, on the other hand, Dingle and the other supporters of the empirico-inductivist First Way to do cosmology. As we have shown, the methodological debate in cosmology was a rich, thorough, and sometimes forceful one; most importantly, the debate shaped the course of subsequent cosmological theorizing through its influence upon the form and content of Bondi's steady-state theory. One could hardly ask for a clearer demonstration of connections between two theories.

Acknowledgements

Presentations covering some of this material were given at Virginia Tech, Pittsburgh, Notre Dame, LSE, and the 1993 annual meeting of the History of Science Society. We have received some excellent suggestions from colleagues and students at these various venues, for which we offer grateful (and sometimes *very* grateful) thanks. Research funded in part by NSF grant DIR-8810699, and the UMKC Office of Research Administration, and the University of Missouri Research Board. G. G. has especially benefitted from conversations with Hank Frankel concerning the material in this essay. Eliseo Fernandez used his Linda Hall Library connection to rescue us more than once. Other essential information was most kindly provided in interviews with Herman Bondi, William Bonnor, Stephen Hawking, William McCrea, Dennis Sciama, A. G. Walker, and G. J. Whitrow. We thank all these helpful colleagues most vigorously, and hope none is disappointed by the use to which we put his graceful assistance!

References

Barrow, John D. & Tipler, Frank J. (1986). *The Anthropic Cosmological Principle*. Oxford: Clarendon Press.

Bondi, Hermann. (1948). "Review of Cosmology." *Monthly Notices of the Royal Astronomical Society* **108**: 104–120.

——— (1955). "Fact and Inference in Theory and in Observation." In *Vistas in Astronomy*. Vol. 1. Arthur Beer, ed. 155–162. London: Pergamon.

——— (1957). "Some Philosophical Problems in Cosmology." In *British Philosophy in the Mid-Century*. C. A. Mace, ed. 195–201. London: Allen & Unwin.

——— (1960). *Cosmology*. 2nd ed. Cambridge: Cambridge University Press.

——— (1975). "Interview with H. Bondi." In *Interviews with Astrophysicists*. 1–48. New York: American Institute of Physics.

——— (1990). "The Cosmological Scene 1945–1952." In *Modern Cosmology in Retrospect*. B. Bertotti, R. Balbinot, S. Bergia and A. Messina, eds. 189–196. Cambridge: Cambridge University Press.

Bondi, Hermann & Kilmister, Clive W. (1959). "The Impact of *Logik der Forschung*." *British Journal for the Philosophy of Science* **10**: 55–57.

Broad, Charlie Dunbar. (1940). "Sir Arthur Eddington's *The Philosophy of Physical Science*." *Philosophy* **15**: 301–312.

Crowther, James G. (1970). *Fifty Years with Science*. London: Barrie & Jenkins.

Dingle, Herbert. (1931). *The Evolution of the Universe*. London: Nature.

——— (1937). "Modern Aristotelianism." *Nature* **139**: 784–786.

——— (1953). "The President's Address." *Monthly Notices of the Royal Astronomical Society* **113**: 53.

Dirac, Paul A. M. (1937). "The Cosmological Constants." *Nature* **139**: 323.

Eddington, Arthur Stanley. (1931). "Contribution." In *The Evolution of the Universe*. Herbert Dingle, ed. 709. London: Nature.

——— (1933). *The Expanding Universe*. Cambridge: Cambridge University Press [Reprint: Ann Arbor: Ann Arbor Paperbacks: Univ. of Michigan Press (1958)].

——— (1935). *New Pathways in Science*. Cambridge: Cambridge University Press [Reprint: Ann Arbor Paperbacks. Univ. of Michigan Press (1959)].

——— (1939). *The Philosophy of Physical Science*. New York: Macmillan.

Gale, George. (1991). "Philosophical Aspects of the Origin of Modern Cosmology." In *Encyclopedia of Cosmology*. N. Heatherington, ed. 381–495. New York: Hackett.

——— (1992). "Rationalist Programmes in Early Modern Cosmology." *Astronomy Quarterly* **8**: 4–16.

Gale, George & Shanks, Niall. (1996). "Methodology and the Birth of Modern Cosmology." *Studies in the History and Philosophy of Modern Physics* **27**: 279–296.

Gale, George & Urani, John R. (1994). "Philosophical Midwifery and the Birth Pangs of Modern Cosmology." *American Journal of Physics* **61**: 66–73.

McCrea, William H. (1939). "The Evolution of Theories of Space-Time and Mechanics." *Philosophy of Science* **6**: 137–162.

——— (1950). "Edward Arthur Milne." *Obituary Notices of Fellows of the Royal Society* **7**: 421–443.

——— (1987). "Clustering of Astronomers." *Annual Review of Astronomy and Astrophysics* **25**: 1–22.

McVittie, George C. (1940). "Kinematical Relativity." *Observatory* **63**: 271–281.

——— (1990). "Interview, 21 March 1978." In *Interviews with Astrophysicists*. New York: American Institute of Physics. Microfilm.

Milne, Edward A. (1929). "The Aims of Mathematical Physics." 19 November 1929. Modern Manuscripts, Bodleian Library, Oxford.

——— (1934). "World-models and the World-picture." *Observatory* **57**: 24–27.

——— (1935a). *Relativity, Gravitation and World-Structure*. Oxford: Clarendon Press.

——— (1935b). "Reviews, Reviewers and Reviewed." *Observatory* **58**: 124–126.

——— (1940). "Cosmological Theories." *Astrophysical Journal* **91**: 129–158.

——— (1941). "Remarks on the Philosophical Status of Physics." *Philosophy* **16**: 356–370.

——— (1943). "The Fundamental Concepts of Natural Philosophy." *Proceedings of the Royal Society* (Edinburgh) **63**: 10–24.

——— (*Corr. C*). [Correspondence with Subrahmanyan Chandrasekhar, 1929–1949]. This correspondence is in Modern Manuscripts, Bodleian Library, Oxford and is used with permission.

——— (*Corr. W*). [Correspondence with A. G. Walker]. This correspondence is in the possession of Professor Walker and is used with permission.

Milne, Edward A., et al. (1937). "On the Origin of Laws of Nature." *Nature* **139**: 997–1008.

Robertson, Howard P. (1935). "Kinematics and World-Structure." *Astrophysical Journal* **82**: 284–301.

——— (1936a). "Kinematics and World-Structure II." *Astrophysical Journal* **83**: 187–201.

——— (1936b). "Kinematics and World-Structure III." *Astrophysical Journal* **83**: 257–271.

Robinson, Richard. (1953). "Necessary Propositions." *Mind* **67**: 209.

Sciama, Dennis. (1990). "Interview with George Gale." 9 January 1990. Taped interview. [Copies of the tape are available for transcription from George Gale].

Singer, Charles Joseph. (1921). *Studies in the History and Method of Science*. Vol. 2. Oxford: Clarendon Press.

Singh, Jagjit. (1970). *Great Ideas and Theories of Modern Cosmology*. New York: Dover.

Smith, Norman Kemp, trans. (1929). *Kant. Critique of Pure Reason*. New York: St. Martin's Press.

Tolman, Richard C. (1932). "Models of the Physical Universe." *Science* **75**: 367–373.

Urani, John R. & Gale, George. (1994). "E. A. Milne and the Origins of Modern Cosmology: An Ubiquitous Presence." In *The Attraction of Gravitation: New Studies in the History of General Relativity* (Einstein Studies, vol. 5). J. Earman, M. Janssen and J. D. Norton, eds. 390–419. Boston: Birkhäuser.

Whitrow, Gerald J. (1990). "Interview with George Gale." 5 January 1990. Unpublished. [Copies of the interview tape are available for transcription from George Gale].

Whittaker, Edmund T. (1941). "Some Disputed Questions in the Philosophy of the Physical Sciences." *Procceedings of the Royal Society* (Edinburgh) **61**: 160–175.

——— (1951). *Eddington's Principle in the Philosophy of Science*. Cambridge: Cambridge University Press.

Steady-State Cosmology and General Relativity: Reconciliation or Conflict?

Helge Kragh

DURING THE LATE NINETEEN-FORTIES Einstein's general theory of relativity was still considered an esoteric theory with almost no connection to experimental reality. Although appreciated by mathematical physicists because of its logical beauty and challenging theoretical problems, in general it seemed irrelevant to empirical physics and was neither appreciated nor well known by the majority of physicists (Eisenstaedt 1986). Cosmology was one of the few areas in which general relativity held a high status, but even in this area it was challenged by rival theories, most of which had their origin in the 1930s (North 1965; Merleau-Ponty 1965). Among these were neo-Newtonian theories, Paul Dirac's varying-G theory and its elaboration by Pascual Jordan, and, in particular, Edward Milne's theory of kinematic relativity. However, by 1948 Milne's alternative was rapidly running out of steam and none of the other alternatives were considered very serious rivals. It looked as if general relativistic cosmology was on its way to attain a kind of paradigmatic status within the small world of cosmologists.

Relativistic cosmology derived its high status from three sources. For one thing, general relativity was held in high esteem because of its comprehensiveness and appealing logical structure. But one thing is to accept the validity of the general theory of relativity, another is to accept its unrestricted validity and applicability to the universe at large. Why should the theory provide the one and only correct framework for cosmology? Historical reasons undoubtedly played a role, for modern cosmology was after all founded on Einstein's theory and its success in accounting for—even predicting—the expansion of the universe added much to the reputation of relativistic cosmology. As a third, important reason was the flexibility and *lack* of unique predictions that characterized the cosmology based on general relativity.

H. Goenner, J. Renn, J. Ritter, T. Sauer (eds.)
The Expanding Worlds of General Relativity
(Einstein Studies, volume 7) pp. 377–402

Relativistic evolution cosmology anno 1948 (or, for that matter, today) was not a theory in the ordinary sense of the word. It was rather a supermarket of theories which had in common that they were all solutions of the same fundamental equations, but otherwise described very different model universes. The basis was Einstein's cosmological field equations of 1917

$$R_{\mu\nu} - \frac{1}{2} g_{\mu\nu} R - \Lambda g_{\mu\nu} = -\kappa T_{\mu\nu}, \tag{1}$$

where Λ denotes the cosmological constant, often taken to be zero. It was generally assumed that the cosmologically relevant solutions were restricted to those satisfying the Cosmological Principle, that is, the solutions of the Friedmann-Lemaître equations[1]

$$\begin{cases} 3\left(\frac{\dot{R}}{R}\right)^2 + 3\frac{kc^2}{R^2} = \left(\Lambda + \kappa\rho c^2\right)c^2, \\ \left(\frac{\dot{R}}{R}\right)^2 + 2\frac{\ddot{R}}{R} + \frac{kc^2}{R^2} = \left(\Lambda - \kappa p\right)c^2. \end{cases} \tag{2}$$

Some of the solutions describe models with a beginning in time ('big bang' models), while others, such as the Lemaître-Eddington model, represent universes of an infinite past age. Although there was no particular model favored about 1948, most relativist cosmologists had become accustomed to the idea of a big bang universe in Lemaître's sense. At least tacitly, it was generally agreed among mainstream relativists that the world has a finite age and its evolution is described by the laws of general relativity.

Even within the class of Friedmann-Lemaître solutions, the number of expanding model universes is very large, including, for example, the Lemaître-Eddington model (with $\Lambda > 0$), the Lemaître model ($\Lambda > 0$), and the Einstein-De Sitter model ($\Lambda = p = k = 0$), not to speak of a range of oscillatory models. The richness of models may seem a methodological embarrassment, but it also meant that relativistic cosmology was able to accomodate new observations, precisely because of the free parameters appearing in the theory. Simply put: If one of the models did not agree with observations, there was always another candidate to fall back on.

The flexibility, and the relativists' willingness to exploit it, was demonstrated in connection with the time-scale difficulty, i.e., the problem that the age of the earth (or stars, or galaxies) appeared to exceed the age of the universe as inferred from the Hubble constant. With the accepted value of the Hubble parameter, the age of the universe would be less than 1.8 billion years according to all Friedmann-Lemaître models with a zero cosmological constant. The age of the earth, based on reliable methods of radioactive dating, was known to be more than 3 billion years. Einstein was deeply worried about the discrepancy and indicated that it

[1] The symbols have their standard meaning: R = cosmic scale factor, $\dot{R}$ = the time derivative of R, k = space curvature parameter, κ = Einstein gravitation contant, ρ = density, p = pressure, Λ = cosmological constant, H = Hubble parameter.

might be necessary to give up relativistic cosmology (Einstein 1945: 132), but other relativist cosmologists took the problem much less seriously. By exploiting the flexibility of the relativistic equations they were able to avoid the dilemma, for example by introducing a positive cosmological constant as in the Lemaître model (e.g., Gamow 1949)—although in all other connections they would put $\Lambda = 0$. Another possibility was to introduce inhomogeneities in the relativistic equations, such as considered by Richard Tolman and his student Guy Omer in 1948. After having designed an inhomogeneous model universe with the age 3.6 billion years, Omer concluded optimistically that "[there is] a very real flexibility of the general theory of relativity. It is not necessary to invent any new physical principles to rationalize the currently accepted observational data" (Omer 1949: 176). Tolman agreed and summarily dismissed the possibility that the time-scale difficulty might indicate a failure in relativistic cosmology. As he saw it, the general theory of relativity was the rock on which scientific cosmology was solidly founded; without this foundation, cosmology would be open to all kinds of "unbridled fancy" (Tolman 1949: 377).

The cosmological fancies Tolman referred to included the theories of Milne and Dirac as well as various tired-light alternatives to the cosmic expansion. Had he known about the new steady-state theory, he would undoubtedly have included it too. Far from considering the "very real flexibility" of the relativistic theory a virtue, to Fred Hoyle, Hermann Bondi and Thomas Gold it signified that relativistic cosmology was methodologically unsound and in need of replacement. It was one among several reasons why the three Cambridge physicists developed the steady-state theory of the universe. The "new cosmology," as it was often called, appeared a few months after Tolman had so briskly dismissed non-relativistic alternatives (Bondi and Gold 1948; Hoyle 1948).[2]

1. The steady-state challenge

The basic motivation of Bondi, Gold and Hoyle was dissatisfaction with the relativistic cosmology, not with the theory of general relativity itself. Bondi and Hoyle (but less so Gold) mastered and appreciated Einstein's theory, but when it came to cosmology they denied its validity and wanted to replace it with a more satisfactory theory which allowed for a universe without the dubious 'beginning' of the big bang models. The result of their considerations was two different versions of the steady-state theory which led to the same observational consequences, the most important of which was the continual creation of matter postulated in order to harmonize the Perfect Cosmological Principle with the expansion of the universe. (The Perfect Cosmological Principle is the requirement that the large-scale features of the universe shall be independent of the spatial as well as temporal location of the observer.) Since matter creation was an essential element of the

[2] A comprehensive treatment of the history of steady-state cosmology is included in Kragh 1996. Earlier accounts of the theory can be found in North 1965 and Merleau-Ponty 1965, and a more modern discussion in Balashov 1994. Parts of the present paper relies on Kragh 1996.

new cosmology, and the phenomenon flatly disagreed with general relativity, so did the steady-state theory.

Einstein never commented in public on the steady-state theory, but it is doubtful if the theory interested him much, and unlikely that he would find it appealing. The only evidence I know of is reported by Peter Michelmore and relates to a lecture Hoyle gave in Princeton in 1952. Einstein was asked what he thought of the new cosmology, and is to have dismissed the theory as a "romantic speculation."[3] If that were Einstein's reaction, he was not alone. The steady-state theory faced stiff and often emotional opposition ever since it was proposed in 1948. Pauli was one of many physicists who rejected the theory because of its violation of the principle of energy conservation. "I consider his [Hoyle's] continuous creation of matter from nothing to be pure madness," he wrote in a letter of 1951, "I see no reason to doubt the conservation of physical energy. I realize that this kind of cosmogonies are *not physics at all, but projections of the unconscious*."[4]

The Bondi-Gold theory was founded deductively on the Perfect Cosmological Principle and differed drastically in style, method and theoretical content from relativistic cosmology. As Bondi emphasized in 1950, "I do not regard cosmology as a minor branch of general relativity" (Bondi 1952: Preface). Bondi and Gold realized of course that the steady-state theory was in conflict with the law of energy conservation, and then with general relativity, but saw no reason to accept energy conservation as an absolute law (see also below). What mattered, according to Bondi, was which principle led to the simplest description of nature: "The principle resulting in greatest overall simplicity is then seen to be not the principle of conservation of matter but the perfect cosmological principle with its consequence of continual creation. From this point of view continual creation is the simplest and hence the most scientific extrapolation from the observations" (Bondi 1952: 144). Bondi and Gold dismissed relativistic cosmology as being scarcely scientific—they judged it "utterly unsatisfactory"—and unequivocally distanced themselves from all attempts to harmonize the steady-state theory with the general theory of relativity.

From a methodological point of view Bondi and Gold objected strongly to relativistic cosmology's flexibility and adaptability to empirical facts. This feature stands in marked contrast to the structure of the general theory of relativity itself. As Einstein emphasized in 1948, his theory is a "theory of principle" (rather than a "constructive theory") the merit of which is its "logical perfection, and the security of [its] foundation." Therefore, he wrote, "If any deduction from it should prove untenable, it must be given up. A modification of it seems impossible without destruction of the whole" (Einstein 1950: 58). It is interesting to note that Bondi

[3] Michelmore 1962: 253. The remark was made in a conversation with Manfred Clynes, a young pianist and acquaintance of Einstein, who apparently reported it to Michelmore.

[4] "Seine continuous creation der Materie aus Nichts halte ich für reinen Unsinn. Ich sehe keinen Grund, an der Erhaltung der physikalischen Energie zu zweifeln. Ich bin mir darüber klar, daß diese Art von Kosmogonien gar keine Physik, sondern Projektionen des Unbewußten sind." Pauli to Aniela Jaffé, 3 December 1951, as reproduced in Meyenn 1996: 447. Pauli's reference to "physical energy" should be seen in relation to his deep interest in Jung's depth psychology, where psychical energy appear. For other early responses to the steady-state theory, see Kragh 1996: 186–201.

and Gold presented their steady-state alternative as precisely a "theory of principle" in Einstein's sense. Indeed, in their methodological arguments for the theory they repeated Einstein's words almost verbatim, only with "it" referring to the new theory of the steady-state universe.

Hoyle's theory was closer in spirit to relativistic cosmology than the more radical Bondi-Gold version; it had less character of a 'theory of principle' and more of a 'constructive theory,' to use Einstein's terminology. Although Hoyle too criticized standard relativistic cosmology, and although he accepted the Perfect Cosmological Principle, he wanted to establish an alternative without leaving completely the framework of general relativity. This wish to modify, rather than reject, the relativistic framework was evident already in the 1948 theory and became even more evident in Hoyle's later versions of the steady-state theory. In 1948 he expressed his program as follows: "Using continuous creation of matter, we shall attempt to obtain, within the framework of the general theory of relativity, but without introducing a cosmical constant, a universe satisfying the wide cosmological principle that shows the required expansion properties and in which localized condensations are continually being formed" (Hoyle 1948: 375).

First of all, Hoyle's theory was a field theory with a mathematical formulation as close to relativistic cosmology as its results were far from it. In order to make the field equations accomodate continual creation of matter, Hoyle in effect replaced the cosmological term $-\Lambda\, g_{\mu\nu}$ in Equation (1) with a symmetric tensor of non-vanishing divergence, the "creation tensor" $C_{\mu\nu}$:

$$R_{\mu\nu} - \frac{1}{2} g_{\mu\nu} R + C_{\mu\nu} = -\kappa S_{\mu\nu}. \tag{3}$$

The reason for denoting the energy-stress tensor $S_{\mu\nu}$, and not $T_{\mu\nu}$, will be clear from what follows. As it soon turned out, "creation" was an explosive term that would invite much controversy. However, Hoyle innocently used it in the neutral sense of 'formation' or 'origin,' simply as expressing the existence of matter where none had been before. He found creation an appropriate term because pair creation was already in widespread use among quantum physicists (and Hoyle was trained in quantum theory before he changed to astronomy).

From the start given by Equation (3) Hoyle was led to a line element of the De Sitter type, namely

$$ds^2 = c^2\, dt^2 - e^{\frac{2ct}{a}} \left[dr^2 + r^2(d\theta^2 + \sin^2\theta\, d\phi^2)\right].$$

For the constant a he deduced that $3/a^2 = \kappa\rho$ and further, by comparing with Hubble's law, that $a = c/H$. In De Sitter's original form for the metric, or rather the form given by Lemaître in 1925, the expression is

$$ds^2 = c^2\, dt^2 - e^{2ct\sqrt{\frac{\Lambda}{3}}} \left[dr^2 + r^2(d\theta^2 + \sin^2\theta\, d\phi^2)\right].$$

Hoyle's steady-state metric can thus be formally derived from the De Sitter metric by the substitution

$$\frac{\Lambda}{3} \longrightarrow \frac{c}{H} = \sqrt{\frac{3}{\kappa\rho}}$$

For the rate of mass creation, Hoyle found the same expression as Bondi and Gold, namely $3\rho H$.

Hoyle did not explicitly point out the analogy in the previous equation, but he did mention that "the $C_{\mu\nu}$ in [Eq. (3)] plays a role similar to that of the cosmical constant in the De Sitter model" (Hoyle 1948: 376). Neither Hoyle nor Bondi and Gold considered retaining the cosmological constant, which they, in agreement with Einstein, dismissed for aesthetical reasons.

Bondi and Gold rejected Hoyle's field-theoretic formulation, which they found "unsatisfactory and unacceptable" and much too close in spirit to the cosmology based on general relativity. Hoyle agreed that "the discrepancies [between astronomical data and relativistic cosmology] can only be resolved by a modification of [Einstein's field equations], or by an entirely new theory," but he preferred the first choice, to keep as closely as possible to existing theory and to try to incorporate the idea of matter creation in it. As he wrote:

> The question of the continuous creation of matter should be discussed by a modification of existing theory rather than by attempting simultaneously to build an entirely new description of the universe. That Einstein's general theory of relativity may ultimately be replaced by a theory of wider scope is no argument against this procedure. (Hoyle 1949: 365, 370)

Hoyle's choice was in part pragmatic, for he realized that a field theory was much better adapted for further development than the closed, deductive system of Bondi and Gold. During the controversy that followed, the three Cambridge physicists fought united and the disagreement between the two versions played no significant role. Almost all of the technical development of the steady-state theory took place within Hoyle's version and for this reason we shall have nothing more to say about the Bondi-Gold version.

Much of the work on steady-state theory that took place during the 1950s was motivated by a wish to modify the theory in such a way that it might become acceptable to the majority of physicists and astronomers who regarded general relativity the natural framework for cosmology. The concept of continual creation of matter and the relationship to the general theory of relativity were of focal interest to those scientists who sympathized with the aspirations of the new cosmology, but hesitated because of its controversial features. If the steady-state theory could somehow be harmonized with general relativity, and the objectionable creation of matter be accounted for within the framework of existing physical theory, then the steady-state universe would become a much more acceptable idea. But could the magic be done?

2. McCrea's theory

The most important contribution to the steady-state theory was a new version presented by McCrea in 1951 (McCrea 1951; North 1965: 215–217). William Hunter McCrea, then a 47-year-old professor of mathematics at the Royal Holloway College of the University of London and a former collaborator of Eddington and Milne, was favorably inclined to the steady-state theory even before it was published; as Secretary for the Royal Astronomical Society and responsible for its publications he had been instrumental in getting the two papers published in the summer of 1948. In 1950 he published a review of the new cosmology, which he found to be highly interesting and worth developing further (McCrea 1950a).

McCrea wanted to develop Hoyle's theory so as to make it accord better with general relativity and, in particular, to satisfy some kind of energy conservation. As mentioned, in Hoyle's work Einstein's field equations were modified by adding the C-tensor in much the same way as the cosmological term was added by Einstein in 1917. The procedure was widely seen as ad hoc and as destroying much of the aesthetic appeal of the original field equations. Furthermore, the addition of a term (other than $\Lambda g_{\mu\nu}$) is not unique and thus opens up a variety of additions and a corresponding variety of cosmological models—which destroys the unique feature of steady-state cosmology which its advocates found so appealing. For these reasons McCrea argued that if modifications had to be made, they should not be modifications in the equations, but in the physical interpretation of the energy-stress tensor $T_{\mu\nu}$.

Ever since Einstein started his search for a unified field theory he had thought that his cosmological field equations (1) were "somehow in bad taste," to use Leopold Infeld's expression (Infeld 1957). This was in part because of the mixing of a geometrical and a physical tensor, which appeared unaesthetical to a monist thinker like Einstein; and in part because of the arbitrariness in the energy-stress tensor. It was this freedom in the interpretation of the energy-stress tensor which McCrea exploited. In Hoyle's Equation (3) the tensor $R_{\mu\nu} - (1/2)\, g_{\mu\nu}\, R$ is of zero divergence, just as in Einstein's Equation (1) with $\Lambda = 0$. That is, $\kappa\, S_{\mu\nu} + C_{\mu\nu}$ is a conserved quantity in the sense of having a vanishing divergence and corresponds, in this respect, to $\kappa\, T_{\mu\nu}$ in the relativistic theory. As noted by McCrea, Hoyle's field equations would in a formal sense become identical with those of general relativity if the substitution

$$\kappa\, T_{\mu\nu} = \kappa\, S_{\mu\nu} + C_{\mu\nu} \tag{4}$$

was made. The idea was thus to subsume the creation term and the ordinary energy-stress tensor $S_{\mu\nu}$, appearing separately in Hoyle's equation, under a new tensor with a different physical interpretation. Then $T^{\mu\nu}{}_{;\nu} = 0$, but $S^{\mu\nu}{}_{;\nu} \neq 0$.

McCrea relied on a result obtained by Edmund Whittaker in 1935, namely, that the density of gravitational mass in general relativity can be expressed as

$$\sigma = \rho + \frac{3p}{c^2} \tag{5}$$

where ρ is the 'relative density' of mass. He further employed a new kind of Newtonian analogy in which terms of the order c^{-2} were included as significant in classical theory, contrary to what was the case in the 1934 McCrea-Milne theory. The strict Newtonian analogy was only valid for $p = 0$, but with the new method McCrea obtained formulae which corresponded completely to those of general relativity also for $p \neq 0$, i.e., the Friedmann-Lemaître equations (2). The extension amounted to recovering the continuity equation

$$\frac{d}{dt}(\rho c^2) + \frac{3p}{c^2} R^2 \dot{R} = 0,$$

in which $\rho > 0$ and p is assumed to be ≥ 0, which for an expanding universe leads to a decreasing density, $d\rho/dt < 0$.[5]

Considering a steady-state universe as given by $k = \Lambda = 0$ and $R = \exp(ct/a)$, or $\dot{R}/R = c/a$, McCrea found by insertion in Equations (2) that

$$\begin{cases} \rho = \dfrac{3}{\kappa a^2}, \\ p = -\dfrac{3c^2}{\kappa a^2}, \end{cases} \tag{6}$$

where a has the same meaning as in Hoyle's work, i.e., $a = c/H$. That is, whereas in Hoyle's theory the pressure was taken to be zero, McCrea found that

$$p = -\rho c^2. \tag{7}$$

By inserting into Equation (5) he then obtained

$$\sigma = -\frac{6}{\kappa a^2} = -\frac{3H^2}{4\pi G}.$$

Up to this point, McCrea had merely rederived results already known from the standard relativistic analysis of the De Sitter universe. Equation (7) was well known from De Sitter's 1917 work, but in this and other classical analyses the pressure was assumed to consist of a matter and a radiation term and thus being positive or zero. McCrea now noted that this interpretation of $T_{\mu\nu}$ was not compulsory. What would happen, he asked, if a uniform *negative* pressure was admitted in accordance with Equation (6)?

The suggested reinterpretation of $T_{\mu\nu}$ corresponded to the introduction of a zero-point stress as an alternative to Hoyle's C-tensor, namely, a change in the zero-level in accordance with Equation (4). In Hoyle's model the expansion was caused by an outward pressure produced by the created matter, whereas McCrea explained the expansion as a result of the negative pressure which corresponds

---

[5] As McCrea noted, these results were well known from relativistic cosmology. See, e.g., Tolman 1934: 381–383.

to the gravitational mass between the observer and a galaxy being negative. He found the work done by the negative pressure (per unit volume and time) to be

$$\frac{d^2W}{dV\,dt} = \frac{3\rho c^3}{a} = 3\rho H c^2.$$

It was this work which, if transformed by means of $E = mc^2$, reappeared in a rate of increase in mass density of $3\rho H$. Consider with McCrea any sphere of fixed radius centred around an observer. The observer will assert that matter is continually receding across the surface of the sphere and being replaced by newly created matter. In 1953 McCrea explained the mechanism as follows:

> By virtue of the negative pressure in the moving medium, the medium outside the sphere is doing work upon the medium inside. . . . *The net flow of energy across the surface is zero.* We can think of the medium as expanding under such a tension that it continually replenishes itself by the work it does upon itself. Therefore the process is energetically self-supporting. The only novel feature in the creation process is the supposition that the energy represented by the work done makes itself apparent in the form of, or is continually being converted into, matter of the same sort as the matter already present. (McCrea 1953: 50)

The pressure appearing in McCrea's theory, but not in Hoyle's, was of no direct physical significance because its gradient vanishes everywhere and thus will have no mechanical effects: "The pressure which it is most natural to invoke happens to be negative, but in any case has no direct accelerating effect because it is uniform, but has such an effect indirectly because of its (negative) contribution to the gravitational mass" (McCrea 1951: 572). One may speculate that McCrea was inspired by Dirac's 1931 theory of 'holes,' in which an analogous feature appeared.[6]

McCrea's reinterpretation remained within the steady-state theory and led to the very same observational results as Hoyle's theory. But instead of making use of the creation process itself as the primary postulate, McCrea's introduction of a zero-point stress in space shifted the focus of the theory away from the mysterious creation of matter, which was no longer seen as a genuine *creatio ex nihilo* process, but rather as a kind of transmutation. In the new interpretation the creation process was a consequence of the existence of a space endowed with negative pressure.

Methodologically, McCrea's work was closely related to Hoyle's, which it developed further in a direction away from the radical deductivism favored by Bondi and Gold, and towards conformity with the general theory of relativity. For example, McCrea did not consider the Perfect Cosmological Principle a fundamental principle of nature because it was not applicable to small-scale physics. One of

---

[6] In his theory of 1931 Dirac introduced an infinite 'sea' of uniformly distributed negative-energy electrons. Because of the uniformity the hypothetical particles would produce no electromagnetic fields and thus be unobservable, merely serving to define a state of normal electrification. Dirac hypothesized that vacant negative-energy states, appearing as particles of positive energy, were identical with protons. (There is no analogous feature in McCrea's cosmic negative-energy substratum.) For details, see Kragh 1990: 89–103.

his ultimate aims was to bridge cosmology and quantum theory, and for this purpose field theory, and not cosmological principles, would have to be used. (Like Hoyle, McCrea had come to astronomy via quantum theory.) Similarly, in McCrea's steady-state theory, Mach's principle did not have the important function that it had in the Bondi-Gold theory. He emphasized as an advantage that in his theory the behavior of any large part of the universe was ascribed to conditions in that part. This element of 'localism' was contrary to the 'holistic' features in the Bondi-Gold theory. "It is only a matter of mathematical convenience to treat a model of the whole universe," McCrea wrote in 1951. "The essential features of the physical interpretation do not depend upon assumptions about the whole universe." (McCrea 1951: 574) With regard to these questions McCrea was in full agreement with Hoyle. The kind of steady-state theory developed by McCrea and others during the 1950s may appropriately be called the Hoyle-McCrea theory.

3. McVittie, Pirani, and others

McCrea's alternative interpretation of the steady-state theory inspired other British scientists to take up similar investigations, including George McVittie, Felix Pirani, P. Roman, and William Davidson. McVittie soon became an ardent antagonist of steady-state cosmology and an advocate of relativistic orthodoxy, but for a brief period in the early 1950s he seems to have found appealing features in the new cosmology. In 1952 he suggested a new world model which incorporated a revised form of McCrea's theory in a framework entirely within the realm of general relativity. "It will be shown," he wrote,

> that a 'continuous' creation process can exist in a suitably chosen general relativity model of the universe. The method employed is that of a priori cosmology in which the model universe is established, not from observational data, but by postulating certain principles to which the universe is supposed to conform. These principles are, in effect, restrictions on the model universe imposed by the investigator because he believes them to be reasonable, or because they are in agreement with his epistemological or philosophical views. (McVittie 1952: 296)

This was a view entirely in agreement with the metaphysical commitments of the steady-state theoreticians. However, McVittie's attitude to the steady-state theory and other rationalistic cosmologies changed to open hostility during the 1950s.

McVittie's model was what he called a "gravitationally steady state" in which the gravitational mass density σ is constant and the acceleration is given by

$$\frac{\ddot{R}}{R} = \frac{\Lambda c^2}{3} - \frac{\kappa \sigma c^4}{6}.$$

McVittie studied in particular the hyperbolic solution

$$R = A \cosh \omega t, \quad \text{with} \quad \omega^2 \equiv \frac{\Lambda c^2}{3} - \frac{\kappa \sigma c^4}{6} > 0.$$

Such a universe has an open past and an open future (i.e., existing for all times), but with a scale factor R which has a minimum at some time $t = 0$. For $t < 0$ the universe is contracting and for $t > 0$ it is expanding. The stress is time-dependent, but negative for all t and with the same meaning as in McCrea's theory, i.e., as a zero-point stress. Following McCrea, McVittie showed that in the expanding phase stress would be converted into matter, and in the contracting phase the reverse process would take place. The conversion follows from $s =$ constant, which is the same as

$$\frac{d\rho}{dt} = -\frac{3}{c^2}\frac{dp}{dt}.$$

"A conversion of energy in the form of stress into energy in terms of matter may seem strange to those accustomed to think in terms of the concepts of Newtonian hydrodynamics, but it is really implicit in general relativity" (McVittie 1951: 298). McVittie showed that under certain assumptions the model would lead to a present average density of matter of $\rho = 2.4 \times 10^{-29}$ g cm^{-3}. This he considered more satisfactory than the much higher value of the Hoyle-McCrea theory (5×10^{-28} g cm^{-3}) because it deviated less drastically from the observed value.

The important feature in McVittie's work was that it presented, for the first time, a world model with matter creation in complete conformity with general relativity. Energy was conserved in the same sense as in McCrea's theory, namely, that $T^{\mu\nu}{}_{;\nu} = 0$, but $S^{\mu\nu}{}_{;\nu} \neq 0$. McVittie considered a perfect fluid with 4-velocity v^μ and energy-stress tensor

$$T^{\mu\nu} = (\rho + \frac{p}{c^2})\, v^\mu v^\nu - g^{\mu\nu}\frac{p}{c^2}.$$

Within standard cosmological models it follows from $T^{\mu\nu}{}_{;\nu} = 0$ that

$$\frac{d\rho}{dt} + 3\frac{\dot{R}}{R}\left(\rho + \frac{p}{c^2}\right) = 0,$$

which is a generalized continuity equation. For $p = 0$ and $\rho = \rho_0$, where ρ_0 is the rest-mass density, the result is

$$\frac{d\rho_0}{dt} + 3\frac{\dot{R}}{R}\rho_0 = 0 \quad \text{or} \quad d(\rho_0 R^3) = 0.$$

It was this equation which McVittie and most other cosmologists in the 1950s took to represent 'matter conservation.'

In many respects, McVittie's world model was a hybrid between steady-state and relativistic cosmology, but in spite of having matter creation and no beginning in time it was not really a steady-state model, at least not in the classical sense. With a creation rate depending on the epoch, it did not satisfy the Perfect Cosmological Principle. McVittie's interest in McCrea's theory did not indicate any support of the steady-state theory, and he seems not to have considered his model with matter creation as a candidate for the real universe. His "suitably chosen general relativity model of the universe" is rather to be considered a demonstration of how powerful and flexible a theory general relativity is. McVittie in fact considered his model to disagree with observations and pointed out "a disadvantage which it shares with other a priori models," namely "its vulnerability to observational tests." In regard of how highly Bondi and his allies praised the steady-state theory's methodological virtues exactly in terms of vulnerability à la Popper, it is remarkable that McVittie considered the same feature a disadvantage.

The works of McCrea and McVittie helped to demystify the concept of continuous creation of matter in the sense that it could now be argued that the process was compatible with general relativity and had its source in a universal stress of negative energy. However, the demystification had curiously little impact on the controversy. Most opponents of the steady-state theory ignored McCrea's idea of a negative intergalactic pressure and maintained that steady-state cosmology and general relativity were irreconcilable quantities. The British physicist William Bonnor was one of the few relativist cosmologists who publicly admitted the validity of the idea—but without therefore admitting the soundness of the steady-state program. As he wrote in 1955, granted the existence of McCrea's negative pressure, then "indeed, the steady-state model does satisfy Einstein's field equations" (Bonnor 1955: 23).

Negative pressure or not, the steady-state creation process was completely obscure from a physical point of view, and the lack of a more definite mechanism made it impossible to discuss the hypothetical process within the framework of quantum physics. Even the sort of created particles—hydrogen atoms, neutrons, or protons and electrons—was pure guesswork. This was a problem which was often discussed among the steady-state theoreticians, but it was usually assumed that a solution would have to wait for some future quantum theory of gravity. Yet even without such a theory it was possible to suggest mechanisms which would bring continual creation of matter within the ambit of more conventional microphysics.

The most interesting of these was suggested by Felix Pirani, who as a young postdoctoral student in Cambridge developed a strong interest in the new cosmology. Pirani had previously worked with Alfred Schild, as his first graduate student, on the quantization of Einstein's field equations using Dirac's new (1950) dynamics of constrained systems. After having received his doctorate in 1951, he went to Bondi in Cambridge and was inevitably exposed to cosmology. Pirani felt the steady-state theory appealing, not only because it resolved the time-scale paradox, but also because it minimised the scope of divine intervention in nature. Realizing that there was a need for connecting continual creation with conventional physics, he wanted to formulate an alternative to Hoyle's C-tensor, which he considered an arbitrary addition to the field equations.

Pirani's idea was to regard the creation of particles in steady-state theory as a kind of collision process in which new entities were produced in addition to the ordinary material particles (Pirani 1955). The new entities, which he named "gravitinos," would exert a pressure of the same kind as an ordinary radiation pressure. They were strange particles indeed. Like the neutrino, a gravitino would have zero rest-mass and be uncharged, but it would also have negative energy and momentum.

"Rather than proliferate elementary particles," Pirani speculated that gravitinos might in fact be negative-energy neutrinos.[7] From this point of view a β-decay process would be described as an absorption of a gravitino. He suggested that the proposed mechanism "may be susceptible to description in quantum-mechanical terms," possibly within the framework of the theory of weak interactions. If gravitinos were a kind of neutrinos this was not an unreasonable suggestion, but it remained undeveloped. Pirani expressed the hope that his hypothesis might be susceptible to experimental test, but unlike the neutrino (which was detected in 1956), the gravitino seems to have been met with silence by experimental physicists. According to *Science Citation Index*, Pirani's paper was only cited twice in the period 1955–1963, and none of the citing authors were concerned with the gravitinos in particular.

The advantage of the hypothesis was that energy and momentum conservation could be upheld even in the process of creation itself. This could be done if it was assumed that gravitinos were produced in such a way that the total 4-momentum in the creation process is zero. If gravitinos are produced together with material particles the reverse process, annihilation of gravitinos and particles, must presumably also take place. However, Pirani gave a heuristic argument that in an expanding universe the annihilation rate would be much smaller than the creation rate and hence could be ignored. On the basis of the gravitino hypothesis Pirani managed to build up a cosmological theory largely identical with the ordinary steady-state theory. There was the difference, though, that in Pirani's theory the mean density of matter would be four times larger than in Hoyle's theory (namely given by $\rho = 12H^2/\kappa c^2$).

4. Matter creation and energy conservation

The works of McCrea, McVittie and Pirani showed that continual creation of matter could be dealt with in a rational, scientific way and thus countered some of the objections against the steady-state theory. Matter creation was hotly debated at the time both among physicists and philosophers, but the crucial point under discussion was the spontaneous creation of matter *out of nothing*, and in this regard the new works did not change much. For example, the American philosopher Milton Munitz attacked the theory precisely because Bondi, Gold and Hoyle introduced particle creation *ex nihilo*, which he claimed was a meaningless concept. Yet this was not at all essential to the steady-state theory, which merely claimed that matter appeared continually where none had been before. As shown by McCrea and Pirani, one could even devise a physical mechanism or cause for the creation. Munitz admitted this to be a methodological improvement, but argued that with

[7] The 'gravitinos' appearing in the supergravity theories of the late 1970s are fermionic partners of the graviton boson. The name of the supergravity gravitino seems to have been suggested in ignorance of Pirani's earlier particle. Apart from the name, the two hypothetical particles have nothing in common. I am not sure when negative-energy particles were first discussed as realistic objects in physics, but in the early 1960s they entered in some theories of tachyons (that is, hypothetical faster-than-light particles).

such a change in the interpretation of the theory it was no longer justified to speak of 'creation' of matter (Munitz 1957). Much of the heated debate over matter creation boiled down to semantics.

Munitz demanded that creation of matter be explained, or at least explainable, in order to qualify as a scientific concept. Hoyle, on the other hand, considered matter creation to be inexplainable in the sense that it is the natural state of affairs of the world, the background situation upon which all phenomena would have to be judged. There is an elementary force in this argument. After all, which state of nature is the natural and unperturbed one, and hence not in need of explanation, is a question which cannot be answered by reference to tradition. About 1600, Aristotelians maintained that the natural state of sublunar bodies were rest or motion towards the centre of the earth and that any other motion needed explanation; but according to Galileo's principle of inertia (or rather to Newton's later version of the principle) uniform motion is the natural state of affairs which requires no explanation in terms of external causes. The only 'explanation' of uniform motion the new physics could offer was that it followed from Newton's second law in the absence of an external force. It was in much the same way that Hoyle, Bondi and Gold suggested density conservation to be 'more natural' than conservation of matter.

Another version of the same argument was given by Bonnor, in a review of Munitz's *Space, Time and Creation.* As Bonnor noted, Munitz demanded an explanation for creation of matter, but took matter conservation to be unproblematic and in no need of explanation. Is there an explanation of matter conservation? If so, the explanation lies in the general theory of relativity from the equations of which matter conservation follows.[8] But Hoyle's modified Equations (3) have the very same structure as Einstein's, and so these equations 'explain' creation of matter (through the C-term) in the same abstract sense that the field equations (1) explain matter conservation. As Bonnor remarked: "Does Professor Munitz demand that creation of matter be given *more* explanation than the conservation of matter?" (Bonnor 1957b). Hoyle made a similar point when he argued that physicists do not try to explain electricity, gravitation and magnetism causally, by asking why there are such forces or where they come from. They 'explain' such phenomena by working out testable consequences based on the hypothesis that they exist, by asking *how* the phenomena operate and not *why* they exist. "Exactly the same situation applies to the creation of matter," Hoyle wrote. "We cannot say why matter is created or where it comes from, but we can say, 'If matter is created continuously, then it is created in such and such a way'" (Hoyle 1952: 51).

McCrea's theory freed steady-state cosmology from the troublesome concept of creation out of nothing, but the alternatives offered were unlikely to appeal to most physicists and in fact they did not. Instead of 'nothingness,' new hypothetical fields or particles were proposed. One could ascribe the creation to some hypo-

[8] In modern parlance, matter conservation means conservation of particle number. Except for a universe with incoherent matter (dust), it is not a consequence of Einstein's field equations, but has to be added as a further assumption. In the debate of the 1950s conservation of particle number was rarely discussed.

thetical creation field, such as Hoyle did; postulate with McCrea a state of negative pressure; or introduce with Pirani the gravitinos. The explanations rested in any case on purely hypothetical entities. Whereas McCrea's uniform negative pressure was unobservable even in principle, there was no a priori reason why gravitinos should not be detectable. However, this was an issue that Pirani did not highlight and from a practical point of view his hypothetical particles might seem as unobservable as the universal negative pressure. There was one bold attempt to explain the stress in terms of physical sources, namely the short-lived electrical universe hypothesis proposed by Bondi and Raymond Lyttleton in 1959 and developed by Hoyle the following year (Lyttleton & Bondi 1959; Hoyle 1960a). According to this hypothesis, McCrea's stress was a manifestation of the electromagnetic field due to a slight inequality between the numerical charges of a proton and an electron. The ingenious theory had to be abandoned when experiments made in 1960–1962 proved that any difference between the two charges was considerably smaller than required theoretically.[9]

In the summer of 1959 Bondi suggested that the electrical universe hypothesis might possibly be connected with an idea forwarded by V. A. Bailey from the University of Sydney. Bailey supported the steady-state theory, but worried that it was associated with "the primitive belief in creation." He therefore suggested that the *apparent* continual supply of matter came from a five-dimensional hyperspace (U_5) in which the ordinary four-dimensional space-time was embedded. Associating the idea with the old Kaluza-Klein theory he imagined that the law of energy conservation would hold in U_5, but not manifestly in U_4. Bailey's motive for proposing his speculation was quite clear:

> It seems desirable to direct attention without delay to the fact that a steady-state universe is possible without the "formation of matter ... out of nothing" [Bondi 1952: 144]. The knowledge of this fact will undoubtedly cause relief in the minds of many persons who would otherwise be unable to accept the steady-state theory. For the old dictum *ex nihilo nihil fit* seems to be one of the few things about which philosophers, scientists and the common man agree. (Bailey 1959: 537)

William Davidson wrote his Ph.D. thesis under the supervision and inspiration of McCrea, whose steady-state theory he investigated within the framework of general relativity (Davidson 1959). As a new feature he concluded that the creation of matter would not be statistically uniform or random in space, but depend on the gravitational influence of local masses. Davidson's analysis suggested that the state given by $\rho c^2 = -p = 3c^2/\kappa a^2$ constituted a natural state of relative stability of the intergalactic medium. By examining the effects of slight disturbances of this state, he found that the expansion tended to restore the steady-state

---

[9] According to Bondi and Lyttleton, the charge excess should be about $2 \times 10^{-18}e$, and according to Hoyle's version $5 \times 10^{-19}e$. Already in 1959 experiments indicated that the excess was smaller than $10^{-20}e$, a result which was substantiated in later experiments. Interestingly, as early as 1924, at a meeting of the Swiss Physical Society, Einstein suggested a charge excess of about $3 \times 10^{-19}e$ in order to explain the magnetic fields of the earth and the sun, but he never published the suggestion. The case of the electrical universe is analysed in Kragh 1996: 214–218, where further references can be found.

conditions. A somewhat similar result was obtained in 1957 by William Bonnor, who, however, used it to argue against the steady-state theory (Bonnor 1957a). Bonnor concluded that condensations could not form in a steady-state universe from small perturbations and that galaxies were therefore unlikely to be formed. Davidson, on the other hand, pointed out the possibility of condensations forming from large disturbances and hence the possibility of the formation of galaxies also in the steady-state theory. The discussion concerning galaxy formation in the rival cosmological models continued through the 1960s, but with no clear conclusion.

McCrea's steady-state model operated with Newtonian or relativistic dynamics in a way which was not very clear. The confidence in this model, whether in its original 1951 formulation or in its later versions, was diminished by an examination made in 1960 by Bonnor, who as an orthodox relativist had no sympathy for the steady-state theory. Although he found McCrea's procedure to be "at once simple and ingeneous," he denied that it led to a description of the real universe. According to Bonnor's analysis, the model led to results that made it physically unrealistic or, at least, drastically reduced its physical credibility. Bonnor showed that within McCrea's framework matter satisfying the equation of state Equation (7) does not respond to gravitation: both the inertial and the passive gravitational densities will be zero. (The passive mass refers to the response of a body placed in a gravitational field.) This implies that the motion of matter is indeterminate, so that, for example, there is no reason to assume that matter moves on a geodesic. The velocity of matter would be wholly indeterminate, that is, unpredictable. "This lack of determination weakens those versions [Newtonian or relativistic steady-state theories] to such an extent that they seem to be of little value as instruments of prediction in cosmology," he concluded (Bonnor 1960: 480).

At the same time Bonnor noted that since the steady-state theory did not rely in itself upon either the general theory of relativity or Newtonian mechanics, then his criticism left, in principle, the theory unaffected. Nonetheless, Bonnor's work weakened its appeal: "What now seems clear, however, if my arguments are correct, is that those who wish to work with the steady-state theory must use a dynamics specifically designed for it, since they are not free to use the existing models of general relativity or Newtonian theory with any degree of confidence." Of course no such special dynamics existed. Bonnor's theoretical analysis of 1960 thus resulted in conclusions very different from those of Davidson and left the impression that McCrea's model might not, after all, be acceptable.

In the absence of a quantum theory of gravitational fields, phenomenological hypotheses or crude analogies were all that could be offered with respect to explaining the continual creation of matter. In his presentation of the steady-state theory at the Solvay Congress in 1958, Hoyle suggested a simple way in which "the inherent plausibility of the creation of matter" could be demonstrated. Hoyle started his plausibility argument by pointing out that the observable part of a steady-state universe is given by the horizon associated with the De Sitter line element, which is c/H. Although the universe is infinite, no signal emitted by a source at a distance larger than c/H will ever be received by the observer. According to Hoyle's 1948 theory the density of matter was given by

$$\rho = \frac{3H^2}{8\pi G}$$

and the mass of the observable universe would thus be

$$\frac{4}{3}\pi\rho\left(\frac{c}{H}\right)^3 = \frac{c^3}{2GH}. \tag{8}$$

Hoyle now noted that the gravitational potential of a mass m at distance r is Gm/r. By dividing the mass of the observable universe by c/H and multiplying by G a quantity corresponding to the gravitational potential will thus appear. The result is $c^2/2$ or, what is the same in this context, about c^2. According to Hoyle, this meant that every particle in the universe exists in a potential well of a depth which is comparable to its rest energy mc^2. "The process of creation can accordingly be thought of as involving no energy expenditure—a particle is created at a negative potential that compensates for its rest mass. Accordingly [*sic*] to quantum theory, particle creation might well be expected under these circumstances" (Hoyle 1958: 57). However, Hoyle's picture was no more than a crude plausibility argument that sceptics might not, and in fact did not, find convincing.

In one way or another, energy conservation was at the heart of the cosmological controversy.[10] This is a complicated issue with which I cannot deal in detail, but a few comments may illustrate the positions. The debate was to some extent obscured by disagreements concerning the proper formulation of the law of energy conservation when applied cosmologically. The steady-state physicists admitted that their theory violated the ordinary formulation, the continuity equation, but questioned if this was the appropriate form of the law. Bondi pointed out that according to the relativistic models the energy of a part of the universe is constant in comoving coordinates, whereas according to the steady-state theory energy is conserved in any proper volume of space. "It may well be correct to speak of conservation of mass in the steady-state model rather than in relativity, since proper volume is more fundamental than coordinate volume" (Bondi 1952: 144). Although considering energy conservation for an infinite universe to be "a meaningless notion," Hoyle agreed with the distinction proposed by Bondi: "When there is creation of matter, the conservation equations require energy to be constant in a box of fixed proper volume. When there is no creation of matter, the conservation equations require energy to be constant in a box that expands with the universe" (Hoyle 1960a: 257).

However, most mainstream relativists had no patience for such "monstrous play on words," as the Soviet physicist Yakov Zel'dovich called it in 1962 (Zel'dovich 1963: 947). And for the majority of physicists the attempts to justify a kind

[10] In the modern interpretation of general relativity, the equation $T^{\mu\nu}{}_{;\nu} = 0$ stands for energy conservation only in the exceptional case when space-time exhibits time-translation symmetry. However, I follow the way of speaking characteristic of the physicists involved in the controversy in the 1950s and 1960s.

of energy conservation in the steady-state theory were wasted. A theory with continual creation of matter was unavoidably associated with violation of energy conservation (whatever that meant) and for that reason unacceptable; "for if there is any law which has withstood all changes and revolutions in physics, it is the law of conservation of energy" (Max Born in Introduction to Jordan 1949).[11] In an interview of 1978, McVittie expressed the same attitude more colorfully. Referring to the creation of matter in steady-state theory, he said:

> It's like breaking the rules when you are playing a game. If you allow yourself in the game of American football to take knives on board with you and stab your oppenent, now and again, of course the results will be very remarkable, particularly if one side only has the knives and the other is merely the recipient. (McVittie 1978)

It is understandable that John North, reviewing the development in 1965, was reminded of the dictum that the principle of energy conservation has survived because energy is defined as that which is conserved (North 1965: 215).

5. Some attempts of the 1960s

The attempts to 'normalize' (not renormalize!) the steady-state theory, i.e., to make it more acceptable from the point of view of general relativity, continued through most of the 1960s, either by developing McCrea's theory or by following other routes. The most important of these reformulations was undoubtedly the program that started with Hoyle's covariant *C*-field theory of 1960, which was inspired by McCrea's approach. Hoyle wrote approvingly that this approach was "preferable because it emphasises the crucial point that the logical structure of the Einstein equations is not thereby changed" (Hoyle 1960b: 257). Hoyle developed his *C*-field theory in different versions in collaboration with Jayant Narlikar, but we shall not be concerned with this rather complex system of theories, aspects of which have previously been subjected to historical review (Sánchez-Ron 1990; Kragh 1996: 358–372).

P. Roman at the University of Manchester took up the steady-state theory in McCrea's version and argued, as other authors had done previously, that matter creation in an expanding universe could be accomodated within a slightly modified law of energy conservation. The mechanism of matter production, clearly adapted from that suggested by McCrea and McVittie, was this:

> The work done by the gravitational force is ... the locally created energy. On the other hand, gravitation is necessary to keep the universe in a steady state. ... In a sense, we may say that gravitation works against a change in rate of expansion and keeps balance, thereby creating mass. However, the

[11] While there are many historical works on the the law of energy conservation, the history of energy nonconservation remains to be written. Violations of the energy law has a fascinating history which goes back to the 1880s, if not earlier. Before the steady-state theory made the issue controversial, energy nonconservation, both locally and cosmologically, was proposed by many physicists. Among them were authorities such as Niels Bohr, Lev Landau, Erwin Schrödinger, Paul Dirac, and Pascual Jordan. Given the long tradition of ideas about energy nonconservation it is surprising that the discussion in the 1950s aroused so much passion.

> expansion is not brought about by the pressure directly, because p is constant and its gradient does not appear in the equation of motion. The pressure, through its negative contribution to the gravitational mass and hence to the gravitational force would tend to bring the system to an accelerated flow. (Roman 1960: 17)

In Roman's hydrodynamical steady-state model the mass density of the universe was $\rho = 24H^2/\kappa c^2$ and the rate of matter creation about 7×10^{-46} g cm^{-3} sec^{-1}, both values being empirically reasonable, although the density was rather too high. The total energy density was given by $-p$, which he interpreted as a form of elastic energy. Comparing his model with the one of relativistic theory he obtained for the cosmological constant the expression

$$\Lambda = 3H^2 + \frac{8\pi G}{c^2} p.$$

With Equation (7) and the mentioned value for the density, this gives $\Lambda = -21H^2 \approx -1.3 \times 10^{-36}$ s^{-2}, which Roman considered "a very reasonable value."

As Roman pointed out, the steady-state theory may seem to be conceptually inconsistent because it includes both continual creation of matter and the Perfect Cosmological principle. The reason is that according to Emmy Noether's theorem of 1918 energy conservation is equivalent to invariance under the time-displacement transformation $t \longrightarrow t + \delta t$. But then a universe with violation of energy conservation will not be homogeneous in time, contrary to the steady-state foundation. However, having mentioned the problem Roman argued that in fact it was apparent only. His modified conservation law, allowing for matter creation, was based on Noether's theorem and the Perfect Cosmological principle.

One of McCrea's aims in suggesting his interpretation of 1951 was that it might open up for an understanding of continual creation in terms of quantum mechanics (see also McCrea 1950b). In the conclusion of his 1951 paper he suggested that the cosmical zero-point stress might be connected with the zero-point fluctuations appearing in quantum field theory. "If any of these can be interpreted as producing a stress, it appears that the connexion might be established," he wrote. Unnoticed by McCrea, Hendrik Casimir had recently demonstrated the reality of zero-point energies and thus opened up for a new understanding of the physics of empty space (Sarlemijn & Sparnaay 1989: 238–272). However, neither the Casimir effect nor other of the early versions of the quantum vacuum were of a kind that could be used in justifying the notion of a universal stress in space.[12]

Yet the dream of introducing quantum mechanics into steady-state theory remained an important motive also in the later developments of Hoyle and Narlikar.

[12] The Casimir effect dates from 1948. It may be relevant to note that in 1953 Casimir applied his theory of zero-point energy in order to account for the Poincaré stress in classical electron models. Although Casimir thus interpreted the zero-point energy as a stress, his result was limited to electromagnetic fields and went at any rate unnoticed by cosmologists. See Carazza & Guidetti 1986 for details and references. As pointed out in the 1980s, the Poincaré stress can be understood in terms of relativistic vacuum cosmology. If the vacuum energy-momentum tensor is included in the Einstein-Maxwell equations, the result becomes an electron model where the Poincaré stress is explained as arising from the vacuum properties. The electron is pictured as a charged spherical shell around a negative gravitational mass in the form of vacuum polarization (Grøn 1985).

In the 1960s, following new developments in quantum field theory, a few attempts were made to give a quantum foundation of continual creation and McCrea's negative-pressure substratum. However, the results were not encouraging (Dowker 1964).

The Czechoslovakian physicist Jaroslav Pachner was another contributor to hybrid forms of steady-state and relativistic cosmologies. In 1960–1961 he proposed a new relativistic theory in which the cosmological constant was related to creation of matter for $\Lambda > 0$, and to annihilation of matter for $\Lambda < 0$. However, because "the hypothesis of creation or annihilation of matter is strange to any scientific theory" he decided that $\Lambda = 0$ (Pachner 1960). In a later version, building on McCrea's idea of a negative universal pressure, he examined a model in which the stress is proportional to the fourth power of the space curvature, i.e., $-p/c^2 = bR^{-4}$. With this assumption Pachner found also the energy to depend on the curvature and was led to an oscillating, singularity-free model which was practically indistinguishable from the Friedman-Lemaître model as far as the present epoch is concerned (Pachner 1965).

An idea somewhat related to McVittie's and Pachner's, but not building on standard relativity, was proposed by António Gião, a Portuguese mathematical physicist, in 1963 (Gião 1964). Based on a generalization of the relativistic field equations, Gião claimed that none of the cosmological models usually discussed, the steady-state model included, were admissible. His alternative was what he called a "generalized steady-state model" in which the constancy of the density of matter was replaced by the constancy of the density of proper energy. From this starting point (which contradicts the Perfect Cosmological Principle) Gião was led to an oscillating cosmological model with a period of about 16×10^9 years and including both creation and destruction of matter. During the phases of contraction creation of matter would take place, during those of expansion a corresponding destruction. In Gião's model the Hubble parameter varied in time, obtaining its maximum during the expansion and its minimum during the contraction. Although the model was ignored by other researchers, certain of its features reappeared in the theories developed by Hoyle and Narlikar a couple of years later. The elaborate and ambitious theory illustrates that even in 1963, when the steady-state theory was under strong attack, the era of heterodoxy in cosmology had not ended.

As illustrated by a comparison of the model universes of McVittie, Pachner and Gião, relativistic cosmology was a really wonderful theory; and it appeared even more wonderful if supplied with unorthodox features related to the steady-state tradition. Indeed, in these versions it seemed able to result in any conceivable model of the universe.

Unnoticed by the participants in the cosmological controversy, some versions of continual matter creation predated the steady-state theory. In fact, the idea can be traced back to the 1920s (Kragh 1995). For example, in 1939 Schrödinger investigated the proper vibrations of quantum waves in expanding cosmological models and concluded that in some cases particles might be formed out of the vacuum—and without violating energy conservation (Schrödinger 1939; Urbantke 1992). Apparently without knowing of Schrödinger's work, Bryce DeWitt argued in 1953 that a varying curved metric can produce pairs of particles through its interaction with the vacuum fluctuations of a scalar field (DeWitt 1953). This was an idea that would be taken up much later, but it played no role in cosmology in the 1950s.

We finally mention a work from 1966 in which the American mathematician Tracy Y. Thomas, without relying on arguments à la McCrea, argued that continual creation of matter is possible within the standard framework of the general theory of relativity. He concluded that "this result strongly supports the Bondi-Gold postulate of the continual creation of mass in space" (Thomas 1966). Note that Thomas found agreement with general relativity necessary in order to accept continual creation of matter and thus gave the theory of general relativity priority over the steady-state theory. This was of course contrary to what Bondi and Gold had argued in 1948; but much had changed since then and in 1966 the authority of general relativity was practically unquestioned.

6. Digression on the cosmological constant

In his work of 1951 McCrea did not notice that his idea of introducing a negative pressure had historical antecedents, albeit belonging to a tradition very different from that of the steady-state theory. In 1933 Lemaître, who always believed that the cosmological constant was an indispensable part of relativistic cosmology, suggested that the constant might be understood as a negative vacuum density (Lemaître 1934). "Everything happens as though the energy in vacuo would be different from zero," he wrote, referring to general relativity applied to regions of space of extremely low density. He then argued that in order to avoid a non-relativistic vacuum or ether—a medium in which absolute motion is detectable—a negative pressure $p = -\rho c^2$ must be introduced, the vacuum density being related to the cosmological constant by

$$\rho_v = \frac{\Lambda c^2}{4\pi G}.$$

This negative pressure is responsible for the exponential (De Sitter) expansion of Lemaître's universe during its last phase, where it contributes a repulsive cosmic force of $\Lambda c^2 r/3$. There is thus a nice equivalence between Lemaître's idea of 1933 and McCrea's of 1951. In fact, the interpretation of the cosmological constant as relating to a negative pressure can be traced back to Einstein's earliest thinking about a universe governed by the laws of general relativity. Einstein in part justified his introduction of the cosmological constant as a means of avoiding a cosmic negative pressure – "*for experience teaches us that the energy density does not become negative*," as he wrote to Michele Besso in December 1916 (Speziali 1972: 97; Kerzberg 1989: 160–163).[13] Almost thirty years later he repeated that

[13] "Geschlossenheit der Welt ... wird auch dadurch gelegt, dass die Krümmung überall dasselbe Vorzeichen hat, weil die Energiedichte erfahrungsgemäss nicht negativ wird." (Speziali 1972: 97).

"I originally introduced a new member [Λ] into the equation instead of the above mentioned [negative] pressure" (Einstein 1945: 111). Einstein reconsidered the connection between the cosmological constant and negative pressure in 1919, but then in a more special case, namely his attempt to provide a link between general relativity and the structure of electrical particles (Einstein 1919).

The interpretation of the cosmological constant as a vacuum energy follows naturally from the field equations (1), where Λ appears as an additional term in the energy-stress tensor $T_{\mu\nu}$. From Equations (2) it follows that the relevant quantities are

$$\begin{cases} \rho_v = \dfrac{\Lambda c^2}{8\pi G}, \\ p_v = -\dfrac{\Lambda c^4}{8\pi G}. \end{cases}$$

Although this must have been evident to cosmologists ever since 1917, Lemaître seems to have been the first to point out explicitly the interpretation. (It is unclear why he wrote the denominator as $4\pi G$ instead of $8\pi G$.) The American physicist Christopher Gregory made use of the same interpretation in 1945, when discussing a cosmological model with a superficial similarity to the later steady-state theory (Gregory 1945).

It was only much later, after the cosmological controversy had died out, that Lemaître's suggestion was taken up. One of the first to do so was Zel'dovich, who showed that a non-zero cosmological constant produces the same gravitational field as space containing matter with a density given by Equation (8) and a pressure $p = -\rho c^2$. "In this sense we can speak of an energy density of the vacuum and a pressure of vacuum," he wrote (Zel'dovich 1968: 390). Apparently Zel'dovich was unaware of Lemaître's old result, which he did not quote, and also unaware of the analogy to McCrea's theory. The same was the case when Alan Guth, Andrei Linde, Paul Steinhardt and Andreas Albrecht proposed the first inflationary universe models in the early 1980s.[14] Indeed, it is a characteristic feature of much of postwar cosmology that the involved physicists and astronomers have lacked knowledge about, or interest in, the history of their science. But then this is not a peculiarity of cosmologists, of course, but a general phenomenon of modern science.

7. Conclusion: The victorious theory of general relativity

The cosmological controversy was not primarily concerned with the general theory of relativity, but since cosmology was perhaps the area which was taken to support Einstein's theory most strongly, the steady-state challenge was also a challenge to general relativity as an exact and universal theory. Yet the authority of this theory

[14] The relativistic McCrea-Hoyle theory is closely related to the inflation universe model which arises from Einstein's field equations when solved with a vacuum energy-stress tensor as its only source. The gravitational mass becomes negative and the vacuum shows a tendency to expand exponentially under the action of its own repulsive gravitation. For the connection between McCrea's theory and the inflation theory, see Grøn 1986.

is clearly featured in the controversy, where, ironically, Einstein's field equations served as a model for much of the opposition against relativistic cosmology. This was not the case with the Bondi-Gold theory, but Hoyle's and McCrea's versions were specifically designed so as to mimic the mathematical structure of the relativistic theory, and it was within these versions that most of the development took place. The wish to conform with general relativity, rather than reject it, or to modify the theory so as to admit unorthodox features such as matter creation, witness to the fascination of Einstein's theory. Whatever models were constructed, however strange the proposed cosmological theories might appear, there was felt a need to justify them theoretically; and theoretical justification meant in practice conformity with general relativity.

If astronomers and physicists had hesitated in admitting the complete validity of general relativity, by the early 1960s they were reassured by a remarkable series of direct and indirect confirmations of the theory. First, general relativity was brilliantly vindicated by laboratory experiments using new microwave technology (Will 1986). Second, Einstein's theory was eagerly explored in new areas of astrophysics, perhaps most notably in the physics of black holes where 1964 marked the beginning of "the golden age" (Thorne 1994). Third, steady-state cosmology suffered its first clear defeat in the form of its inability (without ad hoc hypotheses) to account for radioastronomical measurements, and by 1965 the discovery of the cosmic microwave background sealed the fate of the steady-state theory; as a result, general relativistic cosmology came out of the controversy as the undisputed winner (Kragh 1996). The three developments, all taking place between 1960 and 1965, acted as a forceful and coordinated affirmation of the theory of general relativity. Yet, what may look like coordination was in fact the fortuitous coincidence of three independent lines of development.

At any rate, the experimental and theoretical successes led to an atmosphere of confidence among specialists that Einstein's theory was universally true and no longer a subject of discussion. The renewed faith in the theory of general relativity was given enthusiastical expression by John Wheeler, who in 1962 identified two trends in current gravitation physics. First, he claimed, "Interest has fallen in inventing new theories of gravitation, space-time, and the expanding universe;" and, second, "Increasing numbers of investigators share the conviction that Einstein's 1915–1916 analysis of the curvature of space by energy is a unique theory, of unrivalled scope and reasonableness against which no objection in principle has ever been sustained, and out of which one should now try to read the deeper meaning and consequences" (Wheeler 1962: 40).

There were exceptions to the obedience to the spirit of general relativity, of course, such as exemplified by Milne and also by Bondi and Gold; but most theoretical cosmologists were under the spell of the relativistic magic. For example, in this respect the theories of Jordan and Dirac represented basically the same feature as the Hoyle-McCrea steady-state program. Dirac's theory of 1937 was not only founded on a gravitational constant varying with the epoch, but also on matter being continually created throughout space. In his 1938 version he dropped matter creation and expressed his wish to develop his theory in conformity with the theory of general relativity he valued so highly. This was a persistent theme in Dirac's later cosmological work. At least one physicist argued that it was perfectly possible to accomodate the $G \sim t^{-1}$ hypothesis within standard general relativity (Gilbert 1956)—just as some physicists at the same time argued that the steady-state hypothesis could be reconciled with Einstein's theory.

Acknowledgements

I would like to thank Øyvind Grøn and an anonymous referee for clarifying some of the technical issues in this article.

References

Bailey, V. A. (1959). "The Steady-State Universe and the Deduction of Continual Creation of Matter." *Nature* **184**: 537.

Balashov, Yuri. (1994). "Uniformitarianism in Cosmology: Background and Philosophical Implications of the Steady-State Theory." *Studies in the History and Philosophy of Science* **25**: 933–958.

Bondi, Hermann. (1952). *Cosmology*. Cambridge: Cambridge University Press.

Bondi, Hermann & Gold, Thomas. (1948). "The Steady-State Theory of the Expanding Universe." *Monthly Notices of the Royal Astronomical Society* **108**: 252–270.

Bonnor, William B. (1955). "Fifty Years of Relativity." *Science News* **37**: 1–24.

——— (1957a). "Jeans' Formula of the Nebulae." *Monthly Notices of the Royal Astronomical Society* **117**: 104–117.

——— (1957b). [Review of Munitz 1957]. *The Observatory* **77**: 248.

——— (1960). "The Relativistic Model of the Steady-State Universe." *Monthly Notices of the Royal Astronomical Society* **121**: 475–481.

Carazza, Bruno & Guidetti, Gian P. (1986). "The Casimir Electron Model." *Archive for History of Exact Sciences* **35**: 273–279.

Davidson, William. (1959). "Steady-State Cosmology Treated According to General Relativity." *Monthly Notices of the Royal Astronomical Society* **119**: 309–324.

DeWitt, Bryce. (1953). "Pair Production by a Curved Metric." *Physical Review* **90**: 357.

Dowker, J. S. (1964). "Cosmology and Weak Interactions." *Nuovo Cimento* **32**: 1816–1818.

Einstein, Albert. (1919). "Spielen Gravitationsfelder im Aufbau der materiellen Elementarteilchen eine wesentliche Rolle?" *Sitzungsberichte der Preussischen Akademie der Wissenschaften* (Berlin): 349–356.

——— (1945). *The Meaning of Relativity*. 2nd ed. Princeton: Princeton University Press.

——— (1950). *Out of My Later Years*. London: Thames and Hudson.

Eisenstaedt, Jean. (1986). "La Relativité génerale à l'étiage: 1925–1955." *Archive for History of Exact Sciences* **35**: 15–85.

Gamow, George. (1949). "On Relativistic Cosmogony." *Reviews of Modern Physics* **21**: 367–373.

Gião, António. (1964). "On the Theory of the Cosmological Models with Special Reference to a Generalized Steady-State Model." In *Cosmological Models*. A. Gião, ed. 5–100. Lisbon: Instituto Gulbenkian de Ciência.

Gilbert, C. (1956). "Dirac's Cosmology and the General Theory of Relativity." *Monthly Notices of the Royal Astronomical Society* **116**: 684–690.

Gregory, Christopher. (1945). "On a Supplement to the Field Equations with an Application to Cosmology." *Physical Review* **67**: 179–184.

GRØN, Øyvind. (1985). "Repulsive Gravitation and Electron Models." *Physical Review D* **31**: 2129–2131.

——— (1986). "Repulsive Gravitation and Inflationary Universe Models," *American Journal of Physics* **54**: 46–52.

HOYLE, Fred. (1948). "A New Model for the Expanding Universe." *Monthly Notices of the Royal Astronomical Society* **108**: 372–382.

——— (1949). "On the Cosmological Problem." *Monthly Notices of the Royal Astronomical Society* **109**: 365–371.

——— (1952). "Concepts of the Universe." *New York Times Magazine* 1 June: 11–12, 50–51.

——— (1958). "The Steady State Theory." In *La Structure et l'évolution de l'univers* (Solvay Conference 11). Jules Géhéniau, ed. 53–80. Brussels: R. Stoops.

——— (1960a). "On the Possible Consequences of a Variability of the Elementary Charge." *Proceedings of the Royal Society A* **257**: 431–444.

——— (1960b). "A Covariant Formulation of the Law of Creation of Matter." *Monthly Notices of the Royal Astronomical Society* **120**: 256–262.

INFELD, Leopold. (1957). "Equations of Motion in General Relativity and the Action Principle." *Reviews of Modern Physics* **29**: 398–411.

JORDAN, Pascual. (1949). "Formation of Stars and Development of the Universe." *Nature* **164**: 637–640.

KERZBERG, Pierre. (1989). *The Invented Universe: The Einstein-De Sitter Controversy (1916–17) and the Rise of Relativistic Cosmology*. Oxford: Clarendon Press.

KRAGH, Helge. (1990). *Dirac: A Scientific Biography*. Cambridge: Cambridge University Press.

——— (1995). "Cosmology Between the Wars: The Nernst-MacMillan Alternative." *Journal for the History of Astronomy* **26**: 93–115.

——— (1996). *Cosmology and Controversy: The Historical Development of Two Theories of the Universe*. Princeton: Princeton University Press.

LEMAÎTRE, Georges. (1934). "Evolution of the Expanding Universe." *Proceedings of the National Academy of Science* **20**: 12–17.

LYTTLETON, Raymond A. & BONDI, Hermann. (1959). "On the Physical Consequences of a General Excess of Charge." *Proceedings of the Royal Society A* **252**: 313–333.

MCCREA, William H. (1950a). "The Steady-State Theory of the Expanding Universe." *Endeavour* **9**: 3–10.

——— (1950b). "Quantum Mechanics and Astrophysics." *Nature* **166**: 884–886.

——— (1951). "Relativity Theory and the Creation of Matter." *Proceedings of the Royal Society A* **206**: 562–575.

——— (1953). "Cosmology." *Reports on Progress in Physics* **16**: 321–363.

MCVITTIE, George C. (1952). "A Model Universe Admitting the Interchangeability of Stress and Matter." *Proceedings of the Royal Society A* **211**: 295–301.

——— (1978). Interview conducted by David DeVorkin for the American Institute of Physics (Sources for History of Modern Astrophysics). Transcript held at the Center for History of Physics, College Park, Maryland.

MERLEAU-PONTY, Jacques. (1965). *Cosmologie du XXe siècle: étude épistémologique et historique des théories de la cosmologie contemporaine*. Paris: Gallimard.

VON MEYENN, Karl, ed. (1996). *Wolfgang Pauli. Wissenschaftlicher Briefwechsel mit Bohr, Einstein, Heisenberg u.a.* Vol. 4, part 1. Berlin: Springer.

MICHELMORE, Peter. (1962). *Einstein: Profile of the Man*. New York: Dodd.

MUNITZ, Milton K. (1957). *Space, Time and Creation: Philosophical Aspects of Scientific Cosmology*. Glencoe (IL): The Free Press.

North, John D. (1965). *The Measure of the Universe: A History of Modern Cosmology.* London: Oxford University Press.

Omer, Guy C. (1949). "A Nonhomogeneous Cosmological Model." *The Astrophysical Journal* **109**: 164–176.

Pachner, Jaroslav. (1960). "Dynamics of the Universe." *Acta Physica Polonica* **19**: 662–673.

——— (1965). "An Oscillating Isotropic Universe Without Singularity." *Monthly Notices of the Royal Astronomical Society* **131**: 173–176.

Pirani, Felix A. E. (1954). "On the Energy-Momentum Tensor and the Creation of Matter in Relativistic Cosmology." *Proceedings of the Royal Society A* **228**: 455–462.

Roman, P. (1960). "On the Steady-State Theory of Cosmology." *Nuovo Cimento* **18**: 9–20.

Sánchez-Ron, José-Manuel. (1990). "Steady-State Cosmology, the Arrow of Time, and Hoyle and Narlikar's Theories." In *Modern Cosmology in Retrospect.* B. Bertotti, R. Balbinot, S. Bergia & A. Messina, eds. 233–243. Cambridge & New York: Cambridge University Press.

Sarlemijn, Andries & Sparnaay, Marcus J., eds. (1989). *Physics in the Making: Essays on Developments in Twentieth-Century Physics in Honour of H. B. G. Casimir.* Amsterdam: North-Holland.

Schrödinger, Erwin. (1939). "The Proper Vibrations of the Expanding Universe." *Physica* **6**: 899–912.

Speziali, Pierre, ed. (1972). *Albert Einstein–Michele Beso. Correspondance. 1903–1955.* Paris: Hermann.

Thomas, Tracy Y. (1966). "On the Creation of Mass in an Expanding Universe." *Proceedings of the National Academy of Science* **56**: 1349–1353.

Thorne, Kip S. (1994). *Black Holes and Time Warps: Einstein's Outrageous Legacy.* New York: Norton.

Tolman, Richard C. (1934). *Relativity, Thermodynamics and Cosmology.* Oxford: Oxford University Press.

——— (1949). "The Age of the Universe." *Reviews of Modern Physics* **21**: 374–378.

Urbantke, Helmuth. (1992). "Schrödinger and Cosmology." In *Studies in the History of General Relativity* (Einstein Studies, vol. 3). J. Eisenstaedt and A. J. Kox, eds. 453–459. Boston: Birkhäuser.

Wheeler, John A. (1962). "The Universe in the Light of General Relativity," *The Monist* **47**: 40–76.

Will, Clifford M. (1986). *Was Einstein Right?* New York: Basic Books.

Whittaker, Edmund T. (1935). "Gauss' Theorem and the Concept of Mass in General Relativity." *Proceedings of the Royal Society A* **149**: 384–395.

Zel'dovich, Yakov B. (1963). "Problems of Present-day Physics and Astronomy." *Soviet Physics – Uspekhi* **5**: 931–950.

——— (1968). "The Cosmological Constant and the Theory of Elementary Particles." *Soviet Physics – Uspekhi* **11**: 381–393.

Part IV

Relativity in Debate

Larmor versus General Relativity

José M. Sánchez-Ron

When Thomas Kuhn interviewed John Cockcroft on 2 May 1963, as part of the project "Sources for the History of Quantum Physics,"[1] they began to talk about the Cambridge the famous British physicist had known when he was a young student, early in the 1920s. Referring to Arthur S. Eddington, Cockcroft had no doubt that "he was a big figure," and that "naturally, we went to [his] lectures." Edward A. Milne, also, was well considered, especially because of his work on the internal constitution of the stars; and in the field of quantum physics, there was the great Ernest Rutherford, and others, like Ralph H. Fowler, his son-in-law.

But what about the older generation? J. J. Thomson was still around, and the youngsters "used to go to [his] lectures," not really to learn new physics, but "for the physics of the early twentieth century, ... electrons and all this kind of things; discharge in gases, and isotopes. It was very nice to have him there and to be able to go and listen to the great man." It seems, therefore, that the old 'J. J.' had some audience among Cambridge students; indeed, the instruction he gave did not conflict at all with the direction atomic physics had taken, in as much as he did not express himself against any new ideas, at least according to Cockcroft's recollections.

Another "great man" of the University at the time was Joseph Larmor. Thomson's contemporary (indeed, both took the Mathematical Tripos the same year (1880), Larmor coming out Senior Wrangler and Thomson Second; 'J. J.' also won the Smith's prize), Larmor's career was outstanding by any possible standard: he succeeded Stokes as Lucasian professor in 1903, a chair that he retained until 1932;

[1] For a description of this project, see Kuhn, Heilbron, Forman & Allen 1967.

H. Goenner, J. Renn, J. Ritter, T. Sauer (eds.)
The Expanding Worlds of General Relativity
(Einstein Studies, volume 7) pp. 405–430

president of the London Mathematical Society in 1914–1915, the Royal Society awarded him its Royal Medal in 1915 and its Copley Medal in 1921; knighted in 1909, he also represented Cambridge University in Parliament from 1911 to 1922. Cockcroft, who seems to have been a diligent student, attended Larmor's lectures, but there he would find only two or three more students. "He represented," Cockcroft recalled, "the physics of the 1900s. ... He was a grand old man and it was worth going to hear him. Sometimes he would come out with some new ideas, such as the transmission of electromagnetic waves by the ionosphere and he would come out with those in his lectures before they became generally known. It was interesting to go to but he really did not represent the new age in theoretical physics at all."[2]

Larmor's scientific work centered on electromagnetic theory, optics, analytical mechanics, and geodynamics.[3] Among his most famous works figures his book, *Aether and Matter* (Larmor 1900), significantly subtitled: "A development of the dynamical relations of the aether to material systems on the basis of the atomic constitution of matter, including a discussion of the influence of the earth's motion on optical phenomena." According to Eddington (1942/4: 198), this work "ranks among the great scientific books."

In *Aether and Matter*, Larmor showed that what we now call 'Lorentz transformations' were correct to second order, making thus an important contribution to the development of twentieth-century physics, and to the prehistory of special relativity in particular.[4] However, he did not really feel at home with twentieth-century physics. In this sense, it is worth quoting Eddington again:

> He seemed a man whose heart was in the nineteenth century, with the names of Faraday, Maxwell, Kelvin, Hamilton, Stokes ever on his lips—as though he mentally consulted their judgement on all the modern problems that arose. He would often say that all true scientific progress ceased about 1900—or even earlier, for his own fin de siècle effusion was only dubiously judged by the looser standards of these times; but that was as far as he would go—except when he forgot his pose. There was, of course, a great deal of exaggeration in this pose; but he adopted it so systematically that perhaps he himself could scarcely distinguish it from his natural opinions.[5] (Eddington 1942/4: 205)

---

[2] Larmor himself seems to have recognized such feelings. Thus, on 9 September 1924, he wrote to Oliver Lodge that "I have acquired the idea that I am suddenly old and passé, through being bombarded by the officials (older than me) with formal returns to fill up as to the conditions on which I must retire on a pension now, at 70, and what they can offer." However, he was not prepared to leave Cambridge: "I really would retire if I could only retain a fellowship." Then he added: "They will make Cambridge into a Government bureaucracy yet. The younger generation takes no interest, lets it all slide." Unless otherwise stated, all the letters mentioned in this paper are deposited in the 'Lodge Collection', The Library, University College London, MS Add 89. I am grateful to the librarian for allowing me to have access to this depository.

[3] Larmor's papers are included, with a few exceptions (among which are, as I will commment later on, his papers on relativity) in Larmor 1929.

[4] On Larmor and the FitzGerald-Lorentz contraction, see Warwick 1991, 1995. Other aspects of Larmor's contributions to the development of the electron theory are discussed in Buchwald 1981a, 1981b, 1985, Darrigol 1994 and Hunt 1991.

[5] Other commentators have put forward similar opinions. Thus, Buchwald (1981b: 374) wrote: "In his later years he was reknowned as a staunch conservative in his scientific views, often delighting in

But what were "his natural opinions"? In this paper, I will try to discover his opinions concerning Einstein's general theory of relativity, by using mainly—though not only—his correspondence with Oliver Lodge, whose points of view will also be studied. At the same time, further comments will be made about the points of view of Arthur Eddington, James Jeans and Edmund Whittaker, who corresponded with Lodge.

Before I begin my discussion, let me say that unlike many of the case studies historians of science pay attention to, the present one is, quite clearly, a story not of successes, but of frustations and final defeat. It is the story of how an able and intelligent physicist, brought up during the nineteenth century, reacted when confronted with such a radical new theory as general relativity.

1. The introduction of relativity in Great Britain before 1919

We know now that the early reception of relativity in Britain was a complicated process.[6] Certainly, it was so before 1919, the year of the famous eclipse expedition, led by Frank Dyson and Arthur Eddington, to prove whether Einstein's prediction of the bending of light was true or not. Sometimes it has been argued as if very few people knew of Einstein's new theory of gravitation, general relativity, before that momentous year. However, the more we learn about the history of relativity in Great Britain, the more we find that this was far from being true.[7] And it was not only the new generation of physicists and mathematicians who learnt about relativity quite early on; the 'old guard,' people like Larmor and Lodge, were quite well informed too.

Thus, on 20 October 1916 Larmor wrote to Lodge:

> Einstein is so dissatisfied with an aether and motion relative to it that he makes a gravitational field a modification of surrounding space from the Euclidian form which a body carries about in rigid connexion with it! And our friend Lorentz is fascinated by it—rather as a mathematical exercise I think than a philosophical theory.

That is, by 1916 he knew of the existence of general relativity. To which extent he understood that theory is something we shall consider later on.

Of course, to some extent it is not surprising that both Larmor and Lodge were interested in anything new Einstein had to say, in as much as their own work had dealt in the past with questions that Einstein's previous theory, special relativity, had illuminated in a different manner to what they had expected. This is the case, for instance, with Larmor's already mentioned *Aether and Matter*, as well as with

---

the role of the older scientist defending the Olympian past against the immodest pretensions of the present." Similarly Woodruff (1981: 39) stated that "unlike Lorentz, Larmor did not participate to a larger extent as a guide to the newer generation of physicists developing quantum theory and relativity. In general, he maintained a conservative, critical attitude toward the new ideas, particularly examining the possible limitations of the relativity theories."

6 See, Goldberg 1970, Sánchez-Ron 1987, 1992 and Warwick 1992, 1993.

7 One of the first to show this was John Stachel (1986).

Lodge's 1892 experiments in Liverpool aiming at determining whether the ether could be set in motion.[8]

Contrary to what one could have perhaps expected, it seems that Larmor and Lodge did respect Einstein, although they clearly did not share his points of view (more about this later). Writing to a friend, J. Arthur Hill, Lodge—the less knowledgeable and more hostile in the end of the two—stated (21 January 1915):

> In speaking of Einstein and the other upholders of the doctrine, I ought to add that they are brilliant mathematicians and learned people, and not to be sniffed at lightly—that is what has caused their doctrine to catch hold. They also appear to be sustained by experiment to this extent, that all those who attempted to attack the question of Matter through Ether—and they have been numerous—have had a negative result. The Principle of Relativity assumes that a negative result they always will and must have, no matter how far they are pushed. (Hill 1932: 61–62.)

2. Problems with the mathematics of general relativity

There is no doubt that, contrary to Larmor's case, Lodge found general relativity rather difficult to understand from the very beginning, specially because of his limited grasp of the mathematics involved. And this contrast between two antagonists to relativity deserves our attention, because it allows us to understand better the nature of Larmor's opposition to relativity, rooted in physical and philosophical considerations, and not on mathematical difficulties.

A decade after the eclipse expedition, on 27 May 1929, Lodge confessed to Edmund Whittaker his mathematical shortcomings:

> I was grateful to you for sending me your lecture on "What Energy Is."[9] But I am rather horrified to find that I cannot follow, i.e., understand, it all. It rather shocks me that the tensors have to be introduced in connexion with so fundamental a thing as energy. I don't even know what a tensor is. I know that a vector is a scalar with direction as well as magnitude. One has got used to vectors. I suppose a tensor is a vector with something added. But what? Is it a twist, or what Robert Ball called a wrench? At my age I am not going to learn the tensor calculus whatever happens. And I am rather surprised that conservation of energy has to be muddled up with conservation of momentum in order to make a complete statement. ...
>
> It is the concluding part of your lecture which makes me rather hopeless about understanding the modern view. I feel as if there must sooner or later be some simpler way of specifying fundamental things. Matrices and tensors are not the kind of mathematical weapons which I can imagine posterity using with any satisfaction, even though they are interim necessities. But I look to you to go beyond interim necessities and not to leave us in the quagmire prepared by Dirac and others.[10] It is very little confort to admit

[8] Lodge 1892. For an historical analysis of Lodge's experiments, see Hunt 1986.

[9] Whittaker 1929. This article is based on a lecture Whittaker gave at the Edinburgh University Physical Society on 16 March 1929.

[10] Perhaps it is worth recalling that, one year before, Paul Dirac (1928) had used matrices in his relativistic equation of the electron.

> that matter is a form of energy, if energy turns out to be nothing concrete but only a mathematical abstraction. (Lodge to Whittaker, 27 May 1929)

Lodge's statements represent, of course, a minor historical problem. Why was he so ignorant in, at least, those branches of mathematics? Was it because he went to University College, London, instead of going to Cambridge and taking the Mathematical Tripos? However, in London he had received instruction in "Riemann surfaces and many other parts of higher mathematics" from Henrici, the Professor of Pure Mathematics of UCL, and there was also W. K. Clifford, Professor of Applied Mathematics (Lodge 1932: chap. 6). Obviously, sooner or later he simply forgot those teachings, if he ever learnt them. Why he forgot (if that was the case) is a matter of guesswork, but one possibility is because the London method of examination did not put the same emphasis on the resolution of problems as did the Mathematical Tripos (the aspirants to wranglers had to solve hundreds of problems to sustain any hope of obtaining a good position in the Tripos). But, of course, this is only a possibility, sustained also, it is true, by the fact that people like Larmor or Eddington, who could understand the mathematics of general relativity, were former wranglers.[11]

Whatever problems Larmor had with general relativity, its mathematics was not one of them.

Apart from mathematical ignorance, there was also a rather British dislike for 'merely mathematical' theories, based on the belief that physics was not the same as mathematics, even though the former uses the latter. Like the majority of nineteenth-century natural scientists, Lodge thought that one should understand physical phenomena dynamically. Thus, he confided to Larmor:

> Mathematical analysis rightly leaves the world behind and soars into a region of its own. A writer in The Literary Supplement to *The Times* has recently been saying that music does the same, and rather luminously remarks that this is the reason why one can have musical and mathematical prodigies in infants without the possibility of actual experience. Relativity confronts us with the attempt to link these mathematical soarings with physical fact. It is strange that physical phenomena allow themselves to be expressed in this way. But it cannot be the only way of expressing them. There must be a dynamical method too. But until we have evolved the dynamics of the ether we find that we can proceed by undynamical methods, just as MacCullagh found long ago.[12] (Lodge to Larmor, 9 September 1922)

---

[11] Here is what Buchwald (1981b: 374) said concerning Larmor's education at Cambridge: "In many respects Larmor was a typical, if uncommonly intelligent, product of the Cambridge-centered educational system in physics of the 1880s."

[12] Lodge is referring here to James MacCullagh (1809–1847), who in 1839 developed a dynamical theory of the ether in which he introduced a new type of elastic solid (MacCullagh 1848). MacCullagh's contribution is studied in Schaffner 1972: 59–68, Buchwald 1985: 283–284, Stein 1981: 310–315, Whittaker 1951: 137–145, and Hunt 1991: 8–9, 16–19. Larmor was also a great admirer of MacCullagh, and in particular of his 1839 contribution, of which he wrote (Larmor 1900: 26): "The assumption of elasticity of some kind in the ether is of couse absolutely essential to its optical functions: and the elucidation from the optical phenomena, as a purely abstract problem in analytical dynamics, of the mathematical type of this elasticity, was acomplished in 1839 by MacCullagh in an investigation which may fairly claim to rank amongst the classical achievements of mathematical physics."

In some sense, he saw himself and Larmor as missionaries in a strange land, the land of mathematical physics and mathematically-expressed relativity; their obligation being to make physical sense of the theory: "The attempted marriage between genuine Physics and hyper-Geometry," he wrote also in the last mentioned letter, "was sure to lead to confusions. They will doubtless straighten themselves out, largely perhaps by our aid. For I know no one else who can wield both weapons with equal ease."

3. Debating about relativity: Lodge, Jeans and Eddington

In spite of his mathematical lacunae and ignorance of the physical content of general relativity, Lodge did not mind entering in discussions with other people as to what the theory meant or what its consequences were. It is fortunate that he was that sort of man, for it helps us now to find out what other British scientists thought of relativity as early as 1917. James Jeans and Arthur Eddington were two of those scientists. Let us begin with Jeans.

Replying on 14 August 1917, to a letter of Lodge, Jeans stated:[13]

> I do not regard relativity at all as a settled question but I feel it is much the most hopeful possibility. The only alternative seems to me to be the Fitzgerald contraction which in my opinion leads to endless difficulties. It seems almost to require all atomic motions to be governed by the old-fashioned electromagnetic equations, ... indeed under them the atom would necessarily shrink and collapse, and they seem quite inadequate to account for line spectra etc. I feel reluctant to make such great sacrifices to keep light-propagation through the old-fashioned ether, and then to accept a gravitation which is not propagated through the ether and a gravitational field which does not contract with motion through the ether.
>
> I agree entirely about the complexity of the presentation of relativity. Einstein is not a trained mathematician or I suspect he could have put the whole matter much better. (Jeans to Lodge, 14 August 1917)

At this point, Jeans put forward a problem for considering gravitation along the lines of special relativity. Consider, he said, the equilibrium of the Sun moving with velocity v. Obviously, it is acted on by its own gravitation and by electromagnetic forces from its own electrons. If only the latter existed, then the Sun would contract into a $(1 - v^2/c^2)$ spheroid. "But," Jeans went on, "if the gravitational spheroid is spherical, this prevents complete contraction. A compromise is effected between electromagnetic forces, striving for a Fitzgerald spheroid, and the gravitational forces striving for a strictly spherical shape. The result is a spheroid, less flat than the Fitzgerald one. From this difference between the contraction of the Sun and his measuring rod, an observer could determine v."

What Jeans seems to be saying is that if one has in mind Lorentz contraction plus classical (i.e., infinite speed of propagation) gravitation, then one should get a non-spherical geometry for the Sun, which would allow the determination of its motion through the ether. With Einstein's new theory (general relativity), he added, it was different:

---

[13] Deposited, as all the letters I am using, at the "Lodge Collection," UCL, MS Add 89.

> Einstein suggests that a simpler scheme is got by adjusting the gravitational field also, so that everything contracts uniformly. In this case an observer can not measure v. (Jeans to Lodge, 14 August 1917)

The difference between both approaches was that the former one "required two mechanisms of propagation, apparently unconnected; the latter [general relativity] virtually one."

It is not clear to what extent Jeans really understood then the meaning of Einstein's general theory; there are indications that he thought that general relativity was just special relativity plus gravitation, perhaps something like Poincaré's theory (Poincaré 1906). Indeed, in his August letter to Lodge, he pointed out that his analysis "does not agree in form with Einstein's presentation because of the different ways of measuring length."

Apparently, Jeans was getting his information about general relativity from Willem De Sitter's articles (De Sitter 1916a, 1916b, 1917), although his opinion of them was not very good: "I fear," he confessed to Lodge, "De Sitter has rather prejudged the reception of Einstein's theory by a too abstruse presentation."

Less than two years later, Jeans was still discussing relativity matters with Lodge, although on this occasion the subject of the discussion was the special theory. According to Jeans:

> There are two distinct questions:
> (A) The system of laws in which we believe
> (B) The mechanism which produces ... these laws.
>
> (A) is a question of observation which can be definitely settled in the laboratory; (B) is a question of speculation which probably cannot be finally settled at all. Indeed it seems to me that it is only a guess that (A) may result from mechanism at all. We English being a practical race regard every theme as an engineering problem.
>
> To my mind, if I see things rightly, relativity has only to do with (A). It is simply a guiding principle to suggest laws for trial, and so far these laws have always been confirmed by observation. ...[14]

---

[14] It is interesting to quote what Jeans said here to illustrate his point:

> One law which is suggested by relativity, but cannot be tested observationally, is that a Fitzgerald spheroid ought to be a figure of equilibrium. If this ever can be tested and is found to fail we shall be compelled to give up relativity. But you say "I see no reason why a Fitzgerald spheroid should not be a figure of equilibrium; ... it is the simplest assumption." So far you are with the relativists entirely. Again you say "I feel that the oblate spheroid ... to all practical purposes ... is a sphere; I would not expect any precession of the equinoxes, ... nor lunar perturbations ... " This, if I see the matter rightly, is all pure relativity. So far you agree entirely with Einstein, and to this extent believe in relativity.
>
> Einstein suggests the further step—let us try the effect of introducing similar simplifications everywhere, and examine to what laws they lead. This tentative examination is all that relativity amounts to. The simplification must be defined more precisely, and is put in the form that an actual velocity v in the ether can never enter into observable quantities. It follows that for a moving body $(1 - v^2/c^2)$ must measure all contractions equally—length, gravitational field, etc.
>
> You, Einstein and myself are, I think, all in agreement that "this is the simplest assumption"—purely as a trial assumption of course. I would state this in the form that we all believe equally in relativity, as I think we do.
>
> Now Einstein says, and I think you must agree,—"we have no longer any means of measuring absolute lengths in different directions, so that for all practical purposes we cannot distinguish between a sphere and a spheroid, and instead of saying, as you said

> Question (B) is another matter. Relativity finds the laws, and we may then try to find a mechanism to account for them. My own feeling is that the ether is not likely to account for them and I doubt if any 'mechanism' in the engineering sense, can do so. But certainly I cannot prove this, and there can be as many opinions as men. I rather think Lorentz agrees with you and Einstein with me. I remember at Birmingham Lorentz wrote on the board (in the radiation discussion)
>
> Matter–resonators–ether
>
> and then said "You see I have written 'ether' so I believe in it still; those of you who do not may substitute the word 'vacuum': it will make no difference."[15] The last 5 words are of course the principle of relativity in a nutshell. (Jeans to Lodge, 21 March 1919)

Jeans saw clearly Lodge's position, and, in the end, Lodge had to agree that relativity (special relativity, the theory he felt competent to judge) was a valuable help when trying to know how nature behaves; but what he could never accept is that his cherished ether was not, somehow, a more fundamental concept.[16] However, he found some comfort in this regard because he thought relativity did not enter into conflict with ether. In this sense, he wrote to Edmund Whittaker:[17]

> I know that Relativity theory deals with Physics hyper-geometrically, and does not need ordinary mechanical or kinematical properties for their explication. It truly ignores the ether but by no means denies it. Nor should the possibility of arriving at results by other means prevent the attempt to arrive at them also in more dynamical fashion, with a better understanding of what is really happening. Relativity, like the second law of Thermodynamics, and to some extent the conservation of energy, enables us to obtain results without clearly exhibiting the details of the process. There

---

"I feel that the oblate spheroid ... to all practical purposes ... is a sphere." Einstein says "Let us call it a sphere, since we are at liberty to redefine our measurement of length."

15 Jeans was referring here to the meeting of the British Association for the Advancement of Science held at Birmingham on 10–17 September 1913. A discussion took place on 12 September on the theory of radiation (Report 1913), with the participation of Jeans, E. Pringsheim, A. E. H. Love and Larmor (Lodge, who did not participate, attended the meeting; in fact he was its president, delivering the inaugural conference [Lodge 1913]). Lorentz's comments, mentioned by Jeans, are the following (Report 1913: 381): "we may apply ... the time-honoured name of 'ether,' without discussing the fitness, in the present state of science, of maintaining the ideas that were originally associated with the word. The principle of relativity leads us to consider the question whether the ether has so much of substantiality that one can speak of the notion of a body relatively to it. For the present purpose this is irrelevant, and 'ether' is merely a word, for which we might as well substitute 'vacuum.'" These words do not give quite the same impression about the firmness of Lorentz's opinions on the 'absolute' reality of the ether (circa 1913) as is offered by Jeans's reconstruction.

16 It was Lodge (1933: 5), remember, who wrote: "The Ether of Space has been my life study, and I have constantly urged its claims to attention. I have lived through the time of Lord Kelvin with his mechanical models of an ether, down to a day when the universe by some physicists seems resolved into mathematics, and the idea of an ether is by them considered superfluous, if not contemptible. I always meant some day to write a scientific treatise about the Ether of Space; but when in my old age I came to write this book, I found that the Ether pervaded all my ideas, both of this world and the next. I could no longer keep my treatise within the proposed scientific confines; it escaped in every direction, and now I find has grown into a comprehensive statement of my philosophy."

17 Lodge was prompted to make these remaks by the following comment of Whittaker (in a letter to Lodge, 30 August 1922): "I am not quite sure that Einstein would fully endorse what you say in the first line of page 8 [of Lodge's Silvanus Thomson Memorial Lecture]: in his Leyden lecture of 5 May 1920 he says 'The action of the theory of general relativity is a medium possessing no mechanical or kinematical properties.'"

> must be a dynamical theory, even if we are able to dodge it for a time; though it is true it does not follow that the dynamics of the ether is identical with the dynamics of ordinary matter. But the principle of least action, which is subterraneously the principle of Relativity, cannot be regarded as truly dynamical. (Lodge to Whittaker, 1 September 1922)

Although some specific problems were considered, Lodge's interchange with Jeans remained on an almost epistemological, rather general, level. It was different with the debate which Lodge carried on with Eddington in 1917.

Even though Lodge found general relativity too hard (i. e. too mathematical) to understand, he did not refrain from considering—from his own point of view, of course—the sort of problems that Einstein's new theory had opened. He felt that his world, his classical and 'ethereal' world, was in danger, and he did not hesitate to come to the fore in its defence. This he did in the pages of the *Philosophical Magazine* in 1917 and 1918, as well as corresponding with Arthur Eddington, who also used the pages of the same journal to answer Lodge's claims.[18]

In his first paper, Lodge (1917a) analysed some astronomical consequences of what he called "the electrical theory of matter." As one would have expected of him, Lodge's starting points were expressions of the "electromagnetic worldview." Thus, he supposed that: (1) the motion of matter through the ether has a definite meaning; (2) an extra inertia due to this motion is to be expected at high speeds, "in accordance with the FitzGerald-Lorentz contraction"; and (3) this extra or high-speed inertia is not part of the mass, but is dependent on the ether and hence not subject to gravity. He thought that from these "reasonable hypotheses," all "quite independent of the theory of relativity," some astronomical consequences would follow; in particular, that "the outstanding discrepancy in the theory of the perihelion of Mercury would be accounted for by attributing a certain value to a component of the true solar motion through the aether in the direction of the planet's aphelion path." That is, he wanted to show that his 'electrical theory of matter' could cope also with astronomical problems, with the same success as general relativity. Thus, we read in his paper:

> Professor Einstein's genius enabled him in 1915 to deduce astronomical and optical consequences (some not yet verified) from the Principle of Relativity. ... I wish to show that one of them [the anomalous motion of the perihelion of Mercury] at least can be deduced without reference to that principle. (Lodge 1917a)

Once the paper was published, Lodge sent it to Eddington, who thanked Sir Oliver:

> I was very glad to receive your Phil. Mag. paper and have been much interested in reading it.
>
> The criticism that rises in my mind is that it seems unlikely that we can find any solar motion which will set Mercury right without upsetting something that is already accordant. Einstein's formula had the great advantage

[18] George W. Walker (1918) also made a small contribution to this discussion. I mentioned, without recourse to the correspondence I am using in the present paper, some aspects of that discussion in a previous article (Sánchez-Ron 1992: 62–63).

> of having *d* where you have *ed*, consequently his correction to Venus & the Earth was negligible compared with observation, but I am afraid yours may not be. (Eddington to Lodge, 2 August 1917)

Lodge's points aroused Eddington's interest quite rapidly, and four days later he was writing to Lodge again:

> After writing to you I worried out the matter to what seems a definite conclusion. I have written a paper for the Phil. Mag. and enclose it as I think you would like to see it. (Eddington to Lodge, 6 August 1917)

Indeed, Eddington asked Lodge, who was one of the editors of the journal, to communicate the paper, pointing out that he would "be extremely glad as it would remove any possibility of an appearance of hostility, which was, of course, far from my thoughts, for (although I am somewhat of an Einstenian) I regarded the suggestion as particularly interesting."

In his paper, Eddington (1917a) began by pointing out that Lodge's explanation of the motion of the perihelion of Mercury was "comparatively simple, and on that account will be widely preferred to the recent theory of Einstein, which introduces very revolutionary conceptions, provided that it meets certain other astronomical requirements which seem necessary." To see if those "astronomical requirements" were really fulfilled, Eddington wanted to discover whether or not in removing the discordances for Mercury, Lodge's theory introduced discordances for Venus and the Earth, as he had suggested in his letter of 2 August. "If the explanation breaks down under this," he pointed out, "the discussion will make prominent a feature of the success of Einstein's theory which has perhaps not been sufficiently emphasized." Fortunately for general relativity, Eddington's results did not favor the electrical theory of matter, and the Plumian Professor of Astronomy at the University of Cambridge finally concluded politely that:

> It is disappointing to find that this interesting suggestion [of Lodge], which gives a simple explanation of the most celebrated discordance of gravitational theory, is apparently unable to satisfy the most stringent test proposed. (Eddington 1917a: 167)

However, the discussion did not end at this point. In the course of their correspondence, Lodge pointed out to Eddington that some basic equation the astronomer had been using required amendment when the mass was taken as variable. Eddington (1917b) took the criticism into consideration and looked into it, but he concluded in the October issue of *Philosophical Magazine* that "So far as I can see, the conclusions of my previous paper are not materially modified by this more rigorous calculation."

Confronted with the technicalities displayed by Eddington, Lodge could but declare that, although he did not see how to disprove Eddington's conclusion, he felt "that the last word has not been said on the subject" (Lodge 1917b: 519). His own results for Mercury and Mars were too good "to be readily abandoned." He insisted on the hypothesis that the additional inertia due to motion is not part of the body's true mass, and so is not subject to gravity, offering a further argument:

> In favour of the hypothesis of gravitation independence, I adduce the analogy—admittedly not coercive—of a solid moving through a fluid. The

> apparent inertia of such a body is increased by an amount depending on the fluid displaced, but its floating or sinking properties remain unaffected: the extra inertia is not part of its mass, and is not subject to gravity. If the extra electrical inertia of moving matter is not part of the true mass, but represents only aetherial reactions, which is what I expect, then an astronomical perturbation is bound to be caused in rapidly moving planets; and whether this perturbation can be adjusted to agree with observation, i.e., whether a solar drift can be chosen which shall give a result neither in excess for one planet nor in defect for another, becomes a matter for further detailed calculation. (Lodge 1917b: 519)

On 8 December, Eddington received Lodge's new paper and wrote to him immediately:

> I was very pleased to receive the Phil. Mag. from you this morning, and to read your article, which gives a most excellent presentation of the position we have reached.
>
> The point you now rise had not ocurred to me before, but I feel no doubt it is quite true. If the extra mass is subject to gravitation and the Newtonian law holds unmodified, the perturbations are just doubled. (At any rate the main perturbations—I have not examined those involving the eccentricity). (Eddington to Lodge, 8 December 1917)

However, no matter how "satisfactory" it might be "to have this point, on which there may be legitimate difference of opinion," Eddington thought that it should be put in the background, because for him the real question was "does the strict Newtonian law hold good for systems moving through the aether?"

> In thinking over the matter in the last two months, I had felt that this discussion has produced valuable evidence (*evidence*—not *proof*) for the principle of relativity in its Minkowskian form, with or without the recent extensions of Einstein. We know that optical and electrical laws have entered into a strange conspiracy to prevent us determining our motion through the aether, by methods which at first sight seemed almost certain of success. In generalising this, it has been assumed as an hypothesis (I believe without a vestige of proof) that gravitation conforms to the same principle. By working out your theory, we obtain the first definite indication that gravitation has actually joined the conspiracy. The observed variation of perihelia & eccentricities of Venus and the Earth are just what they would be if the sun were at rest in the aether; hence the Newtonian law has modified itself so as to conceal the effect of the sun's motion (which calculations lead us to expect to be easily observable), and we add one more to the long series of experiments giving null results. (Eddington to Lodge, 8 December 1917)

Indeed, it can be said that through his interchange with Lodge, Eddington's faith in relativity had been strenghthened, while Lodge had been forced to retreat by lack of arguments.

4. Larmor and Leigh Page on radiation reaction

At the same time that Lodge was debating with Eddington, Larmor was entering into a discussion with Leigh Page, who as a young instructor at Yale had been, in

1912, together with R. D. Carmichael, professor of Mathematics at the University of Indiana, one of the first American commentators of special relativity.[19]

The interchange arose from a note which Larmor (1917) published in *Nature*. In it, Larmor emphasized what he considered a contradiction to the principle of relativity involved in the treatment that, following Poynting, he had given (Larmor 1913: 207) at the Fifth International Congress of Mathematics (Cambridge, August 1912) to the radiation reaction experienced by a moving electron on account of its own emission of radiant energy.[20] The problem was that for an isolated radiator that force "was," Larmor (1917) argued, "proportional to its velocity through the aether, and this is said to violate the principle of relativity." The nature of such violation was that according to the relativity principle the velocity of an isolated body does not have any meaning.

Soon afterwards, Page (1918a) tried to refute Larmor's result. His statement of the problem was somewhat clearer than Larmor's: "A question," he wrote (1918a: 47), "of some interest to the astronomer is whether or not a body in motion, such as a star, is retarded by its own radiation. For, on the electromagnetic theory of radiation as developed by Maxwell and his followers, a beam of radiant energy is supposed to have a quasi-momentum, such that if a body emits energy in a single direction it will lose momentum and in consequence suffer a reaction tending to push it in the opposite direction. Now if a star is at rest, and in thermal equilibrium, it follows from symmetry that it will radiate equally in all directions, and there will be no resultant impulse. If, however, the star is in motion, classical electrodynamics leads to a greater emission in the forward direction than in the backward, and consequently it would appear at first sight as though there should be a retardation which would ultimately bring the star to rest." But in this case there would be a problem for special relativity, because a system in uniform motion would be different from another 'at rest' (at rest with respect to the ether, according to Larmor).

Page considered the case of a single oscillator (i.e., a single vibrating electron), which although simpler than a star constituted a "perfectly general test of Larmor's expression," and showed that a "rigorous solution of the problem for this relatively simple case shows the existence of no retarding force. Larmor's result is found to be invalid because of a tacit assumption underlying his reasoning."[21]

Larmor, however, did not give up, and he insisted on his ideas through the pages of the same journal Page had used: the *Proceedings of the National Academy*

---

[19] Page 1912, Carmichael 1912. In his 1912 paper, Page showed that it is possible to use only the principle of relativity and the laws of electrostatics (e.g., Coulomb's law) to derive both Ampère's law of electric currents and Faraday's law of induction—both laws which depend on the motion of charges relative to the observer. One peculiarity of Page's approach was that he considered that the invariance of the velocity of light was a consequence of the principle of relativity and not itself a premise. Page's 1912 contribution has been discussed by Goldberg 1984: 273, Purcell 1980: 108, and Whittaker 1953: 247.

[20] It is interesting to point out that in 1912 neither the principle of relativity nor Einstein's theories or name were subjects of consideration—or even mention—for Larmor.

[21] Page's calculations were not included in his *Proceedings of the National Academy of Sciences* paper (Page 1918a), but in another article he published in *Physical Review* (Page 1918b).

of Sciences (Larmor 1918). We can easily understand his interest: what was at stake was the substantiality of the electromagnetic ether. "It has always seemed to me," he wrote (Larmor 1918: 334), "that this subject, which may be described as that of interaction of the aether with uniform motion, though of slight account phenomenally, is theoretically of high significance, in that it is destined perhaps to throw light on the nature of the forward momentum that is convected by radiation, and thence on the intimate dynamical nature of radiation itself and the physical function of the aether."

As to the mathematical expression derived by Page for the radiation reaction, he did not question its correctness, only Page's interpretation of it: "The expression contains terms involving the acceleration of the system and its time-gradients, but no term involving its velocity except in combination. . . . But the formula obtained seems to leave [the question of whether it challenges the principle of relativity] as it was: for equally the acceleration of the isolated body could have no meaning; and morever, though the thrust of the radiation is, in this case compensated, the velocity appears to be actually involved in the formula in the same manner as the impugned result would require for this particular problem of a convected electrostatic system, for which the radiation is extremely transient and very slight in the absence of extraneous force" (Larmor 1918: 335). To the Lucasian professor, there was no doubt that "However we may clothe our thought in language of relativity, it would appear that this issuing radiation [that "shot out from the electron while it is in varying motion"] does effectively possess an absolute velocity c, and therefore an absolute velocity of its source also is theoretically determinable from observations made in relation to it."

But behind this apparent security there were also doubts. The 'power' of relativity, its capacity to solve problems and to illuminate the physicists' analytical understanding of nature, had undermined perhaps Larmor's conviction not in the truth of the 'ether view of Nature,' but of its status as the unique point of view; maybe a 'relativity view of Nature' was also possible. Thus, he asked himself questions like:

> The special question now in evidence is—Is it now expedient to exchange this frame of reference, corresponding to c infinite, for another far more complex but very slightly different continuum having a finite space-time modulus c? The more fundamental question is—Are we to assign to either frame dynamical properties, typified by propagation of physical effect in space in terms of undulations sustained by stress and inertia, or are we to assign to it properties solely geometrical and regard all physical effect as merely project in duration across space? (Larmor 1918: 336–337)

And knowing that the "forms of special unrestricted relativity which have been recently current ultimately demand and perhaps prefer the latter course," he concluded that: "There is no absolute criterion to decide between the two ideals. The first order of ideas has proved itself as the foundation on which the interlaced fabric of electric and optical science has been actually constructed: the other seems to offer as yet only somehat ingenuous and disjointed though significant expression for certain striking features of recent discovery which the former has not yet succeded in assimilating, and seems to require us to obliterate the course of evolution of the science or perhaps to retain it as a mere historical survival."

The problem—better perhaps the virtue, because it offered to classical physics a possible way out—was that

> all these modes of restatement of departments of physical science in more expressly relative terms may be comprehended as partial analytic developments of the far wider principle of the purely relational character of our external knowledge, which was advanced and systematically fortified with great abstract force in the general metaphysical domain by Bishop Berkeley. ... Thus in these matter we are hardly concerned with refuting any theory, for all are relative: it is fruitless to traverse any proposition, unless we take into account the definitions and context in which it subsists. The question is, as to which scheme of formulation gives as a whole the closest and most expressive representation of the complex of natural knowledge, and affords the most promising clue to its future elaboration and extension. But a choice does not by any means preclude development along other promising but provisional lines for which an interpretation has yet to be found. (Larmor 1918: 337)

Of interest are also the last sentences of Larmor's paper, in which he acknowledged Page's contributions to special relativity, by which one could "translate the usual so-called classical electrodynamic theory into the order of ideas, formulation under relativity of the *c* type being an essential feature, ... effecting this translation in more intuitive and fundamental terms than had previously been attained to; and also that by [its] conciseness and geometrical directness [it] facilitates that comparison and contrast with the alternative order of ideas which is the essential matter." That is, it welcome Page's contribution because it made it possible to express results of classical electrodynamics on special relativistic terms.

5. Larmor and general relativity

Larmor's first publication (Larmor 1919a) which is connected with general relativity came out the following year, 1919, that is, during 'the year of the eclipse.' Previously, he had told Lodge of his interest in, and difficulties with, Einstein's gravitational theory. Thus, on 10 March 1918 he wrote to his friend that having "been compelled by Indian young men and Newnham young women, the only audience, to lecture on relativity, of which they know as little as I do, I have hit upon the precise way in which both gravitational and kinetic mass must be altered on a planet or the Sun to secure it locally. If I could get settled I would contribute to your investigations—but this war!"[22]

A week later he wrote again, now in a rather desperate mood:

> It is hopeless to *explain* their relativity. It does not mean that space and time are abolished; but that many kinds of space and time are undistinguishable

[22] The reference to "Indian young men and Newnham young women"—not really a nice comment—should be understood as one of the consequences of the war, then in its last moments, which was responsible for a substantial reduction in the number of British males who followed courses at Cambridge University. Newnham was one of the two colleges for women which existed then in Cambridge (the other was Girton).

> because to any *set of equations* in one system there is a corresponding *set of equations* of like form in the other system—a covariant set, as the mathematicians say. You can't put that into words. There are a vast number of possible forms of relativity on this basis, short of David Hume's original form that we only know impressions on our organs of sense, and everything else is padding which can be filled in our pleasure. Einstein still keeps an external extension in space & time as well as our sense organs; but in time he will doubtless eliminate them. (Larmor to Lodge, 18 March 1918)

But in spite of such difficulties, he entered the 'relativity arena.' And his 1919 paper is, in one sense, an outstanding publication, as it has the merit of having introduced the first five-dimensional unified field theory in the history of the several efforts to find a unitary framework for the electromagnetic and gravitational interactions. The often mentioned five-dimensional unified theory developed by Theodor Kaluza (1921) was proposed two years after Larmor's, although it is true that it was more comprehensible than that of the Lucasian professor of Cambridge.[23]

Larmor's paper is, in several respects, rather chaotic, and the physical ideas behind it quite confused, but it is nevertheless interesting, as it reveals some aspects of its author's relationship with Einstein's general relativity. However, initially the paper was not related to this theory, but to the "electrodyamics scheme of relativity," or "electrodynamic relativity;" that is, it was closer to special relativity. The idea was to use the symbolic calculus developed by W. J. Johnston, which, as Larmor points out, was a sort of reverse procedure of the one followed by Hermann Minkowski in his four-dimensional interpretation of special relativity:

> The procedure of Minkowski seems to have been, having identified electrodynamic relativity with invariance of the system as regards position in the fourfold continuum, to group and identify the physical quantities of the Maxwellian field as components of various 4-vectors and 6-vectors. ... The procedure of the present calculus is the reverse. The system of invariants natural to a four-dimensional flat continuum are immediately manifest in Mr. Johnston's application of Clifford's calculus; and the totality of them are identified precisely with the vectors of the electrodynamic scheme of Maxwell, with which they are co-extensive. (Larmor 1919a: 336)

Although Larmor was working essentially within the framework of electrodynamics, he was aware of the problem that this meant not having included gravitation:

> But, whatever be the critical obstacles, the problem of probable interaction between gravitation and electrodynamic fields, including rays of light, of course, remains urgent; and if such connection is actually detected by the various astronomical determinations now in progress, data such as hitherto have been entirely non-existent will have been supplied for an attack on this deep-seated question. (Larmor 1919a: 342)

---

[23] On Kaluza's theory, see Vizgin 1994. It is interesting to point out that Vizgin does not mention Larmor, whose contribution seems to have been almost forgotten by most historians; one exception is Jim Ritter (See Goldstein & Ritter A). Thus, and referring of course to Kaluza, one finds in Vizgin's book such sentences as the following: "In 1921 there arose one more new direction in the development of unified geometrized field theories" (p. 149); and "the idea of five dimensions determined one of the main paths of realization of the program of unified geometrized field theories for many years" (p. 160).

Larmor's paper had been received by the Royal Society on 28 August 1919,[24] but the previous comment was included in pages added on 20 October. In the meantime, the news of the results obtained by the British eclipse expedition began to arrive, and Larmor added a note to his statement: "This was written before it became known that the Greenwich and Cambridge astronomers, in their recent eclipse expeditions, had confirmed Einstein's prediction for the amount of the deflection of a ray of light by the influence of the Sun. It must be recognized that the theory has come to stay in some form or other. Its main implication, of instantaneous propagation of change in the constitution of space,[25] seems to be avoidable only on a psychological point of view which would assert that a portion of space is existent only while attention is concentrated on it. It can be managed, however, by including the varying space in a uniform space of higher dimensions, just as the deformation of a two-dimensional surface can be visualized as a whole in uniform space of three dimensions."

This "accommodation"—in a "uniform space of higher dimensions"—of the general relativity world favored by the results of the eclipse expedition, was performed by Larmor in a new addition to his paper (this one dated 20 November; that is, after the 6 November joint meeting of the Royal Astronomical Society and Royal Society). It was there that he developed the five-dimensional unified theory mentioned above. His purpose was no other than to maintain the symbolic calculus à la Johnston he had developed in the previous pages. That calculus was based on a four-dimensional flat continuum (a "homaloid" in Larmor's terminology); but now that Einstein had included the "phenomenon of gravitation" in that Minkowskian scheme "by altering slightly" the expression of the metric (of course, this is not a correct expression of the mathematical and physical structure of general relativity), Larmor (1919a: 353) thought that his previous ideas could be still maintained, and so he pointed out that that "generalization can still be brought within the range of the Clifford geometry by introducing into the analysis a new dimension preferably of space." "Now," he continued (pp. 353–354), "any continuum of four dimensions, having a quadratic line-element, however complex, is expressible as a hypersurface in this homaloid continuum of five dimensions."

---

24 Shortly afterwards, on 12 September, during one of the sessions of the meeting of the British Association for the Advancement of Science held at Bournemouth (9–13 September), Johnston and Larmor had signed jointly a brief note on "The Limitations of Relativity." Among the points they stated there, figures the following, of an evident Newtonian tone: "If time were linked with space after the manner of the fourth dimension, relativity in electrodynamic fields would be secured as above, but the sources of the field could not be permanent particles or electrons. If physical science is to evolve on the basis of relations of permanent matter and its motions, time must be maintained distinct from space, and the effect of convection must continue to be thrown on to material observing system in the form of slight modification of its structure" (Johnston & Larmor 1920: 159). It is interesting to remember that during the presidential address of section A ("Mathematical and physical science") of that meeting, Andrew Gray, professor of Natural Philosophy at the University of Glasgow, voiced some misgivings toward relativity: "I have attacked Minkowski's paper more than once, but have felt repelled, not by the difficulties of his analysis, but by that of marshalling and keeping track of all his results. . . . Some relativists would abolish the ether,—I hope they will not be successful. I am convinced that the whole subject requires much more consideration from the physical point of view than it has yet received from relativists" (Gray 1920: 143–144).

25 This is, of course, a blatant misunderstanding of general relativity, which reveals some of the limitations in Larmor's capacity to understand Einstein's gravity theory.

Like many others, in Britain and elsewhere, Larmor seems to have been stimulated by the general atmosphere aroused by the results of the eclipse expedition. He sent, for instance, a communication (Larmor 1919b) to one of the two main meetings organized in Great Britain soon after the one held on 6 November, to discuss Einstein's 1915 theory: the session hosted by the Royal Astronomical Society on 12 December.[26] As one could have expected, Larmor expressed again on that occasion his doubts concerning Einstein's theory, of which he said: "How far it is from being a determinate theory of the universe will appear to anyone who dips into the writings of its developers."

He might have been critical, but not indifferent. Thus, during the following decade, roughly until the late 20s, Larmor's main fight in the relativity arena was to interpret Einstein's relativistic contributions 'rightly,' developing, if possible, a gravitational theory of his own, as in August 1922, when he wrote to Lodge from his summer house in Dhu Varenn, Portrush:

> I thought I was going to write out my theory of Gravitation; but the sea air is too strong and sleepy for me. Moreover it is off colour. Gravitation is not an essential property of the Electron as I thought. In fact there is no answer to the fundamental question, why should there be Gravitation at all? Only if there is, considerations of isotropic form and symmetry restrict it to the Einstein field-form, if it is to fit in with the null effects of uniform translation (Michelson ...). Thus there is no explanation, but only a definite restriction of possibilities. (Larmor to Lodge, 12 August 1922)

In September, he was still in a depressed mood, still trying to arrive at a satisfactory—satisfactory to him—interpretation of general relativity:

> I have been worrying in the Einstein mazes. Why were they ever invented? So long as you stick to Algebra and don't try to interpret it you can sail away—goodness knows where. But to find out how it relates to external physical science is the rub. (Larmor to Lodge, 1 September 1922)

It was a problem of two worlds, the new, relativistic one, and the old Newtonian and Maxwellian, being mixed up. Larmor tried to follow Einstein's general relativity arguments, but his world was a peculiar mixture of the Newtonian and Maxwellian worlds. And so he found problems in that strange land:

> But in the fourfold the analyst hunts tensors and leaves the world behind. There you may interpret the world in the fourfold frame: you can also do it in the Newtonian frame—either frame is adequate for that content, though they are not equally natural and coincident always. But when you mix the two, and use the developments of Newtonian dynamics in a fourfold pseudo-spatial frame, it is confusion of tongues.
>
> The Maxwell stress is the mother of tensors. Its inner (very remarkable) signification is a most elusive business: I have thought I had it pinned down many a time for years past—and indeed have in a way. It is a case of one medium existing in another—matter in aether—and both in a frame of reference, and the interrelations take some disentangling. I fancy I see it all clear, but may not do so next week. (Larmor to Lodge, 1 September 1922)

---

[26] See Discussion 1919. I have mentioned this meeting in Sánchez-Ron 1992: 65 and nn. 26–27.

One of the problems Larmor tried hard to solve was the interaction between gravitation and light, a problem especially attractive to those who wanted to reproduce the deflection of light effect predicted by general relativity, from a theory based on classical electrodynamics. He tried in 1920 (Larmor 1920) and tried again during 1923.[27] From his summer house he wrote to Lodge:

> I have been here two months—trying to annex what of good there can be in Einstein. In his latest expositions there is one sensible thing latent, viz. Newton's absolute time, though he does not seem to know it.[28] Generally the process of 'saving one's face' seems to have begun.
>
> I can include a propagated gravitation quite comfortably were it not for the now demostrated interaction of light and gravitation, which is a fundamental experimental result. (Larmor to Lodge, 23 August 1923)

In October he thought he had found the solution:

> I have I think fathomed the deflection of light, after many trials, (at any rate to my present satisfaction,) expressed as a result of bringing gravitation, taken as we find it but not explained, within the ambit of the electrodynamic relativity as confirmed by experiment.
>
> The universe is full of absolute atoms, which on account of local relativity each tick out through their spectra the same absolute time to observers at rest relative to them. These local times are connected up physically into a universal time, for the absolute time a ray takes to pass, from a point P to another Q in the aether, is proportional to its length PQ, whether straight or curved. This universal absolute time T in the astronomical time postulated by Newton is no more unintelligible than absolute vibrating atoms identical everywhere in the universe.
>
> But the aether-field is locally disturbed in constitution by adjacent large masses, as the deflection of light shows. If the disturbance is such that the equations of vibrations (and of electrodynamics generally) can be reduced back to the simple form by changing from astronomical time T to an auxiliary time-variable t, namely
>
> $$dT = dt\left(1 - \frac{v}{c^2}\right),$$
>
> depending on locality, the difference being sensible only near large areas, the deflection of light will be as observed, and all the demands of local-relativity as observed will be met, provided this transformation is accompanied by one representating slight radial warping of the space of influence near each mass.

---

27 The following sentences from Larmor 1920: 333 contain some of the main ideas behind his efforts on this direction: "But we now pass from kinematic discussion of frames of reference to physical considerations. If we are to assert, in agreement with the doctrine of relativity plus Least Action, that inertia is a property of organised energy and proportional to it, therefore not solely of matter, and if we are to admit with Einstein, in the same and other connexions, that light is made up of small discrete bundles or quanta of energy, it would appear that each bundle is subjected to gravitation. Therefore if a bundle comes on from infinite distance ... it will swing round the Sun in a concave hyperbolic orbit, and as the result, the direction of its motion will suffer deflection away from the Sun by half the amount that has been astronomically observed."

28 Here, of course, was another of Larmor's problems with relativity. To Lodge, he wrote on 30 September 1925: "In the fourfold continuum, which is a natural geometric theme, there is no time nor motion nor direction; it is there pure length. Succession in time comes to the observer through vision by successively (directionless) rays."

> But this—a fixed ether except as locally affected by masses moving in it, and resulting universal atomic time,—is of course anathema to the relativist philosophers. The universe can however curl up into a ball if they so wish it. (Larmor to Lodge, 11 October 1923)

One paper, entitled "On the Nature and Amount of the Gravitational Deflection of Light" (Larmor 1923b), was the result of those efforts. However, he came to realize that its content was unsatisfactory:

> I am worried over adapting a scheme of gravitation to the experimental deviations of rays and spectral lines (if they are final) now for two years without success. But what I am absolutely certain about, however I turn it over, is that Einstein's theory gives only half results. (As in Phil. Mag. Dec. 1923.) (Larmor to Lodge, 30 September 1925)

However, he kept studying Einstein's works, perhaps with more interest and attention than most British scientists at the time. By the end of 1924 it appears he had restricted his main differences with Einstein to points of interpretation. He did not accept that it was a "final theory" (whatever this may mean), a satisfactorily physical explanation of nature, but he began to appreciate some of its values, as well as Einstein's powers:

> I have been reading Einstein—several times over—and begin to fathom his mentality. It is most picturesque and acute: he gets to where he wants without great regard for coherence. ... He never gives any references (except to his patron Lorentz,) or mentions the people who must be in his mind in his "reflections".
>
> I agree (with Lorentz) that he is a very powerful phenomenon: his "reflections" on quanta of light etc. are much in the same boat—not a logical system at all, but what he adopts as the best choice between alternatives all imperfect; and the choice once made, his business is to bolster it up the most plausible way he can. Is it the oriental imaginative mind of [*sic*; at?] play? Ask any Anglo-Indian.[29] (Larmor to Lodge, 9 September 1924)

Even so, he could not avoid entering into complex and obscure disquisitions:

> The spectrum test is *really* a test of absolute intervals of time! As such I accept its results. The frame of reference is *not* of no consequence. Every system has its absolute frame; namely the frame convected with the system as a whole. Thus each atom has its own frame, each solar system has its own frame; intervals of time in these frames are absolute, because the atoms are such. The function of relativity is to piece them together into a universal frame. If it is to be a space-like extension, this strictly cannot be done. We must therefore discuss *all* cosmic history as one unit, as pieced out by memory, which is astro-physics. The defect of the gravitation scheme is that time in it is entirely illusory; it is not possible to divide the cosmic history into slices so that the following ones are determined by the prior ones. There could be no causation, *no evolution* of the cosmos, if gravitation were 10^4 times as intense as it is, etc. etc. Yet there it is, in possession.
>
> The fundamental fact is that you exist and that you are intellectually identical with me; and I to verify that. Intellectual atomic theory.
>
> You see I am bitten by the metaphysic mosquito. (Larmor to Lodge, 9 September 1924)

[29] Perhaps we should consider the last sentences as an indication of anti-Semitism in Larmor.

6. Larmor, general relativity and the least action principle

According to Eddington (1942/4: 204), Larmor became convinced of the truth of general relativity around 1924; at that time he said to him: "I have been reading continental writers on relativity, and I find it is all least action. I begin to see now." However, his letters to Lodge do not give quite the same impression. Not that he did not emphasize the role of his beloved Least Action Principle. On the contrary, he thought that there laid the real truth, but that general relativity did not incorporate it in an appropiate manner. Thus, he told Lodge as late as 1930:

> Eddington is facetious about the Universe being governed by least action. But the Einstein geodetic orbit is only a case of Least Action *wrongly applied*, as I feel certain. From Vienna they admit to me that there is something wrong. (Larmor to Lodge, 30 September 1930)

And he added: "The view seems to be that the Mach (sensational) philosophy which fascinates Einstein is so inevitable that innate contradictions are merely provisional imperfections."

These sentences seem to indicate that Larmor had some philosophical interests and knowledge of Ernst Mach's works (certainly of the influence that Mach's ideas exerted on Einstein during part of his career), and perhaps also of the Vienna Circle contributions, although one of the expressions he used, "they admit to me that there is something wrong," makes this last possibility rather improbable, in as much as most of the members of the Vienna Circle (in particular Moritz Schlick and Rudolf Carnap) were sympathetic to Einstein's approaches and theories. Further evidence would be needed, however, to understand the real meaning of those sentences of Larmor, as well as his philosophical interests, if any.

What is clear is that he tried to use Least Action as a tool, or help, to undermine general relativity, as well as to develop his own "physical relativity." As when, late in 1922 and early in 1923, he found attractive the criticisms of general relativity made by the French physicist Jean Le Roux (1922a, 1922b, 1922c), which led him to think "that the Einstein *philosophy* is near its end" (Larmor to Lodge, 12 December 1922).[30] A month later he wrote to Lodge in this same sense:

> I have finished my proof that though the Le Roux paradox damns all Einsteinism beyond repair, it leaves my Action in a medium alone. You must be educated into Action. See my edition of Maxwell's "Matter and Motion" [Maxwell 1920]. (Larmor to Lodge, 10 January 1923)

Le Roux's criticisms of Einstein's general relativity were as radical as obscure, offsprings of a mind that mingled together different worlds and concepts—those of Newton's mechanics, Euclidean and Riemann's geometries, and Einstein's gravitational theory—and that could not understand the essence of the approximation procedures which connect in general relativity those worlds and concepts (not to mention wrong assumptions). The sort of problems which were also responsible

---

30 Lodge was also attracted by the Frenchman's work. In a letter he wrote to Larmor on 13 January 1923 he mentioned that he had already read "the first of Le Roux's papers in the C[omptes] R[endus]," and that he was "much tickled with it."

for many of the difficulties found by Larmor and Lodge when confronted with Einstein's relativistic physics.

Specifically, what Le Roux argued against general relativity was: (1) To arrive at the notion of curvature of surfaces it is necessary to consider simultaneously two quadratic forms, one of them having a fixed physical meaning. As in the general theory of relativity, Le Roux went on, one considers only one form, the invariants derived from it cannot be associated to properties of space; thus, the question of whether the space(-time) is finite or infinite cannot be answered (Le Roux 1922a). (2) The success of general relativity in explaining effects like the perihelial shift of Mercury is only apparent, as Einstein's theory does not take into account the perturbations due to mutual actions between the planets of the solar system; and if one suppresses such perturbations, "the concordance disappears," leaving an unexplained shift of 531″ (Le Roux 1922b). (3) Newton's mechanics is not an approximation of Einstein's theory, because both "are based on completely different principles" (Le Roux 1922c).

The fact that a physicist like Larmor was attracted almost immediately (Le Roux's papers were communicated to the Académie des Sciences in November and December 1922) by such ideas, due to a little-known French scientist, shows, perhaps better than in any other way, how prone he was to react in favor of arguments against relativity, or, what is the same, how much he disliked general relativity.

Coming back to the Least Action Principle, we have seen that Lodge was not as impressed with the Principle as was his friend, as can be seen from a letter he sent Larmor:

> The present fashionable mode of writing down equations without specifically referring to an ether, or to any form of mechanism, leaves everything in the air, and prevents the formation of clear conceptions, and evades everything of the nature of explanation or real understanding.
>
> I feel something of the same sort in connexion with your Action theories. They are no doubt powerful and necessary; but none of these main force methods—like the Second Law of Thermodynamics, for instance, or even the Conservation of Energy—can be the last word. They enable one to arrive at a result, but they don't explain it. And until something like a real explanation is given, the ether will not come into its own. (Lodge to Larmor, 9 January 1923)

Finally, Eddington's assertion is misleading in another sense. When he put on Larmor's lips the words: "I have been reading the continental writers and I find it is all least action," one might think that the Lucasian Professor had found Least Action in the works of those continental writers. Read, however, what Larmor wrote to Lodge in 1922:

> None of our foreign savants nowadays know what Hamilton's principle of Varying Action means. They talk glibly of the Hamiltonian principle as if it were a thing to play with at random, instead of the precise kernel of all dynamics. If only they would read their own Helmholtz if they can't read English. I suppose Jacobi had something to do with perverting it off the physical track into general algebraic analysis. (Larmor to Lodge, 1 September 1922)

What is certainly true of what Eddington pointed out in the obituary is that, as I have remarked above, Larmor "wavered very much over Einstein's theory of gravitation" (Eddington 1942/4: 205).

This is one of the reasons why his papers dealing with—or related to—general relativity are so difficult to understand; one never knows what he was aiming at. In the end, he came to value so little his "physical relativity" papers that he decided not to include them in the edition he prepared of his *Mathematical and Physical Papers*.[31]

Explaining this omission in the preface he added to volume II, Larmor referred to general relativity in the following terms:

> In this auxiliary cosmos space and time and motion do not occur; yet it can be of great value in a geographical sense, after the manner of a spherical map of the Earth, for unravelling the intricacies of relations connecting regions in our actual world, which astronomers are permitted to formulate only in terms of the complications of delay in the messages of the informing rays of light. The paralelism of relations between the world of actual perceptions and memory and this particular conglomerated fourfold assembly of permanent ray-models, so to say, extends over a prominent yet necessarily limited range whose boundaries have hardly yet been very closely examined. There may be, however, people who aim at transferring their whole life into this new cosmos. The papers now omitted are concerned largely with various general aspects of this correlation: naturally much of their contents has now become transcended, or modified into improved presentations. (Larmor 1929: vii)

7. Conclusion

To which "improved presentations" Larmor was referring in the last quotation it is not really worth asking; what is clear is that, when he wrote these sentences, relativity had, if not convinced, at least begun to win him over. Were we to think in terms of Lakatos's scientific research programs (Lakatos 1970), we would conclude that the Newtonian-Maxwellian research program that Larmor (and Lodge) tried to develop had finally arrived at a clearly degenerative stage, the progressiveness lying on the side of the Einstenian research program. Ever since he considered that the special and general theories of relativity had won enough attention and consideration from the scientific community, Larmor had tried to criticise their most novel concepts simply because it was not possible to understand—or reduce—them to classical, i.e., Newtonian and Maxwellan, terms ("The main difficulty about the acceptance of the relativity theory," Max Planck once wrote, "was not merely a question of its objective merits but rather the question of how far it would upset the Newtonian structure of theoretical dynamics" [Planck 1930; English translation: Planck 1932: 44].) At the same time, Larmor, or, again, Lodge, had kept trying to accommodate results emanating from the relativistic theories to the framework of classical Newtonian and Maxwellian physics. He, they, (the

---

31 Apart from those already mentioned in the preceding pages, these papers are Larmor 1921, 1923a, and 1927.

'old guard'), were, in short, always one step behind the facts and developments; their efforts were not to uncover a new phenomenal or conceptual world, but to accommodate the world that relativity had discovered—was discovering still—in the physics they loved, the physics of Newton and Maxwell. Not only were they unable to accomplish such a program; while pursing it they were constantly losing the simplicity and coherence that their beloved physics had originally had. No doubt, finally, Larmor understood this, and so abandoned his efforts. His decision to exclude his "physical relativity" articles from his published *Papers* constitutes clear evidence in this sense.

References

BUCHWALD, Jed Z. (1981a). "The Abandonment of Maxwellian Electrodynamics: Joseph Larmor's Theory of the Electron." *Archives Internationales d'Histoire des Sciences* **31**: 135–180.

——— (1981b). "The Abandonment of Maxwellian Electrodynamics: Joseph Larmor's Theory of the Electron. Part II." *Archives Internationales d'Histoire des Sciences* **31**: 373–438.

——— (1985). *From Maxwell to Microphysics*. Chicago & London: The University of Chicago Press.

CARMICHAEL, R. D. (1912). *The Theory of Relativity*. New York.

DARRIGOL, Olivier. (1994). "The Electron Theories of Larmor and Lorentz: A Comparative Study." *Historical Studies in the Physical and Biological Sciences* **24**: 265–336.

DE SITTER, Willem. (1916a). "On Einstein's Theory of Gravitation and Its Astronomical Consequences." *Monthly Notices of the Royal Astronomical Society* **76**: 699–728.

——— (1916b). "On Einstein's Theory of Gravitation and Its Astronomical Consequences. Second Paper." *Monthly Notices of the Royal Astronomical Society* **77**: 155–184, 481.

——— (1917). "On Einstein's Theory of Gravitation and Its Astronomical Consequences. Third Paper." *Monthly Notices of the Royal Astronomical Society* **78**: 3–28.

DIRAC, Paul A. M. (1928). "The Quantum Theory of the Electron." *Proceedings of the Royal Society* (London) *A* **117**: 610–624.

DISCUSSION (1919). "Discussion on the Theory of Relativity." *Monthly Notices of the Royal Astronomical Society* **80**: 96–119.

EDDINGTON, Arthur Stanley. (1917a). "Astronomical Consequences of the Electrical Theory of Matter. Note on Sir Oliver Lodge's Suggestions." *Philosophical Magazine* **34**: 163–167.

——— (1917b). "Astronomical Consequences of the Electrical Theory of Matter. Note on Sir Oliver Lodge's Suggestions, II." *Philosophical Magazine* **34**: 321–327.

——— (1942/4). "Joseph Larmor, 1857–1942." *Obituary Notices of Fellows of the Royal Society* **4**: 197–207.

GOLDBERG, Stanley. (1970). "In Defense of Ether: The British Response to Einstein's Special Theory of Relativity, 1905–1911." *Historical Studies in the Physical Sciences* **2**: 89–125.

——— (1984). *Understanding Relativity*. Oxford: Clarendon Press.

GOLDSTEIN, Catherine & RITTER, Jim. (1998). "Geometry and Physics in Unified Field Theories: 1920–1930." To appear.

GRAY, Andrew. (1920). "Presidential Address." In *Report of the Eighty-Seventh Meeting of the British Association for the Advancement of Science, Bournemouth 1919*. 135–146. London: John Murray.

HILL, J. Arthur, ed. (1932). *Letters from Sir Oliver Lodge*. London: Cassell.

HUNT, Bruce. (1986). "Experimenting on the Ether: Oliver J. Lodge and the Great Whirling Machine." *Historical Studies in the Physical and Biological Sciences* **16**: 111–134.

——— (1991). *The Maxwellians*. Ithaca & London: Cornell University Press.

JOHNSTON, W. J. & LARMOR, Joseph. (1920). "The Limitations of Relativity." In *Report of the Eighty-Seventh Meeting of the British Association for the Advancement of Science, Bournemouth 1919*. 158–159. London: John Murray.

KALUZA, Theodor. (1921). "Zum Unitätsproblem der Physik," *Sitzungsberichte der Preussischen Akademie der Wissenschaften* (Berlin): 966–972.

KUHN, Thomas S., HEILBRON, John L., FORMAN, Paul & ALLEN, Lini. (1967). *Sources for History of Quantum Physics. An Inventory and Report*. Philadelphia: American Philosophical Society.

LAKATOS, Imre. (1970). "Falsification and the Methodology of Scientific Research Programmes." In *Criticism and the Growth of Knowledge*. I. Lakatos and A. Musgrave. eds. 91–195. Cambridge: Cambridge University Press. [Reprinted: *Philosophical Papers*. J. Worrall and G. Currie, eds. Vol. 1: 8–101. Cambridge: Cambridge University Press (1978)].

LARMOR, Joseph. (1900). *Aether and Matter*. Cambridge: Cambridge University Press.

——— (1913). "On the Dynamics of Radiation." In *Proceedings of the Fifth International Congress of Mathematics (Cambridge, 22–28 August 1912)*. E. W. Hobson and A. E. H. Love, eds. Vol. I: 197–216. Cambridge: Cambridge University Press.

——— (1917). "Radiation-Pressure, Astrophysical Retardation, and Relativity." *Nature* **99**: 404.

——— (1918). "On the Essence of Physical Relativity." *Proceedings of the National Academy of Sciences* **4**: 334–337.

——— (1919a). "On Generalized Relativity in Connexion with Mr. W. J. Johnston's Calculus." *Proceedings of the Royal Society* **96**: 334–363.

——— (1919b). "The Relativity of the Forces of Nature." *Monthly Notices of the Royal Astronomical Society* **80**: 119–138.

——— (1920). "Gravitation and Light." *Proceedings Cambridge Philosophical Society* **19**: 324–344.

——— (1921). "Questions in Physical Inter-determination." In *Comptes Rendus du Congrès International des Mathématiciens à Strasbourg*. 3–40. Toulouse: E. Privat.

——— (1923a). "Can Gravitation Really be Absorbed into the Frame of Space and Time?" *Nature* (10 February): 200.

——— (1923b). "On the Nature and Amount of the Gravitational Deflection of Light." *Philosophical Magazine* **45**: 243–256.

——— (1927). "Newtonian Time Essential to Astronomy." *Nature. Supplement* (9 April): 1–12.

——— (1929). *Mathematical and Physical Papers*. 2 vols. Cambridge: Cambridge University Press.

LE ROUX, Jean. (1922a). "La Courbure de l'espace." *Comptes Rendus de l'Académie des Sciences* **174**: 924–927.

——— (1922b). "Sur la gravitation dans la mécanique classique et dans la théorie d'Einstein." *Comptes Rendus de l'Académie des Sciences* **175**: 809–811.

——— (1922c). "La Mécanique de Newton n'est pas une approximation de celle d'Einstein." *Comptes Rendus de l'Académie des Sciences* **175**: 1395–1397.

LODGE, Oliver. (1892). "The Motion of the Ether Near the Earth." *Proceedings of the Royal Institution* **13**: 565–580.

——— (1913). "Continuity." In *Report of the Eighty-Third Meeting of the British Association for the Advancement of Science.* 3–42. London: John Murray.

——— (1917a). "Astronomical Consequences of the Electrical Theory of Matter." *Philosophical Magazine* **34**: 81–94.

——— (1917b). "Astronomical Consequences of the Electrical Theory of Matter. Supplementary Note." *Philosophical Magazine* **34**: 517–521.

——— (1932). *Past Years. An Autobiography.* New York: Charles Scribner's.

——— (1933). *My Philosophy.* London: Ernest Benn.

MAXWELL, James Clerk. (1920). *Matter and Motion.* Notes and Appendices by Joseph Larmor. London: Society for Promoting Christian Knowledge. [Reprinted: New York: Dover (1952)].

MACCULLAGH, James. (1848). "An Essay Towards a Dynamical Theory of Crystalline Reflexion and Refraction." *Transactions of the Royal Irish Academy* **21**: 17–50. This paper is dated 9 December 1839.

PAGE, Leigh. (1912). "A Derivation of the Fundamental Relations of Electrodynamics from those of Electrostatics." *American Journal of Science* **34**: 57–68.

——— (1918a). "Is a Moving Star Retarded by the Reaction of Its Own Radiation?." *Proceedings of the National Academy of Sciences* **4**: 47–49.

——— (1918b). "Is a Moving Mass Retarded by the Reaction of Its Own Radiation?." *Physical Review* **11**: 376–400.

PLANCK, Max. (1930). "Theoretische Physik." In *Aus 50 Jahren deutscher Wissenschaft. Die Entwicklung ihrer Fachgebiete in Einseldarstellungen (Schmidt-Ott Festschrift).* 300–309. Berlin: W. de Gruyter.

——— (1932). "Fifty Years of Science." In *Where is Science Going?.* J. Murphy, trans. 41–63. New York: W. W. Norton.

POINCARÉ, Henri. (1906). "Sur la dynamique de l'électron." *Rendiconti del Circolo Matematico di Palermo* **21**: 129–175 [Reprinted: *Œuvres de Henri Poincaré.* Vol. 9: 494—550. G. Petiau, ed. Paris: Gauthier-Villars (1954)].

PURCELL, Edward M. (1980). "Comments on Special Relativity Theory in Engineering." In *Some Strangeness in the Proportion.* H. Woolf, ed. 106–108. Reading (MA): Addison-Wesley.

REPORT (1913). "Discussion on Radiation." In *Report of the Eighty-Third Meeting of the British Association for the Advancement of Science.* 376–386. London: John Murray.

SÁNCHEZ-RON, José M. (1987). "The Reception of Special Relativity in Great Britain." In *The Comparative Reception of Relativity.* T. F. Glick, ed. 27–58. Boston & Dordrecht: Reidel.

——— (1992). "The Reception of General Relativity among British Physicists and Mathematicians (1915–1930)." In *Studies in the History of General Relativity* (Einstein Studies, vol. 3). J. Eisenstaedt and A. J. Kox, eds. 57–88. Boston: Birkhäuser.

SCHAFFNER, Kenneth F. (1972). *Nineteenth-Century Aether Theories.* Oxford: Pergamon Press.

STACHEL, John. (1986). "Eddington and Einstein." In *The Prism of Science.* E. Ullmann-Margalit, ed. 225–250. Boston & Dordrecht: Reidel.

STEIN, Howard. (1981). "'Subtler Forms of Matter' in the Period Following Maxwell." In *Conceptions of Ether.* G. N. Cantor and M. J. S. Hodge, eds. 309–340. Cambridge: Cambridge University Press.

VIZGIN, Vladimir P. (1994). *Unified Field Theories* (Science Networks, vol. 6). Basel: Birkhäuser.

WALKER, George W. (1918). "Relativity and Electrodynamics." *Philosophical Magazine* **35**: 327–338.

WARWICK, Andrew. (1991). "On the Role of the FitzGerald-Lorentz Contraction Hypothesis in the Development of Joseph Larmor's Electronic Theory of Matter." *Archive for History of Exact Sciences* **43**: 29–91.

——— (1992). "Cambridge Mathematics and Cavendish Physics: Cunningham, Campbell and Einstein's Relativity 1905–1911. Part I: The Uses of Theory." *Studies in History and Philosophy of Science* **23**: 625–656.

——— (1993). "Cambridge Mathematics and Cavendish Physics: Cunningham, Campbell and Einstein's Relativity 1905–1911. Part II: Comparing Traditions in Cambridge Physics." *Studies in History and Philosophy of Science* **24**: 1–25.

——— (1995). "The Sturdy Protestants of Science: Larmor, Trouton, and the Earth's Motion Through the Ether." In *Scientific Practice*. J. Z. Buchwald, ed. 300–343. Chicago: Chicago University Press.

WHITTAKER, Edmund T. (1929). "What is Energy?." *Mathematical Gazette* (April): 401–406.

——— (1951). *A History of Aether and Electricity. I. The Classical Theories*. Edinburgh: Thomas Nelson and Sons.

——— (1953). *A History of Aether and Electricity. II. The Modern Theories, 1900–1926*. Edinburgh: Thomas Nelson and Sons.

WOODRUFF, A. E. (1981). "Larmor, Joseph." In *Dictionary of Scientific Biography*. Vol. 8: 39–41. C. Gillespie, ed. New York: Scribner & Sons.

Kretschmann's Analysis of Covariance and Relativity Principles

Robert Rynasiewicz

For over thirty years now, Erich Kretschmann has been ritually cited in the relativity literature for having shown that the general covariance of the equations of Einstein's general theory of relativity does not entail that the theory automatically satisfies a generalized principle of relativity (Kretschmann 1917). The argument is now a familiar one. Special relativity can be readily cast in generally covariant form. And so can a wide variety of 'pre-relativistic' theories, such as Newtonian gravitational theory, as subsequently shown by Cartan (1923/4). This throws open both the general question, just what a relativity principle is if not the requirement of covariance under certain coordinate transformations, as well as the particular question, whether there is a meaningful sense in which the general theory of relativity is generally relativistic. But rarely is it mentioned that Kretschmann's purpose was to address precisely these questions, much less discussed what answers he proposed.

There are, to be sure, notable exceptions. James Anderson, who should probably be credited for having made the existence of Kretschmann's 1917 paper as well known as it currently is, uses it primarily as a foil for his own views, arguing that according to Kretschmann's position the relativity groups of classical mechanics and special relativity reduce trivially to the identity transformation (Anderson 1966). In several recent historical papers, John Norton reads Kretschmann more sympathetically, interpreting him as equating the relativity group of a theory with the symmetry group of the "geometric structure" posited by the theory, which includes its conformal and affine structure (Norton 1992, 1993; see also Howard & Norton 1993). As it turns out, the relativity groups Kretschmann assigns to special and general relativity, respectively, do happen to be the symmetry groups of the geodesic structure of space-time, but one then becomes curious to

H. Goenner, J. Renn, J. Ritter, T. Sauer (eds.)
The Expanding Worlds of General Relativity
(Einstein Studies, volume 7) pp. 431–462

know Kretschmann's grounds for distinguishing the putative "geometric structure" from "non-geometric structure," particularly given Anderson's position that the relativity group of general relativity (but not of special relativity) is the general covariance group because the affine connection in general relativity is a *dynamical* object, whereas the affine connection in special relativity is not a dynamical, but an *absolute* object (Anderson 1964, 1967).

No doubt, one major reason why the contents of Kretschmann's 1917 paper are not more widely appreciated is, as Norton notes, the convoluted prolixity of the German prose, which deters even the native German speaker. An English translation I produced in collaboration with Frank Döring helped to some extent but proved to be no panacea. Kretschmann's style of thought remains somewhat obscure, no matter what the language. Moreover, there are other significant sources of difficulty. One is the conceptual gap between the extrinsic methods current at the inception of relativity and the more refined intrinsic formulations available today. Norton (1989) has drawn attention to the potential difficulties this presents for understanding Einstein's views on relativity principles and covariance. For interpreting Kretschmann, they can be especially frustrating. Another is the simple fact that the issues tapped, as will be seen, are themselves subtle and difficult.

Nonetheless, I think the thrust of Kretschmann's ideas can be made to stand out reasonably clearly, even if many of the details resist tidy reformulation. My goal here is to provide what hopefully are the essential clues needed. The effort of bringing to life this historical footnote is worthwhile not only on its own terms, but also for the light it potentially sheds on the intricacies of the hole argument and its relation to the point-coincidence argument.

1. The issue

As Kretschmann explains at the very beginning of his 1917 paper, the received view, due largely to Einstein, is that a theory satisfies a given relativity principle just in case the equations expressing that theory are covariant under the group of coordinate transformations associated with the principle. Einstein devotes the first several sections of his 1916 *Grundlage* paper to motivating the need to generalize the restricted principle, valid in the special theory of relativity for inertial systems, to the general principle that the laws of physics must hold good in reference systems in arbitrary states of motion.[1] The general principle of relativity, Einstein explains, will be satisfied by any physics whose laws are expressed by equations that are covariant under arbitrary coordinate substitutions, since included among the latter are those that correspond to all relative motions of (three-dimensional) coordinate systems. Moreover, Einstein argues, there are independent grounds for the requirement of general covariance: all physical experience can be reduced ultimately to purely topological relations or coincidences between material objects or processes, and these are preserved under arbitrary coordinate substitutions.

[1] "Die Gesetze der Physik müssen so beschaffen sein, daß sie in bezug auf beliebig bewegte Bezugssysteme gelten." (Einstein 1916: 772.)

Since this consideration—which, since Stachel 1980, has come to be referred to as the 'point-coincident argument'—is in fact Kretschmann's point of departure, it deserves to be quoted in full:

> That this requirement of general co-variance, which takes away from space and time the last remnant of physical objectivity, is a natural one, will be seen from the following reflexion. All our space-time verifications invariably amount to a determination of space-time coincidences. If, for example, events consisted merely in the motion of material points, then ultimately nothing would be observable but the meetings of two or more of these points. Moreover, the results of our measurings are nothing but verifications of such meetings of the material points of our measuring instruments with other material points, coincidences between the hands of a clock and points on the clock dial, and observed point-events happening at the same place at the same time. The introduction of a system of reference serves no other purpose than to facilitate the description of the totality of such coincidences. We allot to the universe four space-time variables x_1, x_2, x_3, x_4 in such a way that for every point-event there is a corresponding system of values of the variables $x_1 \ldots x_4$. To two coincident point-events there corresponds one system of values of the variables $x_1 \ldots x_4$, i.e., coincidence is characterized by the identity of the co-ordinates. If, in place of the variables $x_1 \ldots x_4$, we introduce functions of them, x_1', x_2', x_3', x_4', as a new system of co-ordinates, so that the systems of values are made to correspond to one another without ambiguity, the equality of all four co-ordinates in the new system will also serve as an expression for the space-time coincidence of the two point-events. As all our physical experience can be ultimately reduced to such coincidences, there is no immediate reason for preferring certain systems of co-ordinates to others, that is to say, we arrive at the requirement of general co-variance. (Einstein 1916: 776–777; translation from Lorentz et. al. 1923: 117–118.)

What is remarkable about this, at least from Kretschmann's point of view, is that there is no mention here of a two-part article that had appeared in the *Annalen* the preceding year, Kretschmann's first publication, in which he had argued at length a thesis similar to the main premise of Einstein's point-coincidence argument, namely, that the factual content of a physical theory is completely exhausted by the totality of topological relations among spatially-temporally extended objects it permits (Kretschmann 1915). It is more likely than not that Einstein was simply unaware of that paper. For one thing, there is a systematic difference in their choices of terminology—*Koinzidenz* for Einstein vs. *Zusammenfall* for Kretschmann. More substantially, though, it would have been evident to Einstein that, for Kretschmann, the thesis applies to any physical theory whatsoever, no matter what relativity principle it is traditionally thought to satisfy. And if it does so apply, then either the point-coincidence argument is a non sequitur or else the general principle of relativity, at least as Einstein intends it, imposes no further physical requirement that would distinguish the so-called general theory of relativity from any other theory.

This is the occasion for Kretschmann's 1917 investigation. As he sees it, the point-coincidence argument shows only that it should be possible, given enough mathematical ingenuity, to cast any theory in generally covariant form, using the absolute differential calculus of Ricci and Levi-Civita. Hence, if a relativity principle is a physical requirement, and not just a formal constraint on how a theory is expressed, it must mean something other than covariance under an associated group of coordinate transformations.

2. A preliminary paradox

Now Kretschmann agrees with Einstein that, if all physical observations are reducible to topological relations and coincidences, then no coordinate systems can be granted privileged status over any others in virtue of these observations, and thus, if the physical content of a theory, is exhausted by the totality of all observations that are physically possible in principle, then it cannot grant a privileged status to any class of reference systems. But certainly special relativity, for example, draws a distinction between inertial and accelerating reference systems. How is this possible, if the content of that theory is completely exhausted by the topological coincidences allowed by it?

Kretschmann's answer, in essence, is that a theory can, in virtue of its physical content, classify reference systems into equivalence classes without thereby conferring a privileged status on any one of the equivalence classes. Whatever preferred status might appear to attach to a given class over any another is entirely an artifact of the particular mathematical formulation of the theory. Kretschmann illustrates this for the case of the law of light propagation in special relativity. Since this example also serves to motivate his eventual proposal concerning the meaning of relativity principles, it is worth going over it in some detail. This will also broach some of the difficulties of recasting Kretschmann's meaning in a modern idiom.

One interpretive problem to be faced at the outset is how to render Kretschmann's notion of a reference system [*Bezugssytem*]. The terminology is not peculiar to Kretschmann. Einstein speaks of "reference systems" and "coordinate systems" [*Koordinatensystem*] interchangeably (Einstein & Grossmann 1913; Einstein 1916). Since Kretschmann treats coordinate systems as devices for assigning quadruples of numbers to point-like physical events, the closest modern cousin is the notion of a coordinate chart.[2] However, the identification of the two notions is problematic for several of reasons. One is that Kretschmann is never clear whether he regards such assignments always to be one-to-one and continuous. Nor does he sense any need to specify precisely the domain and range of the assignments. It is thus unclear what should be assumed, for example, when he deals with transformations from rectilinear reference systems to systems in polar coordinates. The main difficulty, however, is that Kretschmann implicitly supposes, first, that between any two reference systems there is a well-defined coordinate transformation, and second, that the set of all coordinate transformations acts as a transformation *group* on the collection of all reference systems. (So, given the assumption that any two

[2] Kretschmann uses quadruples of the form $(x_1, x_2, x_3, x_4) = (x_1, x_2, x_3, ict)$, where $x_1, x_2, x_3, t \in \mathbf{R}$ and c is the velocity of light.

reference systems are related by a coordinate transformation, there is a single orbit under the group action.) Eventually we will identify this transformation group with the diffeomorphism group of a distinguished manifold.

Kretschmann begins his discussion of special relativity by presenting the usual non-covariant formulation of the law of light propagation[3]

$$(x_1 - x_1{}^0)^2 + \cdots + (x_4 - x_4{}^0)^2 = 0, \tag{1}$$

expressed in terms of the coordinates for an orthogonal, rectilinear reference system Σ. The coordinate transformations that leave (1) unchanged in form are said to be "admissible" [*berechtigt*] transformations[4] and are assumed to form a subgroup $G_{(1)}$ which contains, in addition to the transformations of the Lorentz group, constant translations and uniform dilations. The class of reference systems obtainable from Σ by application of transformations in $G_{(1)}$ is called a class of "co-admissible" [*gleichberechtigt*] frames.[5]

Suppose now that ϕ is a coordinate transformation not in the group of admissible transformations for (1), so that (1) fails to retain its form under ϕ.[6] Let $\phi * \Sigma$ be the reference system to which ϕ carries Σ, and let $\phi * (1)$ be the transformed version of Equation (1). Now consider the class of transformations that leave $\phi * (1)$ unchanged in form, that is, the admissible transformations for $\phi * (1)$. This again forms a group $G_{\phi*(1)}$, which, though distinct from the group of admissible transformations for (1), is nonetheless isomorphic to it.[7] Moreover, the class of reference systems obtainable from $\phi * \Sigma$ by application of the members of $G_{\phi*(1)}$ is a one-to-one image of the reference systems co-admissible with Σ, and constitutes the class of reference systems co-admissible with $\phi * \Sigma$.[8] Thus, the class of all reference systems is partitioned into co-admissibility classes, the elements of any one of which are related by a group of transformations isomorphic to the group of admissible transformations for Equation (1). No one of these co-admissibility classes is structurally distinguished from any other, and hence no single class has a

---

[3] In order to assist the reader who is attempting to relate what I say to Kretschmann's text, I have attempted to retain wherever possible Kretschmann's original notation and numbering of equations.

[4] This terminology appears in Einstein & Grossmann 1914.

[5] This is just the orbit of Σ under the action of $G_{(1)}$.

[6] Kretschmann uses the example of a transformation to polar coordinates. This would be fine except for the difficulties mentioned above. It is nonetheless useful for the reader to recognize that the transformation ϕ need not be one to an accelerated system.

[7] The two groups are conjugate to one another. Think of the group G of all coordinate transformations as also acting on the space of equations (modulo mathematical equivalence in the sense of having identical real-valued solutions). The group of admissible transformations for an (equivalence class of) equation(s) E is then the stabilizer G_E of E. So, if E' is in the orbit of E under the action of G, that is, if there exists a coordinate transformation ϕ such that $E' = \phi * E$, then the stabilizer $G_{E'}$ of E' is the conjugate of G_E by ϕ.

[8] Kretschmann's co-admissibility relation can be defined as follows. For any two reference systems Σ_1 and Σ_2, there are transformations $\phi_1, \phi_2 \in G$ such that $\phi_1 * \Sigma = \Sigma_1$ and $\phi_2 * \Sigma = \Sigma_2$. So, Σ_1 is co-admissible with Σ_2 just in case $\phi_1{}^{-1} \circ \phi_2 \in G_{(1)}$, that is ϕ_1 and ϕ_2 are in the same left coset of $G_{(1)}$. (The relation is well defined without assuming ϕ_1 and ϕ_2 to be unique.)

privileged status *except* in relation to an arbitrary choice of one of the transformed versions of Equation (1).

Consequently, if the group $G_{(1)}$ of admissible transformations of (1) is identified (at least up to isomorphism) as the relativity group of the law of light propagation in special relativity, then it is clear why the law itself picks out no preferred class of reference systems even though its relativity group is not the group of all transformations. It is only the choice of the particular mathematical formulation of the law (Equation (1) or one of its transformed forms) that does so. If this is right, then the point-coincidence argument entails nothing about the satisfaction of relativity principles.

3. Finding the relativity group of special relativity

But on what grounds are we to identify (up to isomorphism) the transformation group $G_{(1)}$ of Equation (1) as the relativity group of the law of light propagation? By adopting the 'manner of expression' of Einstein's general theory of relativity, the same law can be expressed in generally covariant form:

$$\begin{cases} \delta \int ds = 0, \\ ds^2 \equiv \sum_{\mu\nu} g_{\mu\nu} \cdot dx_\mu \cdot dx_\nu = 0, \\ (\lambda\nu, \mu\tau) \equiv 0 \qquad (\lambda, \nu, \mu, \tau = 1 \ldots 4), \end{cases} \tag{2}$$

where $(\lambda\nu, \mu\tau)$ are the components of the Riemann curvature tensor. The general covariance of these equations indicates that the group of admissible transformations of this formulation of the law of light propagation is the group of all coordinate transformations. If a given law or set of laws can be equivalently expressed by different sets of equations having different admissibility groups, then which, if any, of these is the relativity group of the theory?

Kretschmann answers this by considering a variety of equivalent formulations of the law of light propagation having admissibility groups intermediate between those of Equations (1) and (2). For example, the group of admissible transformations of the equation[9]

$$\begin{cases} (x_1 - x_1{}^0)^2 + \cdots + (x_3 - x_3{}^0)^2 + F^2 \cdot (x_4 - x_4{}^0)^2 = 0, \\ F = \text{Const.} \end{cases} \tag{3}$$

includes, in addition to those admissible for Equation (1), all transformations of the form

$$x_4' = \mu \cdot x_4, \quad x_1' = x_1, \quad x_2' = x_2, \quad x_3' = x_3,$$

[9] In the following equation Kretschmann uses the symbol 'c' rather than 'F' even though he has previously used it for the speed of light in vacuo, which is obviously not the intended meaning here. To be more precise, F should be understood to be any function of $x_1, \ldots, x_4$ such that $\partial F/\partial x_\mu = 0$ for $\mu = 1, \ldots, 4$.

for any positive real number μ. Kretschmann's technique now is to compare the "geometrical pictures" [*geometrischen Bilder*] corresponding to the various solutions of Equations (1), (2) and various intermediate forms such as (3).

The terminology "geometric picture" suggests that Kretschmann has in mind some sort of intrinsic space-time model. On closer inspection, though, it is evident that these "geometric pictures" are coordinate-dependent representations and not intrinsic models. However, since *we* are unlikely to be satisfied unless we can understand Kretschmann in intrinsic terms, I will introduce the notion of a *coincidence model* to serve this purpose. It then turns out that a "geometric picture" is an $\mathbf{R}^4$-image of a coincidence model in a coordinate chart.

Although variant formulations are possible, the following characterization of a coincidence model for the law of light propagation in special relativity will suffice. Let M be a fixed manifold diffeomorphic to $\mathbf{R}^4$. A *coincidence structure* is then a pair (M, Λ), such that Λ is a a set of (images of) curves on M, to be thought of as the world-lines of light rays.[10] Now if EQN is some system of equations expressing the law of light propagation in special relativity, we want to be able to state the conditions under which a given coincidence structure $\mathfrak{A} = (M, \Lambda)$ satisfies EQN in a reference system Σ, or equivalently what it means for EQN to be true of $\mathfrak{A}$ in Σ. Since this will involve reference to the image of $\mathfrak{A}$ under some coordinate chart appropriately associated with Σ, the definition in fact proceeds in terms of what it is for a "geometric picture" to satisfy EQN.

The notion of a "geometric picture" $\mathfrak{B}$ in this context can be taken to be a pair (U, Δ), where U is an open neighborhood of $\mathbf{R}^4$ and Δ again is a set of (images of) curves on U. Thus, if $\mathfrak{A} = (M, \Lambda)$ is a coincidence structure, ψ is a coordinate chart in the atlas of M, and U the range of ψ, then $\mathfrak{B} = (U, \psi[\Lambda])$ is the "geometric picture" that is the $\mathbf{R}^4$-image of $\mathfrak{A}$.[11]

Now, since geometric pictures are already set in $\mathbf{R}^4$ rather than in some abstract manifold, a given geometric picture either satisfies or fails to satisfy the system EQN of equations in question without any need to specify a coordinate chart or reference system. In order to state the satisfaction conditions, one needs only to make the quantificational structure explicit. For example, a geometric picture (U, Δ) satisfies (1) just in case the following obtains:

For any two quadruples of real numbers $r = (r_1, \ldots, r_4)$ and $r' = (r'_1, \ldots, r'_4)$ in U and any $C \in \Delta$, if $r, r' \in C$, then

$$(r_1 - r'_1)^2 + \ldots + (r_3 - r'_3)^2 - c^2(r_4 - r'_4)^2 = 0.$$

The satisfaction conditions for Equations (2) and (3) are a bit trickier because of the occurrence of real-valued functions that have no explicit counterparts in the

[10] It might be deemed more faithful to define a coincidence model to have the form (M, S, Λ), where now S is a distinguished subset of M representing those space-time points at which there is an emitting light source, and Λ assigns to each of these points a three dimensional surface representing the future light cone [*Nachkegel*] emanating from it. This, however, leads to complications that are inessential to Kretschmann's proposal.

[11] Here $\psi[\Lambda]$ is defined to be the set $\{\psi[C] \mid C \in \Lambda\}$.

geometrical pictures themselves. Kretschmann's implicit manner of dealing with this is by appeal to higher order existential quantification. Thus, for example, we are to understand that (U, Δ) satisfies (3) just in case the following holds:

> There exists a constant function $F : \mathbf{R}^4 \to \mathbf{R}$ such that for any two quadruples of real numbers $r = (r_1, \ldots, r_4)$ and $r' = (r'_1, \ldots, r'_4)$ in U and any $C \in \Delta$, if $r, r' \in C$, then
>
> $$(r_1 - r'_1)^2 + \ldots + (r_3 - r'_3)^2 - F \cdot c^2(r_4 - r'_4)^2 = 0.$$

Similarly, for Equation (2) we have:

> There exist sixteen functions $g_{\mu\nu} : \mathbf{R}^4 \to \mathbf{R}$ (for $\mu, \nu = 1, 2, 3, 4$) such that ...

where the ellipsis is to be filled in by an appropriate transcription of (2) along with additional formal constraints on the $g_{\mu\nu}$'s, such as symmetry under interchange of indices.

With these examples under our belt, we suppose it is well enough understood in any given case what it means for a geometric picture to satisfy a given system of equations EQN. A coincidence model can then be said to satisfy EQN in reference system Σ just in case its $\mathbf{R}^4$-image under any coordinate chart appropriately associated with Σ satisfies EQN.[12]

Before proceding with Kretschmann's explanation of how the relativity principle of the law of light propagation in special relativity is determined, I want to discuss certain formal features of coincidence models and their associated geometric pictures that will play a role in what follows. There is a natural partial order relation on the class of all coincidence structures (including also those that do not satisfy the law of light propagation in any reference system) under which it forms a complete lattice. This partial ordering is induced by the set inclusion relation between the classes of world-lines of light rays of the coincidence structures. Thus, if $\mathfrak{A}_1 = (M, \Lambda_1)$ and $\mathfrak{A}_2 = (M, \Lambda_2)$, then $\mathfrak{A}_1 \leq \mathfrak{A}_2$ just in case $\Lambda_1 \subseteq \Lambda_2$. The join of $\mathfrak{A}_1$ and $\mathfrak{A}_2$ is the smallest coincidence structure containing all the world-lines in either, i.e., $(M, \Lambda_1 \cup \Lambda_2)$. The meet is defined similarly in terms of intersection. The lattice is complete, since any non-empty set $K = \{\mathfrak{A}_i \mid i \in I\}$ of coincidence structures has a least upper bound, whose collection of world-lines is $\Lambda = \bigcup_{i \in I} \{\Lambda_i \mid (M, \Lambda_i) \in K\}$ and a greatest lower bound given by replacing set union with set intersection. Intuitively, the least upper bound is just the result of superimposing all the world-lines of light rays from the respective coincidence structures in the set K.

Quite obviously, the class of all possible geometric pictures obtained by taking the $\mathbf{R}^4$-images under some fixed coordinate chart inherits this lattice structure. Now what Kretschmann does in effect is to fix a global coordinate chart and then, for each of the Equations (1), (2), and (3), to look at the geometric picture that is the least upper bound of the class of geometric pictures satisfying the equation.

---

[12] In the absence of a precise definition of the notion of a reference system, it remains vague what it means for a chart to be appropriately associated with a reference system. However it is sufficiently clear in enough cases in order to be useful.

One is led directly to [the relativity group] if, for each system of equations, one imagines all the world-lines of light impulses consistent with them drawn out in the coordinate manifold and compares with one another the four-dimensional geometric pictures thus generated.

All the light rays that are possible according to (1) are obtained if, starting from each world point $x_1^0 \dots x_4^0$, one draws all sides of the [light] cone[13] [in conformity] with equation (1). Likewise, each equation obtained by a [coordinate] transformation of (1) describes an infinite class of world lines of light rays which differ in general from the first in relations of measure, but apart from this are completely equivalent in all their topological properties. According to Equation (3), however, from each world point emanates not one, but instead, an infinite number of cones, namely one for each value of the undetermined constant [F] remaining. For (2) ... the multiplicity of cones with the same vertex will be correspondingly greater. Here, of course, it is the meaning of none of these equations that even as many as two light waves forming spatially disjoint and closed surfaces can actually originate from the same point in the ether at the same time. Rather, in each case the single class of world-lines which one obtains by a complete (numerical) specification of the coordinate functions $g_{\mu\nu}$ ... or, respectively, of the constant [F], is a complete picture of the simultaneous and successive light trajectories possible in actuality. Each of these individual pictures, of course, is in perfect topological agreement with the class of world-lines determined by (1), and, consequently, the totality of pictures given by the other [two] systems of equations are compilations of the infinitely many pictures given by (1) and the equations obtainable from (1) by coordinate transformations. As mentioned, any one of these suffices for the representation of what in actuality is possible according to the law in question. Any further one is entirely superfluous from a physical point of view.[14] (p. 582)

In assessing this, it will help enormously to make the simplifying assumption that the transformation group acting on the class of reference systems is isomorphic to the diffeomorphism group of $\mathbf{R}^4$, and thus to the diffeomorphism group of the

13 Kretschmann's terminology, *Nachkegel*, suggests he intends to consider only the future lobe in each case. However, strict faithfulness to this intention would introduce extraordinary complications completely irrelevant to the spirit of his proposal.

14 "Alle nach (1) möglichen Lichtweltlinien erhält man, wenn man von jedem Weltpunkte $x_1{}^0 \dots x_4{}^0$ aus alle Seiten des Nachkegels mit der Gleichung (1) zeichnet. Jede aus (1) transformierte Gleichung stellt ebenfalls eine unendliche Schar von Lichtweltlinien dar, die von der ersten in ihren Maßverhältnissen im allgemeinen verschieden, aber in allen davon unabhängigen topologischen Eigenschaften ihr völlig gleich ist. Nach Gleichung (3) dagegen gehen von jedem Weltpunkte statt eines Nachkegels deren unendlich viele aus, nämlich je einer für jeden Wert der unbestimmt gelassenen Konstanten c. In (2) und der daraus durch den Zusatz $g_{\mu\nu} = 0$ für $\mu \neq \nu$ entstandenen Darstellungsform treten sogar unbestimmte Funktionen—die zehn $g_{\mu\mu}$ [*sic*] bzw. die vier $g_{\mu\mu}$—statt der unbestimmten Konstanten c auf, und die Mannigfaltigkeit der Nachkegel mit gleicher Spitze wird eine dementsprechend größere. Dabei ist es natürlich keiner dieser Gleichungen mit Sinn, daß in der Wirklichkeit auch nur zwei räumlich sich trennende und geschlossene Flächen bildende Lichtwellen von demselben Punkte im Äther zu gleicher Zeit ausgehen könnten. Vielmehr soll in jedem Falle schon jede einzelne der Weltlinienscharen, die man durch vollständige (zahlenmäßige) Festlegung der zwar unbestimmten, doch eindeutigen Koordinatenfunktionen $g_{\mu\nu}$ bzw. der Konstanten c erhält, ein vollständiges Bild der in der Wirklichkeit neben- und nacheinander möglichen Lichtbewegungen sein. Jedes dieser Einzelbilder stimmt natürlich topologisch mit der durch (1) bestimmten Weltlinienschar vollkommen überein, und die Gesamtbilder der anderen drei Gleichungssysteme sind hiernach Zusammenfassungen unendlich vieler der durch (1) und die aus (1) tranformierten Gleichungen gegebenen Bilder. Zur Darstellung des in der Wirklichkeit nach dem betrachteten Gesetze Möglichen genügt, wie gesagt, jedes einzelne von diesen. Jedes weitere ist in physikalischer Hinsicht vollkommen überflüssig."

base manifold M of the coincidence structures.[15] In fact, if we simply identify these groups with one another, then we can speak of a single transformation group D acting on a variety of different spaces. First, is the action of D on the class of reference systems as originally supposed. Second, is the action of D on the space of all coincidence structures. If $\mathfrak{A} = (M, \Lambda)$ and $\phi \in D$, then the action $\phi * \mathfrak{A}$ of ϕ on $\mathfrak{A}$ is $(M, \phi * \Lambda)$, where $\phi * \Lambda = \{\phi[C] \mid C \in \Lambda\}$.[16] Third is the action of D on the collection of all (mathematically) possible geometric pictures. The action of ϕ on a geometric picture $\mathfrak{B} = (U, \Delta)$ is defined as $\phi * \Sigma = (\phi[U], \phi * \Delta)$, where $\phi * \Delta$ is defined analogously to $\phi * \Lambda$. Finally, D can be seen to act on the space of equations of relevant form (considered as linguistic expressions) such that $\phi *$ EQN is the result of transforming EQN under the coordinate expression of the diffeomorphism ϕ.[17]

Given this simplifying assumption, the geometric pictures satisfying the different systems of equations stack up as follows. In general, let Bild[EQN] be the class of geometric pictures satisfying EQN and $\mathfrak{B}_{\text{EQN}}$ the least upper bound of Bild[EQN]. Obviously, $\mathfrak{B}_{(1)}$ corresponds to the usual picture we carry around in our heads of Minkowski space-time with all the light-cones sketched in. Now for each $\phi \in D$, $\text{Bild}[\phi*(1)] = \{\phi*\mathfrak{B} \mid \mathfrak{B} \in \text{Bild}[(1)]\}$ and $\mathfrak{B}_{\phi*(1)} = \phi*\mathfrak{B}_{(1)}$. That is, the totality of geometric pictures satisfying the transformed version $\phi * (1)$ of Equation (1) can be obtained by letting ϕ act on the class of geometric pictures satisfying (1), or equivalently, on the least upper bound of that class. To see this, just imagine the way that ϕ shifts and/or deforms the light cones given by the solutions of (1). Thus, the totality of geometric pictures satisfying the one equation stands in a one-to-one correspondence under the action of ϕ with the totality of geometric pictures satisfying the other equation in such a way that paired pictures are topologically equivalent. The class of geometric pictures satisfying (3), however, is strictly wider than that for (1), and in fact contains a continuum of copies of the family of geometric pictures satisfying (1) according to the continuum of possible values for the constant function F. Moreover, each of these copies is identical to the family of solutions of one of the equations $\phi_k * (1)$, where k ranges over the values the constant function F is allowed to take. In this sense, the collection of geometric pictures satisfying (3) is just the union over this one-parameter family of solutions of transformed versions of (1). Similar remarks apply to the solutions of the general covariant system (2) by letting the diffeomorphism group D play the role that the one-parameter subgroup plays with respect to (3).

Kretschmann's substantive contention here is that each of the copies $\text{Bild}[\phi*(1)]$ of the collection of geometric pictures satisfying (1) represents precisely the same spectrum of physical possibilities as represented by the original collection Bild[(1)]. For Equations (2) and (3), however, the collection of geometric pictures

---

[15] Accordingly, the example invoked by Kretschmann of a transformation to polar coordinates becomes problematic. But this I take to be artifact.

[16] Recall that, in this capacity, ϕ is a diffeomorphism of M. Since C is just a subset of M, $\phi[C]$ is just the image of this set under ϕ.

[17] Actually, D acts on the quotient space of equations modulo mathematical equivalence.

satisfying them, respectively, can be decomposed into a family of subcollections[18] such that each subcollection is the collection Bild[$\phi * (1)$] of geometric pictures satisfying a transformed version $\phi * (1)$ of Equation (1). And since each of these subcollections represents the same spectrum of physical situations, the wider covariance of Equations (2) and (3) serves only to introduce duplicate representations of what is already admitted as possible according to (1) or any of its transformed forms.

Underlying this is a tacit appeal to a criterion of what I shall call the *compossibility of representation* for geometric pictures (and for models in general). By way of explanation, consider two geometric pictures satisfying Equation (3) that involve different choices of value for the parameter k, which, intuitively, determines the steepness of the sides of the light cone at all points. If one thinks of both these pictures as representing possible trajectories of light rays, it then follows that different velocities of light propagation (in vacuo) are possible at the same space-time point, contrary to what is intended. It is true that either picture considered in isolation from the other can be imagined to describe, from the point of view of some reference system or other, a physically possible situation. But from the point of view of one and the same reference system, they cannot both represent physical possiblities. If the one is taken to represent something possible, then the other cannot. Nor is there some fact of the matter as to which picture *really* represents a possibility unless the reference system is specified, not just hypothetically, but in actuality. (See Section 4 for further explanation.)

Thus, Kretschmann's conclusions can be seen to rest on the tacit introduction of a reflexive and symmetric relation of compossibilty of representation on the class of geometric pictures satisfying the equations in question.[19] Let us say that a family of geometric pictures from the collection of all geometric pictures satisfying a given equation is *gauge consistent* just in case the geometric pictures in the family are pairwise compossible representations. Furthermore, say that such a family is maximally gauge consistent just in case no strictly more inclusive family is gauge consistent. Thus, Kretschmann claims that the collections of geometric pictures satisfying (3) and (2), respectively, are not gauge consistent. Rather any maximally gauge consistent family of geometric pictures is identical to one of the collections satisfying Equation (1) or one of its transformed versions. This carries with it the intimation that the covariance groups of (3) and (2) are wider than the relativity group of the law they express, since the additional covariance serves only to introduce families of geometric pictures that are representationally redundant. In this sense, they can be reduced to the less covariant formulation (1).

But can Equation (1) be similarly reduced? Immediately following the quoted passage, Kretschmann answers this in the negative:

> But on the other hand, the particular class of world-lines of light rays which is represented in (1), or an equation transformed from (1), cannot in general

[18] These are pairwise disjoint in the case of (3), but not (2).

[19] The relation happens to be transitive on the classes of geometric pictures of (1) and (3), respectively, but transitivity fails for the class of (2). This is directly related to the point made in the preceding note.

> be reduced without thereby introducing a new law of light propagation going beyond (1). For among the light phenomena at all possible according to (1), there is obviously none which, solely in virtue of Equation (1), excludes any other as incompatible with it.[20] (p. 583)

So, unlike the case with Equations (3) and (2), the collection of geometric pictures satisfying (1) is itself gauge consistent. Hence, any further conditions that would reduce this collection would express a more restrictive physical law.

Kretschmann continues:

> It follows from this that each system of equations which gives expression, in whatever form, to exactly the laws contained in (1) must describe in full at least one of the class of world-lines of light rays belonging to equation (1) and its transformed forms as an image of possible and mutually consistent light motions, and accordingly [each such system of equations] must be satisfied in every reference system in which this picture, i.e., the pertinent equation transformed from (1), is valid. Consequently, the invariance group of transformations which connects these reference systems is the narrowest which is physically distinguishable from all others by means of any manner of representation of the of the laws contained in (1). It is, as required, determined solely by the physical content of the laws independently of the manner of their expression, and in fact it is, according to what was shown, obviously the unique group for which this holds.[21] (p. 583)

The upshot is this. Since the collection of geometric pictures satisfying Equation (1) is maximally gauge consistent, any system of equations EQN expressing the same law must contain one or more diffeomorphic copies of it. Hence, the admissibility group of EQN must contain an isomorphic copy of the group of admissible transformations of (1). Thus, (the isomorphism type of) this latter group is uniquely distinguished by the physical content of the law expressed.

An equivalent way of viewing the distinctive character of this group is this. Take any one of the maximally gauge consistent families of geometric pictures. For each one, there is a subgroup of the group of all transformations (in the capacity of diffeomorphisms of $\mathbf{R}^4$) that carries each geometric picture in the family to a geometric picture also in the family. This invariance group is just the automorphism group of the geometric picture that is the supremum of the family. And since the suprema of the various maximally gauge consistent families are diffeomorphic

---

[20] "Andererseits aber läßt sich die einzelne Schar von Lichtweltlinien, die in (1) oder einer aus (1) transformierten Gleichung dargestellt ist, nicht allgemein vermindern, ohne daß damit ein neues über (1) hinausgehendes Gesetz der Lichtausbreitung eingeführt wäre. Denn unter den nach (1) überhaupt möglichen Lichterscheinungen gibt es offensichtlich keine, die allein vermöge der Gleichung (1) irgendeine andere als nicht mit ihr zusammen möglich ausschlösse."

[21] "Daraus folgt, daß jedes Gleichungssystem, das die in (1) enthaltenen Gesetze und nur diese in irgendeiner Form zum Ausdrucke bringt, mindestens eine der zu Gleichung (1) und ihren transformierten Formen gehörenden Weltlinienscharen vollständig als Abbild möglicher und miteinander verträglicher Lichtbewegungen darstellen muß und demnach auch in allen den Bezugssystemen erfüllt sein muß, in denen dieses Bild, d.h. die zugehörige, aus (1) tranformierte Gleichung, gilt. Die invariante Transformationsgruppe, die diese Bezugssysteme verbindet, ist folglich die engste, die sich durch irgendeine Darstellungsform der in (1) enthaltenen Gesetze physikalisch vor allen anderen auszeichnen läßt. Sie ist, wie verlangt, allein durch den physikalischen Inhalt der Gesetze unabhängig von der Art ihres Ausdruckes bestimmt, und zwar ist es nach dem Dargelegten offensichtlich die einzige Gruppe, für die das gilt."

copies of one another, these invariance groups are isomorphic to one another. As Kretschmann puts it,

> Geometrically, the group is characterized as the group of transformations which transforms [back] into itself the class of all world lines of light rays which are compossible in a single coordinate manifold according to the laws. Thus each system of equations equivalent in content to (1), (2), and (3) satisfies the relativity principle belonging to this group and no additional [relativity principle].[22] (p. 583)

At this point, Kretschmann claims that what he has just done generalizes immediately and straightforwardly to the arbitrary case. Before following him in this regard, though, it will help to clarify the physical meaning of gauge consistency if we examine the relation of the foregoing to Einstein's hole argument. Once we understand the differences between Kretschmann and Einstein on this matter, we can better appreciate the rationale for the body of technical results that Kretschmann labors to produce in order to determine the relativity principle fulfilled by the general theory of relativity.

4. Gauge consistency, point coincidences, and holes

Einstein had convinced himself in late 1913 that the requirement of general covariance leads to unacceptable physical consequences because of the following consideration, known as the hole argument [*Lochbetrachtung*]. (For historical details, see Stachel 1980 and Norton 1987.) Suppose $\mathfrak{S}$ is a solution to some arbitrarily given generally covariant system of field equations relating the metric tensor and the stress-energy tensor, i.e., $\mathfrak{S}$ is a collection of real-valued functions $g_{\mu\nu}$ and $T_{\mu\nu}$ for $\mu, \nu = 1, 2, 3, 4$ defined on (a region of) $\mathbf{R}^4$ satisfying the equations in question. Suppose further that $\mathfrak{S}$ contains a 'matter hole' L in the sense that L is a region of $\mathbf{R}^4$ in which the functions $T_{\mu\nu}$ assume the value zero. Since the equations are generally covariant, any coordinate transformation applied to $\mathfrak{S}$ yields a solution to the same equations. In particular, choose a coordinate transformation that differs from the identity transformation only on the inside of L. This yields a solution $\mathfrak{S}'$ consisting of functions $g'_{\mu\nu}$ and $T'_{\mu\nu}$ such that $T'_{\mu\nu} = T_{\mu\nu}$ everywhere and $g'_{\mu\nu} = g_{\mu\nu}$ outside of L but not on the inside of L. Now if one thinks of the two solutions as describing the same pair of fields only from the points of view of two different reference systems related by the coordinate transformation used to generate $\mathfrak{S}'$, no anomaly emerges. However, suppose you imagine that a reference system Σ has been fixed in advance, that in Σ the values of the stress-energy field has been measured to be $T_{\mu\nu} = T'_{\mu\nu}$ everywhere, that the metric field has been determined to be $g_{\mu\nu} = g'_{\mu\nu}$ outside of L, and that one wants to predict the values that will be obtained for the components in Σ of the metric field for the interior

---

22 "Geometrisch ist die Gruppe gekennzeichnet als die Gruppe der Transformationen, welche die Schar aller Lichtweltlinien, die nach den Gesetzen zusammen in derselben Koordinatenmannigfaltigkeit möglich sind, in sich selbst überführen. Dem zu dieser Gruppe gehörenden Relativitätspostulate, und keinem weiteren, genügt also in physikalischer Hinsicht jedes den Gleichungen (1), (2), und (3) inhaltlich gleichwertige Gleichungssystem."

of L. Since both $\mathfrak{G}$ and $\mathfrak{G}'$ are solutions of the field equations, it appears one will be unable to do so. Hence, it appears that the metric field in the interior of L remains physically underdetermined, even given the field on the exterior of L and the stress-energy field everywhere, and thus that there is a breakdown of causality.

In terms of geometric models and active point transformations the puzzle is this. If (M, g, T) is a model of the field equations in question and ϕ a diffeomorphism of the manifold M, then the triple that results from dragging along the geometric object fields g and T, i.e., the structure $(M, \phi * g, \phi * T)$, is also a model of the equations. So, suppose (M, g, T) has a matter hole $L \subset M$ and ϕ is a non-trivial diffeomorphism that agrees with the identity map on the exterior of L. Then $\phi * g$ differs from g on the inside of L, but $\phi * T = T$ everywhere since T vanishes identically on L. Hence, $(M, \phi * g, T)$ is a distinct model that agrees with (M, g, T) everywhere but for the interior of L. If these models represent distinct physical situations, the alleged breakdown in causality follows. (One can recover the original formulation of the argument by selecting some coordinate chart ψ and letting $\mathfrak{G}$ and $\mathfrak{G}'$ be the images of (M, g, T) and $(M, \phi * g, T)$ under ψ, respectively.)

The dilemma arises, however, only if one assumes, as did Einstein, that, in a fixed reference system, $\mathfrak{G}$ and $\mathfrak{G}'$ really do describe distinct physical fields, or equivalently, that (M, g, T) and $(M, \phi * g, T)$ represent distinct physical situations. This assumption is what I have elsewhere referred to as *"Model Literalism"*: *Distinct solutions or models invariably represent distinct physical realities.* (See Rynasiewicz 1994, 1996.) The threatened breakdown of causality can be avoided either by rejecting the requirement of general covariance or by denying Model Literalism. When in 1915 Einstein re-embraced the requirement of general covariance, he saw his way clear of the dilemma posed by the hole argument by rejecting Model Literalism and adopting instead what Earman and Norton have called *"Leibniz Equivalence"*: *Diffeomorphic models invariably represent the same physical reality.* (See Earman & Norton 1987; Norton 1987; Earman 1989.) But, what independent grounds are available to justify this choice?

Stachel (1980) and Norton (1987) have suggested that such grounds are to be found in the guise of the point-coincidence argument of the 1916 *Grundlage* paper, and have found earlier versions of this argument in Einstein's correspondence that unmistakably link it with the hole argument. For our purposes, though, it is necessary to distinguish between the version of the argument as presented to Ehrenfest and Besso in late 1915 and early 1916, which I shall call the *private* point-coincidence argument, from the *public* point-coincidence argument that appeared in print. The significant difference is that Leibniz Equivalence figures explicitly in the former but not in the latter. The private argument has the following form:[23]

[23] "Everything in the hole argument was correct up to the final conclusion. It has no physical content if, with respect to *the same* coordinate system K, two different solutions $G(x)$ and $G'(x)$ exist. To imagine two solutions simultaneously on the same manifold has no meaning and indeed the system K has no physical reality. The hole argument is replaced by the following consideration. Nothing is physically *real* but the totality of space-time coincidences. If, for example, all physical happenings were to be built up from the motions of material points alone, then the meetings of these points, i.e., the points of intersection of their world-lines, would be the only real things, i.e., observable in principle.

A. Physical reality comprises no more than what is in-principle observable.

B. Only the spatio-temporal coincidences of (or topological relations between) material objects or processes are in-principle observable.

C. Hence, physical reality consists in no more than the spatio-temporal coincidences of (or topological relations between) material objects or processes.

D. One-to-one continuous transformations preserve all coincidences (and topological relations).

E. Hence, topologically equivalent solutions describe the same physical reality (i.e., Leibniz Equivalence).

In contrast, the *public* point-coincidence argument contains no explicit statement of the thesis of Leibniz Equivalence.[24] In the place of **E** it concludes instead:

E′. Hence, there are no grounds for granting privilege to certain coordinate systems over others.

F. Therefore, we arrive at the requirement of general covariance, viz., that the general laws of nature are to be expressed by equations that hold for all coordinate systems.

The difference is important for understanding not only what Kretschmann says in the 1917 paper, but also what he might have said but didn't. Even had he sensed a direct connection between the public point-coincidence argument and the hole argument, he could only have guessed at the hidden role played by Leibniz Equivalence.[25] But as matters stood, Kretschmann had occasion to address only

---

These points of intersection naturally are preserved during all transformations (and no new ones occur) if only certain uniqueness conditions are observed. It is therefore most natural to demand of the laws that they determine no *more* than the totality of the space-time coincidences. From what has been said, this is already attained through the use of generally covariant equations." (Einstein to Besso, 3 January 1916. Quoted from Stachel 1980: 86.)

"In § 12 of my paper of last year, everything is correct (in the first three paragraphs) up to the italicized part at the end of the third paragraph. Absolutely no contradiction to the uniqueness of events follows from the fact that both systems $G(x)$ and $G'(x)$ satisfy the conditions for the gravitational field with respect to the same reference system. The apparent compulsion of this argument disappears at once if one considers that 1) the reference system signifies nothing real, 2) that the (simultaneous) realization of two different g-systems (better said, of two different gravitational fields)in the same region of the continuum is impossible by the nature of the theory.

The following consideration should replace § 12. The physically real, [concerning] what happens in the world (as opposed to what depends on the choice of the reference system), consists of *spatio-temporal coincidences.* For example, the points of intersection of two world-lines, or the assertion that they *do not* intersect is real. Such assertions referring to the physically real are thus not lost because of any (single-valued) coordinate transformations. If two systems of $g_{\mu\nu}$ (or generally, of any variables applied in the description of the world) are so constituted that the second can be obtained from the first by a pure space-time transformation, then they are fully equivalent. For they have all spatio-temporal point coincidences in common, that is, everything that is observable. This argument shows at the same time how natural the demand for general covariance is." (Einstein to Ehrenfest, 26 December 1915. Quoted from Stachel 1980: 86–87.)

[24] One might take the conclusion to be implicit in that this would appear to be the grounds for Einstein's comment that general covariance "takes away from space and time the last residue of physical facticity." ["dem Raum und Zeit den letzten Rest physikalischer Gegenständlichkeit nehmen."] (Einstein 1916: 776.)

[25] It is clear from his 1915 article, which cites several publications (Einstein 1914; Einstein & Gross-

the public point-coincidence argument, although for our purposes it would be most elucidating to know what he might have said to the stronger, private version.

Although such questions, as a matter of history, are unanswerable, I do think that there is a coherent response that would have been available to Kretschmann, one which is not only consistent with what he does say, but is in fact suggested by it. The key issue concerns the role played by the reference system in the semantics of physical theories.

Given his contention that the physical content of a theory is exhausted by the topological coincidences and relations it permits, Kretschmann would certainly agree with Einstein's assertion in the letters to Ehrenfest and Besso mentioned above that the reference system is not something real.[26] Nonetheless, it is possible to draw a distinction between the events to which the reference system assigns quadruples of real numbers in virtue of the stipulation of the reference system, and those events which inherit their coordinates in virtue of topological coincidence with events that have already been tagged with coordinates. The directly tagged events are the physical means by which we make spatio-temporal measurements of other events. This distinction surfaces when Kretschmann speaks of *actual* reference systems. For example, in connection with the law of light propagation in special relativity, he writes:

> If we begin again with form (1) of the law, we see first of all that among the actual reference systems, which are to be thought of as given by measuring rods and clocks with completely precise length and time assignments, those [systems] in which equation (1) is everywhere and always satisfies are distinguished from all those in which this is not the case by the observational results expected in them. . . . [27] (p. 580)

Thus, only if it is presumed that the solutions of the system of equations in question supply exhaustive specifications of the totality of events in a possible world, does it follow that topologically equivalent solutions describe the same possible world. As long as the solutions do not preclude the existence of events other than those that are described by the solutions, these other events provide a potential means for distinguishing between the situations described by topologically equivalent solutions.

Moreover, it need not be presumed that these other background events are given *absolutely*, prior to and independently of the events governed by the law. In fact, the individuation of these background events in general proceeds in tandem with the events described by the solutions of the equations. No doubt this sounds a bit

---

mann 1914) in which the hole argument is presented, that he had followed Einstein's efforts with the *Entwurf* theory (Einstein & Grossmann 1913) and that he would have taken note of the subsequent reversal in regard to the physical acceptability of general covariance in any case, even apart from the thematic connection between the point-coincidence argument and his uncited 1915 paper.

[26] See n. 23 above.

[27] "Geht man wieder von der Form (1) des Gesetzes aus, so sieht man zunächst, daß unter den Bezugssystemen der Wirklichkeit, die durch Maßstäbe und Uhren mit vollkommen scharfen Längen- und Zeitangaben gegeben zu denken sind, die, in denen die Gleichung (1) überall und stets erfüllt ist, von allen, in denen das nicht der Fall ist, durch die prinzipiell, d.h. abgesehen von allen technischen Schwierigkeiten, in ihnen zu erwartenden Beobachtungsergebnisse unterschieden sind."

mysterious, so let me elaborate in terms of the coincidence models for the law of light propagation in special relativity.

The following puzzle may have already occurred to some readers. Suppose the coincidence structure $\mathfrak{A} = (M, \Lambda)$ satisfies Equation (1) in reference system Σ. For the sake of simplification, suppose further that $\mathfrak{A}$ is maximal in the sense that there is no Λ' properly containing Λ such that (M, Λ') also satisfies (1) in Σ. Now let ϕ be a diffeomorphism of M whose coordinate expression in Σ is not in the admissibility group of Equation (1). (Since $\mathfrak{A}$ is maximal, this is equivalent to the requirement that $\phi * \Lambda \neq \Lambda$.) Consider the coincidence structure $\phi * \mathfrak{A} := (M, \phi * \Lambda)$. Although $\phi * \mathfrak{A}$ obviously does not satisfy (1) in Σ, nonetheless there is a reference system in which it does satisfy (1), namely the reference system $\phi * \Sigma$. The puzzle, now, is this. Which reference system, Σ or $\phi * \Sigma$, should qualify as inertial? Four possible answers suggest themselves:

I. Adopt Model Literalism and claim that there is an absolute fact of the matter given beforehand. It may be that Σ is inertial absolutely, or it may be that $\phi * \Sigma$ is, but it cannot be that both are. But if Σ is absolutely inertial, then $\mathfrak{A}$ represents a physically possible situation while $\phi * \mathfrak{A}$ represents a physically impossible situation. If $\phi * \Sigma$ is absolutely inertial, then vice-versa.

II. Adopt Model Literalism and claim that there is no absolute fact of the matter. Both $\mathfrak{A}$ and $\phi * \mathfrak{A}$ represent physically possible situations. Relative to $\mathfrak{A}$, Σ is inertial but $\phi * \Sigma$ is not. Relative to $\phi * \mathfrak{A}$, just the reverse is the case.

III. Adopt Leibniz Equivalence. Then, since $\mathfrak{A}$ and $\phi * \mathfrak{A}$ are isomorphic, they represent the same physical situation, and hence there is no paradox in saying that both Σ and $\phi * \Sigma$ are inertial, since they are really the same reference system.

IV. Reject both Model Literalism and Leibniz Equivalence. Claim there is no fact of the matter independent of a specification of 'gauge.' In the gauge of $\mathfrak{A}$, Σ but not $\phi * \Sigma$ is inertial. In the gauge of $\phi * \mathfrak{A}$, just the reverse. This differs from **II** above, since in the gauge of $\mathfrak{A}$, the model $\phi * \mathfrak{A}$ does not coherently represent any physical situation at all, either physically possible or impossible; and vice-versa in the gauge of $\phi * \mathfrak{A}$.

The first response, I hazard to say, is the only alternative to **III** that Einstein could imagine. I know of no historical figures who would have endorsed response **II**, but it is an instance of the naive absolutist position against which the revamped hole argument of Earman and Norton (1987) can be made to work. The fourth response I take to be the one available to Kretschmann. More expansively, it diagnoses the situation as follows. The behavior of light rays in conformity with Equation (1) is (at least in part) constitutive of what it means for a reference system to be inertial, so that if the individuation of the points of M were something given absolutely, independently of both the specification of the world-lines of light rays and the reference systems Σ and $\phi * \Sigma$, then response (II) would be the appropriate one. However, the possession of a particular method for individuating the points of M amounts to having set in place some reference system or other. Since $\mathfrak{A}$ and $\phi * \mathfrak{A}$ disagree on which reference systems are inertial, it follows that there is no criterion of individuation for the points of M common to the two that also conforms to the law of light propagation. This is what it means to say that $\mathfrak{A}$ and $\phi * \mathfrak{A}$ are in different gauges and why it is that in the gauge of the one the other fails to represent anything at all. In this sense, the individuation of the background events to which the coordinates of the reference system directly attach occurs in tandem with the individuation of the events specifically described by the law.

This also clarifies the sense in which, on Kretschmann's proposal, a relativity principle renders certain reference systems equivalent or indistinguishable, namely *qua* instruments of individuation. Suppose K is a maximally gauge consistent class of models of the law expressed. This means that there exists a single reference system Σ that can coherently serve as a common individuating principle across the models of K. If ϕ is a diffeomorphism whose action on the space of all coincidence structures maps K onto itself, then $\phi * \Sigma$ may also serve as a common principle of individuation across this class, and, in fact, the law itself cannot be appealed to in order to distinguish between Σ and $\phi * \Sigma$. To be precise, two reference systems Σ and Σ' are equivalent according to the law just in case for any diffeomorphism θ such that $\Sigma' = \theta * \Sigma$, any coincidence model $\mathfrak{A}$ is gauge consistent with $\theta * \mathfrak{A}$ according to the law.[28]

5. Generalization to an arbitrary system of laws

Immediately after his discussion of the law of light propagation in special relativity, Kretschmann asserts that the pattern of analysis employed

> can be applied straightforwardly to arbitrary systems of physical laws. When doing so, in order to do justice to all possibilities, one needs only to consider that the size of the physically distinguished transformation group depends also on "contingent" physical circumstances [that are] not lawfully determined.[29] (pp. 583–584)

A footnote clarifies this otherwise opaque remark.

> Such a case arises in Einstein's gravitational theory insofar as, according to it, with certain particular [*singulären*] curvature conditions of the space-time manifold, which, according to Einstein, depend on the contingent distribution of matter, the group that is physically distinguished includes exceptionally more transformations than is usual.[30] (p. 584, n. 1)

---

[28] An equivalent formulation is the following. Σ is equivalent to Σ' just in case for any coincidence model $\mathfrak{A}$ and any charts ψ and ψ' associated with Σ and Σ', respectively, the solutions given by the $\mathbf{R}^4$ images of $\mathfrak{A}$ under ψ and ψ', respectively, are gauge consistent according to the law.

[29] "Die im vorhergehenden gegebene Bestimmung des von den angenommenen Lichtausbreitungsgesetzen erfüllten Relativitätspostulates läßt sich ohne weiteres auf beliebige Systeme physikalischer Gesetze übertragen. Dabei hat man nur, um allen Möglichkeiten gerecht zu werden, auch den Fall zu berücksichtigen, daß der Umfang der physikalisch ausgezeichneten Transformationsgruppe auch von gesetzmäßig nicht bestimmten, "zufälligen", physikalischen Umständen abhängt."

[30] "Ein solcher Fall liegt bei der Einsteinschen Gravitationstheorie insofern vor, als nach ihr bei

The body of the text continues:

> Thereby, it is clear that a relativity principle can hold as (generally) fulfilled only if it is satisfied independently of such circumstances, which in principle can be altered at will.[31] (p. 584)

Thus, the relativity group of a theory should not depend on the existence of any 'accidental' symmetries of its solutions.

Kretschmann then presents his general definition:

> *A system of physical laws satisfies the relativity principle of an invariant transformation group G just in case, if for any arbitrarily formed representation of all the laws of the system, and only these laws, the reference systems physically admissible—and thereby those which are in principle distinguishable from all the rest by observation—form, under all circumstances physically possible according to the laws, a class of such a size that the class of transformations connecting them properly contains or is equal to the group G in some form.*[32] (p. 584)

There are three issues that need to be addressed before we can attempt to restate this in, hopefully, more intelligible terms.

First, up to this point, I have portrayed Kretschmann as working exclusively with the triadic relation: systems Σ and Σ' are co-admissible for equation EQN. However, the above definition employs the dyadic relation: reference system Σ is admissible for EQN. Just prior to stating the definition, Kretschmann characterizes a reference system as

> "physically admissible" for a given form of representation of a system of laws if it is compatible with it in regard to all observations which are not excluded by the system of laws given and are thus regarded as possible in principle.[33] (p. 584)

In fact Kretschmann earlier devotes roughly a page to this idea as it applies to the law of light propagation in special relativity. (See § 4, p. 580.) This is in effect an operational definition indicating how we are to determine whether an actually given reference system is admissible (assuming that the laws in question in fact obtain). It does not (by itself) lead to a classification of reference systems as admissible or inadmissible for a given system of equations on formal grounds alone. This might seem like a shortcoming, but keep in mind that if the laws expressed by

---

gewissen singulären Krümmungsverhältnissen der Raum-Zeit-Mannigfaltigkeit, die ja nach Einstein von der zufälligen Verteilung und Bewegung der Materie abhängen, die physikalisch ausgezeichnete Gruppe ausnahmsweise mehr Transformationen umfaßt als sonst."

31 "Als (allgemein) erfüllt kann dabei offensichtlich nur ein Relativitätspostulat gelten, dem unabhängig von solchen Umständen, die sich prinzipiell nach Willkür ändern lassen, genügt ist."

32 *"Ein System physikalischer Gesetze erfüllt dann und nur dann das Relativitätspostulat einer invarianten Transformationsgruppe G, wenn bei jeder beliebig gestalteten Darstellung aller Gesetze des Systems und nur dieser Gesetze die für die Darstellung physickalisch berechtigten—und dadurch von allen übrigen durch Beobachtung prinzipiell unterscheidbaren—Bezugssysteme unter allen nach den Gesetzen möglichen physikalischen Umständen eine so große Schar bilden, daß die Schar der sie verbindenden Transformationen die Gruppe G in irgendeiner Form als Teil enthält oder ihr gleich ist."*

33 "Nennt man ein Bezugssystem für eine gegebene Darstellungsform eines Gesetzsystems 'physikalisch berechtigt', wenn es bezüglich aller Beobachtungen mit ihr verträglich ist, die durch das gegebene Gesetzsystem nicht ausgeschlossen, also als prinzipiell möglich zu betrachten sind, so . . . "

the equations in question do in fact obtain, then they will be true in at least *some* reference system. So, in applying the definition for theoretical purposes, one can arbitrarily select any formally specified reference system Σ_0 as one in which the laws are hypothetically true, and then any other reference system will count as physically admissible for these equations just in case it is co-admissible with Σ_0.

This suffices at any rate with the understanding that the relevant coincidence structures all have the same base manifold. This raises the second issue, whether Kretschmann perceived the possibility of topologies other than $\mathbf{R}^4$ for the standard field equations of general relativity. Later on he does cite Einstein's 1917 cosmological paper, which introduces a spatially finite, unbounded universe in the static case (Einstein 1917). But he mentions that paper only to say that "unfortunately the most recent modification of Einstein's gravitational field equations cannot be considered further." (p. 595, n. 2.)[34] I shall assume that Kretschmann's analysis is restricted to $\mathbf{R}^4$ topologies and leave it to the reader to see how the approach might be generalized.

The third issue concerns how one is to determine when two systems of equations (or other means of expressing laws of nature) are formulations of one and the same set of laws. Kretschmann has not given explicit criteria, and so we are left to generalize for ourselves from the example at hand. Charity dictates that we suppose that the formulations in question share a common framework in the sense that their coincidence models have a common generic form, so that essentially the issue reduces to how determinations of gauge consistency are to be made. Kretschmann appears to use the criterion that two geometric pictures satisfying an equation EQN are gauge consistent just in case their join also satisfies EQN. Thus, a class of geometric pictures is gauge consistent just in case its least upper bound satisfies EQN. Physically, the criterion requires that geometric pictures, or models, are gauge compatible according to EQN just in case the world lines of light rays from the respective geometric pictures, or models, are literally compossible according to EQN. For general relativity, the coincidence structures implicit in Kretschmann's treatment have the form (M, Λ, Ξ), where M and Λ are as before and Ξ is a collection of world-lines of test particles subject to no external forces. Thus, the same lattice-theoretic criterion of gauge consistency carries over straightforwardly.

Given a criterion of gauge consistency that appeals only to the equations themselves and not the laws they express, we can state when it is that two systems of equations express the same laws. As we have seen, if K is a maximally gauge consistent class of coincidence models or geometric pictures, then certain diffeomorphisms of the base manifold map K back to itself. Call the group of such diffeomorphisms the *invariance group* of K. If the diffeomorphism ϕ is not in the invariance group of K, then the image of K under ϕ is a distinct maximally gauge consistent class K', but which represents, as an entire class, the same spectrum of physical situations. Hence, it does not matter for the physical content of the laws expressed whether the members of K' satisfy the equations expressing those laws.

[34] "Die neueste Änderung der Einsteinschen Quellgleichungen des Schwerefeldes ... konnte leider nicht mehr brücksichtigt werdern."

The essential thing is that there be *some* maximally gauge consistent class, isomorphic in this respect to K, whose members satisfy the equations in question. Thus, two systems of equations express the same system of physical laws just in case any maximally consistent gauge class of the one can be mapped to a maximally consistent gauge class of the other in such a fashion.

Returning now to Kretschmann's general definition of the relativity principle fulfilled by a theory, we can reformulate that definition as follows:

> **Kretschmann's Condition.** The relativity group of a theory is the largest group (up to isomorphism) that is isomorphically embeddable in the invariance group of each maximally consistent gauge class of models of the theory.

In the case of the law of light propagation in special relativity, all maximally consistent gauge classes are isomorphic, and hence the relativity group (up to isomorphism) is the invariance group of any one of these classes. In the case of the law of light propagation together with the law of inertia in general relativity, it turns out, there are structurally distinct maximally consistent gauge classes whose invariance groups are sensitive to the contingent distribution of matter. Kretschmann's project is next to determine whether there is a non-trivial transformation group isomorphically embeddable in the invariance group of each of these gauge classes.

6. Application to general relativity

One way to pursue this end is to take inspiration from the fact that in special relativity one can find the relativity group by imposing non-covariant constraints on the generally covariant Equations (3) without changing the physical content of the theory so as to arrive at Equation (1), whose covariance group coincides with the relativity group. The attempt to do this in the case of general relativity occupies the bulk of Kretschmann's article. But as a preliminary, he needs to establish to what extent a maximally gauge consistent class of coincidence models fixes the metric functions $g_{\mu\nu}$. As he tells us:

> In order to find the covariance properties that are associated in an essential way with Einstein's theory, independently of the choice of expression, an attempt will be made to cast it into the least covariant form possible without altering its physical content. It suffices if this is done with a portion of the Einstein equations, and for this purpose the laws of motion for light rays and point masses in a gravitational field, on which the whole theory is based, will be chosen. In the same way in which the light propagation equation of the original theory of relativity could be put into generally covariant form by introducing the undetermined coefficients $g_{\mu\nu}$ into the expression for the line element, so conversely the conversion of the Einstein equations into a less covariant form of equivalent physical content is a matter of fixing either all or some of those parts of the determination of the coordinate functions $g_{\mu\nu}$ which depend solely on the choice of reference system. In order to know, in turn, to what extent the reference systems which satisfy the relevant conditions on the $g_{\mu\nu}$'s are distinguished from the rest not only

> mathematically, but also observationally, which alone decides the validity of relativity principles as they are understood here, one must above all investigate which assertions concerning the values of $g_{\mu\nu}$ can be verified by observations in an empirically given reference system, according to Einstein's theory.[35] (pp. 585–586)

Kretschmann's worry is that there might be constraints one can impose on the $g_{\mu\nu}$'s that result in a covariance group strictly narrower (up to isomorphism) than the invariance group of any maximally gauge consistent class of models. In order to appreciate this more fully, let's lay out explicitly how the framework developed for special relativity carries over to general relativity.

As mentioned earlier, the coincidence models of the theory now take the form $\mathfrak{A} = (M, \Lambda, \Xi)$, where Ξ represents a class of world-lines of freely moving test particles. A geometric picture, again, is an $\mathbf{R}^4$-image of a coincidence structure under some chart (which for convenience sake we will take to be global). The relevant equations for the motions of light rays and test particles, respectively, are rendered by Kretschmann as[36]

$$0 = \sum_{\mu,\nu}^{1\ldots4} g_{\mu\nu} \frac{dx_\mu}{dx_4} \cdot \frac{dx_\nu}{dx_4} \tag{4}$$

and

$$\sum_\alpha g_{\alpha\rho} \frac{d^2x_\alpha}{ds^2} + \sum_{\mu\nu} \begin{bmatrix} \mu\nu \\ \rho \end{bmatrix}_g \frac{dx_\mu}{ds} \frac{dx_\nu}{ds} = 0, \tag{5}$$

where

$$\begin{bmatrix} \mu\nu \\ \rho \end{bmatrix} \equiv \frac{1}{2} \left(\frac{\partial g_{\mu\rho}}{\partial x_\nu} + \frac{\partial g_{\nu\rho}}{\partial x_\mu} - \frac{\partial g_{\mu\nu}}{\partial x_\rho} \right).$$

As before, the conditions under which a geometric picture satisfies these equations involves existential quantification over real-valued functions for the $g_{\mu\nu}$'s. Let us say that $\mathfrak{B}$ satisfies EQN with $[f_{\mu\nu}]$, if the specific set of functions $[f_{\mu\nu}]$ fits the existential requirement. Kretschmann's concern is this. If $\mathfrak{B}$ satisfies (4, 5), it may be that there exist distinct systems of functions $[f_{\mu\nu}]$ and $[f'_{\mu\nu}]$ such that $\mathfrak{B}$

---

[35] "Um die der Einsteinschen Theorie notwendigerweise und unabhängig von der Wahl des Ausdruckes anhaftenden Kovarianzeigenschaften zu finden, wird man versuchen, sie ohne Änderung ihres physikalischen Inhaltes auf eine möglichst wenig kovariante Form zu bringen. Es genügt, wenn dies mit einem Teil der Einsteinschen Gleichungen geschieht, und zwar sollen die Bewegungsgesetze des Lichtes und der Massenpunkte im Schwerefelde hierzu gewählt werden, auf denen die ganze Theorie ruht. Wie nun oben die Lichtausbreitungsgleichung der ursprünglichen Relativitätstheorie durch Einführung der unbestimmten Koeffizienten $g_{\mu\nu}$ in den Ausdruck des Linienelementes auf allgemeiner kovariante Formen gebracht werden konnte, so kommt umgekehrt jede Verwandlung der Einsteinschen Gleichungen in weniger kovariante gleichen physikalischen Inhaltes auf Festlegung aller oder eines Teiles der nur von der Wahl des Bezugssystems abhängigen Bestimmungsstücke der Koordinatenfunktionen $g_{\mu\nu}$ hinaus. Um aber zu wissen, wie weit die Bezugssysteme, in denen die getroffenen Festsetzungen über die $g_{\mu\nu}$ gelten, vor den übrigen nicht nur mathematisch, sondern auch für die Beobachtung ausgezeichnet sind, was allein über die Gültigkeit von Relativitätspostulaten, wie sie hier verstanden werden, entscheidet, muß man vor allem untersuchen, welche Angaben über die Werte der $g_{\mu\nu}$ in einem empirisch gegebenen Bezugssysteme nach der Einsteinschen Theorie durch Beobachtungen nachgeprüft werden können."

[36] Equation (5) below appears as equation (7) in (Kretschmann 1917).

satisfies (4, 5) both with $[f_{\mu\nu}]$ and with $[f'_{\mu\nu}]$. If $\mathfrak{B}$ is maximal, that is, $\mathfrak{B}$ is the least upper bound of a maximally consistent gauge class of geometric pictures, and the two systems of functions are related by a coordinate transformation under the rules of the tensor calculus, then that transformation is an automorphism of $\mathfrak{B}$.[37] So, it needs to be known whether there are such sets of functions related in this way. The danger is that if there are, then a constraint on the $g_{\mu\nu}$'s that is not covariant under the coordinate transformation in question will lead to a narrower group than the relativity group.

Kretschmann establishes a pair of results in this connection, which together are noteworthy enough to be called *Kretschmann's Lemma*. The first is that if $\mathfrak{B}$ satisfies (4) with $[f_{\mu\nu}]$ and also with $[f'_{\mu\nu}]$, then there is a real-valued function λ such that $f'_{\mu\nu} = \lambda \cdot f_{\mu\nu}$, for each μ, $\nu = 1, 2, 3, 4$. Second if $\mathfrak{B}$ satisfies (4, 5) with $[f_{\mu\nu}]$ and also with $[f'_{\mu\nu}]$, then λ is a positive constant. In other words, the class of null geodesics determines the metric up to a conformal factor, and the time-like geodesics further force this conformal factor to be constant.[38]

The question now is, if $g_{\mu\nu} = f_{\mu\nu}(x_1, \ldots, x_4)$, then for what transformations from $(x_1, \ldots, x_4)$ to $(x'_1, \ldots, x'_4)$ is it the case that $g'_{\mu\nu} = \lambda \cdot f_{\mu\nu}(x'_1, \ldots, x'_4)$ for constant λ? Kretschmann argues that if there are more than four independent scalar invariants homogenous in $g_{\mu\nu}$ and its derivatives, then the only such transformation is the identity. The exceptional cases are thus ones of "peculiarly extensive functional dependence" of the scalar invariants on one another.

> *In general, then, there are no further reference systems that would be equivalent, even from a purely physical point of view, to those that are mathematically distinguished by the constraints on the $g_{\mu\nu}$'s*—since those further reference systems would have to belong to the same actual manifold, i.e., would have to be transformable into the latter systems.[39] (p. 591)

---

[37] The argument is this. If ϕ is the transformation in question, then automatically $\phi * \mathfrak{B}$ satisfies (4) and (5) with $[f'_{\mu\nu}]$ as well. Hence, $\phi * \mathfrak{B}$ is gauge consistent with $\mathfrak{B}$. But $\mathfrak{B}$ is maximal. So $\phi * \mathfrak{B} = \mathfrak{B}$.

[38] As Norton mentions, the result is "usually attributed to a later tradition initiated by Weyl." (Norton 1992: 300)

[39] *"Im allgemeinen sind demnach den durch irgendwelche Bedingungen für die $g_{\mu\nu}$ mathematisch hervorgehobenen Bezugssystemen auch in rein phyikalischer Hinsicht keine weiteren*—die ja stets derselben wirklichen Mannigfaltigkeit angehören, d.h. in jene transformierbar sein müßten—*gleichberechtigt."*

Kretschmann goes on to say: "In the exceptional cases just mentioned, a one-parameter group of transformations, which take $g_{\mu\nu} = f_{\mu\nu}(x_1 \ldots x_4)$ into $g'_{\mu\nu} = \lambda . f_{\mu\nu}(x'_1 \ldots x'_4)$, suffices in order to obtain all physically privileged reference systems from the ones mathematically distinguished." ["In den eben genannten Ausnahmefällen genügt offenbar eine einparametrige Gruppe von Transformationen, die $g_{\mu\nu} = f_{\mu\nu}(x_1 \ldots x_4)$ in $g'_{\mu\nu} = \lambda . f_{\mu\nu}(x'_1 \ldots x'_4)$ überführt, um aus den mathematisch hervorgehobenen alle physikalisch berechtigten Bezugssysteme zu erzeugen."] There is a footnote to this that reads:"For the determination of relativity principles actually satisfied in the sense of § 7, this exception becomes insignificant, insofar as it arises only with [a] quite special configuration of the contingent, i.e., not lawfully determined, physical conditions (distribution of matter), which according to Einstein's theory contribute to the determination of the nature of the curvature of the universe, and consequently can be disrupted by an (arbitrary) change of these contingencies." ["Für die Bestimmung der im Sinne von § 7 wirklich erfüllten Relativitätspostulate wird diese Ausnahme insofern bedeutungslos, als sie nur bei ganz besonderer Gestaltung der zufälligen, d.h. gesetzlich nicht bestimmten, physikalischen Umstände (Verteilung der Materie), die ja nach Einsteins Theorie den Krümmungscharackter der Welt mitbestimmen, besteht und daher durch eine (willkürliche) Änderung dieser Zufälligkeiten aufgehoben werden kann."] (p. 591, n. 2.)

Thus it is permissible to impose any constraints on the $g_{\mu\nu}$s that do not exclude in its entirety any isomorphism class of maximally gauge consistent classes of models of the original equations.

In what follows, though, Kretschmann is unable to find any such clean minimally covariant formulation. He pursues a maze of technical considerations so subject to repeated qualifications and restrictions that in the end one is left wondering what has been accomplished. The upshot for the overall dialectic seems to be to demonstrate that the principle barrier to finding such a formulation is the existence of the exceptional cases mentioned. Otherwise, one can use the curvature properties of the space-time to distinguish, physically, certain coordinate systems uniquely. Consequently, Kretschmann turns to a "geometric determination" of the relativity principle satisfied by general relativity.

This demonstration proceeds by comparing the maximal geometric pictures corresponding to the various possible solutions for $g_{\mu\nu}$. Each such solution $g_{\mu\nu} = f_{\mu\nu}(x_1, \ldots, x_4)$ determines an infinite set of world-lines representing all the possible motions of light rays and free point masses in that space-time. (The terms *Weltlinienshar* and *Extremalshar* are used interchangeably to refer to a maximal set of such world-lines.) Kretschmann wants to classify these geometric pictures first into topological equivalence classes, which he calls *Teilmengen*, and then into further subclasses, called *Untermengen*, in virtue of the choice of constant conformal factor λ. However, the ensuing discussion makes sense only if one assumes that the maximal set of geodesics of the solution $g_{\mu\nu} = f_{\mu\nu}(x_1, \ldots, x_4)$ is numerically distinct from that of the solution $g_{\mu\nu} = \lambda \cdot f_{\mu\nu}(x_1, \ldots, x_4)$, even though the two sets of geodesics coincide exactly. So, in effect, what Kretschmann does is to classify, not the geometric pictures (i.e., the *Weltlininiensharen*) corresponding to the possible solutions for $g_{\mu\nu}$, but the solutions themselves. Kretschmann's carelessness in this regard is particularly frustrating since, as he tells us later in a footnote, he wants to leave open the possibility that the constant conformal factor λ has an actual physical significance. (See p. 609, n. 2.) But in that event, the physical content of the theory cannot be assessed in terms of the geometric pictures he in fact uses, but instead these would have to be supplemented so as to reflect differences in λ in such a way that the correspondence between solutions and maximal geometric pictures is one-to-one.

What Kretschmann wants to say, or should have said, is this. The set of all possible solutions for $g_{\mu\nu}$ can be partitioned into equivalence classes (*Teilmengen*) such that any two members of the same class are topologically equivalent in the sense that there is a smooth mapping that carries the geodesics of the one onto those of the other. However, given Kretschmann's result that the solutions related by a constant conformal factor are geodesically identical even if they cannot be related by a coordinate transformation, it follows that one cannot necessarily generate an entire *Teilmenge* by the action of the group of all possible transformations on a single solution in the class. Thus, each *Teilmenge* can be partitioned further into subclasses (*Untermengen*) such that any solution can be obtained from any other of the same subclass by some transformation. Generally, a *Teilmenge* consists of a continuum of *Untermengen*, one for each value of the positive constant λ.

In the language of space-time models, this classification works as follows. Consider the collection of all relativistic space-times (M, g) for some fixed manifold M. The unique connection ∇_g compatible with g does the same duty as one of Kretschmann's geometric pictures or *Weltliniensharen.*[40] The space-times (M, g) and (M, g') belong to the same *Teilmenge* just in case there is a diffeomorphism ϕ of M such that $\nabla_{g'} = \phi * \nabla_g$ and to the same *Untermenge* just in case $g' = \phi * g$ for some diffeomorphism ϕ.

The 'geometric' determination of the relativity principle satisfied by a set of laws presupposes that, in principle, those laws can be expressed just as well in strictly extensional terms by a set of solutions as it can by a set of equations. The basic idea then is to choose a minimal subset of solutions for $g_{\mu\nu}$ having the same physical content as the entire class of solutions of the covariant equations. In order to do this, it suffices to select arbitrarily a single member from each *Untermenge.*[41] The rationale is presumably the same as in the case of special relativity. Corresponding to each space-time model (M, g) there is a maximally gauge consistent set of coincidence models K_g such that (M, Σ, Ξ) is a member of K_g just in case each curve in Λ is a null geodesic of g and each curve in Ξ a time-like geodesic. If ϕ is a diffeomorphism of M such that $\phi * g \neq \lambda g$, then $K_{\phi * g}$ is a maximally gauge consistent class, which represents precisely the same spectrum of physical situations as represented by K_g, and hence is redundant for the representation of the physical content of the laws in question.

Having selected one space-time model (M, g) from each *Untermenge*, we then ask what is the group of diffeomorphisms that preserves K_g. By Kretschmann's lemma, ϕ is a member of this group just in case $\phi * g = \lambda g$ for some λ. But, as Kretschmann has already argued, in the general case this is the trivial group consisting of the identity transformation alone. "*Consequently, Einstein's theory physically satisfies no relativity principle whatsoever ... ; it is a completely absolute theory in regard to its content.*"[42] (p. 610). Not only is general relativity *less* relativistic than special relativity, it follows there is no theory in the same category that embodies a relativity principle more general than that of special relativity.[43]

7. General discussion

Is any sort of theory conceivable that would satisfy a completely general principle of relativity? In answer to this Kretschmann invokes, without explanation, a dis-

---

[40] So, Kretschmann's Lemma states that if $g' = \lambda g$ for some positive constant λ, then $\nabla_{g'} = \nabla_g$.

[41] Or, if the conformal factor λ has no physical significance, then a single member from each *Teilmenge*.

[42] "*Die Einsteinsche Theorie genügt demnach physikalisch* im Sinne des in § 7 Ausgefürten *überhaupt keinem Relativitätspostulate; sie ist ihrem Inhalte nach eine vollkommene Absoluttheorie.*"

[43] A theory in the same category is one "*which confers on the extremal curves of a space-time manifold, with line elements of Minkowski normal form, a significance accessible to observation, or represents in some other way the invariant metrical character of the manifold as observable in principle to the same extent*" ["*welche den Extremalen ... einer Raum-Zeit-Mannigfaltigkeit mit Minkowskischer Normalform des Linienelements eine der Beobachtung zugängliche Bedeutung zuerkennt oder den invarianten metrischen Charakter der Mannigfaltigkeit sonstwie in gleichem Umfange als prinzipiell beobachtbar hinstellt*"] (p. 610).

tinction between laws that are "unconditionally affirmative" [*unbedingt bejahend*] and those that are "negative and conditional" [*verneinenden und bedingten*] and asserts that *"it is thus the negative content of the kinematical laws known to us, i.e., the restricting and denying of possibilities of coincidence, which makes impossible the physical satisfaction of the general principle of relativity."*[44] (p. 613)

At first blush, it might seem that any distinction between "affirmative" and "negative" laws must be completely illusory, since if a law is to have any physical content whatsoever, then it must prohibit the occurrence of certain logically conceivable situations. However, the distinction Kretschmann has in mind appears to be related to a distinction between existential and universal claims. There is a lore amongst logicians as to how to capture this distinction syntactically for the first-order predicate calculus in terms of prenex normal form. (See, for example, Chang & Keisler 1973: 34, exercise 1.3.4.) But it is more instructive to think of how the distinction is reflected semantically. An existential sentence continues to hold true no matter how a model of the sentence is extended to a larger structure. In contrast, a universal sentence fails to remain true in arbitrary extensions of its models, although it holds in all substructures of its models. For coincidence structures of the sort entertained above, natural correlates of the notions of extension and submodel are given in terms of the partial order relation. That is, a law is existential (affirmative) just in case for any coincidence structures $\mathfrak{A}_1$ and $\mathfrak{A}_2$ such that $\mathfrak{A}_1 \leq \mathfrak{A}_2$, if $\mathfrak{A}_1$ is a model of the law, then so is $\mathfrak{A}_2$. Conversely, the law is universal (negative) if $\mathfrak{A}_1$ is a model whenever $\mathfrak{A}_2$ is a model.

According to this characterization, the laws of motion from special and general relativity considered by Kretschmann qualify as universal.[45] What then might an "unconditional and affirmative" law satisfying the general principle of relativity look like? Kretschmann suggests:

> As an example of an unconditional, purely affirmative kinematical law, one could perhaps advance the proposition that in the course of time each point mass collides with at least one other and then must remain joined with it forever after. The geometrical picture which contains the totality of world-lines not contradicting this fictitious law must transform into itself under each continuous coordinate transformation, for its overlap with an arbitrarily distorted picture can never yield a singly standing world-line running together with no other, and thus no world-line not already contained

---

44 *"Demnach ist es also der verneinende Inhalt der uns bekannten kinematischen Gesetze, d.h. die in ihnen enthaltene Beschränkung und Verneinung von Koinzidenzmöglichkeiten, welche die physikalische Erfüllung des allgemeinen Relativitätspostulates unmöglich machen."*

45 Kretschmann locates the negative content for the laws of motion of light and inertial point masses in the assertion "that through two distinct world points there can never pass two different world-lines of the sort in question." ["daß durch zwei vershiedene Weltpuntkte niemals zwei verschiedene der genannten Weltlinien gehen können."] (p. 613). (Note, however, that this proposition is generally valid only locally.) He then cites the failure of this assertion to be preserved under arbitrary extensions as the crucial property that limits the relativity group. "In fact, it is precisely this proposition that, in each coordinate manifold, limits the class of mutually consistent world-lines and thereby also the group of transformations that transform it [this maximal class of world-lines] into itself ... since it would be violated by adding any further world-line to the class." ["In der Tat ist es gerade dieser Satz, der in jeder Koordinatenmannigfaltigkeit die Schar der miteinander verträglichen Weltlinien und damit zugleich die Gruppe der Transformationen, die sie sich überführen ... begrenzt, da er durch jede weitere der Schar zugefügte Weltlinie verletzt würde."] (p. 613)

in the original picture.[46] (p. 613)

Note that the fictitious law here is not strictly existential according to the above characterization. In general, if we have a collection of world-lines satisfying the stated requirement, we can always add other world-lines that violate it. However, it does meet a weaker condition that, as it turns out, is by itself sufficient for the purposes at hand. The condition is that the least upper bound of any collection of models of the law is again a model. Thus, if it is stipulated that a class of models is gauge consistent just in case its least upper bound is also a model, then the general principle of relativity follows if this weaker condition is met, since then the entire class of models is a maximally gauge consistent class.

So far so good. But can we always assess gauge consistency in the indicated fashion? Imagine the world-lines corresponding to a collection of charged particles in Newtonian (or neo-Newtonian) space-time obeying Coulomb's law. If we superimpose on this the world-lines of a second collection of such particles, the result in general need not be a collection of world-lines of particles satisfying Coulomb's law. Does this mean that the two models are gauge inconsistent? An affirmative answer leads to a difficulty pointed out by J. L. Anderson (1966). If a finite system of point masses were to constitute a maximally gauge consistent model, then it would follow that classical mechanics fails to satisfy any non-trivial relativity principle, since, e.g., one could use the center of mass of the system and its moments of inertia to define a unique reference system. Similar considerations apply to special relativity as well. It seems to me, though, that an affirmative answer here does not meet the spirit of Kretschmann's proposal, and thus that what is in order is a reconsideration of the formal criterion for gauge consistency rather than a trashing of the the program itself. The least upper bound criterion we have teased out of Kretschmann's exposition appears to capture what is desired only under special circumstances, namely, when the compossibility of representation reduces to the actual compossibility of what is represented. Moreover, up to this point I have called no attention to the fact that Kretschmann speaks of the relativity principles satisfied by *kinematical* laws. Now Kretschmann nowhere gives an explanation of what he means by this, but it is noteworthy that the types of processes governed by the laws he does consider are *ideal processes* in the sense that they are influenced by, but do not themselves influence other physical processes. The least upper bound criterion of gauge consistency is an appropriate necessary and sufficient conditions for processes that are ideal in this sense. In the example of particles interacting via a Coulomb force, we are not dealing with ideal processes. Nonetheless, we can simultaneously entertain the two imagined systems as genuinely possible physical situations in virtue of the fact that we have no difficulty imagining them in turn against the backdrop of one and the same reference system,

[46] "Als Beispiel eines unbedingten rein bejahenden kinematischen Gesetzes könnte man etwa den Satz aufstellen, daß jeder Massenpunkt im Laufe der Zeit mit mindestens einem anderen zusammentreffen und dann dauernd mit ihm vereint bleiben müsse. Das geometrische Bild, das die Gesamtheit der diesem fingierten Gesetze nicht widersprechenden Weltlinien enthält, muß bei jeder stetigen Koordinatentransformation in sich selbst übergehen; denn seine Überdeckung mit einem beliebig verzerrten Bilde kann niemals eine einzeln stehende, mit keiner anderen zusammenfließende Weltlinie ergeben, also keine im urspünglichen Bilde noch nicht enthaltene Weltlinie."

physically instantiated by certain processes and measuring devices, even though the two systems cannot exist together (against this backdrop) without violating Coulomb's law.

We can only guess how Kretschmann might have responded to Anderson's challenge. One possible response is to claim that dynamical laws invariably either presuppose or mandate certain 'kinematical' laws, and that it is the latter alone that dictate the relativity principle satisfied. Alternatively, one might attempt to reformulate the criterion of gauge consistency so as to apply to 'non-ideal' processes as well. A suggestion along these lines might be to 'localize' the least upper bound criterion by requiring only that at each point of the composite system there is a neighborhood that can be extended to a model of the law. But just what if anything along these lines might be workable is an interesting foundational, not historical question.

8. Conclusion

In the following year, 1918, Einstein published a short note in the *Annalen der Physik* (Einstein 1918), once again reconsidering the foundations of general relativity, this time in response to a number of articles, chief among them, Kretschmann's. As he now sees it, the theory rests on three main principles: a) the principle of relativity, b) the principle of equivalence, and c) Mach's Principle. He formulates the first of these as: "the laws of nature are only statements concerning spatio-temporal coincidences; they therefore find their only natural expression in generally covariant equations."[47] (p. 241).

He then responds to Kretschmann as follows:

> Concerning a) Hr. Kretschmann remarks that the principle of relativity thus formulated is not a statement about physical reality, i.e., about the *content* of the laws of nature, but only a condition concerning the mathematical *formulation*. To wit, since the totality of physical experience refers only to coincidences, it must always be possible to represent experiences concerning the lawful dependencies of these coincidences by means of generally covariant equations. For this reason he maintains it necessary to associate with the relativity requirement another meaning. I maintain that Hr. Kretschmann's argument is correct, but that the proposal put forth by him is inadvisable. For although it is correct that one must be able to put any empirical law in generally covariant form, nonetheless principle a) possesses an important heuristic force, which already has brilliantly proved itself in the problem of gravitation and is based on the following: Of two theoretical systems consistent with experience, the one which is the simpler and more transparent from the standpoint of the absolute differential calculus is [the one] to be preferred. If one were to put Newtonian gravitational mechanics into the form of absolutely covariant (four-dimensional) equations, one would undoubtedly be convinced that principle a) rules them out, granted not theoretically, but practically![48] (p. 242)

47 "Die Naturgesetze sind nur Aussagen über zeiträumliche Koinzidenzen; sie finden deshalb ihren einzig natürlichen Ausdruck in allgemein kovarianten Gleichungen."

48 "Zu a) bemerkt Hr. Kretschmann, das so formulierte Relativitätsprinzip sei keine Aussage über die

As far as I know, Kretschmann never replied. Perhaps he was satisfied that Einstein had conceded the main point: that the requirement of general covariance is not a physical requirement, but only a formal constraint on the mathematical form of the equations. Unfortunately, Einstein's comments leave open significant room for substantive disagreement, particularly in regard to his dismissal of Kretschmann's "proposal" as "inadvisable." If Einstein meant by this that searching to discover the laws of nature by seeking equations in "least covariant form" is inadvisable, then what follows is to the point. However, if he meant that, *pace* Kretschmann, there is no intelligible or interesting sense in which relativity principles can be understood to reflect something about the content of the laws of nature, then the explanation that follows fails to indicate why, since the heuristic value of general covariance does not obviously preclude this. In general, it is disappointing that Einstein ignores entirely Kretschmann's contention that there is a perfectly good sense in which Lorentz invariance reflects something important about the physical content of special relativity. Moreover, although Einstein, in response to Kretschmann, clearly softens the conclusion he draws from the public point-coincidence argument (it now follows only that laws find their *natural* expression in generally covariant equations), the bearing of Kretschmann's arguments on the private point-coincidence argument remains unaddressed.

Only much later, in Einstein's "Autobiographical Notes" (Einstein 1949), is there a hint of a response. There he mentions again that one might deny that "the principle of general covariance by itself would embody an actual restriction on the laws of physics, since it will always be possible to reformulate a law postulated initially only for certain coordinate systems so that the new formulation of the form becomes afterwards generally covariant."[49] (Einstein 1949: 64)

And again he invokes "the eminent heuristic significance of the general principle of relativity."[50] But then, without either preliminary or subsequent discussion, he asserts the thesis of Leibniz Equivalence: "Fields that can be transformed into each other by such transformations describe the same real state of affairs."[51] (p. 64).

---

physikalische Realität, d.h. über den *Inhalt* der Naturgesetze, sondern nur eine Forderung bezüglich der mathematischen *Formulierung*. Da nämlich die gesamte physikalische Erfahrung sich nur auf Koinzidenzen beziehe, müsse es stets möglich sein, Erfahrungen über die gesetzlichen Zusammenhänge dieser Koinzidenzen durch allgemein kovariante Gleichungen darzustellen. Er hält es deshalb für nötig, einen anderen Sinn mit der Relativitätsforderung zu verbinden. Ich halte Hrn. Kretschmanns Argument für richtig, die von ihm vorgeschlagene Neuerung jedoch nicht für empfehlenswert. Wenn es nämlich auch richtig ist, daß man jedes empirische Gesetz in allgemein kovariante Form muß bringen können, so besitzt das Prinzip a) doch eine bedeutende heuristische Kraft, die sich am Gravitationsproblem ja schon glänzend bewährt hat und auf folgendem beruht. Von zwei mit der Erfahrung vereinbarten theoretischen Systemen wird dasjenige zu bevorzugen sein, welches vom Standpunkte des absoluten Differentialkalküls das einfachere und durchsichtigere ist. Man bringe einmal die Newtonsche Gravitationsmechanik in die Form von absolut kovarianten Gleichungen (vierdimensional) und man wird sicherlich überzeugt sein, daß das Prinzip a) diese Theorie zwar nicht theoretisch, aber praktisch ausschließt!

49 "Zunächst kann man sogar bestreiten, dass die Forderung allein eine wirkliche Beschränkung für die physikalischen Gesetze enthalte; denn es wird stets möglich sein, ein zunächst nur für gewisse Koordinatensysteme postuliertes Gesetz so umzuformulieren, dass die neue Formulierung der Form nach allgemein kovariant wird."

50 "Die eminente heuristische Bedeutung des allgemeinen Relativitätsprinzips. . . . "

51 "Felder, die durch solche Transformationen ineinander übergeführt werden können, beschreiben

This sentence is significant in connection with the claim several pages later, "We have stated *physical* reasons for the fact that invariance with respect to the wider group must be demanded in physics."[52] (p. 72, emphasis in original). What are these physical reasons? In the preceding paragraph Einstein explains:

> But even if, from all the conceivable Lorentz-invariant laws, one had accidently guessed exactly the law belonging to the wider group, one would still not be at the level of understanding attained by the general principle of relativity. For, from the standpoint of the Lorentz group, two solutions would be regarded incorrectly as physically distinct from one another, if they are transformable into one another by a non-linear coordinate transformation, i.e., are only different representations of the same field from the standpoint of the wider group.[53] (p. 72)

Einstein appears to commit himself here to the claim that the physical significance of the relativity group is to partition the solutions of the equations into equivalence classes, the members of which represent the same physical state of affairs. In other words,

> **Einstein Equivalence.** Two solutions represent the same physical circumstance if they can be transformed into one another by a member of the relativity group.

Leibniz Equivalence then follows on the assumption that the relativity group is the general covariance group.

The question that remains is how, according to Einstein, the correct relativity group is determined. According to Kretschmann, the correct group is the relativity group of the correct theory. It seems, though, that according to Einstein, the correct group is determined by means of the sorts of general considerations that enter into the point-coincidence argument. Thus, had Einstein and Kretschmann debated their differences, it is possible that the same ground would have been covered over and over due to a fundamental difference over the upshot of the published version of that argument. On the other hand, if Einstein were to have made explicit his commitment to Leibniz Equivalence, or more exactly, to Einstein Equivalence, it is conceivable that Kretschmann's response might have led to interesting developments.

Acknowledgements

This work was supported by NSF grant number SBR-9511796. I am deeply indebted to Frank Döring for his generous assistance in deciphering Kretschmann's prose. Thanks also to John Norton for much advice and encouragement.

---

denselben realen Sachverhalt."

52 "Wir haben *physikalische* Gründe dafür angegeben, dass Invarianz gegenüber der weiteren Gruppe in der Physik gefordert werden muss; ... "

53 "Selbst aber wenn man von all den denkbaren Lorentz-invarianten Gesetzen zufällig gerade das zu der weiteren Gruppe gehörige Gesetz erraten hätte, so wäre man immer noch nicht auf der durch das allgemeine Relativitätsprinzip erlangten Stufe der Erkenntnis. Denn vom Standpunkt der Lorentz-Gruppe wären zwei Lösungen fälschlich als physikalisch voneinander verschieden zu betrachten, wenn sie durch eine nichtlineare Koordinaten-Transformation ineinander transformierbar sind, d.h. vom Standpunkt der weiteren Gruppe nur verschiedene Darstellungen desselben Feldes sind."

REFERENCES

ANDERSON, James L. (1964). "Relativity Principles and the Role of Coordinates in Physics." In *Gravitation and Relativity*. H.-Y. Chui and W. F. Hoffmann, eds. 175–194. New York: Benjamin.

——— (1966). "Maximal Covariance Conditions and Kretschmann's Relativity Group." In *Perspectives in Geometry and Relativity*. B. Hoffmann, ed. 16–27. Bloomington & London: Indiana University Press.

——— (1967). *Principles of Relativity Physics*. New York & London: Academic Press.

CARTAN, Élie. (1923/4). "Sur les variétés à connexion affine et la théorie de la relativité généralisée (première partie)." *Annales scientifiques de l'École normale supérieure* **40**: 325–412; **41**: 1–25. [Reprinted: *Œuvres complètes*. Part 3. 659–746; 799–823. Paris: Gauthier-Villars (1955)].

CHANG, Chen C. & KEISLER, H. Jerome. (1973). *Model Theory*. Amsterdam & London: North-Holland.

EARMAN, John. (1989). *World Enough and Spacetime: Absolute versus Relational Theories of Space and Time*. Cambridge (MA) & London: MIT Press.

EARMAN, John & NORTON, John D. (1987). "What Price Spacetime Substantivalism? The Hole Argument." *British Journal for the Philosophy of Science* 38: 515–525.

EINSTEIN, Albert. (1914). "Prinzipielles zur verallgemeinerten Relativitätstheorie." *Physikalische Zeitschrift* **15**: 176–180. [= EINSTEIN *CP4*: doc. 25].

——— (1916). "Die Grundlage der allgemeinen Relativitätstheorie." *Annalen der Physik* **49**: 769–822. [= EINSTEIN *CP6*: doc. 30].

——— (1917). "Kosmologische Betrachtungen zur allgemeinen Relativitätstheorie." *Sitzungsberichte der Preussischen Akademie der Wissenschaften*: 142–152. [= EINSTEIN *CP6*: doc. 43].

——— (1918). "Prinzipielles zur allgemeinen Relativitätstheorie." *Annalen der Physik* **55**: 241–244.

——— (1949). "Autobiographisches." In *Albert Einstein: Philosopher-Scientist*. P. A. Schilpp, ed. 2–95. LaSalle (IL): Open Court. [Reprinted as *Autobiographical Notes*. LaSalle & Chicago: Open Court (IL) (1979). Page references are to the German text in this edition].

——— (*CP4*). *The Collected Papers of Albert Einstein*. Vol. 4. *The Swiss Years: Writings, 1912–1914*. M. J. Klein, A. J. Kox, J. Renn & R. Schulmann, eds. Princeton: Princeton University Press (1995).

——— (*CP6*). *The Collected Papers of Albert Einstein*. Vol. 6. *The Berlin Years: Writings, 1914–1917*. A. J. Kox, M. J. Klein, & R. Schulmann, eds. Princeton: Princeton University Press (1996).

EINSTEIN, Albert & GROSSMANN, Marcel. (1913). *Entwurf einer verallgemeinerten Relativitätstheorie und einer Theorie der Gravitation*. Leipzig: Teubner. [Reprinted with added "Bemerkungen," *Zeitschrift für Mathematik und Physik* **62** (1914): 225–261 = EINSTEIN *CP4*: doc. 13].

——— (1914). "Kovarianzeigenschaften der Feldgleichungen der auf die verallgemeinerte Relativitätstheorie gegründeten Gravitationstheorie." *Zeitschrift für Mathematik und Physik* **63**: 215–225. [= EINSTEIN *CP6*: doc. 2].

HOWARD, Don & NORTON, John D. (1993). "Out of the Labyrinth? Einstein, Hertz, and the Göttingen Answer to the Hole Argument." In *The Attraction of Gravitation: New Studies in the History of General Relativity* (Einstein Studies, vol. 5). J. Earman, M. Janssen and J. D. Norton, eds. 30–62. Boston: Birkhäuser.

KRETSCHMANN, Erich. (1915). "Über die prinzipielle Bestimmbarkeit der berechtigten Bezugssytemebeliebiger Relativitätstheorien." *Annalen der Physik* **48**: 907–942, 943–982.

——— (1917). "Über den physikalischen Sinn der Relativitätspostulate, A. Einsteins neue und seine ursprügliche Relativitätstheorie." *Annalen der Physik* **53**: 575–614.

Lorentz, Hendrik A., Einstein, Albert, Minkowski, Hermann & Weyl, Hermann. (1923). *The Principle of Relativity: A Collection of Original Memoirs on the Special and General Theory of Relativity*. W. Perrett and G. B. Jeffery, trans. London: Methuen. [Reprint: New York: Dover (1952)].

Norton, John D. (1987). "Einstein, the Hole Argument and the Reality of Space." In *Measurement, Realism and Objectivity*. J. Forge, ed. 153–188. Dordrecht: Reidel.

——— (1989). "Coordinates and Covariance: Einstein's View of Space-Time and the Modern View." *Foundations of Physics* **19**: 1215–1263.

——— (1992). "The Physical Content of General Covariance." In *Studies in the History of General Relativity* (Einstein Studies, vol. 3). J. Eisenstaedt and A. J. Kox, eds. 281–315. Boston: Birkhäuser.

——— (1993). "General Covariance and the Foundations of General Relativity: Eight Decades of Dispute." *Reports on Progress in Physics* **56**: 791–858.

Rynasiewicz, Robert. (1994). "The Lessons of the Hole Argument." *British Journal for the Philosophy of Science* **45**: 407–436.

——— (1996). "Is There a Syntactic Solution to the Hole Argument?" *Philosophy of Science* **63** (Proceedings): S55–S62.

Stachel, John. (1980). "Einstein's Search for General Covariance." Paper delivered at the Ninth International Conference on General Relativity and Gravitation, Jena, Germany (DDR), 17 July 1980. [Quotations and page numbers from revised version in *Einstein and the History of General Relativity* (Einstein Studies, vol. 1). D. Howard and J. Stachel, eds. 63–100. Boston: Birkhäuser (1989)].

Point Coincidences and Pointer Coincidences: Einstein on the Invariant Content of Space-Time Theories

Don Howard

Since their rediscovery some fifteen years ago by John Stachel (Stachel 1980), Einstein's hole and point-coincidence arguments have come to play an important role in several different but not unrelated literatures. Thus, in line with Stachel's original interest, a proper understanding of the two arguments has led historians of general relativity to rewrite the story of Einstein's struggle with the problem of general relativity and gravitation, correcting the older view according to which Einstein's unfortunate turn away from generally covariant field equations in 1913 was the result of perhaps trivial mathematical error.[1] Stachel has also suggested that the hole and point-coincidence arguments play an important role in the history of mathematical physics as the principal stimulus to clarification of the Cauchy problem (see Stachel 1988). In the foundations literature, John Earman and John Norton (1987; see also Norton 1988) have used the hole argument as the basis for a new critique of space-time substantivalism, thereby reopening an old debate but in a manner that significantly advances the discussion by bringing to the fore the connection between the substantivalist position (or the various different versions of substantivalism) and assumptions about the individuation of space-time points. In the physics literature, the point-coincidence argument has been interpreted as demonstrating that we should do general relativity not on manifolds but on fiber bundles (Stachel 1986a, 1986b, 1993). Finally, in the newly developing literature on the history of the philosophy of science, Michael Friedman has claimed to find in Einstein's exposition of the point-coincidence argument

[1] For representative statements of that older view, see Earman & Glymour 1978a, 1978b, and Vizgin & Smorodinskii 1979. Norton 1984 is another important contribution to the revised history of Einstein's struggle with general relativity

H. Goenner, J. Renn, J. Ritter, T. Sauer (eds.)
The Expanding Worlds of General Relativity
(Einstein Studies, volume 7) pp. 463–500

the origin of the "modern observational/theoretical distinction" and therewith "the beginnings of the empiricist and verificationist interpretations of science characteristic of later positivism" (Friedman 1983: 24),[2] a theme that has also been explored by Thomas Ryckman (1992).

Friedman's claim will be disputed in the course of the present paper, which focuses on a pair of problems at the intersection of the ontological and epistemological implications of the point-coincidence argument. In brief, the question is: What kinds of point coincidences determine the invariant content of a generally-covariant space-time theory, and what is it about such coincidences that entitles them to play such a crucial role? More specifically, is the invariant content of a generally-covariant space-time theory determined wholly by the infinitesimal point coincidences that make up the set of all intersections of possible world lines, or is some, at least, of the invariant content determined also (or perhaps only) by what will be termed here 'pointer coincidences,' namely, the spatiotemporally extended intersections of the world tubes of 'observable' objects such as the needle on an electrometer and a mark on its scale? And whichever kind of coincidences play the role of determining the invariant content of a generally-covariant space-time theory, do they do so by virtue of their being invariant under arbitrary transformations or by virtue of their observability? I will argue that, on Einstein's understanding at least, it is only *infinitesimal point coincidences* that play such a privileged role and that they do so by virtue of their invariance, not their observability, such invariance being a necessary, but not a sufficient condition for the relevant kind of observability.

Why are these important questions? They are important, first and foremost, simply as part of our effort to understand Einstein's point-coincidence argument, which remains a rich wellspring of new ideas and insights even now, some fifteen years after the current revival of scholarly interest in it. That more is yet to be done by way of understanding the implications of the argument is evidenced by the fact that one still encounters in some of the literature on general relativity what is, in my opinion, too quick an association between invariance and observability, an association that traces its legacy back to earlier responses to the point-coincidence argument, and thus a mistaking of ontological questions for epistemological ones (see, for example, Komar 1958 and Rovelli 1991a, 1991b).

These questions about invariance and coincidences are important, as well, because of the stories now being told about alleged connections of an epistemological sort between the point-coincidence argument and the origins of logical empiricism. I believe that we have been misreading Einstein's remarks about the alleged observability of coincidences because, as the not-so-innocent victims of several generations of positivist propaganda, we have been too quick to read Einstein's talk of observability in verificationist terms. Whatever later logical empiricists themselves might have read into the argument, it is simply wrong to claim that the point-coincidence argument—as Einstein understood and intended it—contains, in germ, the modern observational/theoretical distinction and thus

---

[2] In fairness, it should be noted that Friedman has since retracted this claim (private communication).

the verificationist interpretation of science so prominent in later logical empiricism. Not only does such a reading of history miss the point (no pun intended) of the point-coincidence argument, it also misses what are, in my opinion, the more important, but also more tangled threads, linking the point-coincidence argument and later logical empiricism. These threads grow out of Einstein's reassertion, in the point-coincidence argument, of a methodological principle that was already asserted in the hole argument, a principle well known in Einstein's day by the name given to it in 1895 by the minor positivist, Joseph Petzoldt, das "Gesetz der Eindeutigkeit," or the principle of univocalness (Petzoldt 1895), which required that an adequate theory determine for itself a unique representation of that which it aimed to describe or depict (see Howard 1992). In brief, it was the growing realization, in the mid- to late 1920s that such a principle could not be satisfied by any reasonably interesting scientific theory when conceived axiomatically—owing the non-categoricity of any theory incorporating a mathematical apparatus as powerful as first-order Peano arithmetic—that led Carnap to adopt the neo-Machian constructionist program of the *Aufbau* (see Howard 1996).[3]

Finally, the question whether the invariant content of a space-time theory is exhaustively determined by point coincidences alone or by pointer coincidences, as well, is important if we are to understand some remarks by Einstein about the degree of underdetermination characteristic of space-time theories at the level of their basic, space-time event ontologies. If presumably directly observable pointer coincidences are accorded an equally fundamental role with point coincidences in determining the event ontology of a space-time theory, then it will prove difficult to understand why Einstein believed that *different* conceptions of 'reality' must be associated with the observable and the unobservable portions of a space-time theory's ontology, this because of the characteristic underdetermination, according to Einstein, of space-time theories at the level of their event ontologies. Making sense of Einstein's view of this matter is essential if we are to make sense of his larger philosophy of science.

The question here being, in the first place, a question of history and the interpretation of texts, let us begin there.

1. The origins of the point-coincidence argument

Since the point-coincidence argument and most of the relevant texts are already well known, I will review them only briefly (for further details, see Stachel 1980 and Norton 1984).

It was most likely in the late fall of 1913 that Einstein first wrongly convinced himself, via the hole argument, that, whatever other failings they were thought to possess, such as allegedly failing to yield the correct Newtonian limit for weak, static gravitational fields (Einstein & Grossmann 1913: 257), fully generally-covariant gravitational field equations also could not be correct because, so Einstein

[3] In a different way, Ryckman (1992) also seeks, commendably, to complicate the story of the connection between the point-coincidence argument and later logical empiricism

believed, they failed to satisfy the *Eindeutigkeit* principle. As he explained in his most lucid statement of the argument (Einstein 1914: 1067), given a 'hole,' a region of space-time, Σ, that is devoid of matter and energy, a coordinatization of the manifold, K, and a solution of the generally-covariant field equations expressed in terms of those coordinates, $g_{\mu\nu}(x_\nu)$, we can express the same solution, the same representation of the gravitational field, in terms of a new set of coordinates, K', that is identical with K everywhere outside Σ but comes smoothly to differ from K inside Σ, this transformed expression of that same solution being written $g'_{\mu\nu}(x'_\nu)$. Einstein then asks us to replace the x'_ν in the functions $g'_{\mu\nu}(x'_\nu)$ by the numerically equivalent values of the x_ν, yielding what is claimed to be a new and, for the region inside Σ, different solution of the field equations, $g'_{\mu\nu}(x_\nu)$.[4] Were the argument correct, the pathology would be severe, since, in general, numerically different values of the metric tensor, $g_{\mu\nu}$ and $g'_{\mu\nu}$, would thereby be assigned to one and the same space-time point, x_ν. Since the generally-covariant field equations would thus fail to determine a unique solution for the gravitational field inside the hole and therefore fail to satisfy the *Eindeutigkeit* principle, they must not be correct:

> *Processes* [das Geschehen] *in the gravitational field cannot be determined uniquely* [eindeutig festgelegt] *by means of generally-covariant differential equations for the gravitational field.* If we demand, therefore, that the course of events [*der Ablauf des Geschehens*] in the gravitational field be completely determined [*vollständig bestimmt*] by means of the laws that are to be established, then we are obliged to restrict the choice of the coordinate system. (Einstein 1914: 1067)

When, in November of 1915, Einstein finally returned to the generally-covariant field equations, having corrected his earlier misunderstanding about the Newtonian limit, he needed also a way around the hole argument's seemingly definitive disqualification of the generally covariant equations. Helped perhaps by some probing, critical questions about the hole argument posed in correspondence from the Göttingen physicist, Paul Hertz, in the summer and early autumn of 1915 (see Howard & Norton 1993), Einstein came to realize that, at the crucial step in the hole argument where he had inferred that, in general, the 'two solutions' of the generally-covariant field equations, $g_{\mu\nu}$ and $g'_{\mu\nu}$, assign different values of the metric to *one and the same point* within the hole, he had illicitly assumed that a mere coordinatization of the space-time manifold was sufficient for individuating the points of the manifold. The problem, of course, is that coordinatization does not yield an invariant scheme of individuation. The question then arose, how to individuate the points of the space-time manifold in an invariant manner. The answer was contained in the point-coincidence argument.

---

[4] That $g'_{\mu\nu}(x_\nu)$ is a solution of the field equations follows from the facts that (a) x_ν and x'_ν agree everywhere outside of Σ, so that in that region, $g'_{\mu\nu}(x_\nu)$ and $g'_{\mu\nu}(x'_\nu)$ are identical, and (b) inside Σ, the $t_{\mu\nu}$, which determine the $g_{\mu\nu}$, are everywhere 0 (Σ is a 'hole'). That, in general, $g'_{\mu\nu}(x_\nu)$ and $g_{\mu\nu}(x_\nu)$ differ follows from the fact that numerically equivalent values of x_ν and x'_ν label different points within Σ, so that the numerical value of $g'_{\mu\nu}$ originally assigned to the point x'_ν and now assigned to the point x_ν, will, in general, differ from the numerical value of $g_{\mu\nu}$ originally assigned to the point x_ν.

Einstein's new way of thinking about individuation and coordinatization was first hinted at in the second and third of the three communications to the Berlin Academy that announced the completion of general theory of relativity and the return to generally-covariant field equations in November 1915 (Einstein 1915a, 1915b, 1915c). Einstein's second paper, communicated on 18 November, contains the explanation of the anomalous precession of the perihelion of Mercury. Its opening paragraph includes what proved, for a short time, to be Einstein's favorite way of expressing his new attitude toward coordinatizations. Without any further explanation, he says that because of the introduction of generally-covariant field equations, "time and space are robbed of the last trace of objective reality" (Einstein 1915b: 831), a formulation echoed a few weeks later when, in announcing the new theory to Moritz Schlick in a letter of 14 December, Einstein says about the covariance properties of the field equations, "time and space thereby lose the last vestige of physical reality" (EA 21-610). The perihelion paper contains one further hint about Einstein's new way of thinking about coordinates and covariance. In introducing a solution of the generally-covariant field equations for a universe containing a single, central mass point like the sun, Einstein comments:

> It is well to bear in mind, however, that for a given solar mass the $g_{\mu\nu}$ are still not completely mathematically determined by equations (1) and (3). This follows from the fact that these equations are covariant with respect to arbitrary transformations with determinant 1. One may be justified, however, in assuming that all of these solutions can be reduced to one another by means of such transformations, and that (for the given boundary conditions) they therefore differ from one another only formally, but not physically. (Einstein 1915b: 832)

But what, exactly, does this imply about coordinatizations? In the concluding paragraph of his third paper, communicated on 25 November, Einstein observed: "The relativity postulate in its most general formulation, which turns the space-time coordinates into physically meaningless parameters, leads with compelling necessity [*zwingender Notwendigkeit*] to a completely determinate theory of gravitation" (Einstein 1915c: 847).

Time and space have lost the last vestige of physical reality. Coordinates have become physically meaningless parameters. All of this was sufficiently puzzling to Einstein's contemporaries to lead them to press him for further explanation, one of the first being Paul Ehrenfest, to whom Einstein proferred a long explanation in a letter of 26 December (referring to Einstein 1914):

> In § 12 of my work of last year, everything is correct (in the first three paragraphs) up to that which is printed with emphasis at the end of the third paragraph. From the fact that the two systems $G(x)$ and $G'(x)$, referred to the same reference system [the same x], satisfy the conditions of the grav. field, no contradiction follows with the uniqueness of processes [*Eindeutigkeit des Geschehens*]. That which was apparently compelling in these reflections founders immediately, if one considers that
>
> 1) the reference system signifies nothing real
>
> 2) that the (simultaneous) realization of two different g-systems (or better, two different grav. fields) in the same region of the continuum is impossible according to the nature of the theory.

> In place of § 12, the following reflections must appear. The physically real in the universe [*Weltgeschehen*] (in contrast to that which is dependent upon the choice of a reference system) consists *in spatiotemporal coincidences.** Real are, e.g., the intersections of two different world lines, or the statement that they *do not* intersect. Those statements that refer to the physically real therefore do not founder on any univocal [*eindeutige*] coordinate transformation. If two systems of the $g_{\mu\nu}$ (or in general the variables employed in the description of the world) are so created that one can obtain the second from the first through mere space-time transformation, then they are completely equivalent [*gleichbedeutend*]. For they have all spatiotemporal point coincidences in common, i.e., everything that is observable.
>
> These reflections show at the same time how natural the demand for general covariance is.
>
> *) and in nothing else! (EA 9-363)

This is the earliest surviving extended statement of the point-coincidence argument. Essentially the same argument is found in a letter of 3 January 1916 from Einstein to Michele Besso:

> In the *Lochbetrachtung*, everything was correct up to the final conclusion. There is no physical content in the existence of two different solutions $G(x)$ and $G'(x)$ with reference to the *same* coordinate system K. Attributing two different solutions to the same manifold is senseless, and the system K has, indeed, no physical reality. The following consideration takes the place of the *Lochbetrachtung*. From a physical point of view, nothing is *real* except the totality of spatiotemporal point coincidences. If, e.g., physical processes [*das physikalische Geschehen*] were to be built up solely out of the movements of material points, then the meetings of the points, i.e., the points of intersection of their world lines, would be the only reality, i.e., observable in principle. These points of intersection are naturally preserved under all transformations (and no new ones are added), if only certain uniqueness conditions [*Eindeutigkeitsbedingungen*] are maintained. Thus, it is most natural to demand of the laws that they do not determine *more* than the totality of the spatiotemporal coincidences. According to what has been said, this is already achieved by generally-covariant equations. (EA 7-272; Speziali 1972: 63–64)

Thus the point-coincidence argument. The central idea is that a generally-covariant theory of gravitation satisfies Petzoldt's principle of univocalness or *Eindeutigkeit* if we assume that the physcial reality it aims to describe consists exclusively of spatiotemporal coincidences or intersections of world lines, since such coincidences are themselves appropriately invariant under arbitrary transformations.

It was the invariance of the coincidences under arbitrary transformations that Einstein sought to illustrate in a second letter to Ehrenfest, which was evidently occasioned by Ehrenfest's continuing difficulty with the point-coincidence argument. Ehrenfest's side of the correspondence is now lost, but he seems to have proposed an example in which light from a star traverses a 'hole' of the kind posited in the hole argument and then passes through a slit in a screen, whereby it is directed to a specific spot on a photographic plate. On 5 January 1916, Einstein wrote to Ehrenfest:

I cannot blame you for the fact that you have not yet understood the admissibility of generally-covariant equations, because I, myself, took so long to achieve clarity on this point. Your difficulty is rooted in your instinctively treating the reference system as something 'real'.

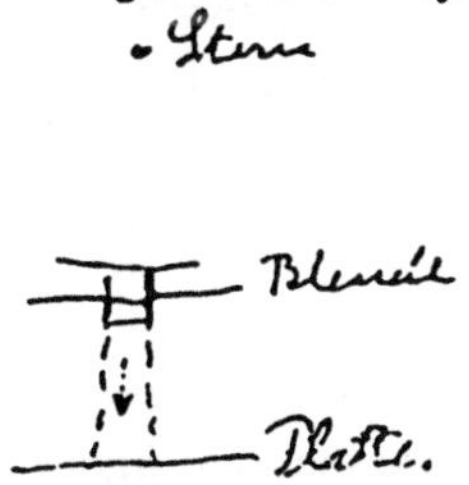

Your somewhat simplified example: You consider two solutions with the same boundary conditions at infinity, in which the coordinates of the star, the material point of the screen, and the plate are the same. You ask whether "the direction of the wave normal" at the screen always turns out the same. As soon as you speak of "the direction of the wave normal *at the screen*," you are treating this space with respect to the functions $g_{\mu\nu}$ as an infinitely small space. *This and the determinateness of the coordinates of the points on the screen entail that the direction of the wave normal at the screen is the same for all solutions.*

This is my claim. The following by way of more detailed explanation. In the following way you obtain all of the solutions that general covariance brings in its train in the above special case. Trace the little figure above on completely deformable tracing paper. Then deform the tracing paper arbitrarily in the paper-plane. Then again make a copy on notepaper. You obtain then, e.g., the figure

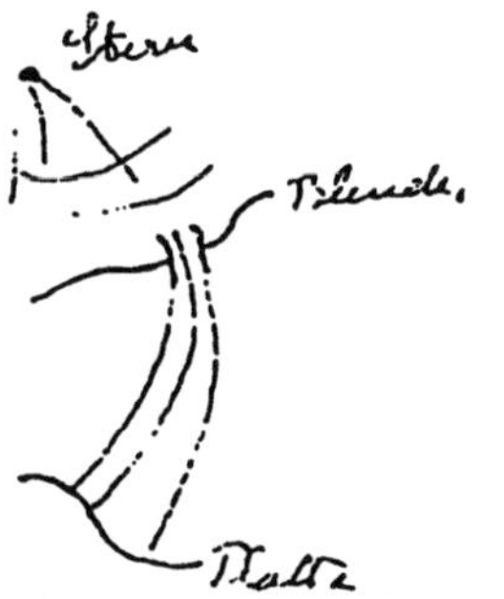

If you now refer the figure again to orthogonal notepaper-coordinates, then the solution is mathematically a different one from before, naturally also with respect to the $g_{\mu\nu}$. But physically it is exactly the same, because even the notepaper-coordinate system is only something imaginary [*etwas eingebildetes*]. The same points of the plate always receive light. If you carry out the distortion of the tracing paper only in a finite region and in such a way that the image of the star, the screen, and the plate remains undisturbed without violating continuity, then you obtain the special case to which your question refers.

What is essential is this: As long as the drawing paper, i.e., 'space,' has no reality, the two figures do not differ at all. It is only a matter of 'coincidences,' e.g., whether points on the plate are struck by light or not. Thus, the difference between your solutions *A* and *B* becomes a mere difference of representation, with physical agreement.

If the equations of physics were not generally covariant, then you could, to be sure, also make the above argument; but the same laws would not hold in the *second* figure, relative to the notepaper-system, as in the first. To that extent the two would not be equally justified. But this difference drops away with general covariance (EA 9-372)

In this construction, space is represented by the various drawing papers—both the tracing paper and the notepaper—upon which we can imagine coordinate charts being drawn. In its original version, the figure on the notepaper has coordinates—those given by the chart on the notepaper—implicitly assigned to the star (or rather the point where the light is emitted from the star), the point where the light from the star strikes the screen, and the point where that light strikes the photographic plate. Project this drawing (trace it) onto the tracing paper, distort the tracing paper as you will (maybe rubberized tracing paper would work best), and then project it (trace it) back onto the notepaper. In this final version of the figure, there are different notepaper coordinates for the star, the point where the light strikes the screen, and the point where the light strikes the plate. In this sense the final figure provides a mathematically different representation of the star, light, screen, and plate. But only the coordinate labels assigned to these point-coincidences have changed. Since the coincidences themselves remain unchanged—light still stikes the same points on the plate, as Einstein says—nothing physical has changed. 'Space' (or space-time), in the form of a coordinatization, has lost "the last vestige of physical reality," because changes in coordinate labels have no physical significance.

Soon we will focus more closely on one essential feature of the point-coincidence argument, namely, the claim that "*nothing* is real except the totality of spatiotemporal point coincidences" (emphasis mine), and on one feature that is arguably less essential but that has, as mentioned, grown in importance in the more recent secondary literature, namely, the assertion that such point coincidences are "observable in principle." But first let us pause to consider another feature of the argument that is not explicitly stressed by Einstein, which is that it yields, implicitly, a new, invariant scheme of individuation for the points of the space-time manifold. For if "nothing is real except the totality of spatiotemporal point coincidences," then the space-time events that constitute the space-time manifold, the fundamental ontology of the general theory of relativity, must be nothing other than such spatiotemporal point coincidences. Two space-time points will be accounted the same if and only if they are defined or constituted by the same spatiotemporal coincidence. Indeed, it is the availability of this alternative, invariant scheme of individuation that enables us to avoid the error of the hole argument. If, for example, a point of the space-time manifold in the 'hole' is defined as the intersection of two world lines, then since these world lines are determined by the metric, it would make no sense to speak of assigning different values of the metric to one and the same point; for if the value of the metric at 'that' point were to be different, then the world lines determined by the metric would be different and would not, in general, intersect as they did given the original assignment of a metric.

So far, I have hewn rather closely to the vocabulary employed by Einstein himself in propounding the hole and point-coincidence arguments. How, if at all, to translate these arguments into modern, coordinate-free language, is a matter of some dispute, into which we need not enter here. But for all that he lacked the modern apparatus of coordinate-free methods in differential geometry and topology, Einstein here again evinces the insight and sound instincts that we encounter in so many other contexts, especially in his deft development of the tracing-paper example, in which, by the seat of his pants, as it were, he comes quite close to a modern formulation of the problem.[5]

2. Point coincidences, pointer coincidences, and observability

The early formulations of the point-coincidence argument in Einstein's letters to Ehrenfest and Besso were not written for the public eye. The argument was first published in Einstein's 1916 review paper, "Die Grundlage der allgemeinen Relativitätstheorie," which was received by the *Annalen der Physik* on 20 March 1916, two and one-half months after Einstein's 6 January letter to Ehrenfest. It is found in § 3, where Einstein seeks to motivate the principle of general covariance. He first introduces the rotating-disk thought experiment for the purpose of showing that space and time cannot be defined in general relativity in such a way as to permit spatial and temporal differences to be directly measured by idealized rods and clocks, since the length of a rod and the rate of a clock depend upon where they are located (for example, at the center or the periphery of the disk). This being the case, there is no alternative, Einstein says, but to allow arbitrary coordinatizations, which comes to the same thing as requiring that "*the general laws of nature are to be expressed by equations which hold good for all systems of co-ordinates, that is, are co-variant with respect to any substitutions whatsoever (generally co-variant)*" (Einstein 1916a: 117). Any physical theory that is generally-covariant will therefore also satisfy the general relativity postulate, says Einstein, since the class of *all* such substitutions (read 'coordinate transformations') will surely include those corresponding to all relative motions of three-dimensional coordinate systems. Then comes the crucial passage:

> That this requirement of general co-variance, which takes away from space and time the last remnant of physical objectivity, is a natural one, will be seen from the following reflexion. All our space-time verifications [*Konstatierungen*] invariably amount to a determination of space-time coincidences. If, for example, events consisted merely in the motions of material points, then ultimately nothing would be observable but the meetings of two or more of these points. Moreover, the results of our measurings are nothing but verifications of such meetings of the material points of our measuring instruments with other material points, coincidences between the hands of a clock and points on the clock dial, and observed point-events happening at the same place at the same time.
>
> The introduction of a system of reference serves no other purpose than to facilitate the description of the totality of such coincidences. We allot to the universe four space-time variables x_1, x_2, x_3, x_4 in such a way that for every point-event there is a corresponding system of values of the variables $x_1 \ldots x_4$. To two coincident point-events there corresponds one system

[5] As a model for the case of a 4-dimensional space-time, the notepaper (graph paper?) plays, in effect, a role analogous to $\mathbf{R}^4$, with the tracing paper playing the role of the bare differentiable manifold. Deformation of the tracing paper is analogous to a diffeomorphic mapping of the manifold onto itself; the tracing operation—tracing from the tracing paper onto the notepaper—is analogous to a mapping of the bare manifold onto $\mathbf{R}^4$. For a more thorough comparison of Einstein's view with the modern coordinate-free view, see Norton 1989; for a different perspective, see Stachel 1993.

> of values of the variables $x_1 \ldots x_4$, i.e., coincidence is characterized by the identity of the co-ordinates. If, in place of the variables $x_1 \ldots x_4$, we introduce functions of them x'_1, x'_2, x'_3, x'_4, as a new system of co-ordinates, so that the systems of values are made to correspond to one another without ambiguity, the equality of all four co-ordinates in the new system will also serve as an expression for the space-time coincidence of the two point-events. As all our physical experiences can be ultimately reduced to such coincidences, there is no immediate reason for preferring certain systems of co-ordinates to others, that is to say, we arrive at the requirement of general covariance. (Einstein 1916a: 117–118)

It is this passage, in particular, that Friedman has in mind in arguing, as mentioned above, that the point-coincidence argument contains the origin of the "modern observational/theoretical distinction and therewith the beginnings of the empiricist and verificationist interpretations of science characteristic of later positivism" (Friedman 1983: 24). As we shall see in a moment, not all of Einstein's contemporaries gave this statement of the point-coincidence argument such a pronouncedly empiricist reading. Most interesting of all is Schlick, who, in fact, gave it an explicitly anti-empiricist interpretation. For now, compare this statement of the point-coincidence argument with those Einstein gave in his letters to Ehrenfest and Besso.

For those who wish to give the point-coincidence argument an empiricist reading it is noteworthy that the emphasis on the observability of coincidences has been there from the start, at least when it comes to coincidences such as the intersections of the world lines of "material points," although it is here that Einstein for the first time speaks explicitly about the meetings of the material points of measuring instruments, such as coincidences between the hands of a clock and points on its dial, the kinds of coincidences that I prefer to distinguish from literal point coincidences by the name 'pointer coincidences.' Otherwise, the earlier statements of the argument had a decidely less *epistemological* and decidedly more *ontological* flavor, with Einstein there stressing not the question of observability, but the question of physical reality. In those earlier versions of the point-coincidence argument, the invariance of coincidences is what qualifies them as part or all of physical *reality*, by contrast with space-time points individuated by coordinatizations, space and time thus understood losing "the last vestige of physical reality" because of the failure of coordinatizations to yield an invariant scheme of individuation. That space-time coincidences are also observable is mentioned on those earlier occasions almost as an afterthought.

Is the difference here merely one of emphasis, questions of ontology being stressed for one audience, questions of epistemology for another? Is there, perhaps, no difference at all? Does Einstein mean to be *defining* 'reality,' in good positivist fashion, in terms of observability? He did, after all, write in the 3 January 1916 letter to Besso: "If, e.g., physical processes were to be built up solely out of the movements of material points, then the meetings of the points, i.e., the points of intersection of their world lines, would be the only reality, i.e., observable in principle."

Consider, to begin with, a small but important terminological point. What most clearly invites a verificationist reading of the version of the point-coincidence argument published in the 1916 review paper in the *Annalen der Physik* is the crucial role of the expression "space-time verifications." But while the term *Konstatierung*, which is here translated as 'verification,' was later appropriated by Schlick in the context of the protocol-sentence debate as his preferred term for designating the confrontation between theory and experience (see Schlick 1934), the term had not acquired those explicitly verificationist connotations in 1916 and could just as well, indeed, might more properly be translated in that earlier context as something like 'determination,' thus removing much of the seeming epistemological content from Einstein's formulation of the argument.

Still, there is no denying the frequent gestures of respect to positivism in Einstein's correspondence and published writings of this period. Thus, the 1916 review paper from which the first published version of the point-coincidence argument was quoted above begins with Einstein explaining an "inherent epistemological defect" in both special relativity and classical mechanics, "which was, perhaps for the first time, clearly pointed out by Ernst Mach" (Einstein 1916a: 112). Einstein asks us to consider two fluid bodies of the same size and nature hovering freely in space, and so far apart that only internal gravitational effects are relevant. One of the masses rotates with respect to the other and one of the masses presents a somewhat flattened, ellipsoidal shape while the other is perfectly spherical. How are we to explain this difference? Einstein writes: "No answer can be admitted as epistemologically satisfactory, unless the reason given is an *observable fact of experience*" (Einstein 1916a: 112–113). Classical mechanics and special relativity fail this test because they proffer "*factitious*," unobservable causes as explanations, rotation relative to absolute space in the case of classical mechanics and rotation relative to an inertial frame in the case of special relativity.

Nor was this the only occasion on which Einstein expressed his respect for Mach's philosophy. Three months earlier, in the 14 December 1915 letter to Schlick from which I quoted above, Einstein complimented Schlick on his recently published essay, "Die philosophische Bedeutung des Relativitätsprinzips" (Schlick 1915), which Einstein had read the day before: "It is among the best that have until now been written about relativity. From the philosophical side, nothing appears to have been written on the subject that is at all so clear" (EA 21-610). He singled out one particular part of the essay for special praise:

> Quite correct also is your explanation of how positivism suggests the relativity theory without, however, requiring it. And you have seen correctly that this line of thought has been of great influence on my endeavors, especially E. Mach and, even more, Hume, whose treatise on the understanding I studied with eagerness and admiration shortly before the discovery of the relativity theory. It is very well possible that I would not have come upon the solution without these philosophical studies. (EA 21-610)

By this time Einstein had also made the acquaintance of Petzoldt, author of "Das Gesetz der Eindeutigkeit" (Petzoldt 1895), organizer in 1912 of the Gesellschaft für positivistische Philosophie (to which Einstein lent his name as a founding member), and for some time famous as the proponent of a "relativistic positivism," which took its life from a willful misreading of the relativity theory's introduction of the concept of the reference system as entailing a necessary reference to the subjective perspective of a sentient observer. Indeed, there is evidence that by 1915 Petzoldt was attending Einstein's Berlin lectures on relativity. In late 1914 or early 1915, Einstein wrote to Petzoldt about his recently-published *Das Weltproblem vom Standpunkte des relativistischen Positivismus aus, historisch-kritisch dargestellt* (Petzoldt 1912): "Today I have with great interest read your book in its entirety, and I happily infer from it that I have for a long time been your companion in your way of thinking" (EA 19-067). However credible the polite sentiments, the spirit of sympathy with positivism is clear.

Earlier still, in 1909, Einstein had begun a warmly respectful correspondence with Mach himself. In what is evidently his first letter to Mach, of 9 August 1909, he writes:

> Naturally, I am well acquainted with your principal works, of which I especially admire the one on mechanics. You have had such an influence on the epistemological views of the younger generation of physicists that even your current opponents, such as, e.g., Herr Planck, would undoubtedly have been declared to be 'Machists' by the kind of physicists that prevailed a few decades ago. (Einstein *CP5*: doc. 174)

Einstein alludes here to Max Planck's recent and widely-discussed attack on Mach in his Leiden lecture, "Die Einheit des physikalischen Weltbildes" (Planck 1908; see also Mach's reply, Mach 1910, and Planck's rejoinder, Planck 1910). In later letters, Einstein expresses sympathy for Mach in the wake of "Planck's unjustified criticism" (Einstein to Mach, 25 June 1913, Einstein *CP5*: doc. 448) and says how pleased he is "that the development of the theory brings to the fore the depth and importance of your investigations on the foundation of classical mechanics" (Einstein to Mach, December 1913, Einstein *CP5*: doc. 495).

But the record of Einstein's sympathies for positivism of the Machian variety by the mid-1910s is not unambiguous. To begin with, recall that later—much later, to be sure—Einstein characterized the aim of his early papers on the foundations of statistical mechanics and Brownian motion (Einstein 1902, 1903, 1904, 1905b, 1905c) as being "to find facts that would guarantee as much as possible the existence of atoms of definite finite size" (Einstein 1946: 45), adding that the empirical success of his theory of Brownian motion, together with Planck's derivation of Avogadro's constant from the radiation law "convinced the skeptics, who were quite numerous at that time (Ostwald, Mach), of the reality of atoms" (Einstein 1946: 45, 47). Einstein went on to explain the source of the skepticism:

> The hostility of these scholars toward atomic theory can undoubtedly be traced back to their positivistic philosophical attitude. This is an interesting example of the fact that even scholars of audacious spirit and fine instinct can be hindered in the interpretation of facts by philosophical prejudices. The prejudice—which has by no means disappeared—consists in the belief that facts by themselves can and should yield scientific knowledge without free conceptual construction. Such a misconception is possible only because one does not easily become aware of the free choice of such concepts, which,

> through success and long usage, appear to be immediately connected with the empirical material. (Einstein 1946: 47)

Can we trust reflections penned forty years later to represent accurately Einstein's attitude toward positivism in 1905 or 1915?

For generations, Einstein's critique of absolute distant simultaneity in his first paper on relativity (Einstein 1905d) was read, especially by logical empiricists eager to find august progenitors, as evincing a commitment to a strong observability requirement: A relationship of absolute distant simultaneity is to be proscribed because there is no way to determine it observationally. That is a persuasive reading. A more straightforward reading finds Einstein objecting not that the relationship is not observationally determinable, but that observation alone (say by simply comparing arrival times of light signals from distant events) does not yield an unambiguous assignment of simultaneity relations to distant events, since "such an assignment has the drawback that it is not independent of the position of the observer" (Einstein 1905d: 893). Einstein goes on to introduce his own unambiguous and "far more practical arrangement" with a procedure that has one "stipulate *by definition* [durch Definition *festsetzt*]" the equality of outbound and inbound light signals. A concern with ambiguities resolved by a definition sounds more like Henri Poincaré's conventionalism than Mach's positivism, a surmise only stregthened by the fact that Poincaré himself had introduced a strikingly similar procedure a few years earlier (Poincaré 1900: 272) in a paper that Einstein cites explicitly in his 1906 paper on the inertia of energy (Einstein 1906: 627; see also Howard 1993).

More to the point of Einstein's attitude toward positivism at the time of the completion of the general theory of relativity is his obituary notice for Mach published in the *Physikalische Zeitschrift* on 1 April 1916. Here, where one would have expected and does find Einstein's most explicit review of his debt to Mach, it is not Mach's phenomenalism nor his antimetaphysical attitude that Einstein praises. Einstein says that it is difficult but also not essential to try to distinguish Mach from John Stuart Mill, Gustav Kirchhoff, Heinrich Hertz, and Hermann von Helmholtz with respect to his "general epistemological standpoint." Mach's real legacy was his historical-critical approach to scientific concepts:

> The fact is that Mach exercised a great influence upon our generation through his historical-critical writings. ... Concepts that have proven useful in ordering things easily achieve such an authority over us that we forget their earthly origins and accept them as unalterable givens. Thus they come to be stamped as 'necessities of thought,' 'a priori givens,' etc. The path of scientific advance is often made impassable for a long time through such errors. For that reason, it is by no means an idle game if we become practiced in analyzing the long commonplace concepts and exhibiting those circumstances upon which their justification and usefulness depend, how they have grown up, individually, out of the givens of experience. By this means, their all-too-great authority will be broken. They will be removed if they cannot be properly legitimated, corrected if their correlation with given things be far too superfluous, replaced by others if a new system can be established that we prefer for whatever reason. (Einstein 1916b: 102)

The manuscript containing these remarks was received by the editors of the *Physikalische Zeitschrift* on 14 March 1916, six days before Einstein's major review article on the general theory of relativity was received by the editors of the *Annalen der Physik*, meaning that Einstein was thus distancing himself from the more problematic side of Mach's positivist epistemology at about the same time he was writing the above-quoted remarks about the observability of coincidences.

The key to understanding Einstein's attitude toward Mach is realizing that since at least 1909, the greater influence on Einstein's own epistemology had been not Mach, but Pierre Duhem (see Howard 1990). My best guess is that Einstein read Duhem's *La Théorie physique: son objet et sa structure* (Duhem 1906) in its German translation (Duhem 1908) by Friedrich Adler, Einstein's old Zurich student friend, in the fall of 1909, shortly after he returned from Bern to take up his first official academic appointment at the University of Zurich and became the upstairs neighbor and close intellectual friend of Adler. When he lectured on electricity and magnetism at the University of Zurich in the Winter semester of 1910/11, he made unmistakably clear that his epistemology was Duhem's rather than Mach's. The question was how to define the concept of the quantity of electricity in the interior of a solid, charged body, a region inaccessible to test bodies of the kind a strict Machian would think necessary for the desired definition. Einstein's answer:

> We set up a conceptual system the individual parts of which do not correspond directly to empirical facts. Only a certain totality of theoretical material corresponds again to a certain totality of experimental facts. ... But the force [on a unit charge inside the body] is no longer directly accessible to exp. It is one part of a theoretical construction that can be correct or false, i.e., consistent or not consistent with experience, only *as a whole*. (Einstein *CP3*: 325)

On my reading, Einstein's frequent remarks to the effect that scientific concepts and theories are the "free creations of the human spirit" (see, for example, Einstein 1921: 5) are rooted also in his commitment to a Duhemian holistic and underdeterminationist version of conventionalism. Not only does the holism and underdeterminationism imply the impossibility of any simple Machian reductionist account of the relation between scientific concepts and the "elements of experience," it also insures, as Duhem himself stressed, a role for the creative intellect, the free imagination of the theoretical physicist, precisely because our theories, not being reducible to experience, are therefore also not unambiguously determined by experience (see, again, Howard 1993).

That for which Einstein was truly indebted to Mach, therefore, was Mach's exhibiting—through historical-critical analysis—the historical contingency of theoretical constructions. Appreciation of such contingency was valuable to Einstein not, as for Mach (or, at least, his more simple-minded disciples), for the purpose of questioning the empirical legitimacy of those constructions, but for showing that, not being forced upon us by experience alone, those constructions are the free creations of the human spirit.

The evidence of Einstein's rather complicated attitude toward positivism by 1915 suggests that one should be slow to read his remarks about the observability of point coincidences and pointer coincidences as implying any straightforward endorsement of a Machian epistemology or an anticipation of the later logical empiricists' verificationist conception of the meaning of scientific terms. One might ask how Einstein's philosophically sophisticated contemporaries read such remarks. Who better with whom to start than Schlick?

3. Schlick on coincidences and reality

Though a philosopher by profession, Schlick was a physicist by training, having taken his doctorate in 1904 in Berlin under Planck. He turned to philosophy in the years 1908 to 1910, studying at the University of Zurich, and then, with encouragement from his friend, Max von Laue, he began studying relativity around 1912, publishing his first essay on the topic in 1915, the paper about which Einstein said such kind things in the above-quoted letter of 14 December. Perhaps emboldened by Einstein's praise, and the close personal and intellectual relationship with Einstein that was growing throughout 1916, Schlick turned his attention to the newly-completed general theory of relativity, publishing in 1917 the first serious philosophical study of the theory, *Raum und Zeit in den gegenwärtigen Physik* (Schlick 1917a and b).

Schlick announced the principal aim of his monograph in a letter to Einstein of 4 February 1917. Schlick was sending Einstein the manuscript of the first version, which was to appear in *Die Naturwissenschaften* in two installments, on 16 March and 23 March, asking for Einstein's critical comments. Schlick wrote that it is to be hoped that the general theory of relativity would soon be widely known and understood, "not merely for physical, but also especially for philosophical reasons." What might those philosophical reasons be? Schlick writes: "The essay is less a presentation of the general theory of relativity itself and more a thoroughgoing exposition of the proposition that, in physics, space and time have now forfeited all objectivity [*alle Gegenständlichkeit eingebüsst haben*]" (EA 21-568). This is just Schlick's paraphrase of Einstein's way of asserting the ontological insignificance of coordinatizations in general relativity. In other words, the overall aim of Schlick's essay is to explore the philosophical implications of the chief corollary to the point-coincidence argument.

Einstein's reaction to reading the manuscript was immediate and warmly positive. On 6 February he wrote to Schlick, "Your exposition is of unsurpassed clarity and perspicuousness" (EA 21-612). He added that Schlick's earlier essay on special relativity (Schlick 1915) was "splendid" and asked for a few more reprints, having already given away those Schlick had earlier given him, so that he could send some to "my friends in Zurich." On 21 March Einstein wrote again to correct "a slight inaccuracy" in Schlick's essay (EA 21-614) and on 1 April he wrote to invite Schlick to visit with him again so that they could discuss "the question of the constitution of space," remarking that "your splendid work has already imparted an understanding of the theory to many people, as I have convinced myself" (EA 21-616).

Soon after the essay's appearance in *Die Naturwissenschaften*, Schlick had turned it into a monograph that was published by Springer, adding a new, concluding section, "Relations to Philosophy." Bear in mind that this was written while Schlick was teaching at Rostock, six years before he made the move to Vienna (1922—there was a one-year interlude at Kiel, 1921–22) and thus long before he took a turn toward the antimetaphysical, logical empiricism for which he is now famous and which led to a break in his relations with Einstein by 1930. In 1915, Schlick was still a realist, but his was a realism of a curious sort.

The argument of the last, philosophical section of the monograph version of *Raum und Zeit in den gegenwärtigen Physik* was premised in important ways upon Schlick's distinctive version of an underdeterminationist view of the relation between theory and reality, a view already developed at length in his 1915 essay on relativity. Schlick's starting point was the semiotic theory of truth—quite influential in its day—that he developed in his 1910 essay on "Das Wesen der Wahrheit nach der modernen Logik" (Schlick 1910). According to Schlick, a proposition is true if and only if it is unambiguously coordinated with the unique fact(s) it aims to represent. Truth requires nothing more. As in the relationship between a sign and what it signifies, no picturing or other material realtionship based on content is needed. The correlation between proposition and fact is everything. If the correlation of proposition to fact is one-to-many, then the proposition is false. But many-to-one correlations of proposition to fact are possible without harm to the truth of the proposition. On the contrary, such many-to-one correlations are perhaps the norm. Schlick explained this point in detail in his 1915 essay on relativity:

> The totality of our scientific propositions, in word and formula, is in fact nothing else but a system of symbols *correlated* to the facts of reality; and that is equally certain, whether we declare reality to be a transcendent being or merely the totality and interconnection of the immediately 'given.' The system of symbols is called 'true,' however, if the correlation is completely univocal. Certain features of this symbol system are left to our arbitrary choice; we can select them in this way or that without damaging the univocal character of the correlation. It is therefore no contradiction, but lies, rather, in the nature of the matter, that under certain circumstances, several theories may be true at the same time, in that they achieve indeed a different, but each for itself completely univocal designation of the facts. (Schlick 1915: 149)

In 1917, Schlick asserted the same idea in similar terms:

> Every theory consists of a structure of concepts and judgments, and it is *correct* or *true* if the system of judgments is *univocally* correlated with the world of facts. ... It is, however, possible to indicate the *same* set of facts by means of *different* systems of judgments; consequently there can be different theories for which the criterion of truth holds in the same way, and which then do justice in equal measure to the observed facts and lead to the same predictions. They are even different systems of symbols that are correlated to the same objective reality, different modes of expression that reproduce the same set of facts. (Schlick 1917b: 61–62)

Having already been led to a kind of underdeterminationism by his reading of Duhem, Einstein would probably have been quite receptive to Schlick's position. But what elicited explicit reaction from Einstein was what Schlick had to say about the kind of reality to which a theory may be coordinated.

The issue first arises in the context of a discussion of Kantian theories of space and time, which for Schlick becomes a discussion of the psycho-genesis of intuitional, psychological space. Here is where one notion of 'coincidences' has its origin. How do we connect an element of optical space with an element of tactual space to form the concept of a 'point' in objective space? Schlick answers:

> Experiences of coincidences must be taken into consideration here. In order to determine a point in space, one must somehow, directly or indirectly, *point* to it [hinzeigen], one must make the point of a compass, or a finger, or a set of cross-hairs, coincide with it [*zur Deckung bringen*], that is, one sets up a spatiotemporal coincidence of two otherwise separated elements. Now it turns out that these coincidences always appear to agree for all the intuitional spaces of the various senses and for various individuals; for just that reason, an objective 'point' is defined by them, i.e., one independent of individual experiences and valid for all. ... Upon more careful reflection, one easily finds that we arrive at the construction of physical space and time exclusively by this method of coincidences and in no other way. The space-time manifold is precisely nothing other than the totality of objective elements defined by this method.
>
> This is the result of the psychological-empiriocritical analysis of the space and time concept, and we see that we encounter precisely *the* significance of space and time that Einstein recognized as alone essential for physics and there gave proper expression. For he repudiated the Newtonian conception ... and instead founded physics on the concept of the coincidence of events. (Schlick 1917b: 57–58)

This analysis of the concept of coincidences would seem wholly consonant with Einstein's talk of the observability of coincidences. Indeed, the space-time manifold has its psychological origin in the notion of observable coincidences.

But is this really the whole story? Hardly. Schlick notes immediately that there is one point in which scientific theory goes beyond the realm of psychological data. Permit me to quote him at length:

> Physics introduces the coinciding [*Zusammenfallen*] of two *events* as an ultimate, indefinable concept; but the psychogenetic analysis of the idea of objective space ends with the concept of the spatiotemporal coincidence [*Koinzidenz*] of two *elements of experience*. Are both simply the same?
>
> The strict positivism of a Mach asserts that they are. According to this view, the immediately experienced elements—colors, tones, pressures, warmths, etc.—are the only real. ... Where physics nevertheless speaks of other coincidences, there it is a question, according to Mach, only of abbreviated ways of speaking, economical auxilliary concepts, not realities in the same sense in which the sensations are realities. ...
>
> But this view is not the only possible interpretation of the state of affairs in science. If distinguished researchers in the exact sciences again and again declare that they are not satisfied by the strict positivistic world view, then the reason for that undoudtedly lies in the fact that none of the quantities appearing in physical laws designate 'elements' in Mach's

> sense; the coincidences that are expressed by the differential equations of physics are not immediately accessible to experience, they do not directly signify a coinciding of sense data, but rather, above all, a coinciding of non-intuitive [*unanschaulichen*] quantities, such as electrical and magnetic field strengths. Now nothing compels us to assert that only the intuitive elements—colors, tones, etc.—exist in this world; one can just as well assume that there are, in addition to them, elements or qualities that are not directly experienced and can likewise be designated as 'real,' whether or not they are comparable to those intuitive ones. (Schlick 1917b: 58–59)

The realistic tendency of Schlick's thinking here is clear.

There is no rigorous proof of the correctness or incorrectness of either of these two opposed points of view, says Schlick, but he cites a couple of reasons for why he is drawn to the "more realistic" view. First, the positivists' circumscription of reality seems to Schlick "arbitrary, indeed dogmatic." Second, the positivistic world view is unsatisfactory because of the way in which it "rips holes in reality" that are to be filled by "mere auxilliary concepts." "The pencil in my hand is supposed to be real, but the molecules constituting it mere fictions" (Schlick 1917b: 60). It is better, says Schlick, to assume that every concept that proves useful in the natural sciences may be regarded as a sign for something real. In this way we can achieve a worldview that satisfies the demands of the "realists," without sacrificing any of the advatages of the positivistic world view.

What are those advantages? Above all else, there is the fact that "the relation of individual theories to one another is properly recognized and evaluated" (Schlick 1917b: 61). What is that relationship? This is where Schlick rehearses his above-quoted exposition of his own version of underdeterminationism: *Different* theories can simultaneously give *true* accounts of the *same* objective reality, the same set of facts. One might wonder in what sense such a view of the relationship between theory and fact was also an advantage of positivism, for the semiotic theory of truth that grounds Schlick's view has so little in common with the reductionist phenomenalism of Mach, which by investing each scientific concept and proposition with its own determinate empirical content seems incompatible with any form of underdeterminationism.

Perhaps the point is that since scientific concepts and theories are, for the Machian, mere auxilliaries, tools, or *Hilfsbegriffe*, there may well be several different sets of tools that can accomplish any one job, considerations of economy giving us the only grounds for choosing among them. In fact, Schlick goes on to stress that even though more than one theory can "do justice in equal measure to the observed facts and lead to the same predictions," among all the theories that contain the same "kernel of truth" one must be "the simplest." But then he once again distances himself from Mach, by asserting that our preferring this simplest theory over its rivals is grounded "not merely on a practical economy, a kind of mental laziness, but ... has a logical basis in the fact that the simplest theory contains a minimum of *arbitrary* factors."

Schlick explains what he means by the simpler theory's being prefereable because it contains fewer arbitrary factors:

> The more complicated views necessarily contain superfluous concepts, with which I can do as I please, and which consequently are not determined by the facts under consideration. In the case of the simplest theory, by contrast, the role of each individual concept is demanded by the facts; it forms a symbol system with no indispensible additions. (Schlick 1917b: 62)

The same idea had been developed at even greater length in Schlick's 1915 paper on relativity. There he made even clearer the ontological implications of this doctrine:

> In very many cases, the greater simplicity of a theory rests upon its containing fewer arbitrary elements. ... But it is clear now that the greater the number of arbitrary elements that a theory contains, the more in it results from my willful choice, the less from what the facts compel. But naturally we must say that a theory represents reality only to the extent that it is determined just by the objective facts. ... Naturally, however, we want to exclude from our theories, as far as possible, not only the false but also the superfluous accessories, our own addition. We do this by choosing those with a minimum of arbitrary assumptions, that is, the simplest. Then we are certain to stray at least no farther from reality than required by the limits of our knowledge. (Schlick 1915: 154–155)

In the 1917 monograph, Schlick gives the example of the special theory of relativity, which is said to be simpler than Lorentz's ether theory, because the latter incorporates the concept of a privileged frame of reference without providing any physical means for picking it out, meaning that the choice of what to count as the absolute frame is arbitrary. Note that, on Schlick's reading, the concept of absolute space is not faulted here for its being unobservable, but for its being arbitrary, which calls to mind Einstein's argument in the 1905 special relativity paper that the concept of absolute distant simultaneity lacks objective significance because there is no non-arbitrary way to establish a system of simultaneity relations.

Are there any other interesting examples of such arbitrary and hence superfluous features of scientific theories?

> The concepts of space and time, in the form in which they have heretofore appeared in physics, are also to be counted among such superfluous factors—we have recorgnized this as a result of the general theory of relativity. They too find no application by themselves alone, but rather only insofar as they enter into the concept of the spatiotemporal coincidence of events. We may thus reiterate that they represent something real only in this combination, and not by themselves alone. (Schlick 1917b: 63)

By the time Schlick wrote these words, Einstein was clearly thinking along similar lines and no doubt educating Schlick. Thus, when he drafted his first popular exposition of relativity in late 1916, Einstein wrote:

> We refer the four-dimensional space-time continuum in an arbitrary manner to Gaussian coordinates. We assign to every point of the continuum (event) four numbers x_1, x_2, x_3, x_4 (coordinates), which have no direct physical significance at all, but only serve to number the points of the continuum in a definite but arbitrary manner. ... A one-dimensional line in the four-dimensional continuum thus corresponds to the material point. Lines of such a kind in our continuum correspond just as well to many points in

> motion. The only statements concerning these points that can lay claim to physical reality are in truth the statements about their encounters. (Einstein 1917: 64)

Einstein makes no explicit connection here with considerations of simplicity, but otherwise the implicit premise is the same as Schlick's: Only that which is not arbitrary in a physical theory can be held to represent physical reality. It is Petzoldt's principle of *Eindeutigkeit*—the univocal is the real in a physical theory's representation of reality. And in a space-time theory it is the point coincidences that are nonarbitrary and hence real.

4. Einstein's "two different peoples"

Einstein read the longer, monograph version of *Raum und Zeit in den gegenwärtigen Physik* sometime in the spring of 1917. He wrote to Schlick on 21 May: "Again and again I take a look at your little book and am delighted by the splendidly clear expositions. And the last section, 'Relations to Philosophy,' appears to me to be excellent." After a short comment on a technical point, Einstein turned his attention to what Schlick had written about both the Machian "elements of sensation" and unobservable physical events having to be counted as equally 'real':

> The second point to which I want to refer concerns the reality concept. Your view stands opposed to Mach's according to the following schema:
>
> Mach: Only impressions are real.
>
> Schlick: Impressions and events (of a phys[ical] nature) are real.
>
> Now it appears to me that the word 'real' is taken in different senses, according to whether impressions or events, that is to say, states of affairs in the physical sense, are spoken of.
>
> If two different peoples pursue physics independently of one another, they will create systems that certainly agree as regards the impressions ("elements" in Mach's sense). The mental constructions that the two devise for connecting these "elements" can be vastly different. And the two constructions need not agree as regards the "events"; for these surely belong to the conceptual constructions. Certainly only the "elements," but not the "events," are real in the sense of being "given unavoidably in experience."
>
> But if we designate as 'real' that which we arrange in the space-time-schema, as you have done in the theory of knowledge, then without doubt the "events," above all, are real.
>
> Now what we designate as 'real' in physics is, no doubt, the 'spatiotemporally-arranged,' not the 'immediately-given.' The immediately-given can be illusion, the spatiotemporally-arranged can be a sterile concept that does not contribute to illuminating the connections between the immediately-given. *I would like to recommend a clean conceptual distinction here.* (EA 21-618)

Schlick had argued that space-time events are just as real as the Machian elements and, presumably, real in just the same sense as the elements. Einstein agrees that both may be regarded as real, but he would have us distinguish two different senses of reality, depending on the degree of underdetermination attaching to each: an empirically underdetermined physical reality of space-time events and a wholly determinate phenomenal reality of sense impressions. Sadly, Einstein does not go on in this letter to explain further why it was important thus to distinguish two kinds of reality, so we must try to reconstruct his reasoning from context.

The broad context is the holistic view of the empirical content of theories and the associated thesis of the underdetermination of theories by evidence that Einstein first learned through his reading of Duhem's *La Théorie physique* and that was then reinforced and perhaps subtly modified by his reading of Schlick's works on the philosophical significance of relativity as well as his discussions of these issues with Schlick, both in person and in correspondence. This holism and underdeterminationism was to find ever more frequent and insistent expression in Einstein's writings during the late 1910s and early 1920s as he sought first to distance himself from neo-Kantianism and then from logical empiricism.[6] Consider, for now, just two representative examples of such expressions: In his address on the occasion of Max Planck's sixtieth birthday in 1918, Einstein wrote: "No logical path leads to these elementary [physical] laws. ... In this state of methodological uncertainty one can think that arbitrarily many, in themselves equally justified systems of theoretical principles were possible; and this opinion is, *in principle*, certainly correct (Einstein 1918b: 31);[7] and in a little-known article in the *Berliner Tageblatt* of 25 December 1919, he wrote: "The *truth* of a theory can never be proven. For one never knows that even in the future no experience will be encountered that contradicts its consequences; and still other systems of thought are always conceivable that are capable of joining together the same given facts" (Einstein 1919: 1; for a broader review of the place of Einstein's holism and underdeterminationism in his philosophy of science, see Howard 1993).

I take it that when Einstein speaks in his letter to Schlick about the "two different peoples" agreeing about the "impressions," but disagreeing about the "mental constructions," he means by this simply to assert, albeit in a poetical manner, the same general idea developed on these later occasions, namely, that in the domain of space-time theories there exist empirically-equivalent theories, a choice among which is underdetermined by the empirical evidence. More specifically,

---

[6] Einstein argued against the neo-Kantians that a holistic view of theories blocked any principled distinction between a priori and posteriori components of a physical theory and thus undercut a Kantian theory of science (see, for example, Einstein 1924a). The problem with logical empiricism was the atomistic conception of the empical content of theories that we find well developed in the writings of Schlick, Hans Reichenbach, and others by the late 1920s and early 1930s. On this latter view, empirical propositions are distinguished from coordinating definitions in such a way that, once the coordinating definitions associated with a theory are fixed by convention, each of the theory's empirical propositions is vested with a determinate empirical content sufficient to allow the truth or falsity of the proposition to be determined unambiguously by experience. As late as 1949, in his reply to Reichenbach's contribution to the Schilpp volume, *Albert Einstein: Philosopher Scientist* (Reichenbach 1949), Einstein was still going out of his way to criticize such a conception, starting from the point of view of his own holistic conception of empirical content (Einstein 1949: 677–678). The irony is that Schlick was one of those who influenced Einstein in the direction of holism and underdeterminationism, but then became a target of Einstein's criticism for his having abandoned that view. For more on the role of Einstein's holism in his dispute with the neo-Kantians and in his split with Schlick, see Howard 1994.

[7] Of course the quoted passage from Einstein 1918b continues with the well-known, if problematic remark that "in practice, the world of perceptions determines the theoretical system unambiguously, even though no logical path leads from the perceptions to the basis principles of the theory." For a more thorough anlaysis of what Einstein meant by underdetermination in principle and determination in practice, see Howard 1990.

since what I have been calling pointer coincidences are to be paradigmatically 'observable,' they must fall on the 'impressions' side of the impressions/events distinction. Thus, if all observations are ultimately reducible to coincidences of the pointer variety, and if space-time events are individuated as point coincidences, then Einstein is saying that the pointer coincidences underdetermine the physical constructions that we make employing point coincidences. In short, pointer coincidences underdetermine point coincidences.

What, however, is the nature of the underdetermination that Einstein has in mind? Surely it is not just the arbitrary choice of a coordinatization itself. For while the choice of a coordinatization is not dictated by empirical considerations and may, instead, be driven by simplicity considerations, as in the case of our choice among empirically-equivalent theories, Einstein and Schlick were both adamant about there being no physical reality attaching to a coordinatization—that was the lesson of the point-coincidence argument—whereas the underdetermination at issue in the quoted letter to Schlick concerns precisely the 'reality' posited in two different 'mental constructions.' As Einstein says, "above all" it is the "events," presumably individuated as point coincidences, that are "real" in the second of the two senses he seeks to distinguish, where what distinguishes this second sense of the "real" is just that "two constructions need not agree as regards the 'events'; for these surely belong to the conceptual constructions."

So the kind of underdetermination that Einstein has in mind affects primarily the space-time events, individuated as point coincidences, that constitute the fundamental manifold ontology of a space-time theory like general relativity. One might find it ironic that the space-time events thus individuated, which, as invariant entities, play the privileged role *within* a space-time theory of representing reality (by contrast with a coordinatization), are at the same time the parts of a theory most affected by underdetermination. Just as Schlick was untroubled by the possibility that "several theories may be true at the same time," so too Einstein seems not to have been troubled, as many contemporary scientific realists would be, by the idea of underdetermined representations of physical reality. Still, what is the nature of this underdetermination? How can it be that "two constructions" do not agree "as regards the events"?

One might imagine radically different ways of doing physics that are not space-time theories in the sense of structures on a space-time manifold. Quantum mechanics would qualify as such an alternative physics. But I do not think that this is what Einstein meant by different "constructions," since the difference is said to be a difference "as regards the events" and there are no space-time 'events' in quantum mechanics because, there being no well-defined trajectories in quantum mechanics, there are no world lines whose intersections could be taken to constitute the 'events.' Moreover, Einstein believed that it was virtually an a priori condition for the very possibility of an objective science that it be based on a space-time representation of nature, since he could imagine no other objective scheme for the individuation of physical systems except on the basis of their being separated by a non-null spatiotemporal interval (see Howard 1997).[8]

One might also imagine two empirically-equivalent space-time theories that differ at the global topological level and hence differ with respect to what one might term, to coin a phrase, the 'global catalogue of coincidences,' or the pattern of individuation of space-time events (see Sklar 1974: 312–314). Global topological underdetermination is not what was being denied by those neo-Kantians of the early 1920s, like Ernst Cassirer and the young Rudolf Carnap, who argued that the lesson of general relativity was that while metrical structure could no longer be accorded a priori status, topological structure must be granted such a status, because topological relations are invariant under the general relativistic class of transformations, for what they were speaking about was the univocal determination of the local topological structure of space-time theories (see Cassirer 1921 and Carnap 1921). But even if the argument did bear upon global topological structure, it likely would not have moved an Einstein whose holism led him to deny the a priori/a posteriori distinction and, with it, a Kantian theory of science. Nevertheless, underdetermination at the level of global topology seems not to have figured prominently in Einstein's thinking, certainly not in the late 1910s and early 1920s, perhaps because at least the standard examples of such underdetermination all have the flavor of ontologically and epistemologically uninteresting mathematical tricks, such as replacing a time-cylindrical world with closed, timelike world lines by an infinitely repeated set of temporal strips, where one has disindividuated the points that now appear at the boundaries of the strips (Sklar 1974: 313–314).

Perhaps a better guess as to what Einstein intended is that he was thinking about the kind of underdetermination of the *metric* that he discussed at length four years later in his essay, *Geometrie und Erfahrung* (Einstein 1921). There he first developed the argument that went on to become standard in all the literature on the conventionality of the metric, even if Einstein's understanding of the argument did not survive in these later rehearsals: The metric that we ascribe to a region of space-time depends upon our choice of a unit measuring rod, a choice that is, in turn, not forced upon us by any empirical considerations. Choose a different meter stick and get a different metric.

What is the significance of the resulting differences in the metric? The answer given by the logical empiricists, following Reichenbach's lead, was: "Not much of a difference at all." On Reichenbach's interpretation, our choice of a meter stick had the status of a conventional coordinating definition (Reichenbach 1928: 23–29). Fix the coordinating definitions by conventional stipulations and the truth or falsity of each remaining empirical proposition in a theory is determined unambiguously and individually by the empirical evidence. If the resulting difference in the metric is thus traceable to a *mere* choice of coordinating definition, then it is no difference at all, no more a difference than that between saying *il pleut* and *es regnet*, or, better still, between *до свидания* and *do svidaniya*. Two theories that differ by nothing

[8] On a few occasions, however, Einstein did speculate about whether or not the problem of the quantum would force us to abandon the mathematics of the continuum as a framework for fundamental physics, in favor of an "algebraic" framework; see Stachel 1986c. Such a change would probably also entail the abandonment of a spatiotemporal scheme of individuation.

more than a conventional coordinating definition share the same empirical content and thus say the same thing.

The holistic conventionalism favored by Einstein does not permit a systematic distinction between coordinating definitions and empirical propositions and consequently takes far more seriously the differences between two empirically-equivalent theories. They might share the same total empirical content, but if one is not a verificationist about meaning, if one does not think that the truth or falsity of each proposition constituting a theory is separately determined by the facts corresponding to its own, individual empirical content, and if one thinks that the meanings of scientific terms and theories extend beyond the empirical to a rich ontology of unobservables, then the differences between empirically-equivalent theories may be vast; indeed, two such theories may tell mutually inconsistent stories about states of affairs at the level of that deep ontology.

Even for Einstein, the difference between two coordinatizations is like the difference between two alphabets, which is to say, no difference at all, but then no physical reality attaches to a coordinatization. The difference between two metrics, consequent on the choice of different meter sticks, is much more of a difference. Is it, however, a difference such as that between two "constructions" that do not agree "as regards the events"?

The assignment of a different metric because of the choice of a different meter stick cannot change the coincidences and thus does not change the topological structure of the manifold. Two world lines that intersected under the original characterization of the metrical structure will still intersect. In this respect, the assignment of a different metric entails no change in the 'reality' represented by the theory, which may be the whole story if the reality we seek to describe were exhausted by the events individuated as point coincidences. A space-time theory requires more, however, than a manifold of space-time events. If we are to *explain* anything with such a theory, we need a structure on that manifold, at the very least a metrical structure, if not other fields as well. But the assignment of a different metric will lead to changes in what count as geodesics and so will lead to changes in how we explain the motion of bodies: A body whose world line was a geodesic under the first assignment of a metric, and thus was represented as freely falling, may be represented under the new assignment of a metric by a non-geodesic world line, and thus be represented as moving under the influence of a force. If 'reality' includes not just the manifold but also the structures defined on the manifold, then the assignment of different metrics because of the choice of different measuring rods does amount to a different picture of reality.

For the moment, then, let us take Einstein's term, "construction," to designate an assignment of a metric in consequence of a choice of a meter stick and the associated division of the structure on the space-time manifold into metrical structure and the structure associated with whatever other fields have to be introduced to explain deviations from what count, given that metric, as geodesic trajectories. Given any one such "construction," there will be an invariant content associated with the structure(s) that that "construction" defines on the manifold. Two such "constructions," however, may divide the metrical and non-metrical structure in different ways. Is this what Einstein meant in the "two different peoples" letter?

Again, I think not. However one might divide the structure on the manifold into metrical and non-metrical pieces, what determines the corresponding invariant content of space-time is still the spatiotemporal coincidences, paradigmatically the coincidences of world lines, for, as Einstein says, in consequence of the point-coincidence argument, "every physical description resolves itself into a number of statements, each of which refers to the space-time coincidence of two events *A* and *B*" (Einstein 1917: 65). But dividing up the structure on the manifold in different ways between metrical and non-metrical structure does nothing to change the catalogue of coincidences. If Einstein's point was to distinguish sense impressions from space-time events on the basis of the high degree of determination of the former and the underdetermination of the latter, then we have still not found that distinction.

My best guess is that Einstein meant something far more subtle than any of the interpretations ventured so far. The underdetermination that we are seeking must concern the space-time events themselves if there is to be an interesting contrast with sense impressions. We have considered space-time events individuated as coincidences, in particular, as intersections of world lines, without, however, having considered carefully the question, "World lines of what"? A common answer is "test particles" and light rays. But how does one decide what counts as a test particle? Ordinarily we imagine a chargeless, point-like particle, with a mass that is finite but small enough so that it does not induce significant local deformations of the metric. Surely there is some vagueness and arbitrariness in this conception. How large, how heavy may a particle grow before it no longer qualifies as a test particle? More importantly, why a chargeless particle? Assumimg that test particles are chargeless means that we are begging important questions of the same kind that were raised just above in connection with different ways of dividing structure between the metrical and the non-metrical as a result of the choice of a different measuring rod. Why not make no assumption at the outset about whether or not our test particles are charged or uncharged? Finally, one must ask about the implicit criteria for the individuation of test particles. How does one decide, when the world lines of two identical particles, *A* and *B* intersect, that it is particle *A*, say, that continues seemingly uninterruptedly, as opposed to particle *B*'s having been deflected along that path?

One might also ask similar questions about the kinds of particles that follow light-like trajectories in space-time. Why are they assumed to be responding only to metrical structure and not non-metrical structure? And how are they individuated?

A different choice of test particles and different assumptions about particle individuation will yield a different set of world lines, a different catalogue of coincidences, and thus a different space-time event ontology. Is this what Einstein had in mind? Before answering that question, let us pause to reiterate an important fact about the underdetermination of space-time events here proposed. It is that *pointer coincidences* are in no way involved here. Since pointer coincidences belong to the realm of the directly observable, they fall on the "impressions" side of the impressions/events distinction that Einstein urges in our conception of the real and thus suffer from no underdetermination whatsoever. Stipulation of a different kind of test particle will not reveal itself in the reading of any voltmeter.

In some respects, the stipulation of what is to count as a test particle is similar in kind to the stipulation of what is to count as one's meter stick. This is true both epistemologically, in that the stipulation is not forced upon us by any empirical considerations, and physically, in that different choices of both test particles and meter sticks lead, among other things, to differing divisions of structure on the manifold between metrical and non-metrical structure. The two kinds of stipulations are similar also in one additional respect, which is that if and when a space-time theory is elaborated into a full-blown theory of the microstructure of matter and radiation, say in the form of Einstein's hoped-for unified field theory, then the need for arbitrary stipulations of meter sticks and test particles may well be obviated by the theory's entailing the structure of such objects. Such theories would still be underdetermined by the totality of the empirical evidence, but in such a way that made the aspect of holism still more evident by the way in which the physics and the geometry or the physics and the topology would be inextricably intertwined.

Einstein said this explicitly about the meter sticks in his *Geometrie und Erfahrung*:[9]

> *Sub specie aeterni* Poincaré is right, in my opinion. The concept of the measuring-rod and the concept of the clock coordinated with it in the theory of relativity do not find an exactly corresponding object in the real world [there are no perfectly rigid rods]. It is also clear that the solid body and the clock do not play the role of irreducible elements in the conceptual edifice of physics, but that of composite structures, which may not play any independent role in theoretical physics. But it is my conviction that in the present stage of development of theoretical physics these concepts must still be employed as independent concepts; for we are still far from possessing such certain knowledge of the theoretical foundations as to be able to give theoretical constructions of such structures. (Einstein 1921: 8)

Exactly the same thing could be said about the role played by the concept of the test body. Lacking a fundamental theory of the structure of matter and radiation in a space-time context, test bodies must be introduced as logically-independent posits. Eventually, however, one would hope to have their structure determined as part of a solution to the field equations of a unified field theory.

Helpful evidence that we are on the right track in seeking an interpretation of Einstein's "two different peoples" letter is to be found, ironically, in the closing pages of Reichenbach's *Philosophie der Raum-Zeit-Lehre*. I say "ironically," because, as noted above, by the time Reichenbach published this in 1928, he and Einstein had diverged considerably in their understanding of the role of convention in science, Reichenbach wanting to restrict the role of convention to the choice of coordinating definitions, Einstein favoring a more holistic view that did not respect a principled distinction between coordinating definitions and empirical propositions. But we should remember that when Reichenbach finished writing

[9] See also Einstein's letter to Walter Dällenbach, November 1916, EA 9-072

his *Philosophie der Raum-Zeit-Lehre* in October of 1927, he had been Einstein's colleague in the physics department at the University of Berlin for a year, occupying a special chair in the philosophy of science that Einstein and Planck had helped to create for him (Hecht & Hoffmann 1982). Einstein and Reichenbach were thus in regular, close contact, making it safe to assume that they had ample opportunity to discuss fundamental questions of the kind under consideration here. The passage in question, which I quote at length, reads almost as if it were an explication of and commentary upon Einstein's "two different peoples" letter.

Reichenbach is at this point concluding his discussion of the role of topology in a space-time theory, arguing, not unlike Cassirer and Carnap before him, that local topological structure, unaffected by any changes in the metric, establishes the order relationships of events in space-time and thus undergirds the causal structure of space-time. He then comments on neo-Kantian explanations of the objectivity of topological structure:

> Other attempts have been made to explain the topology of space and time. The coordinate system assigns to the system of coincidences, of point-events, a mutual order that is independent of any metric. This order of coincidences must therefore be understood as an ultimate fact. The attempt has been made to justify this order as necessary; it has been regarded as a function of the human perceptual apparatus rather than of the objective world. Accordingly, it has been claimed that sense perceptions supply directly only coincidences, and that the ultimate element of space-time order is determined by the character of our sense perceptions. In this connection appeal is made to the experimental methods of the physicist, in which coincidences of dials and pointers play an important role.
>
> This view is untenable. First of all, it is obvious that we cannot regard the order of coincidences as immediately given, since the subjective order of perceptions does not necessarily correspond to the objective order of external events. It can serve only as the basis of a complicated procedure by which the objective order is inferred. ... We must therefore introduce rules for the construction of the objective order; such rules have been formulated in the topological coordinative definitions.
>
> It is a serious mistake to identify a coincidence, in the sense of a point-event of space-time order, with a coincidence in the sense of a sense experience. The latter is *subjective coincidence*, in which sense perceptions are blended. ... The former, on the other hand, is *objective* coincidence, in which physical things, such as atoms, billiard balls or light rays collide and which can take place even when no observer is present. The space-time order deals only with objective coincidences, and we go outside the realm of its problems in asking how the system of objective coincidences is related to the corresponding subjective system. The analysis of this question belongs to that part of epistemology that explains the connection between objective reality, on the one hand, and consciousness and perception on the other. Let us say here only that any statement about objective coincidences has the same epistemological status as any other statement concerning a physical fact.
>
> It is therefore not possible to reduce the topology of space and time to subjective grounds springing from the nature of the observer. On the contrary, we must specify the principles according to which an objective coincidence is to be ascertained. This means that we must indicate a method how to decide whether a physical event is to be considered as one, or as two

> or more separate point events.[10] (Reichenbach 1928: 285–286)

After reminding the reader of how he had earlier illustrated the use of the principle of contact action in deciding whether to regard an event as one point-event in a two-dimensional space or two point-events in a one-dimensional space, Reichenbach continues:

> Objective coincidences are therefore physical events like any others; their occurrence can be confirmed only within the context of theoretical investigation. Since all happenings have until now been reducible to objective coincidences, we must consider it the most general empirical fact that the physical world is a system of coincidences. It is this fact on which all spatio-temporal order is based, even in the most complicated gravitational fields. What kind of physical occurrences are coincidences, however, is not uniquely determined by empirical evidence, but depends again on the totality of our theoretical knowledge. (Reichenbach 1928: 287)

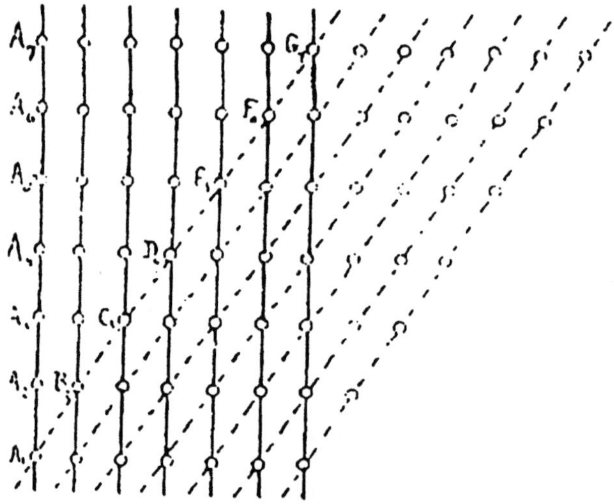

A few pages earlier, in the context of a discussion of the concept of genidentity and causal order, Reichenbach gives a helpful illustration of the empirical underdetermination of coincidences. It concerns what are, in effect, photon trajectories (though, of course, the same argument would hold for any boson). Photons are indistinguishable particles. As in the figure above, if one has a region of space-time continuously filled with a field like an electromagnetic field carried by indistinguishable field particles, how does one decide what count as different points along a single field particle (photon) trajectory as opposed to neighboring points on different field particle (photon) trajectories? One choice is as good as another, as

---

[10] In criticizing those who wanted to conflate spatiotemporal point coincidences with coincidences in sense experience, Reichenbach may also have had in mind Petzoldt's writings of the 1920s, wherein creative misinterpretations of relativity theory were advanced in support of a neo-Machian point of view that he called "relativistic positivism." On the specific subject of coincidences, see for example, Petzoldt 1921: 118.

long as one restricts ones choice to timelike world lines.[11] The arbitrariness in the determination of the field particle or photon world lines, which Reichenbach calls an arbitrariness in the "striation [*Zerspaltung*] of the world lines of the material field" (Reichenbach 1928: 271), would induce an arbitrariness in the determination of coincidences and hence in the event ontology. No such arbitrariness would be encounted with certain other "material fields," such as those corresponding to "atomic matter," for which there is a "natural striation," presumably because the constitutive particles are not indistinguishable from one another. For atomic matter, therefore, the concept of genidentity is still well defined.

Reichenbach attributes this construction involving identical particles to Einstein, who is supposed to have seen in it "the collapse of the old concept of substance," adding that "Einstein thus wishes to characterize the metrical field that propagates gravitational forces" (Reichenbach 1928: 270–271). In fact, Einstein's purpose in the work cited by Reichenbach, his 1920 address, *Äther und Relativitäts-theorie* (Einstein 1920), was, on the face of it, somewhat different. There the point was to argue that the classical conception of the electromagnetic ether had to be abandoned in relativity theory, or rather that it could be maintained only if we did not ascribe to it a definite state of motion in the sense of postulating the existence of a privileged frame of reference, the ether frame. But we do just this—that is, we implicitly assume the existence of a privileged ether frame—if we think of the electromagnetic field as consisting of separate "material" lines of force, each of which can be tracked through the course of time (Einstein 1920: 10). Einstein then elaborates:

> Generalizing, we must say the following: We may conceive of extended physical objects to which the concept of motion cannot be applied. They may not be thought of as consisting of particles that allow themselves to be tracked individually through time. In Minkowski's idiom this is expressed as follows: Not every extended structure can be conceived as being composed of world threads [*Weltfäden*]. The special principle of relativity forbids our assuming that the ether consists of particles that can be tracked through time, but the special theory of relativity is not in conflict with the ether hypothesis in itself. Only one must be careful not to ascribe a state of motion to the ether. (Einstein 1920: 10)

In place of the electromagnetic ether—the "Lorentz ether"—we have the "ether of the general theory of relativity," namely, the gravitational field itself, in the form of the space-time manifold with a metrical field defined upon it. Einstein stressed the fundamental difference between the gravitational field and the electromagnetic field: We can imagine a region of space without an electromagnetic field, but we cannot imagine a region of space without a gravitational field, for the latter is what endows space with its metrical properties, "without which it cannot, after all, be

---

[11] On the restriction to timelike world lines, see Reichenbach 1928: 271. Of course this way of formulating a restriction on the "slope" of the world lines of "material field particles" assumes that the conformal or causal structure on the manifold is given independently of the specification of photon trajectories. How that is to be done if we have not already assumed some metrical structure is not explained by Reichenbach, which is especially puzzling in a program such as his, where theoretical primitives, such as "lightlike world line" acquire empirical content only via explicit coordinating definitions.

conceived" (Einstein 1920: 13–14). But of this ether too Einstein says that it "may not be thought of as being endowed with the characteristic property of ponderable media, which is consisting of parts that are trackable through time; the concept of motion may not be applied to it" (Einstein 1920: 15).

Nowhere in *Äther und Relativitäts-theorie* does one find the specific construction discussed by Reichenbach. The germ of such an idea might, however, be sought in Einstein's assertion that the electromagnetic field and other extended object fields to which the concept of a well-defined state of motion cannot be applied on pain of assuming the existence of a preferred frame therefore "may not be thought of as consisting of particles that allow themselves to be tracked individually through time," since so conceiving such fields would entail ascribing to them a definite state of motion, namely that of particle-like carriers of the field. I take it that what Einstein concluded from this in 1920 was simply that the electromagnetic field (and the gravitational field) should be viewed as a reality in its own right, not reducible to particle-like carriers of the field. He said near the end of *Äther und Relativitäts-theorie*, "thus our current world picture recognizes two realities that are conceptually completely separate from one another, even if they are causally connected with one another, namely, the gravitational ether and the electromagnetic field or—as one might also call them—space and matter." Far from wanting to reduce electromagnetic fields to particle-like carriers of the field, Einstein said that, "according to current views," the elementary particles of matter are essentially "condensations [*Verdichtungen*] of the electromagnetic field" (Einstein 1920: 14). A future unified field theory might reduce these two realities to one, but it cannot portray even this unified field as being carried by temporally trackable field particles.

That was in 1920. By the mid-1920s, Einstein's thinking about the nature of radiation and the way it lives in space-time had changed, thanks largely to the work of Satyendra Nath Bose, who demonstrated that the Planck distribution formula for black-body radiation could be derived by applying to a photon gas a peculiar non-Boltzmannian statistics based upon the assumption that photons are indistinguishable particles (Bose 1924). Einstein himself translated Bose's article for publication in the *Zeitschrift für Physik* and then extended Bose's method to material gases (Einstein 1924b, 1925a, 1925b). It would not be surprising if Einstein saw the connection between the argument of *Äther und Relativitäts-theorie* and Bose's way of regarding the photon gas.

In 1920, Einstein was most likely still assuming that it was in the nature of all particles that they "allow themselves to be tracked individually through time," and thus disallowed a particle interpretation of the electromagnetic and gravitational fields. But after Bose had showed that photons not only could be regarded as indistinguishable particles but must be so regarded in order to derive the Planck formula, Einstein could easily have seen that his 1920 argument could also be used to show that, according to relativity theory, quite independendly of the quantum theory, the only acceptable particle interpretation of the electromagnetic field would be one in which the field particles were identical and indistinguishable. Only thus could one give a particle interpretation to the electromagnetic field without implicitly attributing to it a preferred state of motion, since the underdetermination in the ascription of world lines to the field particles that follows from regarding them as indistinguishable would mean that no *unique* state of motion was being associated with the electromagnetic field.

We have reached a remarkable conclusion: A purely relativistic argument for the indistinguishability of the field particles. The condition for the possibility of giving an acceptibly *relativistic* particle interpretation of the electromagnetic field or any other field is thus that the field particles be indistinguishable and hence obey non-Boltzmannian statistics. Particle indistinguishability is necessary because it induces an underdetermination in the way we lay particle world lines into space-time, meaning that we do not tacitly ascribe a privileged state of motion to the field carried by these particles. Such underdetermination in turn induces an underdetermination in the coincidences of these world lines and thus also in the event ontology of space-time itself.

We see then that the empirical underdetermination of point coincidences that was stressed by Reichenbach and earlier emphasized by Einstein as the distinguishing feature of the kind of reality unique to a space-time theory's event ontology is not merely a trifling special case of a larger genus of empirical underdetermination. Instead, it is a fact about space-time event ontologies of fundamental significance for understanding the kinds of invariant structure that can live in a space-time. Surely not all of this was evident to Einstein in November and December of 1915. Still, if we misread the point-coincidence argument that Einstein invented at that time as concerning finite pointer coincidences, as opposed to infinitesimal point coincidences, then we will overlook some of the most important physical content of general relativity.

5. Conclusion

If, as it has emerged, it is so important to distinguish in principle unobservable, infinitesimal point coincidences from observable, finite pointer coincidences with respect to the degree of underdetermination attaching to each, why in his early presentations of the point-coincidence argument did Einstein lay so much emphasis on the observability of at least pointer coincidences?

Given that his most explicit pronouncements of this kind were in his 1916 review article in the *Annalen der Physik*, Einstein may have been moved by a concern for his audience—one very much influenced by Mach, all the more so in the weeks immediately following Mach's death—to give the point-coincidence argument a more positivistic feeling, by contrast with the more robustly ontological versions of the argument in his correspondence with Ehrenfest and Besso. It may even have been a gesture of respect to Mach himself. On the other hand, given Einstein's polite silence about Mach's phenomenalism in his spring 1916 obituary for Mach, perhaps we should not be so quick to read even that 1916 statement of the argument as being quite so verificationist in tone as it might at first appear.

The stress on the observability of point coincidences may also have been simply a confusion on Einstein's part, a confusion that was clarified in the course of

Einstein's exchange with Schlick. Einstein was probably aware of the old tradition in the foundations of mathematics that linked topological relations of incidence and order with observable properties of space and time. If, as has been conjectured (see Howard & Norton 1993), he read Erich Kretschmann's late 1915 essay on the determination of "justified" coordinate systems at about the time of the point-coincidence argument, he would have certainly been reminded in a clear way about that traditional linkage of the topological with the observable. But the notion of observability that figured in that tradition never carried the kind of proto-logical empiricist epistemological significance that some would read into Einstein's talk of the observability of coincidences, something more like 'objective determinability' being what was actually intended by talk of observability.

Finally, it is important to remind ourselves that when Einstein did assert the observability of point coincidences, as in the quoted letters to Ehrenfest and Besso, he did it as an addendum to the primary assertion of the physical reality of point coincidences, a reality they possess by virtue of their invariance properties. Only a determined positivist would read these remarks as involving the definition of the real via observability. A reading closer to Einstein's would be that point coincidences are observable only because they are real, observability being a characteristic feature of the real, but not part of its very definition. That is the reading that I prefer to give.

Once we stop reading Einstein's point-coincidence argument as an anticipation of verificationism, a very different view of its implications for the development of the philosophy of science in the early twentieth century begins to emerge. Space does not permit telling the whole, somewhat long and complicated story here, but, in brief, it is that the point-coincidence argument is the starting point of at least a decade-long discussion of the extent to which a scientific theory can be expected to satisfy Petzoldt's "Gesetz der Eindeutigkeit." This discussion involves figures as diverse as Petzoldt himself, Schlick, Cassirer, Carnap, and Weyl. The *Eindeutigkeit* principle comes to play an especially important role in mature critical idealism of the Marburg variety, where, in Cassirer's writings, a hoped-for univocal formal-axiomatic representation of the real would take the place of the individual representations that orthodox Kantianism thought it the business of intuition to provide. But, starting in the mid-1920s, the recognition begins to spread, thanks to results like the Löwenheim-Skolem theorems, that non-categoricity will be endemic among reasonably powerful first-order theories, marking an end to the hope of a univocal theoretical representation of reality. In the early 1930s the final word was spoken by Gödel, who showed that such non-categoricity is a corollary to his first incompleteness theorem, and hence will be encountered in any theory as powerful as, or more powerful than, elementary arithmetic, meaning, among others, any physical theory that incorporates more than the most primitive mathematical tools. In Carnap's case, recognition of the phenomenon of non-categoricity was perhaps the determining factor in turning him away from the program of implicit definitions of scientific primitives—à la Hilbert and the early Schlick—and toward the program of explicit definitions featured in the *Aufbau*. The irony here is that, far from the point-coincidence argument's being the origin of verificationism, the recognition that the *Eindeutigkeit* principle upon which the point-coincidence argument rests could *not*, in general, be satisfied is what moved a thinker like Carnap to the doctrines about empirical meaning that we associate with logicial empiricism.[12]

Acknowledgements

Special thanks are owed to John Norton, whose gentle prodding in an e-mail exchange in September of 1994 was the stimulus that led to the writing of this paper, and to John Stachel for a careful, critical reading of the manuscript. My thanks also for their helpful comments to the participants in the Fourth International Conference on the History of General Relativity, Berlin, August 1995 and to audiences at the University of Notre Dame and Boston University, where earlier versions of this paper were read in January and May of 1996, respectively.

References

Bose, Satyendra Nath. (1924). "Plancks Gesetz und Lichtquantenhypothese." *Zeitschrift für Physik* **26**: 178–181.

Carnap, Rudolf. (1921). "Der Raum. Ein Beitrag zur Wissenschaftslehre." Ph.D. dissertation, University of Jena. [Published: Göttingen: Dieterichschen Universität-Buchdruckerei, W. Fr. Kaestner (1921). Reprinted as *Kant-Studien* (Ergänzungshefte, no. 56). Berlin: Reuther & Reichard (1922)].

Cassirer, Ernst. (1921). *Zur Einsteinschen Relativitätstheorie. Erkenntnistheoretische Betrachtungen*. Berlin: Bruno Cassirer.

Duhem, Pierre. (1906). *La Théorie physique: son objet et sa structure*. Paris: Chevalier & Rivière.

——— (1908). *Ziel und Struktur der physikalischen Theorien*. Friedrich Adler, trans. Foreward by Ernst Mach. Leipzig: Johann Ambrosius Barth. [German translation of Duhem 1906].

Earman, John & Glymour, Clark. (1978a). "Lost in the Tensors: Einstein's Struggle with Covariance Principles 1912–1916." *Studies in History and Philosophy of Science* **9**: 251–278.

——— (1978b). "Einstein and Hilbert: Two Months in the History of General Relativity." *Archive for History of Exact Sciences* **19**: 291–308.

Earman, John & Norton, John D. (1987). "What Price Spacetime Substantivalism? The Hole Story." *British Journal for the Philosophy of Science* **38**: 515–525.

Einstein, Albert. (1902). "Kinetische Theorie des Wärmegleichgewichtes und des zweiten Hauptsatzes der Thermodynamik." *Annalen der Physik* **9**: 417-433. [= Einstein *CP2*: doc. 3].

——— (1903). "Eine Theorie der Grundalgen der Thermodynamik." *Annalen der Physik* **11**: 170-187. [= Einstein *CP2*: doc. 4].

——— (1904). "Zur allgemeinen molekularen Theorie der Wärme." *Annalen der Physik* **14**: 354-362. [= Einstein *CP2*: doc. 5].

——— (1905a). "Über einen die Erzeugung und Verwandlung des Lichtes betreffenden heuristischen Gesichtspunkt." *Annalen der Physik* **17**: 132–148. [= Einstein *CP2*: doc. 14].

[12] For more on the philosophical implications of the point-coincidence argument and the *Eindeutigkeit* principle, see Howard 1992 and 1996.

——— (1905b). "Eine neue Bestimmung der Moleküldimensionen." Ph.D. dissertation, University of Zurich. [Published: Bern: Buchdruckerei K. J. Wyss (1906) = EINSTEIN *CP2*: doc. 15].

——— (1905c). "Über die von der molekularkinetischen Theorie der Wärme geforderte Bewegung von in ruhenden Flüssigkeiten suspendierten Teilchen." *Annalen der Physik* **17**: 549–560. [= EINSTEIN *CP2*: doc. 16].

——— (1905d). "Zur Elektrodynamik bewegter Körper." *Annalen der Physik* **17**: 891–921. [= EINSTEIN *CP2*: doc. 23].

——— (1906). "Das Prinzip von der Erhaltung der Schwerpunktsbewegung und die Tragheit der Energie." *Annalen der Physik* **20**: 199–206. [= EINSTEIN *CP2*: doc. 35].

——— (1914). "Die formale Grundlage der allgemeinen Relativitätstheorie." *Sitzungsberichte der Königlichen Preussischen Akademie der Wissenschaften* (Berlin): 1030–1085. [= EINSTEIN *CP6*: doc. 9].

——— (1915a). "Zur allgemeinen Relativitatstheorie." *Sitzungsberichte der Königlichen Preussischen Akademie der Wissenschaften* (Berlin): 778–786, 799–801. [= EINSTEIN *CP6*: doc. 21].

——— (1915b). "Erklärung der Perhelbewegung des Merkur aus der allgemeinen Relativitätstheorie." *Sitzungsberichte der Königlichen Preussischen Akademie der Wissenschaften* (Berlin): 831–839. [= EINSTEIN *CP6*: doc. 24].

——— (1915c). "Die Feldgleichungen der Gravitation." *Sitzungsberichte der Königlichen Preussischen Akademie der Wissenschaften* (Berlin): 844–847. [= EINSTEIN *CP6*: doc. 25].

——— (1916a). "Die Grundlage der allgemeinen Relativitätstheorie." *Annalen der Physik* **49**: 769–822. [= EINSTEIN *CP6*: doc. 30]. Except where otherwise noted, page numbers and quotations from the translation: "The Foundation of the General Theory of Relativity." In H.A. Lorentz et al. *The Principle of Relativity: A Collection of Original Memoirs on the Special and General Theory of Relativity*. W. Perrett and G. B. Jeffery, trans. 109–164. London: Methuen (1923). Reprinted: New York: Dover (1952)].

——— (1916b). "Ernst Mach." *Physikalische Zeitschrift* **17**: 101–104. [= EINSTEIN *CP6*: doc. 29].

——— (1917). *Über die spezielle und die allgemeine Relativitätstheorie (Gemeinverständlich)*. Braunschweig: Friedrich Vieweg und Sohn.

——— (1918a). "Bemerkung zu Herrn Schrödinger's Notiz: Über die Lösungssystem der allgemein kovarianten Gravitationsgleichungen." *Physikalische Zeitschrift* **19**: 165–166.

——— (1918b). "Motive des Forschens." In *Zu Max Plancks sechzigstem Geburtstag. Ansprachen, gehalten am 26. April 1918 in der Deutschen Physikalischen Gesellschaft*. 29–32. Karlsruhe: C. F. Müller.

——— (1919). "Induktion und Deduktion in der Physik." *Berliner Tageblatt* (25 December) Suppl. 4: 1.

——— (1920). *Äther und Relativitäts-Theorie. Rede gehalten am 5. Mai 1920 an der Reichs-Universität zu Leiden*. Berlin: Julius Springer.

——— (1921). *Geometrie und Erfahrung. Erweiterte Fassung des Festvortrages gehalten an der Preussischen Akademie der Wissenschaften zu Belin am 27. Januar 1921*. Berlin: Julius Springer.

——— (1924a). [Review of: *Alfred Elsbach. Kant und Einstein. Untersuchungen über das Verhältnis der modernen Erkenntnistheorie zur Relativitätstheorie*]. *Deutsche Literaturzeitung* **45**: 1688–1689.

——— (1924b). "Quantentheorie des einatomigen idealen Gases." *Sitzungsberichte der Preussischen Akademie der Wissenschaften* (Berlin): 261–267.

——— (1925a). "Quantentheorie des einatomigen idealen Gases. Zweite Abhandlung." *Sitzungsberichte der Preussischen Akademie der Wissenschaften* (Berlin): 3–14.

——— (1925b). "Zur Quantentheorie des idealen Gases." *Sitzungsberichte der Preussischen Akademie der Wissenschaften* (Berlin): 18–25.

——— (1946). "Autobiographical Notes." In Schilpp 1949: 1–94. [Quotations and page numbers are taken from the corrected English translation in *Autobiographical Notes: A Centennial Edition*. P. A. Schilpp, trans. and ed. La Salle (IL): Open Court (1979)].

——— (1949). "Remarks Concerning the Essays Brought together in this Cooperative Volume." In Schilpp 1949: 665–688.

——— (*CP2*). *The Collected Papers of Albert Einstein*. Vol. 2: *The Swiss Years: Writings, 1900–1909*. J. Stachel, D. C. Cassidy, J. Renn, & R. Schulmann, eds. Princeton: Princeton University Press (1989).

——— (*CP3*). *The Collected Papers of Albert Einstein*. Vol. 3. *The Swiss Years: Writings, 1909–1911*. M. J. Klein, A. J. Kox, J. Renn & R. Schulmann, eds. Princeton: Princeton University Press (1993).

——— (*CP4*). *The Collected Papers of Albert Einstein*. Vol. 4. *The Swiss Years: Writings, 1912–1914*. M. J. Klein, A. J. Kox, J. Renn & R. Schulmann, eds. Princeton: Princeton University Press (1995).

——— (*CP5*). *The Collected Papers of Albert Einstein*. Vol. 5. *The Swiss Years: Correspondence, 1902–1914*. M. J. Klein, A. J. Kox & R. Schulmann, eds. Princeton: Princeton University Press (1993).

——— (*CP6*). *The Collected Papers of Albert Einstein*. Vol. 6. *The Berlin Years: Writings, 1914–1917*. A. J. Kox, M. J. Klein, & R. Schulmann, eds. Princeton: Princeton University Press (1996).

EINSTEIN, Albert & GROSSMANN, Marcel.

——— (1913). *Entwurf einer verallgemeinerten Relativitätstheorie und einer Theorie der Gravitation. I. Physikalischer Teil von Albert Einstein. II. Mathematischer Teil von Marcel Grossmann*. Leipzig and Berlin: B. G. Teubner. [Reprinted with added "Bemerkungen," *Zeitschrift für Mathematik und Physik* **62** (1914): 225–261 = EINSTEIN *CP4*: doc. 13].

——— (1914). "Kovarianzeigenschaften der Feldgleichungen der auf die verallgemeinerte Relativitätstheorie gegründeten Gravitationstheorie." *Zeitschrift für Mathematik und Physik* **63**: 215–225. [= EINSTEIN *CP6*: doc. 2].

EISENSTAEDT, Jean & KOX, Anne J., eds. (1992). *Studies in the History of General Relativity* (Einstein Studies, vol. 3). Boston: Birkhäuser.

FRIEDMAN, Michael. (1983). *Foundations of Space-Time Theories: Relativistic Physics and Philosophy of Science*. Princeton, New Jersey: Princeton University Press.

HECHT, Hartmut & HOFFMANN, Dieter. (1982). "Die Berufung Hans Reichenbachs an die Berliner Universität." *Deutsche Zeitschrift für Philosophie* **30**: 651–662.

HOWARD, Don. (1990). "Einstein and Duhem." *Synthese* **83**: 363–384.

——— (1992). "Einstein and *Eindeutigkeit*: A Neglected Theme in the Philosophical Background to General Relativity." In EISENSTAEDT & KOX 1992: 154–243.

——— (1993). "Was Einstein Really a Realist?" *Perspectives on Science: Historical, Philosophical, and Sociological* **1**: 204–251.

——— (1994). "Einstein, Kant, and the Origins of Logical Empiricism." In *Language, Logic, and the Structure of Scientific Theories: The Carnap-Reichenbach Centennial*. Wesley Salmon and Gereon Wolters, eds. 45–105. Pittsburgh: University of Pittsburgh Press; Konstanz: Universitätsverlag.

——— (1996). "Relativity, *Eindeutigkeit*, and Monomorphism: Rudolf Carnap and the Development of the Categoricity Concept in Formal Semantics." In *Origins of Logical Empiricism*. Ron Giere and Alan Richardson, eds. 87–150. Minneapolis: University of Minnesota Press.

——— (1997). "A Peek behind the Veil of Maya: Einstein, Schopenhauer, and the Historical Background of the Conception of Space as a Ground for the Individuation of Physical Systems." In *The Cosmos of Science: Essays of Exploration*. John Earman and John Norton, eds. 115–164. Pittsburgh: University of Pittsburgh Press.

Howard, Don & Norton, John D. (1993). "Out of the Labyrinth? Einstein, Hertz, and the Göttingen Answer to the Hole Argument." In *The Attraction of Gravitation. New Studies in the History of General Relativity* (Einstein Studies, vol. 5). John Earman, Michel Jannsen and John Norton, eds. 30–62. Boston: Birkhäuser.

Howard, Don & Stachel, John, eds. (1989). ***Einstein and the History of General Relativity*** (Einstein Studies, vol. 1). Boston: Birkhäuser.

Komar, Arthur. (1958). "Construction of a Complete Set of Observables in the General Theory of Relativity." *Physical Review* **111**: 1182–1187.

Kretschmann, Erich. (1915). "Über die prinzipielle Bestimmbarkeit der berechtigten Bezugssysteme beliebiger Relativitätstheorien." *Annalen der Physik* **48**: 907–942, 943–982.

——— (1917). "Über den physikalischen Sinn der Relativitätspostulate, A. Einstein's neue und seine ursprüngliche Relativitätstheorie." *Annalen der Physik* **53**: 575–614.

Mach, Ernst. (1910). "Die Leitgedanken meiner naturwissenschaftlichen Erkenntnislehre und ihre Aufhahme durch die Zeitgenossen." *Scientia* **7**: 2ff. [Reprinted: *Physikalische Zeitschrift* **11** (1910): 599–606].

Norton, John D. (1984). "How Einstein Found His Field Equations." *Historical Studies in the Physical Sciences* **14**: 253–316. [Reprinted in Howard & Stachel 1989: 101–159].

——— (1987). "Einstein, the Hole Argument and the Reality of Space." In *Measurement, Realism and Objectivity*. J. Forge, ed. 153–188. Boston & Dordrecht: D. Reidel.

——— (1988). "The Hole Argument." In *PSA 1988*. Vol. 2. Arthur Fine and Jarrett Leplin, eds. 56–64. East Lansing, Michigan: Philosophy of Science Association.

——— (1989). "Coordinates and Covariance: Einstein's View and the Modern View." *Foundations of Physics* **19**: 1215–1263.

Petzoldt, Joseph. (1895). "Das Gesetz der Eindeutigkeit." *Vierteljahrsschrift für wissenschaftliche Philosophie und Soziologie* **19**: 146–203.

——— (1912). *Das Weltproblem vom Standpunkte des relativistischen Positivismus aus, historisch-kritisch dargestellt*. 2nd ed. (Wissenschaft und Hypothese, vol. 14). Leipzig & Berlin: B. G. Teubner.

——— (1921). *Die Stellung der Relativitätstheorie in der Geistigen Entwicklung der Menschheit*. Dresden: Sibyllen-Verlag.

Planck, Max. (1908). "Die Einheit des physikalischen Weltbildes." *Physikalische Zeitschrift* **10**: 62–75.

——— (1910). "Zur Machschen Theorie der physikalischen Erkenntnis. Eine Erwiderung." *Physikalische Zeitschrift* **11**: 1186–1190.

Poincaré, Henri. (1900). "La Théorie de Lorentz et le principe de réaction." In *Recueil de Travaux offerts par les Auteurs à H. A. Lorentz, professeur de Physique à l'Université de Leiden, à l'occasion du 25ᵉ anniversaire de son Doctorat le 11 décembre 1900*. Johannes Bosscha, ed. The Hague: Martinus Nijhoff = *Archives Néerlandaises des Sciences exactes et naturelles* **5** (2): 252–278. [Reprinted: *Œuvres de Henri Poincaré*. Vol. 9: 464—488. G. Petiau, ed. Paris: Gauthier-Villars (1954)].

REICHENBACH, Hans. (1928). *Philosophie der Raum-Zeit-Lehre*. Berlin: Julius Springer. [Page numbers and quotations from the English translation: *The Philosophy of Space & Time*. Maria Reichenbach and John Freund, trans. New York: Dover (1957)].

——— (1949). "The Philosophical Significance of the Theory of Relativity." In Schilpp 1949: 289–311.

ROVELLI, Carlo. (1991a). "What Is Observable in Classical and Quantum Gravity?" *Classical and Quantum Gravity* **8**: 297–316.

——— (1991b). "Quantum Reference Systems." *Classical and Quantum Gravity* **8**: 317–331.

RYCKMAN, Thomas A. (1991). "*Conditio sine qua non*? *Zuordnung* in the Early Epistemologies of Cassirer and Schlick." *Synthese* **88**: 57–95.

——— (1992). "P(oint)-C(oincidence) Thinking: The Ironical Attachment of Logical Empiricism to General Relativity (and Some Lingering Consequences)." *Studies in History and Philosophy of Science* **23**: 471–497.

SCHILPP, Paul Arthur, ed. (1949). *Albert Einstein: Philosopher-Scientist*. The Library of Living Philosophers, vol. 7. Evanston (IL): The Library of Living Philosophers.

SCHLICK, Moritz. (1910). "Das Wesen der Wahrheit nach der modernen Logik." *Vierteljahrsschrift für wissenschaftliche Philosophie und Soziologie* **34**: 386–477.

——— (1915). "Die philosophische Bedeutung des Relativitätsprinzips." *Zeitschrift für Philosophie und philosophische Kritik* **159**: 129–175.

——— (1917a). "Raum und Zeit in der gegenwärtigen Physik. Zur Einführung in das Verständnis der allgemeinen Relativitätstheorie." *Die Naturwissenschaften* **5**: 161–167, 177–186.

——— (1917b). *Raum und Zeit in den gegenwärtigen Physik. Zur Einführung in das Verständnis der allgemeinen Relativitätstheorie*. Berlin: Julius Springer.

——— (1918). *Allgemeine Erkenntnislehre*. Berlin: Julius Springer.

——— (1934). "Über das Fundament der Erkenntnis." *Erkenntnis* **4**: 79–99.

SKLAR, Lawrence. (1974). *Space, Time, and Spacetime*. Berkeley: University of California Press.

SPEZIALI, Pierre, ed. (1972). *Albert Einstein–Michele Besso. Correspondance. 1903–1955*. Paris: Hermann.

STACHEL, John. (1980). "Einstein's Search for General Covariance, 1912–1915." Paper delivered at the Ninth International Conference on General Relativity and Gravitation, Jena, Germany (DDR), 17 July 1980. [Page numbers and quotations from revised version in HOWARD & STACHEL 1989: 63–100.

——— (1986a). "What a Physicist Can Learn from the History of Einstein's Discovery of General Relativity." In *Proceedings of the Fourth Marcel Grossmann Meeting on General Relativity*. Remo Ruffini, ed. 1857–1862. Amsterdam: Elsevier.

——— (1986b). "How Einstein Discovered General Relativity: A Historical Tale with Some Contemporary Morals." In *General Relativity and Gravitation. Proceedings of the 11th International Conference on General Relativity and Gravitation, Stockholm, July 6–12, 1986*. M. A. H. MacCallum, ed. 200–208. Cambridge: Cambridge University Press (1987).

——— (1986c). "Einstein and the Quantum: Fifty Years of Struggle." In *From Quarks to Quasars: Philosophical Problems of Modern Physics* (University of Pittsburgh Series in the Philosophy of Science, vol. 7). Robert G. Colodny ed. 349–385. Pittsburgh: University of Pittsburgh Press.

——— (1988). "The Early History of the Cauchy Problem in General Relativity, 1916–1937." In EISENSTAEDT & KOX 1992: 407–418.

——— (1993). "The Meaning of General Covariance: The Hole Story." In *Philosophical Problems of the Internal and External Worlds: Essays on the Philosophy of Adolf Grünbaum*. John Earman, Allen I. Janis, Gerald J. Massey, and Nicholas Rescher, eds. 129–160. Pittsburgh: University of Pittsburgh Press and Konstanz: Universitätsverlag.

VIZGIN, Vladimir & SMORODINSKIY, Ya. A. (1979). "From the Equivalence Principle to the Equations of Gravitation." *Soviet Physics. Uspekhi* **22**: 489–513.

Contributors

Silvio Bergia, Dipartamento di Fisica, Università degli Studi di Bologna, Via Irnerio 46, I-40126, Bologna, Italy

John Earman, Department of History and Philosophy of Science, University of Pittsburgh, Pittsburgh PA 15260, U.S.A.

George Gale, Department of Philosophy, University of Missouri-Kansas City, 222 Cockefair Hall, Kansas City MO 64110-2499, U.S.A.

Hubert Goenner, Institut für Theoretische Physik, Universität Göttingen, Bunsenstraße 9, D-3400 Göttingen, Germany

Peter Havas, Department of Physics, Temple University, Barton Hall 009-00, Philadelphia PA 19122, U.S.A.

Don Howard, Department of Philosophy, University of Notre Dame, 336 O'Shaughnessy, Notre Dame IN 46556, U.S.A.

Michel Janssen, Department of Philosophy, Boston University, Boston MA 02215, U.S.A.

Daniel Kennefick, SOCAS, University of Wales, 50 Park Place, Cardiff CF1 3AT, Wales, U.K.

Helge Kragh, Institute for the History of the Exact Sciences, Aarhus University, Ny Munkegard, DK-8000C Aarhus, Denmark

Roberto de Andrade Martins, Group of History and Theory of Science, State University of Campinas, Caixa Postal 6165, 13081-970 Campinas, SP, Brazil

Lucia Mazzoni, Dipartamento di Fisica, Università degli Studi di Bologna, Via Irnerio 46, I-40126, Bologna, Italy

John D. Norton, Department of History and Philosophy of Science, University of Pittsburgh, Pittsburgh PA 15260, U.S.A.

Jürgen Renn, Max-Planck-Institut für Wissenschaftsgeschichte, Wilhelmstraße 44, D-10117, Berlin, Germany

Jim Ritter, Département de Mathématiques, Histoire et Philosophie des Sciences, Université de Paris VIII, 2 rue de la Liberté, F-93526 Saint-Denis, France

Robert Rynasiewicz, Department of Philosophy, Johns Hopkins University, 3400 N. Charles St., Baltimore MD 21218, U.S.A.

José M. Sánchez-Ron, Universidad Autónoma de Madrid, Departamento de Física Teórica, Cantoblanco, E-28049 Madrid, Spain

Tilman Sauer, Institut für Wissenschaftsgeschichte, Universität Göttingen, Humboldtallee 11, D-37073 Göttingen, Germany

John Urani, Department of Physics, University of Missouri-Kansas City, Kansas City MO 64110-2499, U.S.A.

Scott Walter, Max-Planck-Institut für Wissenschaftsgeschichte, Wilhelmstraße 44, D-10117, Berlin, Germany

Index

A page number followed by the letter "n" indicates a *note* on that page; by the letter "f" a *figure*; and by the letter "t" a *table*. The letter "*p*" (*passim*) indicates occasional scattered references in the pages cited.

GPSR Compliance

The European Union's (EU) General Product Safety Regulation (GPSR) is a set of rules that requires consumer products to be safe and our obligations to ensure this.

If you have any concerns about our products, you can contact us on ProductSafety@springernature.com

In case Publisher is established outside the EU, the EU authorized representative is:

Springer Nature Customer Service Center GmbH
Europaplatz 3
69115 Heidelberg, Germany

Batch number: 10577874

Printed by Printforce, the Netherlands